WITHDRAWN FROM
KENT STATE UNIVERSITY LIBRARIES

# FOREST GROWTH RESPONSES TO THE POLLUTION CLIMATE OF THE 21st CENTURY

# Forest Growth Responses to the Pollution Climate of the 21st Century

Edited by

LUCY J. SHEPPARD

and

J. NEIL CAPE

*Institute of Terrestrial Ecology,
Edinburgh, Scotland,
United Kingdom*

Reprinted from *Water, Air, and Soil Pollution* 116: 1–2, 1999

KLUWER ACADEMIC PUBLISHERS
Dordrecht/Boston/London

A C.I.P. Catalogue record for this book is available from the Library of Congress.

ISBN 0-7923-5991-7

Published by Kluwer Academic Publishers
P.O. Box 17, 3300 AA Dordrecht, The Netherlands

Sold and distributed in North, Central and Latin America
by Kluwer Academic Publishers
101 Philip Drive, Norwell, MA 02061, U.S.A.

In all other countries, sold and distributed
by Kluwer Academic Publishers
P.O. Box 322, 3300 AH Dordrecht, The Netherlands

*Printed on acid-free paper*

All rights reserved
© 1999 Kluwer Academic Publishers
No part of the material protected by this copyright notice may be reproduced
or utilized in any form or by any means, electronic or mechanical,
including photocopying, recording or by any information storage and
retrieval system, without written permission from the copyright owner.

Printed in the Netherlands

# TABLE OF CONTENTS

| | |
|---|---|
| Preface | 1 |
| Acknowledgements | 3 |
| D. FOWLER, J.N. CAPE, M. COYLE, C. FLECHARD, J. KUYLENSTIERNA, K. HICKS, D. DERWENT, C. JOHNSON and D. STEVENSON / The Global Exposure of Forests to Air Pollutants | 5–32 |
| H. SPIECKER / Overview of Recent Growth Trends in European Forests | 33–46 |
| H. RENNENBERG and A. GESSLER / Consequences of N Deposition to Forest Ecosystems – Recent Results and Future Research Needs | 47–64 |
| B.A. EMMETT / The Impact of Nitrogen on Forest Soils and Feedbacks on Tree Growth | 65–74 |
| M.F. HOVMAND and J. BILLE-HANSSEN / Atmospheric Input to Danish Spruce Forests and Effects on Soil Acidification and Forest Growth Based on 12 Years Measurements | 75–88 |
| U. SKIBA, L.J. SHEPPARD, C.E.R. PITCAIRN, S. VAN DIJK and M.J. ROSSALL / The Effect of N Decomposition on Nitrous Oxide and Nitric Oxide Emissions from Temperate Forest Soils | 89–98 |
| W. FLÜCKIGER and S. BRAUN / Nitrogen and Its Effect on Growth, Nutrient Status and Parasite Attacks in Beech and Norway Spruce | 99–110 |
| M. CARNOL, P. CUDLÍN and P. INESON / Impacts of $(NH_4)_2SO_4$ Deposition on Norway Spruce (*Picea Abies* [L.] Karst) Roots | 111–120 |
| T. STASZEWSKI, S. GODZIK, P. KUBIESA and J. SZDZUJ / Fate of Nitrogen Compounds Deposited to Spruce (*Picea Abies* Karst.) and Pine (*Pinus Silvestris* L.) Forests Located in Different Air Pollution and Climatic Conditions | 121–127 |
| P. SCHLEPPI, I. BUCHER-WALLIN, R. SIEGWOLF, M. SAURER, N. MULLER and J.B. BUCHER / Simulation of Increased Nitrogen Deposition to a Montane Forest Ecosystem: Partitioning of the Added $^{15}N$ | 129–134 |
| J. WÖLLECKE, B. MÜNZENBERGER and R.F. HÜTTL / Some Effects of N on Ectomycorrhizal Diversity of Scots Pine (*Pinus Sylvestris* L.) in Northeastern Germany | 135–140 |
| A. BYTNEROWICZ, S. GODZIK, M. POTH, I. ANDERSON, J. SZDZUJ, C. TOBIAS, S. MACKO, P. KUBIESA, T. STASZEWSKI and M. FENN / Chemical Composition of Air, Soil and Vegetation in Forests of the Silesian Beskid Mountains, Poland | 141–150 |
| S. McLAUGHLIN and K. PERCY / Forest Health in North America: Some Perspectives on Actual and Potential Roles of Climate and Air Pollution | 151–197 |
| R. MATYSSEK and J.L. INNES / Ozone – A Risk Factor for Trees and Forests in Europe? | 199–226 |
| J.M. SKELLY, J.L. INNES, J.E. SAVAGE, K.R. SNYDER, D. VANDERHEYDEN, J. ZHANG and M.J. SANZ / Observation and Confirmation of Foliar Ozone Symptoms of Native Plant Species of Switzerland and Southern Spain | 227–234 |
| N.E. GRULKE and L. BALDUMAN / Deciduous Conifers: High N Deposition and $O_3$ Exposure Effects on Growth and Biomass Allocation in Ponderosa Pine | 235–248 |
| M. TAUSZ, A. BYTNEROWICZ, W. WEIDNER, M.J. ARBAUGH, P. PADGETT and D. GRILL / Changes in Free-Radical Scavengers Describe the Susceptibility of *Pinus Ponderosa* to Ozone in Southern Californian Forests | 249–254 |
| A. CHAPPELKA, G. SOMERS and J. RENFRO / Visible Ozone Injury on Forest Trees in Great Smoky Mountains National Park, USA | 255–260 |
| A. CHAPPELKA, J. SKELLY, G. SOMERS, J. RENFRO and E. HILDEBRAND / Mature Black Cherry Used as a Bioindicator of Ozone Injury | 261–266 |
| E. GRAVANO, M. FERRETTI, F. BUSSOTTI and P. GROSSONI / Foliar Symptoms and Growth Reduction of *Ailanthus altissima* Desf. in an Area with High Ozone and Acidic Deposition in Italy | 267–272 |
| R. INCLÁN, A. RIBAS, J. PEÑUELAS and B.S. GIMENO / The Relative Sensitivity of Different Mediterranean Plant Species to Ozone Exposure | 273–277 |
| R.L. HEATH / Biochemical Processes in an Ecosystem: How Should They Be Measured? | 279–298 |

M.S.J. BROADMEADOW, J. HEATH and T.J. RANDLE / Environmental Limitations to $O_3$ Uptake – Some Key Results from Young Trees Growing at Elevated $CO_2$ Concentrations     299–310

D.F. KARNOSKY, B. MANKOVSKA, K. PERCY, R.E. DICKSON, G.K. PODILA, J. SOBER, A. NOORMETS, G. HENDREY, M.D. COLEMAN, M. KUBISKE, K.S. PREGITZER and J.G. ISEBRANDS / Effects of Tropospheric $O_3$ on Trembling Aspen and Interaction with $CO_2$: Results from an $O_3$-Gradient and a Face Experiment     311–322

M.S. GÜNTHARDT-GOERG, S. MAURER, J. BOLLIGER, A.J. CLARK, W. LANDOLT and J.B. BUCHER / Responses of Young Trees (Five Species in a Chamber Exposure) to Near-Ambient Ozone Concentrations     323–332

S. MANNINEN, D. LE THIEC, C. ROSE, G. NOURRISSON, F. RADNAI, J.P. GARREC and S. HUTTUNEN / Pigment Concentrations and Ratios of Aleppo Pine Seedlings Exposed to Ozone     333–338

R.M. COX and J.W. MALCOLM / Passive Ozone Monitoring for Forest Health Assessment     339–344

G. GEROSA, F. SPINAZZI and A. BALLARIN DENTI / Tropospheric Ozone in Alpine Forest Sites: Air Quality Monitoring and Statistical Data Analysis     345–350

M. FERRETTI, I. BONINI, F. BUSSOTTI, C. CELESTI, E. CENNI, A. CHIARUCCI, A. COZZI, V. DE DOMINICIS, P. GROSSONI and C. LEONZIO / Short-Term Changes of Response Indicators of Ecosystem Status in Broadleaved Forests in Tuscany (Central Italy)     351–356

S. BRAUN, B. RIHM, C. SCHINDLER and W. FLÜCKIGER / Growth of Mature Beech in Relation to Ozone and Nitrogen Deposition: An Epidemiological Approach     357–364

M.V. KOZLOV and P. NIEMELÄ / Difference in Needle Length – A New and Objective Indicator of Pollution Impact on Scots Pine (*Pinus sylvestris*)     365–370

M. FERRETTI, F. BUSSOTTI, E. CENNI and A. COZZI / Implementation of Quality Assurance Procedures in the Italian Programs of Forest Condition Monitoring     371–376

F. BATIČ, P. KALAN, H. KRAIGHER, H. ŠIRCELJ, P. SIMONČIČ, N. VIDERGAR-GORJUP and B. TURK / Bioindication of Different Stresses in Forest Decline Studies in Slovenia     377–382

G. THELIN, U. ROSENGREN-BRINCK and B. NIHLGÅRD / Can Graphical Vector Analysis be Used to Identify Micro Nutrient Deficiency?     383–388

J. NEIRYNCK and P. ROSKAMS / Relationships Between Crown Condition of Beech (*Fagus sylvatica* L.) and Throughfall Chemistry     389–394

W. PRUS-GLOWACKI, A. WOJNICKA-POLTORAK, J. OLEKSYN and P.B. REICH / Industrial Pollutants Tend to Increase Genetic Diversity: Evidence from Field-Grown European Scots Pine Populations     395–402

J.L.J. HOUPIS, P.D. ANDERSON, J.C. PUSHNIK and D.J. ANSCHEL / Among-Provenance Variability of Gas Exchange and Growth in Response to Long-Term Elevated $CO_2$ Exposure     403–412

J.C. PUSHNIK, D. GARCIA-IBILCIETA, S. BAUER, P.D. ANDERSON, J. BELL and J.L.J. HOUPIS / Biochemical Responses and Altered Genetic Expression Patterns in Ponderosa Pine (*Pinus ponderosa* Doug ex P. Laws) Grown under Elevated $CO_2$     413–422

A. WONISCH, M. TAUSZ, M. MÜLLER, W. WEIDNER, L.J. DE KOK and D. GRILL / Treatment of Young Spruce Shoots with $SO_2$ and $H_2S$: Effects on Fine Root Chromosomes in Relation to Changes in the Thiol Content and Redox State     423–428

D.C. GORDON and K.E. PERCY / Effect of UV-B Dose on Biosynthesis of Epicuticular Waxes in Blue Spruce (*Picea Pungens* Engelmann.) Primary Needles: Preliminary Investigation     429–436

A. AUGUSTAITIS / A New Imitative Model to Predict the Impact of Air Pollutants on Scots Pine Health and Radial Increment     437–442

K. PERCY, J. BUCHER, J. CAPE, M. FERRETTI, R. HEATH, H.E. JONES, D. KARNOSKY, R. MATYSSEK, G. MULLER-STARCK, E. PAOLETTI, U. ROSENGREN-BRINCK, L. SHEPPARD, J. SKELLY and G. WEETMAN / State of Science and Knowledge Gaps with Respect to Air Pollution Impacts on Forests: Reports from Concurrent IUFRO 7.04.00 Working Party Sessions     443–448

# PREFACE

This Special Issue of *Water, Air and Soil Pollution* offers contributions from the 18$^{th}$ IUFRO workshop on Air Pollution Stress, Forest Responses to the Pollution Climate of the 21$^{st}$ Century held in Edinburgh, Scotland, from September 21 to 23, 1998.

The meeting was held under the auspices of IUFRO, Research Group 7.04.00 chaired by Dr Kevin Percy of Canada. A new session structure was adopted to stimulate activity within the six working parties and a brief resume of these is presented at the front of this volume. The two, one-day plenary sessions were devoted to the two important air pollution issues, nitrogen deposition and ozone. Invited papers were augmented by a large and excellent contribution of poster papers. The final day comprised parallel Working Party Sessions with pre-arranged speakers to stimulate discussions. One hundred and thirty one scientists attended, representing 20 countries and 7 IUFRO regions: Northern Europe, Central Europe, Eastern Europe, Mediterranean, North America, Asia and the Western Pacific.

*Lucy Sheppard*
*David Fowler*

# ACKNOWLEDGEMENTS

The Edinburgh IUFRO meeting was organised by a few dedicated people and the goodwill of the IUFRO air pollution specialists. We should like to thank members of the Institute of Terrestrial Ecology, especially Kate Potter and Lucy Douglas, who between them undertook the compilation of the proceedings, all the administrative arrangements and the seemingly endless reformatting of papers. Prior to and since the meeting Davis Innes and Lisa Murphy have provided unstinting support, performing any task asked of them. Tom Murray produced the flier and members of the pollution group – Ian Leith, Alan Crossley, Carole Pitcairn, Ute Skiba, Frank Harvey, Alastair Dunster and our station director Melvin Cannell - gave up their Sunday afternoon to show nearly 70 visitors round the acid mist field fumigation experiment at Deepsyke and the open top chamber facility at Bush. The weather was outstanding – it was Edinburgh's hottest day in 1998! The UK forestry commission assisted with the organisation of the meeting and the UK government, represented by Lord Sewell, Minister of Forestry and Agriculture for Scotland, hosted an excellent reception for the meeting, in Edinburgh Castle. Visitors were treated to a private viewing of the Scottish Crown Jewels.

Over 60 manuscripts were submitted and reviewed, and the 41, which appear here, survived a rigorous reviewing and rewriting period. We are very grateful to the authors who responded politely to the requests for redrafts of their papers. We are also very grateful to all our reviewers listed at the end of these proceedings and their institutions, who gave of their time and effort.

The front cover of the hardbound book was produced by Zoe Fowler and Mhairi Coyle.

Mariette de Jong and Astrid Zandee of the staff of Kluwer Academic Publishers have been particularly helpful, responding rapidly to all requests for guidance. Our respective home institution, the Institute of Terrestrial Ecology, has supported us throughout this initiative. The meeting was entirely self-financing.

*Lucy Sheppard*
*Neil Cape*
*David Fowler*

# THE GLOBAL EXPOSURE OF FORESTS TO AIR POLLUTANTS

DAVID FOWLER[1], J. NEIL CAPE[1], MHAIRI COYLE[1], CHRIS FLECHARD[1],
JOHAN KUYLENSTIERNA[2], KEVIN HICKS[2], DICK DERWENT[3], COLIN JOHNSON[3] AND DAVID STEVENSON[3]

[1]*Institute of Terrestrial Ecology, Edinburgh Research Station, Bush estate, Penicuik, Midlothian EH26 0QB.TE Edinburgh, UK;* [2]*Stockholm Environment Institute, University of York,UK;* [3]*Meteorological Office, London Road, Bracknell, Berks. RG12 2SY UK*

(Received 9 April 1999; accepted 23 April 1999)

**Abstract**: The tall, aerodynamically rough surfaces of forests provide for the efficient exchange of heat and momentum between terrestrial surfaces and the atmosphere. The same properties of forests also provide for large potential rates of deposition of pollutant gases, aerosols and cloud droplets. For some reactive pollutant gases, including $SO_2$, $HNO_3$ and $NH_3$, rates of deposition may be large and substantially larger than onto shorter vegetation and is the cause of the so called "filtering effect" of forest canopies. Pollutant inputs to moorland and forest have been compared using measured ambient concentrations from an unpolluted site in southern Scotland and a more polluted site in south eastern Germany. The inputs of S and N to forest at the Scottish site exceed moorland by 16% and 31% respectively with inputs of 7.3 kg S ha$^{-1}$ y and 10.6 kg N ha$^{-1}$ y$^{-1}$. At the continental site inputs to the forest were 43% and 48% larger than over moorland for S and N deposition with totals of 53.6 kg S ha$^{-1}$ y$^{-1}$ and 69.5 kg N ha$^{-1}$ y$^{-1}$ respectively.

The inputs of acidity to global forests show that in 1985 most of the areas receiving > 1 kg H$^+$ ha$^{-1}$ y$^{-1}$ as S are in the temperate latitudes, with 8% of total global forest exceeding this threshold. By 2050, 17% of global forest will be receiving > 1 kg H$^{-1}$ ha$^{-1}$ as S and most of the increase is in tropical and sub-tropical countries.

Forests throughout the world are also exposed to elevated concentrations of ozone. Taking 60 ppb $O_3$ as a concentration likely to be phytotoxic to sensitive forest species, a global model has been used to simulate the global exposure of forests to potentially phytotoxic $O_3$ concentrations for the years 1860, 1950, 1970, 1990 and 2100. The model shows no exposure to concentrations in excess of 60 ppb in 1860, and of the 6% of global forest exposed to concentrations > 60 ppb in 1950, 75% were in temperate latitudes and 25% in the tropics. By 1990 24% of global forest is exposed to $O_3$ concentrates > 60 ppb, and this increases to almost 50% of global forest by 2100. While the uncertainty in the future pollution climate of global forest is considerable, the likely impact of $O_3$ and acid deposition is even more difficult to assess because of interactions between these pollutants and substantial changes in ambient $CO_2$ concentration, N deposition and climate over the same period, but the effects are unlikely to be beneficial overall.

**Keywords**: acid deposition, ozone, forests, nitrogen deposition, pollution climate.

## 1. Introduction

The major regional pollutants associated with acidic deposition (oxidised sulphur and nitrogen compounds and reduced nitrogen, and the photochemical oxidant ($O_3$) represent a threat to a wide range of ecosystems including agricultural crops, wetlands and heathland vegetation. However, much of the interest in pollutant effects during the last three decades has been focussed on forests. Examples include the role of air pollutants in the decline in health of Ponderosa Pine in Southern California (Miller *et al.*, 1963), the dieback of high elevation forests in the Appalachian mountains of Eastern North America (Eager and Adams, 1992) and the 'new type' of forest decline in Germany and other areas of central Europe (Schultze and Freer-Smith, 1991). More recently, observations of increased rates of growth of European forests have focussed on the role of elevated $CO_2$ concentrations and the deposition of fixed nitrogen (Cannell *et al.*1997; Spiecker *et al.*, 1996; Thornley *et al.*, 1991), both oxidised ($NO_x$) and reduced ($NH_x$).

Why should forests become such a clear focus of interest in these pollution problems? Is it simply a consequence of their abundance as land cover or size relative to other vegetation in the non-agricultural landscape and their importance for amenity, or, are there factors, which predispose forests to the influence, positive or negative of air pollutants?

In this paper, the properties of forests as sinks for the major gaseous and particulate pollutants are described and contrasted with shorter vegetation. In this way one of the major characteristics of forests, their aerodynamic interaction with the atmosphere, is highlighted as an important contributor to their sensitivity to the presence of reactive air pollutants. The current and projected trends in emissions of acidifying pollutant and the presence of photochemical oxidants are then used to quantify the spatial and temporal trends in pollutant exposure globally.

## 2. Forests and the Atmosphere

The exchange of mass, sensible heat and momentum between vegetation and the atmosphere is effected by turbulent transfer, molecular processes only becoming important very close (<< 1mm) to terrestrial surfaces. The turbulence generated within the surface layers of the atmosphere occurs as a consequence of frictional drag at the earth's surface and forests, among all natural land cover classes, generate the greatest frictional drag at the surface. Forests are therefore aerodynamically rough surfaces with rates of turbulent exchange between the atmosphere and forests larger by an order of magnitude or more than those over grassland.

The rates of exchange of mass, energy and momentum between forests and the atmosphere are discussed in detail by Jarvis *et al.* (1975) and Thom (1975). The extent to which the larger rates of turbulent transfer above forests lead to larger inputs of pollutants to forests than other vegetation is determined by physical and chemical processes at the foliar surfaces, and within the viscous boundary layer close to foliage. Thus the physical and chemical properties of the pollutants interact strongly with the properties of forests to determine rates of exchange.

## 3. The Pollutants

The pollutants occur as gases, $SO_2$, $NO$, $NO_2$, $HNO_3$, $NH_3$, $O_3$, as aerosol particles mainly in the size range

| | |
|---|---|
| 0.1 to 1.0 µm (diameter) | $SO_4^{2-}$, $NO_3^-$, $NH_4^+$, $H^+$ |
| 0.1 to 0.5 µm (diameter) | heavy metals (e.g. Pb, Cd, Zn, Cu) |
| 1.0 to 3.0 µm (diameter) | base cations $Ca^{2+}$, $Mg^{2+}$ |

and as cloud droplets in the range 3 μm to 20 μm in diameter, containing the major ions ($SO_4^{2-}$, $NO_3^-$ $NH_4^+$, $H^+$, $Cl^-$, $Na^+$, $Ca^{2+}$, $Mg^{2+}$).

Generalising the properties of forests necessarily conceals the variability between forests of different species composition and climates. The alternative, of describing the aerodynamic properties of a range of tree species and canopy structures, is outside the scope of this paper. Furthermore, the extent of published literature to provide the basis for such an exercise is limited to relatively few species.

TABLE I

Potential effects of forests on deposition fluxes

|  |  | Moorland | Forest | % Increase Forest/Moor |
|---|---|---|---|---|
| Canopy height | h (m) | 0.15 | 10 | - |
| Zero plane displacement | d (m) | 0.1 | 7 | - |
| Roughness length | $z_0$ (m) | 0.01 | 1.0 | - |
| Friction velocity | $u_*$ (m s$^{-1}$) | 0.32 | 0.82 | 156% |
| Momentum flux | $\tau$ (N m$^{-2}$) | 131 | 840 | 541% |
| Maximum deposition velocity for $SO_2$ | $V_{max}$ $SO_2$ (mm s$^{-1}$) | 18.6 | 35.1 | 89% |
| Maximum deposition velocity for $NO_2$ | $V_{max}$ $NO_2$ (mm s$^{-1}$) | 20.0 | 43.5 | 118% |
| Maximum deposition velocity for $NH_3$ | $V_{max}$ $NH_3$ (mm s$^{-1}$) | 21.4 | 55.5 | 160% |

In describing and contrasting the micrometeorology of temperate and tropical forests Shuttleworth (1989) showed broad similarities in the rates of exchange of water and energy of boreal conifer forest and tropical rainforests. The forests have had to adapt to common problems of limited water supply and very large potential rates of water loss despite the very different architecture of the forest canopies and the climatologies of these regions. The same processes which create the potential for large rates of exchange of momentum, water and sensible heat also influence the deposition of pollutants.

## 4. Potential deposition rates of pollutants on forests

The extent to which these generalisations apply to the deposition of pollutants on forests varies with the physical and chemical forms of the pollutant. It is therefore necessary initially to quantify first the potential deposition rates onto forests as a consequence of their aerodynamic properties and contrast these for different canopies and atmospheric conditions. This background may then be used to work through each of the major pollutant categories to show the extent to which the 'scavenging effect' of forests occurs in the field.

To demonstrate the potential rates of deposition of pollutants onto forest relative to short vegetation conventional micrometeorological approaches may be used to calculate momentum and gas fluxes in typical conditions. For simplicity, taking an extensive forest of uniform height in a flat landscape, the wind velocity profile within the constant flux layer may be used to estimate the friction or eddy velocity ($u_*$) and the momentum flux ($\tau = \rho u_*^2$, where $\rho$ is air density) from an assumed wind velocity at a reference height above the surface (Monteith and Unsworth, 1990).

Using this simple approach to contrast the turbulent exchange over a 10 m forest and a moorland (height 0.15 m) for a constant windspeed of 3 ms$^{-1}$ at a reference height of 10 m above the effective surface (d), the eddy velocity over the trees is increased by nearly a factor of 3 and the momentum flux is increased by a factor of 6.4 (Table 1). It is easy to see from this simple example the magnitude of the effects of enhanced turbulence over the trees.

The increased momentum flux for the forest results directly from the increased frictional drag of the canopy, quantified as the drag coefficient ($C_m$) and given by

$$C_m = [\frac{u_*}{u_{(z)}}]^2 \quad \text{where } u_{(z)} \text{ is the windspeed at height z.}$$

Thus $C_m$ may be calculated directly from a knowledge of $u_*$ and the wind velocity profile used to produce Table I. The drag coefficient is therefore related directly to the roughness length ($z_o$) within the wind profile equation.

$$u(Z) = \frac{u_*}{k} \ln\left(\frac{z-d}{z_o}\right)$$

In which k is von Karman's constant and d the zero plane displacement (Monteith and Unsworth, 1990).

The direct link between aerodynamic roughness of the canopy and the turbulent transfer of momentum is clear from these simple relationships. The transfer of momentum however, differs from that of trace gases or particles in that momentum transfer occurs both as frictional drag and as form drag, whereas the transfer of gases and particles is restricted to the equivalent of frictional drag, since the gases rely on molecular diffusion at the surface. A convenient method to quantify the reduced rate of transfer of gases relative to momentum is provided within the resistance analogy of mass and energy transfer between the surface and the atmosphere (Thom, 1975). In the analogy the total resistance to transfer of momentum ($r_{am}$), is provided by ($u_{(Z)} / u_*^2$) (Fig. 1). For gases and particles the additional transfer resistance through a viscous sub-layer ($r_b$) completes the transfer pathway from a reference point in the free atmosphere to the external surfaces of vegetation where uptake of the gas or capture of the particle may occur (Fig. 1). The additional resistance ($r_b$) for gases and particles may be regarded as a different level within the plant canopy which represents the effective sink for the gas, and since gas transfer is less efficient than momentum, the height of the effective sink for gases will be deeper within the canopy than the sink for momentum. The maximum rates of transfer for the pollutant gases to canopies of vegetation are therefore somewhat smaller than those of momentum, and differ between gases as a consequence of the different molecular diffusivities of the different gases. The individual values for the maximum rate of deposition ($v_{max}$) for $SO_2$, $NO_2$ and $NH_3$ to a 10m forest and 15cm moorland in Table I differ as a consequence of differences in $r_b$.

The deposition velocity $V_d$ is simply the reciprocal of the total resistance in the deposition pathway ($r_t$) i.e. $r_t = V_d^{-1}$. In the case of $V_{max}$, the maximum rate of deposition is simply the condition in which all molecules transported to the surface are absorbed (or react). In practice, the gas may not react readily at the surface and the extent to which a gas is not removed from the atmosphere by contact with the surface may be expressed as a residual, or surface resistance. The maximum rates of deposition onto a forest, in these conditions, vary between 89% and 160% larger than their maximum rates of deposition onto moorland under the same atmospheric conditions. The largest absolute value of $V_{max}$ and largest forest-moorland difference are for $NH_3$ the pollutant in this group with the largest molecular diffusivity.

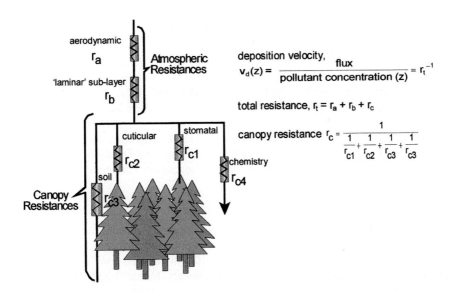

Fig. 1 A simple resistance analogy to simulate the partitioning of pollutant fluxes between the principle sinks in a forest canopy

The bulk of the published deposition velocity data for these gases show the presence of a surface resistance, although the magnitude varies according with surface and atmospheric conditions. The potential enhancement of pollutant deposition by forests is only realised for circumstances in which atmospheric transfer (i.e. $r_a + r_b$ in Figure 1) significantly influences the magnitude of $r_t$ (and hence $V_d$).

The overall deposition process is subject to controls by rates of turbulent diffusion above and within the vegetation and to chemical reaction with a range of surfaces, including epicuticular wax, leaf surface water, senescent vegetation and soil. While rates of turbulent transfer and aerodynamic resistances can be estimated directly from a knowledge of roughness characteristics of the surface and wind velocity profiles, the canopy resistance cannot in general be quantified from first principles or theory and requires empirical values from field or laboratory measurement.

## 5. Measurements of deposition fluxes and quantifying canopy resistance

The measurement of wet deposition fluxes in the field is straightforward and networks of wet deposition samplers provide the basis for wet deposition maps of Europe and North America and parts of Asia. The measurements are more difficult in complex terrain and especially in mountainous regions. However, even in these regions extensive wet deposition measurements have been made (Dore et al. 1990; Mohnen, 1992).

By contrast, the measurement of dry deposition to vegetation requires instrumentation to determine either the co-variance in short term variability in concentration $\chi^1$ and the vertical component of turbulence $w^1$ to provide the flux ($F_s$) as

$$F_s = \rho w^1 \chi^1$$

Alternatively, fluxes may be measured by aerodynamic flux gradient or Bowen ratio methods. The detailed methodology lies outside the scope of the paper and is described by Woodward and Sheehy (1983) and Fowler and Duyzer (1989) while the micrometeorological background for the work is provided by Monteith and Unsworth (1990).

These measurement methods have provided a reasonable understanding of the main sinks for the pollutant gases in forest canopies (Erisman and Draaijers, 1995; Fowler et al., 1989; Duyzer et al., 1992) and in shorter vegetation (Erisman & Wyers, 1993; Flechard and Fowler, 1998).

### 5.1 SO₂ DEPOSITION

The primary control of $SO_2$ deposition occurs at the surface, with canopy resistances generally exceeding aerodynamic resistance for the majority of the time. Typical aerodynamic resistances (rates) being between 10 s m$^{-1}$ and 20 s m$^{-1}$ for conifer forests (Jarvis et al, 1975) and 30 s m$^{-1}$ to 50 s m$^{-1}$ for cereals (Monteith and Unsworth, 1990), whereas the canopy resistances for $SO_2$ deposition range from 10 s m$^{-1}$ to 500 s m$^{-1}$ (Erisman and Draaijers, 1995; Fowler and Unsworth, 1979) with median values in excess of 100 s m$^{-1}$. The large range of canopy resistances arises as a consequence of a combination of different sinks for $SO_2$ within vegetation. For entirely dry canopies the epicuticular waxes on foliage represent a relatively weak sink for $SO_2$ with canopy resistances of 500 s m$^{-1}$. The stomata represent efficient sites of $SO_2$ uptake, limited only by the diffusive resistance to transport to the apoplast fluids within sub-stomatal cavities. Making allowance for the relative molecular diffusivities of $SO_2$ and $H_2O$, canopy conductances for $H_2O$ may be used to calibrate the magnitude of stomatal uptake and the canopy resistance with which it is associated. Thus the minimum values are typically 100 s m$^{-1}$ for the stomatal component of the canopy resistance to $SO_2$ uptake by cereals and grasses. For forests, which are characterised by larger canopy resistances for evapotranspiration, the work of Jarvis et al (1975) can be used to estimate the equivalent values of $r_{c1}$ for $SO_2$ deposition onto forests, of 150 s m$^{-1}$ to 200 s m$^{-1}$ for conifers in the absence of water stress. In practice, the two components of canopy resistance, for stomatal uptake and for uptake onto the external surfaces ($r_{c2}$) (Fig. 1) operate in parallel, and thus provide the total canopy resistance ($r_c$) as

$$rc^{-1} = r_{c1}^{-1} + r_{c2}^{-1}$$

This is the simplest representation of the processes within the canopy in which many additional sinks exist (e.g. soil, dead or senescent plant material). Furthermore, the canopy structure is far from uniform and may be better described following sub-division of the canopy into a series of horizontal layers. The more recent work on $SO_2$ from long-term monitoring of deposition fluxes shows that the overall canopy resistance is strongly linked to the chemistry in surface layers of water on the vegetation (Erisman et al, 1998a; Flechard and Fowler, 1998). The measurements show that the canopy resistance for $SO_2$ is regulated by the supply of $NH_3$, which neutralises the acidity formed by $SO_2$ oxidation to $SO_4^{2-}$ (Flechard and Fowler, 1999). Thus the local chemical climate of the atmosphere regulates the composition of surface water on vegetation and thus the canopy resistance for $SO_2$. Sufficient data have been accumulated within the recent long-term flux measurement studies to parameterise deposition models in which the interaction of pollutants in regulating deposition rates is simulated.

5. 2 $NO_2$ DEPOSITION

The uptake of $NO_2$ by a variety of tree species has been shown in laboratory conditions to be almost exclusively by stomata (Hanson and Lindberg, 1991). For a grassland canopy, Hargreaves et al. (1992) showed in field conditions that $NO_2$ uptake is limited to stomata. Such findings greatly simplify the deposition of $NO_2$ to vegetation, and reduce the differences between forests and other surfaces, since for the aerodynamically rough surfaces ($z_0 > 1$ m) bulk stomatal resistances are generally larger than for short vegetation ($z_0 < 0.2$ m). A complication arises in the measurement of $NO_2$ uptake by forests in particular since soil emissions of NO, which does not deposit at significant rates, are readily oxidised within the canopy to $NO_2$ by $O_3$ (Duyzer et al., 1995). The $NO_2$ may then be absorbed within the canopy by stomata, and such an internal cycle for the soil-plant represents no net uptake by the canopy of vegetation (Fowler et al., 1998a).

5.3 $NH_3$ DEPOSITION

The vegetation-atmosphere exchange of $NH_3$ is complicated by the presence of $NH_4^+$ within apoplast fluids of the leaf mesophyll. Thus with open stomata the net flow of $NH_3$ between the sub-stomatal cavity and the atmosphere is regulated by apoplast pH and [$NH_4^+$] and the ambient concentration of $NH_3$ within the plant canopy (Sutton et al., 1995). The process, illustrated schematically in Fig. 2 has been simulated in a simple canopy compensation point model by Sutton et al. (1993c).

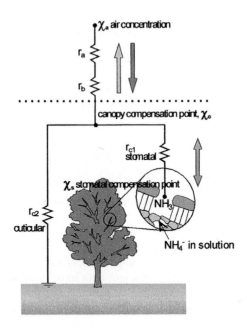

Fig. 2 A schematic representation of the canopy compensation point model to simulate emission and deposition fluxes of $NH_3$

Field measurements of $NH_3$ exchange are therefore influenced by reactions with the leaf surface chemical processes within liquid films on foliage which also $SO_2$ deposition and by the physiological controls of $NH_4$ in the apoplast. In practice, the canopies with the greatest potential for emission fluxes of $NH_3$ are the heavily fertilised agricultural canopies and especially grassland which in intensive production may receive in excess of 300 kg N ha$^{-1}$ annually in fertiliser applications. Forests in contrast, generally receive much smaller quantities of nitrogen, except in the vicinity of large agricultural sources of $NH_3$, such as livestock farms. In long-term flux measurements of $NH_3$ over semi-natural (moorland) vegetation Flechard and Fowler (1998) showed a median canopy resistance of 37 s m$^{-1}$, providing deposition velocities typically in the range 10 mm s$^{-1}$ to 20 mm s$^{-1}$, and net emission fluxes of $NH_3$ were observed only 6% of the time during 13 months of continuous measurement. Similarly, Sutton *et al.* (1993a, b) showed large rates of $NH_3$ deposition onto acidic moorland vegetation.

5.4 CONTRASTING POLLUTANT DEPOSITION ON FORESTS WITH MOORLAND
While far from comprehensive in land surface cover, these extensive measurements provide the necessary data to parameterise a simple model with canopy resistances for a forest and moorland (or other semi-natural short vegetation canopy). This allows the effect of forest on pollution fluxes to be quantified, given the same ambient pollution climate and an identical meteorological and similar surface conditions.

### TABLE II a

Effect of forests on pollutant fluxes: a remote relatively unpolluted site (S. Scotland)

| Pollutant Concentrations ($\mu g\ m^{-3}$) | | | | Dry Deposition of Gases | | |
|---|---|---|---|---|---|---|
| $SO_2$ | $NO_2$ | $NH_3$ | | $SO_2$ | $NO_2$ | $NH_3$ |
| 2.4 | 6.0 | 0.9 | | kg S ha$^{-1}$ a$^{-1}$ | kg N ha$^{-1}$ a$^{-1}$ | kg N ha$^{-1}$ a$^{-1}$ |
| Canopy resistance $R_c$ (s m$^{-1}$) | | | Moorland | 1.9 | 0.6 | 2.7 |
| | | | Forest | 2.1 | 0.6 | 4.2 |
| $SO_2$ | $NO_2$ | $NH_3$ | | | | |
| 150 | 1000 | 37 | | | | |
| | | | | Dry Deposition of Aerosols | | |
| Wet deposition fluxes (kg ha$^{-1}$ a$^{-1}$) | | | | | | |
| $SO_4^{-S}$ | $NO_3^{-N}$ | $NH_4^{-N}$ | | $SO_4^{-S}$ | $NO_3^{-N}$ | $NH_4^{-N}$ |
| 4.1 | 1.9 | 2.4 | | kg S ha$^{-1}$ a$^{-1}$ | kg N ha$^{-1}$ a$^{-1}$ | kg N ha$^{-1}$ a$^{-1}$ |
| | | | Moorland | 0.3 | 0.2 | 0.3 |
| | | | Forest | 1.0 | 0.6 | 1.0 |
| | | | | | 0.6 | 1.0 |

### TABLE IIba

Effect of forests on pollutant fluxes: a polluted site in Central Europe

| Pollutant Concentrations ($\mu g\ m^{-3}$) | | | | Dry Deposition of Gases | | |
|---|---|---|---|---|---|---|
| $SO_2$ | $NO_2$ | $NH_3$ | | $SO_2$ | $NO_2$ | $NH_3$ |
| 13.8 | 19.8 | 7.3 | | kg S ha$^{-1}$ a$^{-1}$ | kg N ha$^{-1}$ a$^{-1}$ | kg N ha$^{-1}$ a$^{-1}$ |
| (5 ppb) | (10 ppb) | (10ppb) | | | | |
| Canopy resistance $R_c$ (s m$^{-1}$) | | | Moorland | 25.3 | 1.8 | 22.6 |
| | | | Forest | 35.9 | 1.9 | 34.5 |
| $SO_2$ | $NO_2$ | $NH_3$ | | | | |
| 32 | 1000 | 37 | | | | |
| | | | | Dry Deposition of Aerosols | | |
| Wet deposition fluxes (kg ha$^{-1}$ a$^{-1}$) | | | | | | |
| $SO_4^{-S}$ | $NO_3^{-N}$ | $NH_4^{-N}$ | | $SO_4^{-S}$ | $NO_3^{-N}$ | $NH_4^{-N}$ |
| 9.3 | 5.2 | 11.3 | | kg S ha$^{-1}$ a$^{-1}$ | kg N ha$^{-1}$ a$^{-1}$ | kg N ha$^{-1}$ a$^{-1}$ |
| | | | Moorland | 3.0 | 4.1 | 1.8 |
| | | | Forest | 8.4 | 11.7 | 4.9 |
| | | | | | 0.6 | 1.0 |

The concentration of the pollutants $SO_2$, $NO_2$ and $NH_3$, the aerosol concentrations of $SO_4^{2-}$, $NO_3^-$, $NH_4^+$, the canopy resistance values for each and the wet deposition fluxes were all obtained from direct measurements at Auchencorth Moss (Fowler et al., 1996,1998b) and summarised in Table IIa and b. Assuming a constant wind speed of 5 ms$^{-1}$ at a height of 10 m (above d + $z_O$), a thermally neutral atmosphere and roughness lengths ($z_O$) of 0.01 and 1.0 m respectively for moorland and forest the annual deposition fluxes were calculated as shown in Table IIa.

The annual input fluxes of sulphur and nitrogen of 6.3 kg. S and 8.1 kg. N (Table III) to moorland and forest are small but are typical of the upland areas in the north and west of the UK (RGAR, 1997). The effect of increasing the aerodynamic roughness of the surface may be seen in the comparison for this relatively unpolluted site between moorland and forest. The change in aerodynamic roughness increases the sulphur and nitrogen deposition by 16% and 31% respectively, the larger increase in N deposition being largely a consequence of $NH_3$ deposition.

The influence of the pollution climate on both the total input and the importance of the aerodynamic effects of the forest are quantified by running the same procedure to calculate annual inputs for data from a more polluted location. In this case the pollutant data were provided by measurements from Melpitz, a pollution flux monitoring station close to the city of Leipzig (Spindler et al., 1996). The canopy resistances used were identical to those in the previous example to allow the interaction of pollutant concentration and forest moorland to be compared without the complication of different canopy resistances. The input data and output fluxes are summarised in Table IIb with a summary of the annual S and N inputs for both sites and canopies in Table III. The annual inputs of sulphur at the polluted site range from 37 kg S ha$^{-1}$ for m moorland to 54 kg S ha$^{-1}$ for forest, while nitrogen inputs range from 47 kg N ha$^{-1}$ for moorland to 70 kg N ha$^{-1}$ for the forest. These values are all large for the polluted site and the effect of the forest in enhancing the deposition is typically + 50%.

TABLE III

Effect of forests on pollutant fluxes: annual deposition fluxes of S and N at contrasting sites

|  | Total Wet + Dry Deposition | | | |
|---|---|---|---|---|
|  | Unpolluted area (S. Scotland) | | Polluted area (central Europe) | |
|  | S | N | S | N |
|  | kg S ha$^{-1}$ a$^{-1}$ | kg S ha$^{-1}$ a$^{-1}$ | kg S ha$^{-1}$ a$^{-1}$ | kg S ha$^{-1}$ a$^{-1}$ |
| Moorland | 6.3 | 8.1 | 37.6 | 46.8 |
| Forest | 7.3 | 10.6 | 53.6 | 69.5 |
| % increase Forests/Moor | 16% | 31% | 43% | 48% |

The magnitude of the effect of forests in enhancing pollutant deposition is therefore very variable, but increases rapidly for the soluble reactive pollutants $SO_2$ and $NH_3$ (and $HNO_3$) with the ambient concentration of these gases. These calculations contrast two lowland locations, one in the windy, relatively unpolluted environment of southern

Scotland and one in the more continental and polluted climate of central Europe. The exercise shows the magnitude of the forest effect on these gases and aerosols and demonstrates that, given adequate monitoring data for the pollutant gases and meteorological variables, the calculation of both total inputs and effects of different land cover is straightforward. Very few sites in rural areas of Europe provide the necessary data for these flux measurements.

## 6. Aerosols and cloud droplets

The processes which lead to the enhancement in deposition by forest relative to other terrestrial surfaces combine the general aerodynamic properties of the forest with surface processes which regulate the rate of uptake. The aerosols and cloud droplets containing pollutants are present over a broad size range from a few nm in diameter to 20 µm in diameter. Over this size range the control over deposition rates changes from processes regulating the atmospheric transfer for the larger particles to processes controlling the capture processes at the surface for the smaller particles (Beswick et al. 1991).

The large particle sizes (> 10 µm), and especially large cloud droplets, have significant rates of gravitational settling. At 20 µm droplet radius, unit density spheres sediment at about 10 mm $s^{-1}$, whereas for all particles smaller than 5 µm the gravitational settling velocity is less than 1 mm $s^{-1}$ and can generally be neglected. The rates of turbulent diffusion above forests provide upper limits for deposition rates in the range 30 mm $s^{-1}$ to 50 mm $s^{-1}$ for the gases (Table I) and also particle deposition. The extent to which these large rates apply to the particles is determined by the transport of particles through the viscous sub-layer of air close to surfaces and to the capture efficiency of the surfaces.

The conventional particle deposition theory, largely based on wind tunnel studies, shows efficient deposition of particles and droplets larger than 10 µm diameter onto vegetation (Chamberlain, 1975). For particles smaller than about 0.1 µm diameter, transport through the viscous sub-layer by Brownian diffusion provides significant deposition rates and for particles larger than 5 µm diameter interception and impaction become increasingly efficient. However, for particles in the size range 0.1 µm diameter to 5 µm diameter, the wind tunnel measurements show very inefficient transport and capture of particles on vegetation. The relationship between particle deposition and particle size from the work of Chamberlain (1975) and others is illustrated in Figure III. Also illustrated in Figure III are results of several recent field experiments summarised by Gallagher et al. (1997). This analysis shows much larger rates of deposition of sub-micron aerosols then earlier work and for which there is no underlying theoretical framework. However, the extent of the direct experimental support for larger rates of sub-micron aerosol deposition on forest is now sufficient to be confident that for these surfaces at least, aerosol deposition rates are substantially larger than the 0.1 mm $s^{-1}$ to 1.0 mm $s^{-1}$ indicated by wind tunnel studies. The potential mechanisms available to provide a mechanistic basis for the observations include electrophoresis and other phoretic mechanisms (Goldsmith et al., 1963). Somehow, the small particles are transported through the viscous sub-layer at rates which exceed rates of Brownian diffusion and for particle sizes which have insufficient inertia for efficient impaction.

However, to date the no unifying theory has been advanced to explain the observed rates of deposition of particles in the size range 0.1 µm diameter to 1.0 µm diameter. For this exercise, and based on the data provided in Figure III, a deposition velocity of 5 mm s$^{-1}$ has been assumed for particles in the size range 0.1 µm diameter to 2.0 µm diameter (i.e. including all the aerosols $SO_4^{2-}$, $NO_3^-$ and $NH_4^+$).

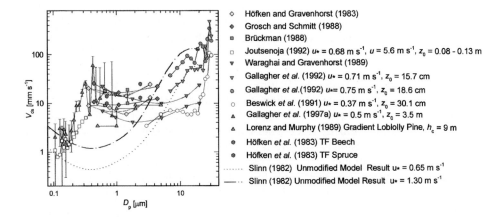

Fig. 3 The relationship between particle diameter and deposition velocity from field measurements compared with the model of Slinn (1982.) (Adapted from Gallagher et al. 1997).

To examine the effect of different canopy heights on the long-term average inputs of aerosols and cloud droplets the inventory of $^{210}$Pb in soil has been very valuable (Fowler et al., 1998d). The method relies on the deposition of radioactive $^{210}$Pb containing aerosols and cloud droplets onto vegetation, and the capture and retention of the $^{210}$Pb by organic matter in soils. The inventory of $^{210}$Pb ($I_s$) then undergoes radioactive decay with a half life (t $^1/_2$) of approximately 22 years and average lifetime ($\lambda = \ln 2/t^{1/2}$), so that the long-term annual deposition flux ($F_D$) is provided by

$$F_D = \frac{I_s}{\lambda}$$

The data in Figure IV from Fowler et al. (1998d) shows the increase in the $^{210}$Pb inventory beneath a forest canopy relative to those beneath a nearby grass canopy in the Scottish Borders. The increase in the inventory occurs largely as a consequence of a greater cloud droplet capture by the trees than by grassland. Measurements of cloud frequency and windspeed at the site allow independent estimates of cloud droplet deposition to be made for the site and these are in good agreement with the observed increase (~ 40%) in the $^{210}$Pb inventory at the site (Fowler et al., 1998d).

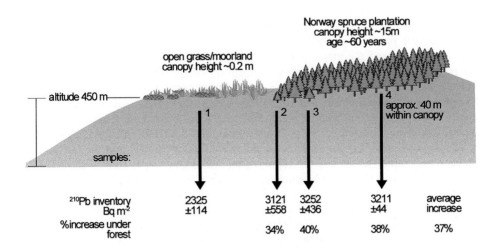

Fig. 4. The soil inventories of $^{210}$Pb under moorland and conifer forest, showing the effect of enhanced capture of cloud droplets and aerosols by the forest canopy (from Fowler et al 1998d)

Overall, the 'filtering effect' of forest canopies for pollutants, relative to shorter vegetation, is therefore supported by current understanding of the deposition processes. It is clear however, that the degree of enhancement is highly variable, depending on chemical interaction at foliar surfaces for the gases and particle size for particle deposition, as well as the meteorological and morphological characteristics of individual sites which regulate turbulent exchange between vegetation and the atmosphere.

## 7. Current and future global patterns of acidic deposition

The period of the 1970s and 1980s provided a series of research programmes investigating effects of acid deposition on forests (e.g. Heij et al., 1991; Eagar and Adams, 1992). The health of forests became an important issue in Europe and North America in particular; during this period air pollutants were believed to be responsible, at least in part, for areas of pronounced decline. No single causal mechanism was found to be responsible for problems of forest decline in the high elevation red spruce forests of the Appalachians of Eastern North America (Johnson, 1987), the fir and spruce decline in Germany and parts of France, in spruce and fir forests of the Czech Republic and Poland and in pine in the South East United States and California.

However, acid deposition and sulphur in particular was shown to be the cause of marked reductions in frost hardiness of red spruce (DeHayes, 1992; Cape et al., 1991; Sheppard et al., 1991). Acid deposition was also shown to be the major contributor to acidification of soils and along with the direct of $SO_2$ was the cause of widespread decline in high elevation conifer stands in the Czech Republic and Poland.

The control measures introduced in Europe and North America were in part influenced by these findings. Peak sulphur emissions in Europe during the late 1970s were followed by a monotonic decline from approximately 29 Tg $Sy^{-1}$ in 1980 to 15 Tg S in 1997, a 48% reduction. In North America the change has been somewhat smaller from 16 TG S $y^{-1}$ to 12 Tg $y^{-1}$ over the same period (Table IV). Sulphur has been the clear focus for political action because it was perceived as the major contributor to acidic deposition and the major cause of terrestrial effects on freshwaters and forests. The marked reductions in S emissions represent the major results of regional pollution control, and also a considerable financial investment in abatement technology throughout Europe and North America. Elsewhere in the world, and especially in the rapidly developing economies of south and east Asia as well as Africa, South and Central America, emissions of all pollutants, including S have been increasing rapidly. In S.E. Asia in particular the increase in sulphur emissions throughout the last decade has been very rapid and already this region of the world emits more sulphur dioxide into the atmosphere than Europe or North America (Galloway and Rodhe, 1991).

The sensitivity of soils and ecosystems to acidification has now been mapped at global scales by the Stockholm Environment Institute (Foell *et al.* 1995). This analysis identifies large areas within the tropics and sub-tropics in which soils are very sensitive to acidification (Rodhe and Herrera, 1988).

TABLE IV

Emissions of sulphur in Europe and N America in 1980 and 1997

|  | 1980 | 1997 |  |
| --- | --- | --- | --- |
| Europe | 29 Tg | 15 Tg | -48% |
| N. America | 16 Tg | 12 Tg | -25% |

The global inventory of sulphur emissions for 1985 and the projected emissions for the year 2050 have been used to model the global deposition of sulphur, using the Moguntia model (Crutzen *et al.*, 1997) based on the emission projections IS92a (IPCC 1995). The results of the model are shown in Figure V, which reveal large areas of Asia, Africa and Central America with increased sulphur deposition. To identify the consequence of the sulphur deposition scenarios for forests throughout the world, land cover from Matthews (1983) has been used to identify the areas of forest in two broad categories, temperate/sub polar and tropical/sub-tropical (Figure VI). The modelled sulphur deposition may then be used to calculate the areas of forest receiving atmospheric sulphur input in excess of 1 keq $ha^{-1}$ annually for 1985 and 2050 (Figure VII). Making the assumption that at an annual sulphur deposition in excess of 1 keq S $ha^{-1}$ the forest is at risk on acid sensitive soils, the areas of potential forest damage may be calculated. The value of 1 keq S $ha^{-1}$ represents a much lower sensitivity than the most sensitive soil (0.25 keq S $ha^{-1}$) applied by Fowler et al (1995) and applies a mid-point of the sensitivity scale applied (0.25 to 2.5 keq S $ha^{-1}$). The exercise, despite its simplifying assumptions, reveals an increase in the area of exceedance by a factor of

2.14 for all forest over the 65 year period (Table Va). If, however, the threshold for damage is taken as 2 kg S ha$^{-1}$ then overall the risk to forests increases by a factor of 7.2, with a 5 fold increase in temperate forests and an increase from zero to 2.3 x 10$^6$ km$^2$ of tropical forest at risk (Table Vb).

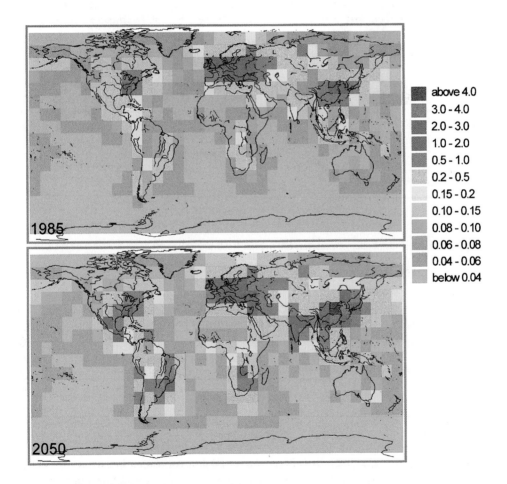

Fig. 5 Total deposition of sulphur at 10$^o$ x10$^o$ (keq H$^+$ -S ha$^{-1}$ yr$^{-1}$) from the MOGUNTIA model based on the IPCC IS92a scenario (data provided by the Stockholm Environment Institute).

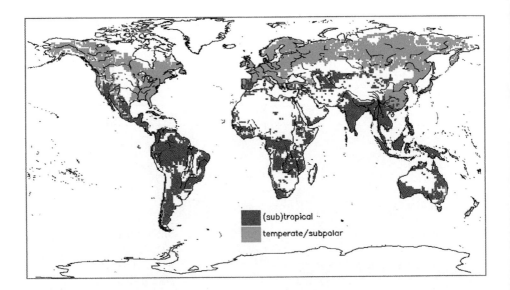

Fig. 6 The global distribution of temperate and tropical (+sub-tropical) forest cover derived from the work of Mathews (1983).

TABLE V a

Assuming forest at risk is that receiving > 1 keq $H^+$ $ha^{-1}$ as S

|  | Potential area of exceedance | | |
| --- | --- | --- | --- |
|  | all forests | tropical + subtropical | temperate + subpolar |
| 1985 | 2.77 x $10^6$ $km^2$ | 0.02 x $10^6$ $km^2$ | 2.75 x $10^6$ $km^2$ |
| % | 8 | 0.2 | 15 |
| 2050 | 5.9 x $10^6$ $km^2$ | 2.3 x $10^6$ $km^2$ | 3.6 x $10^6$ $km^2$ |
| % | 17 | 15 | 20 |
| % change 1985 to 2050 | 114 | 9200 | 32 |

TABLE V b

Assuming forest at risk is that receiving > 2 keq $H^+$ ha$^{-1}$ as S

|  | Potential area of exceedance | | |
|---|---|---|---|
|  | all forests | tropical + subtropical | temperate + subpolar |
| 1985 | $0.28 \times 10^6$ km$^2$ | 0 | $0.28 \times 10^6$ km$^2$ |
| % | 0.8 | 0 | 1.5 |
| 2050 | $5.9 \times 10^6$ km$^2$ | $2.3 \times 10^6$ km$^2$ | $3.6 \times 10^6$ km$^2$ |
| % | 6.0 | 4.4 | 7.3 |
| % change 1985 to 2050 | 624 | - | 376 |

The majority of the increase is however in tropical and sub-tropical countries in which the increase in areas of potential exceedance increases by almost two orders of magnitude. By contrast, the temperate and sub-polar forests change by a modest +32% relative to 1985. There are important uncertainties in this simplistic exercise; however, it is clear that the majority of forest acidification effects will be in developing regions of the world, in the tropics and sub-tropics in particular. There is also a clear requirement for improved understanding of processes, and monitoring of pollution climates and their effects on forests in these regions.

## 8. Eutrophication, nitrogen deposition and forests

The focus on sulphur emissions and deposition and its role in acidification throughout the 1970s and 1980s distracted attention from the other major contributors to acidification, namely $NO_y$ and $NH_x$. However, in the Scandinavian research of the 1970s it was recognised that the four dominant ions of anthropogenic origin in wet deposition were $SO_4^{2-}$, $NO_3^-$, $NH_4^+$ and $H^+$. The $NO_3^-$ and $NH_4^+$ ions are both potentially acidifying depending on their biological and chemical processing within soil and vegetation (Sutton *et al.*, 1993c). Throughout the last four decades, the deposition of nitrogen compounds has contributed a substantial fraction of the acidifying deposition. Furthermore, as sulphur emissions have declined in Europe and North America, gradually the contribution of nitrogen to deposited acidity has increased. In many parts of Europe and North America the atmospheric inputs of $NO_y$ and $NH_3$ already contribute the dominant fraction of acidifying deposition.

The control measures developed to limit sulphur dioxide emissions from large combustion plant have not been accompanied by similarly effective controls on $NO_x$

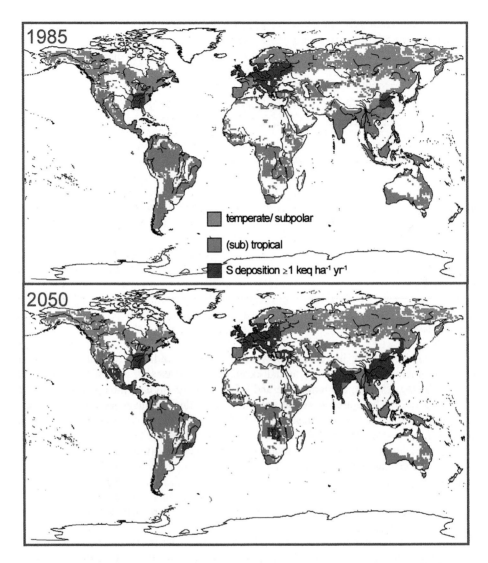

Fig. 7 The global distribution of forest cover on which total sulphur deposition exceeds 1 keq ha$^{-1}$ yr$^{-1}$, for 1985 and 2050

emissions, and there are currently very few effective controls on emissions of ammonia (NH$_3$ other than reductions in livestock numbers or intensity of production (Lekkerkirk, 1998). The greatest concern over NH$_3$ emissions has been in the Netherlands, where NH$_3$ deposition provides most of the acidifying input to terrestrial surfaces throughout the country and to forests in particular (Erisman and Draaijers, 1995). The introduction of control measures has not resulted in the expected reductions in NH$_3$ concentrations or deposition (Erisman et al., 1998b). Current understanding of the effects of different control techniques on the net exchange of NH$_3$ between soil-vegetation and the atmosphere has therefore to be significantly improved before effective controls on $NH_3$ emissions can be implemented. It seems likely therefore that the deposition of nitrogen compounds will increasingly dominate the acidifying atmospheric inputs to forests throughout Europe and North America over the next decade.

Atmospheric inputs of fixed nitrogen to forests have also attracted recent interest as a potential cause of increased forest growth over a wide geographical scale in Europe (Speicker et al., 1996). Other potential causes of the increased forest growth include $CO_2$ fertilisation as a consequence of the increase in atmospheric $CO_2$ concentration which between 1960 and 1998 changed from 320 ppm to 365 ppm (Neftel et al., 1985). The current interest in nitrogen deposited onto forests therefore includes both of the major regional and global environmental problems, acidification and global warming.

Critical loads for nitrogen inputs to forests have been developed from mass balance and empirical approaches (Hornung et al., 1995, Kuylenstierna et al, 1999). While there is still considerable uncertainty in the values selected, both approaches include values in the range of 15 to 20 kg N $ha^{-1}$ $y^{-1}$ for the most sensitive forest ecosystems.

In considering the current interest in deposition of fixed nitrogen onto forests, it is useful to draw a distinction between the regional increases in deposition that have taken place over the last 100 years and local deposition of nitrogen close to point sources in which inputs may exceed 50 kg N $ha^{-1}$ annually. The regional increases from 2 to 5 kg N $ha^{-1}$ to 10 to 20 kg N $ha^{-1}$ $y^{-1}$ have occurred over substantial areas of the industrialised regions of Europe and the USA. Taking the UK as an example, the annual deposition of fixed nitrogen as $NO_y$ and $NH_x$, is approximately 385 kt N (Fowler et al., 1998e), which averages the deposition to terrestrial surfaces in the UK at about 16 kg N $ha^{-1}$ with a range from 5 kg N $ha^{-1}$ to 40 kg N $ha^{-1}$ (Goulding et al., 1998).

The terrestrial effects of this scale of increase in nitrogen input include increased emissions of the radiatively active $N_2O$ to the atmosphere and also enhanced emissions of NO (Skiba et al., 1994). However, the magnitude of these increased emissions is small relative to the potential for terrestrial sequestration of $CO_2$ from the atmosphere by increased forest growth as a consequence of increased N deposition.

The average increase in regional deposition of nitrogen conceals 'hot spots' which result from proximity to large sources. The dry deposition rate of $NH_3$ is substantially larger than that for $SO_2$ and this, along with the ground level sources of $NH_3$ exaggerates the spatial variability in N deposition relative to that in S (Fowler et al. 1998c). These hot spots include the areas close to intensive livestock production units ($NH_3$), hill tops on which cloud droplet deposition may contribute substantial inputs and urban areas with large $NO_2$ concentrations. At such sites, the local N input from the atmosphere may exceed 50 kg N $ha^{-1}$ (Fowler et al., 1998c). The dispersion of $NH_3$ from ground level sources and the typical size of livestock units leads to the individual areas of high deposition close to each farm being small. Taking as an example a poultry farm surrounded by woodland, the woodland between 15m and 50m downward of the source experienced a mean $NH_3$ concentration of 29 $\mu g m^{-3}$ and received approximately 42 kg $NH_3^-$ N $ha^{-1}$ annually from the source. At a distance of 150m from the source the ambient $NH_3$ concentration was 4 $\mu g$ $NH_3$ $m^{-3}$ and the dry deposition was 8 kg $NH_3^-$ N $ha^{-1}$ annually. Thus the broad regional maps smooth the variability due to landscape features and conceal the presence of hot spots of deposition close to local sources.

## 9. Photochemical Oxidants

Among the pollutants recognised as potential causes of forest decline in Europe and North America, $O_3$ was identified as a probable contributor (McLaughlin and Kohut, 1992). From field studies however, in most of the areas in which clear evidence of dieback of forests has been observed, there are several possible causes. In the case of high elevation conifers, these are subjected to prolonged exposure to cloud water containing large concentrations of the major ions $SO_4^{2-}$ $NO_3^-$ $NH_4^+$ and $H^+$ (Mohnen, 1992). At some forest locations, the bulk of the sulphur and nitrogen is deposited as cloud droplets, with concentrations of the major ions between 5 and 7 times those in precipitation at the same site (Crossley et al., 1992). The same high elevation locations experience prolonged exposure to elevated ozone largely because at these altitudes the development of a stable nocturnal boundary layer is suppressed by the larger wind speeds (PORG, 1997). However, laboratory and open-top chamber studies in which ozone has been introduced in controlled experiments show clear effects on the physiology and growth of many of the common tree species of temperate forests (Skarby et al., 1998; Chappelka and Samuelson, 1998). The threshold concentration above which ozone may influence physiological processes in trees is **not** well defined. In practice, a cut-off at 40 ppb has been widely adopted in assessing the potential for ozone damage to both field crops and trees (Karenlampi and Skarby, 1996) even though there may well be important effects at smaller concentrations. The use of the arbitrary cut-off of 40 ppb has also been helpful in developing 'dose'-effect relationships in which biomass yield, especially of cereal crops, has been linearly related to the accumulated exposure to concentrations in excess of 40ppb, the $AOT_{40}$ concept (Fuhrer et al., 1997). However, the threshold is arbitrary and, in practice, there may well be physiologically important effects at concentrations significantly smaller than 40 ppb.

The threshold of 40 ppb is however very close to ambient background concentrations at many exposed locations in mid-northern latitudes during the Spring and Summer months. To contrast the global distribution of ozone concentrations which represent a threat to forest productivity and health, it is helpful to work at higher concentrations. At 60 ppb, surface ozone concentrations are both significantly elevated above the zonal background concentration and are also sufficiently large to represent a potential threat to the productivity of many forests.

Networks of ozone monitoring stations have been established in many of the countries of Europe as well as in the USA, Canada and Japan. These direct measurements are restricted mainly to the last decade and are therefore not adequate to provide a chronology of $O_3$ concentration over the last century, during which the concentrations are believed to have increased and are of course limited to a small area of the global forests. An alternative method of assessing the global exposure of forests to potentially changing $O_3$ concentrations is to simulate the photochemical production of ozone within a global model containing the emission inventories for the ozone precursors, the fluid dynamics which transport the pollutants and the processes that remove soluble, and reactive species through wet and dry deposition. Such models have made great progress in recent years in simulating the concentrations and their seasonal and changes (Collins et al., 1997, Wang et al, 1998).

The model used for this paper is a 3-D chemistry-transport model (CTM) run off line from archived meteorological data generated by the UKMO unified model at climate resolution (3.75° longitude x 2.5° latitude with 19 vertical levels). The archived meteorological information provides winds, pressures, temperatures, humidity, tropopause height, cloud and precipitation and boundary layer and surface properties. The model (STOCHEM) described by Collins et al. (1997) used a Lagrangian method in which the atmosphere comprises 50,000 air parcels in which the chemical processing of pollutants is simulated. The model uses 70 chemical species and 174 reactions. It has been used to simulate the atmospheric transport, and chemistry of the pollutant emissions using 5 scenarios for 1860, 1950, 1970, 1990 and 2100 each for a period of 15 months from which the first 3 months were discarded (Stevenson et al., 1998). Thus, the effect of changes in anthropogenic and biomass burning emissions over 240 year period on the global distribution of tropospheric ozone are simulated.

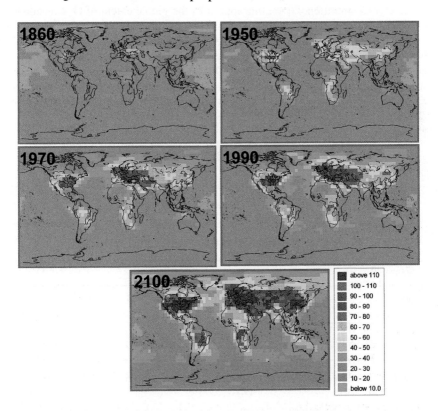

Fig. 8 July average global surface ozone concentrations at 5km x 5km from the STOCHEM model (Stevenson et al 1998).

The results in map form are shown in Figure VIII. The spatial resolution is sufficient to show the regional patterns and the 1860 map shows no areas in which July average surface $O_3$ concentration exceeds 50 ppb and peak concentrations of approximately 25 ppb at a latitude of 40°N. By 1950, substantial areas of Europe, the Middle East and the southern States of the USA experience peak daily $O_3$ concentrations in excess of 60 ppb with small areas reaching average concentrations of 70 ppb. The areas of experiencing elevated $O_3$ concentration expand rapidly between 1950 and 1990 and average concentrations in the most polluted areas reach 80 ppb. The simulation for the year 2100 is based on emissions according to the IS92a scenario (IPCC 1992) using meteorology derived from 2 x $CO_2$ climate.

The land cover maps (Figure IX) have been used to quantify the areas of global forest exposed to peak surface concentrations in excess of 60ppb. The use of 60ppb as the threshold to quantify the aereal and temporal changes in the exposure of phytotoxic ozone overcomes some of the problems of uncertainty in the analysis. However, uncertainty in $O_3$ exposure due to the vertical gradient in $O_3$ concentration in the lowest 10m to 20m of the boundary layer (Grunhage et al. 1999), and the uncertainties in precursor emission and the overall chemical regulation of boundary layer $O_3$ concentration, argue for a cautious approach in quantifying the likely distribution of phytotoxic $O_3$ concentration. Given this approach, the global extent of $O_3$ exceedance is largely restricted to the temperate latitudes in 1950 and extends to $1.7 \times 10^6$ km$^2$ or 9% of the temperate and sub-polar forest (Table VI). For 1970, $6.3 \times 10^6$ km$^2$ or 18% of global forest is exposed to potentially phytotoxic $O_3$ concentrations, of which 2/3 is in the temperate and sub-polar latitudes. By 1990 the aereal extent of the forest exposed to $O_3$ concentrations in excess of 60ppb has grown to $8.3 \times 10^6$ km$^2$, almost a quarter of

Fig. 9 Global areas of forest cover where the July peak surface ozone concentration exceeds 60 ppb from the model output of the STOCHEM model (Stevenson et al.1998).

TABLE VI
Assuming forest at risk is that exposed to surface ozone concentrations >60 ppb

| | Potential area of exceedance | | |
|---|---|---|---|
| | all forests | tropical + subtropical | temperate + subpolar |
| 1860 | no exceedance of 60 ppb | | |
| 1950 % | $2.1 \times 10^6$ km$^2$ 6.3 | $0.5 \times 10^6$ km$^2$ 2.9 | $1.7 \times 10^6$ km$^2$ 9.2 |
| 1970 % | $6.3 \times 10^6$ km$^2$ 18.5 | $1.9 \times 10^6$ km$^2$ 12.1 | $4.4 \times 10^6$ km$^2$ 24.1 |
| 1990 % | $8.3 \times 10^6$ km$^2$ 24.4 | $3.0 \times 10^6$ km$^2$ 19.1 | $5.3 \times 10^6$ km$^2$ 29.1 |
| 2100 % | $17.0 \times 10^6$ km$^2$ 49.8 | $6.0 \times 10^6$ km$^2$ 37.9 | $11.0 \times 10^6$ km$^2$ 60.2 |

The global forest with 60% in the temperate and sub-polar region. During the period 1950 to 1990 as well as the overall increase in area of forest exposed (by a factor of 4), the relative increase in area is almost an order of magnitude over the 40 years while the increase in the temperate zone is only (!) a factor of 3. The modelled forest exposure for 2100 is necessarily dependent on global emissions for this year and is somewhat

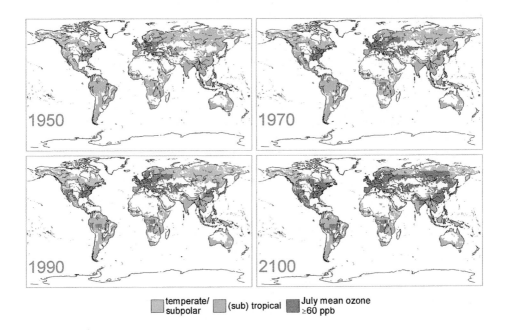

Fig. 9 Global areas of forest cover where the July peak surface ozone concentration exceeds 60 ppb from the model output of the STOCHEM model (Stevenson *et al.*1998).

speculative. However, the very large area (17 x $10^6$ km$^2$), representing 50% of global forest exposed to potentially phytotoxic O$_3$ concentrations, identifies clearly the potential magnitude of the problem. While projected emissions and climate a century in advance introduce considerable uncertainty in the actual exposure to O$_3$, it seems probable from the simulation that photochemical oxidants will remain a major global pollutant threat to forest productivity globally.

Many variables interact to determine the actual exposure to potentially damaging concentrations. Considering first the damage mechanism, which is much more likely to be due to the absorbed dose of O$_3$ exceeding a threshold for toxicity than simply exposure to large ambient concentrations (Fowler *et al.*, 1995). Such a threshold would almost certainly vary between species and with different environmental conditions. The cumulative exposure over a growing season in excess of the threshold may then be used to quantify the likely scale of dry matter lost. A very speculative exercise could be attempted for the global exceedance to 60ppb O$_3$. However, in practice the elevated CO$_2$ concentrations for the year 2100 will influence the photosynthetic process and net carbon exchange such that the water-use efficiency may have changed considerably. Such effects would decrease canopy conductance and both evapotranspiration and the ozone flux would be appreciably smaller. Regional modification of evapotranspiration, and therefore leaf temperatures, as a consequence of a larger fraction of net radiation partitioned into sensible heat loss as a consequence of smaller canopy stomatal conductances, would also feed-back to further decrease the canopy stomatal conductance. These strong feedbacks would be expected to reduce the canopy ozone fluxes, and thus lead to smaller exceedance of fluxes in excess of a threshold for dry matter loss. Therefore, even given accurate forecasts of the ozone concentrations at dates sufficiently far into the future, the climate would probably have changed significantly, and any attempt to quantify the response of forest productivity becomes extremely speculative.

A further, and probable consequence of elevated $CO_2$ in a warmer climate on boundary layer ozone concentration would be a reduction in the rate of removal of ozone at the surface by dry deposition due to reduced canopy conductance, and this would increase boundary layer $O_3$ concentrations.

TABLE VII

% Change in global areas of potential ozone effects

| % Change | All forests | Tropical + subtropical | Temperate + subpolar |
|---|---|---|---|
| 1950 – 1970 | 195 | 318 | 162 |
| 1970 – 1990 | 32 | 59 | 21 |
| 1990 – 2100 | 104 | 98 | 107 |
| 1950 – 2100 | 694 | 1214 | 553 |

## 10. Conclusions

The underlying processes contributing to pollutant deposition on forests are reasonably well understood for the pollutant gases $SO_2$, $NO_2$, $NH_3$, $HNO_3$ and $O_3$. While uncertainties remain on the chemical interaction of pollutants on foliar surfaces and especially within water films on vegetation, rates of deposition of the pollutants on specified forest can be quantified given ambient concentrations of the pollutants and knowledge of the key environmental variables (wind velocity, air temperature, solar radiation and the presence of surface water on foliage). The average rates of deposition are appreciably larger for the gases $SO_2$, $NO_2$ and $NH_3$ onto forests than onto shorter vegetation. This leads to inputs of fixed atmospheric nitrogen into forests in the UK exceeding those for most other land classes by between 30% and 50% depending on the pollution climate of the site. Rates of particle deposition onto forest also exceed those for other land classes, although important uncertainties exist in current understanding of the physical processes which create these differences. In particular, the deposition rates of sub-micron aerosols onto forests appear from recent field studies to exceed those suggested earlier from wind tunnel studies by an order of magnitude. Since rates of turbulent transfer and the associated aerodynamic resistance are reasonably well known, the mechanism(s) leading to efficient transfer through the viscous sub-layer require explanation.

Cloud droplet capture by forest remains a very efficient deposition process for high elevation forest, and in polluted regions this mechanism may lead to very large inputs. The systematic study of chemical composition of orographic cloud composition around the globe may identify important regional pollutants and considerably aid understanding of hydrochemical cycling processes in remote regions.

The current temporal trends of acidifying pollutant emissions show a marked contrast between the total emissions, which are increasing globally, and emissions in Europe and North America, which are declining. These trends will, over the next decade, appreciably reduce the areas of forest in Europe and N. America currently receiving inputs of acidity in excess of 2 keq $H^+$ $ha^{-1}$ annually. However, increasing emissions in rapidly developing countries, and especially in South East Asia, will greatly increase the area of forest in exceedance of 2 keq $H^+$ $ha^{-1}$ annually. By 2050 more than half of the global forest subject to acidifying inputs in excess of 2 keq $H^+$ $ha^{-1}$ annually will be in tropical and sub-tropical countries.

The photochemical oxidants, and $O_3$ in particular, have been shown to be major pollutants for forests globally, and to have increased rapidly throughout the last half century. Model forecasts of the global exposure of forest to $O_3$ indicate continued increases in exceedance of thresholds for physiological effects, and probably also dry matter decreases. The timescale of the projections is sufficiently long that important interactions between the effects of climate, $CO_2$ and $O_3$ on the responses of forests to $O_3$ are likely. The scale of the interactions is also likely to be of the same order as the direct effect of the $O_3$, so that significant positive or negative feedback is likely. Within the same general theme, the steady increase in deposition of nitrogen compounds (as oxidised and reduced compounds $NO_2$, $NH_3$, $NO_3^-$, $HNO_3$, $NH_4^+$) with time is also likely to interact with climate change and enhanced $CO_2$. Thus, the net effect on forest composition, productivity and sensitivity to environmental stresses becomes very difficult to predict. There is little doubt that given the scope for environmental changes in the chemical and physical climates to which forests will be exposed over the next century that current models could simulate dramatic changes to the composition of the afforested landscape. It is much more likely that the actual effects which occur will be unforeseen, and that they may be either positive or negative, and only extensive monitoring and research programmes in this field will provide the opportunity to identify such effects sufficiently early to introduce adequate control measures to prevent important social and economic consequences.

## Acknowledgements

The authors are grateful for the help of colleagues at their respective institutions for the background work necessary to enable this review to be completed. The UK authors gratefully acknowledge support of the UK Department of Environment, Transport and the Regions for support for this work.

## References

Beswick, K.M., Hargreaves, K.J., Gallagher, M.W., Choularton, T.W. and Fowler, D.: 1991, Size resolved measurements of cloud droplet deposition velocity to a forest canopy using an eddy correlation technique. *Q. J. R. Met. Soc.* **117**: 623-645.

Cannell, M.G.R., Thornley, J.H.M., Mobbs, D.C. and Friend, A.D.: 1997, UK conifer forests may be growing faster in response to increased N deposition, atmospheric $CO_2$ and temperature, *Forestry*, Vol. **71**, No. 4.

Cape, J.N., Leith, I.D., Fowler, D., Murray, M.B., Sheppard, L.J., Eamus, D. and Wilson. R.H.F.: 1991, *New Phytol*, **118**: 119-126.

Chamberlain, A.C.: 1960, Aspects of the deposition of radioactive and other gases and particles. *Int. J. Air Pollut.* **3**: 63-88.

Chamberlain, A.C.: 1975, The movement of particles in plant communities. In *Vegetation and the Atmosphere, Volume 1 Principles*, ed. J.L. Monteith, 155-203. Academic Press, London.

Chappelka, A.H. Samuelson, I.J.:1998, Ambient ozone effects on forest trees of the eastern United States: a review. *New Phytol.* **139**:91-108.

Collins, W.J., Stevenson, D.S., Johnson, C.E. and Derwent, R.G..: 1997, Tropospheric ozone in a global-scale three-dimensional Lagrangian model and its response to $NO_x$ emission controls. *Journal of Atmospheric Chemistry* **26**: 223-274.

Crossley, A., Wilson, D.B. & Milne, R.:1992, Pollution in the upland environment. *Environ. Pollut.* **75**, 81-88.

Crutzen, P., Brost, B., Dentener, F., Feichter, H., Hein, R., Kanakidou, M., Lelieveld, J. & Zimmermann, P.: 1997, Development of a time-dependent global tropospheric air chemistry model "GLOMAC" based on the weather forecast of the "ECMWF". In: *Tropospheric Modelling and Emission Estimation*, eds. A.Ebel, R Friedrich & H Rodhe, 380-391. Springer-Verlag Berlin Heidelberg New York.

DeHayes, D.H.: 1992, Winter Injury and Developmental Cold Tolerance of Red Spruce. In: *Ecology and Decline of Red Spruce in the Eastern United States*, eds. Eagar, C. and Adams, M.B., Berlin-Heidelberg, Springer-Verlag.

Dore, A.J., Choularton, T.W., Fowler, D. and Storeton-West, R.: 1990, Field measurements of wet deposition in an extended region of complex topography. *Q. J. Roy. Met. Soc.* **116**: 1193-1212.

Duyzer, J.H., Verhagen, H.L.M., Westrate, J.H. & Bosveld, F.C.: 1992, Measurement of the dry deposition flux of $NH_3$ on to coniferous forest. *Environ. Pollut.*, **75**, 3-14.

Duyzer, J., Deinum, G. and Baak, J.: 1995, The interpretation of measurements of surface exchange of nitrogen oxides; correction for chemical reactions. *Philosophical Transactions of the Royal Society London* **351**: 231-248.

Eagar, C. and Adams, M.B. (eds.) 1992, Ecology and Decline of Red Spruce in the Eastern United States, Berlin-Heidelberg, Springer-Verlag.

Erisman, J.W. and Draaijers, G.P.J.: 1995, Atmospheric Deposition in relation to Acidification and Eutrophication, Elsevier Science B.V., Amsterdam.

Erisman, J.W. and Wyers, G.P: 1993, Continuous measurements of surface exchange of $SO_2$ and $NH_3$: implications for their possible interaction in the deposition process. *Atmos Environ* **A27**: 1937-1949.

Erisman, J.W., Bleeker, A. and Jaarsveld, J.A. van: 1998b, Evaluation of ammonia emission abatement on the basis of measurements and model calculations, *Environ Pollut* **102**, S1: 269-274.

Erisman, J.W., Mennen, M., ., Fowler, D., Flechard, C.R., Spindler, G., Gruner, G., Duyzer, J.H., Ruigrok, W. and Wyers, G.P.:1998a, Deposition monitoring in Europe. *Environ. Monit. Assess.* **53**, 279-295.

Flechard, C.R. and Fowler, D.: 1998, Atmospheric ammonia at a moorland site. II: Long-term surface/atmosphere micrometeorological flux measurements. *Q.J.R. Meteorol. Soc.* **124**: 733-757.

Flechard, C.R., Fowler, D., Sutton, M.A. & Cape, J.N.:1999, A dynamic chemical model of bi-directional ammonia exchange between semi-natural vegetation and the atmosphere. *Q.J.R.Meteorl. Soc.* **125**, 1-33.

Foell, W., Green, C., Amann, M., Bhattacharya, S., Carmichael, G., Chadwick, M., Cinderby, S., Haugland, T., Hettlingh, J-P., Hordijk, L., Kuylenstierna, J., Shah, J., Shrestha, R., Streets, D. & Zhao, D.: 1995, Energy use, emissions, and air pollution reduction strategies in Asia. Water Air and Soil Pollut., 85 2277-2282.

Fowler, D. and Duyzer, J.H.: 1989, Micrometeorological techniques for the measurement of trace gas exchange. In: *Exchange of Trace Gases Between Terrestrial Ecosystems and the Atmosphere* (eds. Andreae, M.O. and Schimel, D.S.): 189-207. Wiley, New York.

Fowler, D. and Unsworth, M.H.: 1979, Turbulent transfer of sulphur dioxide to a wheat crop. *Q. J. R. Met. Soc.* **105**: 767-784.

Fowler, D., Cape, J.N. and Unsworth, M.H.: 1989, Deposition of atmospheric pollutants on forests, *Phil. Trans. R. Soc. Lond.* B **324**: 247-265.

Fowler, D., Smith, R.I., Coyle, M., Weston, K.J., Davies, T.D., Ashmore, M.R. and Brown, M.: 1995, Quantifying the fine scale (1km x 1km) exposure, dose and effects of ozone. Part 1. Methodology and application for effects on forests. *Water Air and Soil Pollution* **85**: 1479-1484.
Fowler, D., Flechard, C.R., Milford, C., Hargreaves, K.J., Storeton-West, R.L., Nemitz, E. and Sutton, M.A.: 1996, *Measurements of pollutant concentrations and deposition fluxes to moorland at Auchencorth Moss in southern Scotland, 1995*. 47 pp. Commisssion of the European Communities.
Fowler, D., Flechard, C., Skiba, U., Coyle, M. and Cape, J.N.: 1998a, The atmospheric budget of oxidized nitrogen and its role in ozone formation and deposition. *New Phytol.* **139**: 11-23.
Fowler, D., Flechard, C., Sutton, M.A. and Storeton-West, R.L.: 1998b, Long term measurements of the land-atmosphere exchange of ammonia over moorland. *Atmos Environ* **32**: 453-459.
Fowler, D., Pitcairn, C.E.R., Sutton, M.A., Flechard, C., Loubet, B., Coyle, M. and Munro, R.C.: 1998c, The mass budget of atmospheric ammonia in woodland within 1 km of livestock buildings, *Environ Pollut* **102**: 343-348.
Fowler, D., Smith, R.I., Leith, I.D., Crossley, A., Mourne, R.W., Branford, D.W. and Moghaddam, M.: 1998d, Quantifying fine-scale variability in pollutant deposition in complex terrain using $^{210}$Pb inventories in soil. *Water, Air and Soil Pollution* **105**: 459-470.
Fowler, D., Sutton, M.A., Smith, R.I., Pitcairn, C.E.R., Coyle, M., Campbell, G. and Stedman, J.: 1998e, Regional mass budgets of oxidized and reduced nitrogen and their relative contribution to the nitrogen inputs of sensitive ecosystems, *Environ Pollut* **102**, S1: 337-342.
Fuhrer, J., Skarby, L. and Ashmore, M.R.: 1997, Critical levels for ozone effects on vegetation in Europe. *Environ Pollut* **97**: 91-106.
Gallagher, M.W., Beswick, K.M., Duyzer, J., Westrate, H., Choularton, T.W. and Hummelshoj, P.: 1997, Measurements of aerosol fluxes to Speulder Forest using a micrometeorological technique. *Atmos Environ* **31**: 359-373.
Galloway, J.N. and Rodhe, H.: 1991, Regional atmospheric budgets of S and N fluxes: how well can they be quantified? *Proceedings of the Royal Society of Edinburgh*, **97B**: 61-80.
Goldsmith, P., Delafield, J.H. and Cox, L.C.: 1963, The role of diffusiophoresis in the scavenging of radioactive particles from the atmosphere. *Quart. J.R. Met. Soc.* **89**: 43-61.
Goulding, K.W.T., Bailey, N.J., Bradbury, N.J., Hargreaves, P., Howe, M., Murphy, D.V., Poulton, P.R. and Willison, T.W.: 1998, Nitrogen deposition and its contribution to nitrogen cycling and associated soil processes, *New Phytol.* **139**: 49-58.
Grünhage, L., Jäger, H.-J., Haenel, H.-D., Löpmeier, F.-J., Hanewald, K.:1999, The European critical levels for ozone: improving their usage. *Environ Pollut.* **105**: 163-174.
Hanson, P.J. and Lindberg, S.E.: 1991, Dry deposition of reactive nitrogen compounds: a review of leaf, canopy and non-foliar measurements. *Atmos Environ* **25A**: 1615-1634.
Hargreaves, K.J., Fowler, D., Storeton-West, R.L. and Duyzer, J.H.: 1992, The exchange of nitric oxide, nitrogen dioxide and ozone between pasture and the atmosphere. *Environ Pollut.* **75**: 53-59.
Heij, G.J. & Schneider, T. Eds.: 1991, *Acidification Research in The Netherlands*. Studes in Environmental Science 46. Elsevier Science.
Hornung, M., Sutton, M.A. & Wilson, R.B.: 1995, *Mapping and Modelling of Critical Loads for Nitrogen – a Workshop Report*, Proceedings of UNECE workshop, Grange-over-Sands, October 1994. Institute of TerrestrialEecology, Edinburgh.
IPCC,: 1992, *Climate Change 1992. The Supplementary Report to the IPCC Scientific Assessment*. Cambridge University Press.
IPCC,: 1995,Climate Change 1994. *Radiative Forcing of Climate Change and An Evaluation of the IPCC 1S92 Emission Scenarios*. Cambridge University Press.
Jarvis, P.G., James, G.B. and Landsberg, J.J.: 1975, *Coniferous Forest*. In: Vegetation and the Atmosphere, ed. Monteith, J.L., Academic Press Inc. (London) Ltd.
Johnson, A.H.: 1987, Deterioration of red spruce in the Northern Appalachian Mountains. In *Effects of atmospheric pollutants on forests, wetlands and agricultural ecosystems* (ed. Hutchison, T.C. and Meema, K.M.) NATO ASI series vol. 916): 83-99. Berlin-Heidelberg: Springer-Verlag.
Kärenlampi, L. & Skärby, L. Eds:1996, *Critical levels for Ozone in Europe: Testing and Finalising the Concepts*. Proceedings of UNECE Workshop, Kuopio, Finland, 1996, University of Kuopio.
Kuylenstierna, J. C.I., Hicks, W.K., Cinderby, S. & Cambridge, H.:,1999 Critical loads for nitrogen deposition and their exceedance at European scale. *Environ. Pollut.* **102**:591-598.
Lekkerkerk, I.J.A.: 1998, Implications of Dutch ammonia policy on the livestock sector. . *Atmos Environ* **32**: 581-588.
Matthews, E.: 1983, Global vegetation and land use:new high-resolution data bases for climate studies. *J. Clim. Appl. Meteorrol.*, **22**, 474-487.

McLaughlin, S.B. and Kohut, R.J.: 1992, The Effects of Atmospheric Deposition and Ozone on Carbon Allocation and Associated Physiological Processes in Red Spruce. In: *Ecology and Decline of Red Spruce in the Eastern United States*, eds. Eagar, C. and Adams, M.B., Berlin-Heidelberg, Springer-Verlag.

Miller, P.R., Parmeter, J.R. Jr., Taylor, O.C. and Cardiff, E.A.: 1963, Ozone injury to the foliage of Ponderosa pine. *Phytopath* **53**: 1072-1076.

Mohnen, V.A.: 1992, Atmospheric Deposition and Pollutant Exposure of Eastern U.S. Forests. In: *Ecology and Decline of Red Spruce in the Eastern United States*, eds. Eagar, C. and Adams, M.B., Berlin-Heidelberg, Springer-Verlag.

Monteith, J.L. and Unsworth, K.H.: 1990, *Principles of Environmental Physics*. Edward Arnold, London.

Neftel, A, Moor, E., Oeschger, H. and Stauffer, B.: 1985, Evidence from polar ice cores for the increase in atmospheric CO2 in the past two centuries. *Nature* **315**: 45-47.

PORG: 1997. *Ozone in the UK*. Fourth report of the Photochemical Oxidants Review Group. 234 pp. Department of the Environment, Transport and the Regions, London. (ITE Edinburgh).

RGAR: 1997, Acid deposition in the United Kingdom 1992-1994. In *The Fourth Report of the United Kingdom review group on acid rain*, DETR.

Rodhe, H. and Herrera, R.(eds): 1988, *Acidification in Tropical Countries, Scope 36*, J. Wiley, Chichester, Great Britain.

Schulze, E.-D. and Freer-Smith, P.H.: 1991, An evaluation of forest decline based on field observations focussed on Norway spruce, Picea abies. *Proceedings of the Royal Society of Edinburgh*, **97B**, 155-168.

Sheppard, L.J.: 1991, Causal mechanisms by which sulphate, nitrate and acidity influence forest hardiness in red spruce: review and hypothesis, *New Phytol*. **127**, 69-82.

Shuttleworth, W.J.: 1989, Micrometeorology of temperate and tropical forest, *Phil. Trans. R. Soc. Lond.* B **324**: 299-334.

Skarby, L., Ro-Poulsen, H., Wellburn, F.A.M. and Sheppard, L.J.: 1998, Impacts of ozone on forests: a European perspective, *New Phytol*, **139**, 109-122.

Skiba, U., Fowler, D. and Smith, K.A.: 1994, Emissions of NO and $N_2O$ from soils. In: Ham, L.J.J.van, Janssen, H.M. and Swart, R.J. (eds). *Non-$CO_2$ Greenhouse Gases*, Dordrecht: Kluwer:153-158.

Slinn,: 1982, Predictions for particle deposition to vegetative caopies. *Atmos. Environ.* 16, 1785-1794.

Spiecker, H., Mielikainen, K., Kohl, M and Skovsgaard, J.P. (eds.): 1996, *Growth Trends in European Forests*, Springer-Verlag, Berlin Heidelberg.

Spindler, G. & Gruner, A.:1996, Towards development of a deposition monitoring network for air pollution of Europe; The Melpitz site in Germany – Results of measurements made in 1995, IFT, Leipzig, Germany.

Stevenson, D.S., Johnson, C.E., Collins, W.J., Derwent, R.G., Shine, K.P. and Edwards, J.M.: 1998, Evolution of tropospheric ozone radiative forcing. *Geophysical Research Letters* **25**: 3819-3822.

Sutton, M.A., Fowler, D. and Moncrieff, J.B.: 1993a, The exchange of atmospheric ammonia with vegetated surfaces. I Unfertilized vegetation. *Q. J. R. Met. Soc.* **119**: 1023-1045.

Sutton, M.A., Fowler, D., Moncrieff, J.B. and Storeton-West, R.L.: 1993b, The exchange of atmospheric ammonia with vegetated surfaces. II: Fertilized vegetation. *Q. J. R. Met. Soc.* **119**: 1047-1070.

Sutton, M.A., Pitcairn, C.E.R. and Fowler, D.: 1993c, The exchange of ammonia between the atmosphere and plant communities. *Advances in Ecological Research*, 24: 301-393.

Sutton, M.A., Schorring, J.K. and Wyers, G.P.: 1995, Plant-atmosphere exchange of ammonia. *Philosophical Transactions of the Royal Society London* A. **351**: 261-278.

Thom, A.S.: 1975, Momentum, mass and heat exchange of plant communities. In *Vegetation and atmosphere* (ed. Monteith, J.L.): 57-109. London: Academic Press.

Thornley, J.H.M., Fowler, D. and Cannell, M.G.R.: 1991, Terrestrial carbon storage resulting from $CO_2$ and nitrogen fertilisation in temperate grasslands. *Plant Cell and Environment*, **14**: 1007-1011.

Wang, Y., Logan, J.A., Jacob, D.J. and Spivakovsky, C.M.: 1998, Global simulation of $O_3$-$NO_x$ – hydrocarbon chemistry, 2. Model evaluation and global ozone budget. *Journal of Geophysical Research* **103 D9**, 10757-10767.

Woodward, F.I., & Sheehy, J.E.: 1983, *Principles and measurements in Environmental Biology*. London; Butterworths.

# OVERVIEW OF RECENT GROWTH TRENDS IN EUROPEAN FORESTS

HEINRICH SPIECKER

*Institute for Forest Growth, Albert-Ludwigs-University, Freiburg, Germany*
*D-79085 Freiburg, Bertoldstr. 17*

(Received September 1998; accepted February 1999)

**Abstract.** Natural environmental changes and human activities have altered forest growth for centuries. Recent long-term growth investigations indicate an increasing growth trend in European forests. The investigations are based on forest inventory, permanent plot and tree analysis data. The observed trends are species specific, locally varying and modified by remarkably large periodic growth variations. On a European scale, species and site specific quantitative information about the extent and spatial as well as temporal variation in growth acceleration is lacking. Future growth development may differ from past observations. A better understanding of changes in site conditions, their causes and consequences is needed to guide sustainable management of European forests.

**Key words:** growth trends, site productivity, European forests, climate change, management

## 1. Forest growth - an indicator of environmental changes

Forests have always been exposed to environmental changes. These can arise naturally, for example, as a result of natural climatic fluctuations or extreme climatic events followed by disturbances such as pests, diseases, fire, storm or snow, or as result of human activities, for example, forest degradation due to pasturing and litter raking, and other historical management practices such as choice of tree species and provenances, site preparation, weed control, regeneration methods, tending, liming, fertilization, amelioration, drainage, prescribed burning and wildlife management. These influences may change forest ecosystems considerably. Today we have to acknowledge that human activities change the environment both locally and globally. We discuss possible effects of atmospheric deposition as for example nitrogen, of changes in tropospheric air chemistry like the concentration of $CO_2$ and Ozone ($O_3$), in radiation, in air temperature and in other site factors on forest ecosystems. Site factors have to be considered as constantly changing.

Forest growth research is asked to provide information not only on growth as a result of site and stand characteristics but on the extent to which growth deviates from earlier observations. In the past it was often assumed that site specific growth patterns do not change much with time. Changing environmental conditions raise the question of whether site conditions, and as a consequence growth patterns have changed. The title of the 18th International Meeting for Specialists in Air Pollution Effects on Forest Ecosystems "Forest Growth Responses to the Pollution Climate of the 21st Century" suggests that the pollution climate may affect forest growth. This paper describes aspects of recent research in possible growth changes.

Growth responses of trees may be used as indicators of environmental changes in their spatial and temporal extent. They are, however, by themselves not a sufficiently powerful tool for the diagnosis of specific causes. When we talk about forest growth we may think of total biomass production, tree growth including growth of roots, trunk, branches and leaves or wood volume growth per hectare. This paper will concentrate on wood volume growth calculated from tree diameter and height growth simply because these are the only data available over long time periods in many European countries. In this paper the following definitions are used: *Site*: inclusive of all environmental conditions existing at a particular location; *site productivity*: Timber production potential of a site for a particular species or forest type and *growth trends*: Long-term site-induced deviations from former growth rates. The paper concentrates on site-induced deviations from expected growth. Other effects on forest growth data such as inconsistencies in inventory methods, changes in forest area over time, changes in species and age class distribution, and changes in silvicultural methods are taken into account. Methods for analyzing growth trends are described, results are presented on a European scale and possible causes of the observed changes are discussed.

## 2. Existing growth data

2.1 FOREST INVENTORY DATA

The first national survey was carried out in Finland in the early 20's. Today national forest surveys are conducted in most European countries. Many inventories are designed on a statistical basis (European Commission 1997). Those inventories cover the forest conditions and the variability in forest stands and sites for the entire inventory area in a representative way. The sampling errors can be given for all estimated parameters. As sample plots are typically laid out in a sample grid, forest inventory data can be regionalized, for example with geostatistical methods. For further large-scale analysis of possible causes, data can be combined with other geo-referenced data such as climatic data, soil data, or atmospheric deposition data (Spiecker *et al.*, 1994).

Sample-based forest surveys are a rather recent concept for European forests. For most European national forest inventories only two or three survey cycles have been accomplished and long-term and medium-term growth data are not available. These national surveys are designed independently and have different traditions and objectives. Besides the differences in the time periods covered and the number of observations, major differences can be found in respect to measurement methods for important attributes such as forest area or stand volume (Päivinen *et al.*, 1998). The measurement methods might have been modified with time, due to changed information needs and new experience. Changes through time might be partially related to the differences in definitions. Inventory data are most appropriate to describe growth of larger forests in a representative way.

2.2 PERMANENT RESEARCH PLOT DATA

Many forest research organizations in Europe installed hundreds of long-term permanent plots. The earliest were established in the middle of the 19th century. The great advantage of long-term permanent plot data is the generally well-known stand history. The trees within the plots have been numbered and the diameters of all trees as well as the height of selected trees have been remeasured over periodic time intervals. Also survivor growth, ingrowth, mortality and cut have been registered. The earliest permanent research plots generally were established to determine wood volume growth potential of a site for a particular species or forest type. Later, new research objectives were added: provenance trials, spacing and thinning trials, fertilizer trials etc.. Disadvantages are non-representativity due to insufficient replications, subjective implementation and only periodic time resolution.

Stand growth can be expressed as basal area, height or volume growth. Basal area growth is easy to measure, but is more influenced by silvicultural measures than height growth. Volume growth combines basal area and height growth, including stem form and its changes. A comparison of long-term volume growth is only valid if consistent methods for tree volume estimates are used. In addition to periodic measurements it is possible to take increment cores or cross sections from trees to be removed from permanent plots.

If stand characteristics are calculated on the basis of models derived from sample tree measurements, the number of sample trees must be big enough. Stand height is usually estimated from 20-40 sample trees per plot. The previous selection of the sample trees has an influence on the results (Magnussen and Penner, 1996). Trees in different social classes may show different growth trends. These trends, arising from natural stand dynamics, must not be misinterpreted. Analyzing site induced growth trends in mixed species stands is problematic due to differences in the growth rhythms of tree species and the continuously changing competition between them. Thus species can show different growth trends in the same stand depending on tree age, species composition, type of mixture and site influences.

## 2.3 TREE ANALYSIS DATA

The measurement of annual rings and shoots allows the reconstruction of tree growth even over long observation periods with yearly time resolution on any site where trees are growing. As a result height growth, diameter growth, development in stem form and volume growth of individual trees can be described retrospectively. This source of data may be used when no long-term permanent plot data are available and/or a yearly time resolution of growth is needed.

For the reconstruction of height growth, trees generally have to be harvested. Annual shoot growth can be analyzed either by measuring the shoot lengths along the stem or by cutting discs at short distances along the stem and counting for each cross section the cambial age. Annual radial growth can be measured by cutting discs at predefined locations along the stem and by measuring annual growth along several radii in order to account for irregular and eccentric radial growth. As an alternative, increment cores can be used to measure annual ring width. Here the trees do not need to be cut but will be damaged and possibly infected by fungi and the cores may be compressed in the process of coring. The main problem with tree analysis data is that the stand history may not be known and it may be impossible to find individual trees which represent stand growth over a long period by their growth. Even all surviving trees in a stand may not represent past growth of the stand properly (Spiecker, 1992). Individual competition dynamics are influenced by the constellation of neighbouring trees which may be modified by thinning operations, genetic and site induced differences in growth dynamics of neighbouring trees, natural mortality or ingrowth. The population representing the stand is changing over time. Thus

Fig. 1. Differences in radial growth trends of Norway spruce (Picea abies [L.] Karst.) in the Black Forest caused by past competition: 1: wide initial spacing, no thinning (Kenk,1990) and 2: heavy release after growing in a rather dense forest (Spiecker ,1992).

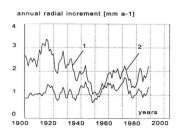

tree analysis does not generally allow a reliable reconstruction of the long-term basal area or volume growth of a stand. An example of management influences on radial growth is shown in Figure 1. This figure also shows the large year to year variability and synchronous periodic variation of radial increment.

Methods of detecting site-related growth trends from radial increment or ring width data of single trees have to include non-site factors, especially stand density and aging (e.g. Van Deusen, 1987; Briffa, 1992; LeBlanc, 1993). To minimizing competition effects, dominant trees are selected. The growth reaction of dominant trees may however differ from the reaction of codominant and suppressed trees. The potential of tree-ring measurement lies in the precise reconstruction of the annual variability of increment. Events with strong effects on tree growth are rather easy to detect (Schweingruber *et al.*, 1990). Tree-ring analysis has been successfully used, especially in short- and medium-term dendroecological studies. Growth variation and their correlation with climatic variation has been described in many studies (f.e. Becker, 1987; Becker *et al.*, 1994; Spiecker, 1986, 1987, 1991; Kahle, 1994). Mortality may correlate with climatic fluctuation as well (Figure 2).

Data from height analysis may be used as indicators of site productivity. However, the reliability of this indicator has to be validated. The combination of tree analysis data with permanent research plot data provides an improved data base for detecting growth trends. Tree analysis data are collected, for example, when applying gradient or exclusion methods especially when rarely available long-term environmental data exist for specific locations.

Combining single tree analysis data with periodically measured plot data helps to overcome some limitations of each of the two data sources. The combination may provide reliable area related annual growth values. In addition, systematic errors of periodic diameter measurements can be detected and eliminated (Spiecker 1991). The results of forest inventories in terms of species composition or other stand characteristics and site conditions can be used as a basis for testing the representativeness of observations on growth and yield plots.

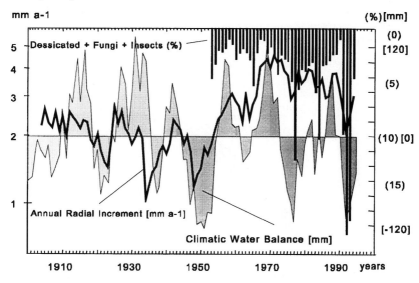

Fig. 2 Growth variation, mortality and climatic variation. Annual radial growth of Norway spruce (Picea abies [L.] Karst.) in the Black Forest varies considerably. Growth (curve) and mortality (bars, expressed in % of volume of allowable cut) are correlated with climatic conditions (here expressed by the climatic water balance based on the method of Thornthwaite and Mather (1955) during the period of May to September of the five preceding years, deviation from average 1900-1990).

## 3. Detecting changes in forest growth

### 3.1 DETECTING CHANGES IN FOREST GROWTH BY COMPARING THE GROWTH OF TREES AND STANDS WITH DIFFERENT GERMINATION DATES

Deviations of actual growth from expected growth may indicate site productivity changes. As a reference for expected growth past growth of stands with similar characteristics on the same location or on sites with similar history and actual site conditions may be used. Yield tables generally do not fulfill these requirements and therefore do not describe expected growth adequately. Growth trend estimation methods are discussed by several authors (Zahner, 1988; Cook *et al.*, 1990; Van Deusen, 1991; Dupouey *et al.*, 1992; Spiecker *et al.*, 1996, c.f. these for further literature). Further methodological information is given in many publications relating to growth effects caused by forest damage, for example by McLaughlin and Bräker (1985), Lorenz (1987), and Deutscher

Verband Forstlicher Forschungsanstalten, Sektion Ertragskunde (1988).

Site induced growth trends may be analyzed by comparing the growth of stands with different germination dates. The following conditions must be fulfilled: site conditions at a given point in time must be the same; management practices must be constant over time, especially when diameter growth is used for comparison; the genetics and stand structure of the compared stands should be similar; and fluctuations in growth conditions such as unfavorable weather conditions have to be taken into account.

Fig. 3. Height growth of Norway spruce (Picea abies [L.] Karst.) with different germination dates growing at the same site unit (Untheim, 1996). Height growth of the younger trees is faster.

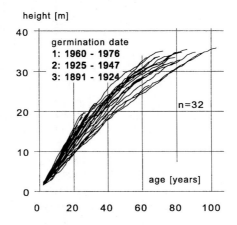

When individual tree data are compared, equal competitive regimes and site distributions in time must be represented by the data. Otherwise a possible trend cannot be found by the analysis. Examples of this method are described by Untheim (1996 see Figure 3) and Wenk and Vogel (1996) using height growth data and by Becker (1989 see Figure 4), Abetz (1984), Bert (1992), Schneider and Hartmann (1996) and Schadauer (1996) using tree-ring data. Site classification at one point in time is a problem in growth trend studies, because possible growth trends may be caused by site changes such as soil succession. These changes are not accounted for when site classification is based on height or on plant communities in the stand.

Comparisons between several site types may give a better insight into possible causes. For example regional comparisons of site types with rich and poor nutritional status could be helpful. Additional information could also be drawn from analyzing tree response to fertilization.

3.2 DETECTING CHANGES IN FOREST GROWTH BY COMPARING THE GROWTH OF SUCCESSIVE GENERATIONS

Growth patterns of successive generations at identical geographical location have been compared by several authors (e.g. Kenk *et al.,* 1991; Keller, 1992; Eriksson and Johansson, 1993; Untheim, 1996). This is a straightforward approach for detecting site productivity changes. In cases where the second generation stand has not been measured, single tree height development data might be compared with research plot data of previous generations (Figure 5).

Fig. 4 Radial growth trend (left: average annual radial growth -ir- over age; right: average radial growth deviation -ir%- from the average annual growth at a given age as shown in the left curve). Annual radial growth has increased in recent decades (Becker, 1989).

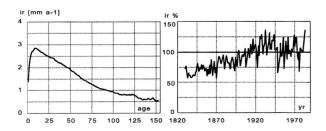

Fig. 5. Height growth of two successive Norway spruce (Picea abies [L.] Karst.) generations on the same plot; $h_{100}$ = mean height of the 100 thickest trees per hectare; in subsequent generation mean height of five dominant trees (Kenk et al., 1991).

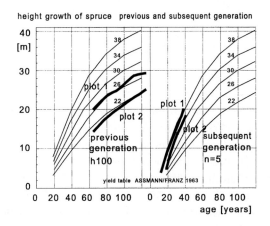

Interpretation has to consider aspects of management practices and genetic structure as mentioned in section 3.1. In addition, several consecutive generations of the same species might alter site productivity (Wiedemann, 1923; Krauss, 1939; Moosmayer, 1957; Genssler, 1959; Holmsgaard et al.. 1961 and Eriksson and Johansson, 1993). If there are enough height data from successive tree generations in a region, some kind of correction could be made by modelling the site index change as a function of time.

3.3 DETECTING CHANGES IN FOREST GROWTH BY COMPARING GROWTH WITH YIELD TABLES OR OTHER GROWTH MODELS

Yield tables and other growth models are often used as growth references. For the interpretation of deviations from the reference the data sources of these references must be known. The requirements listed in 3.1 have to be fulfilled when yield tables or other growth models are used for reference. Modelling single tree growth by using long-term constant factors such as topographic and edaphic characteristics and varying site factors such as weather conditions as predictor variables enable comparisons to be made between former growth and today's growth response. Assuming that tree competition, genetic, and aging effects are removed from the time-series, an altered response must be due to changed site conditions. This method enables the separation of expected reactions from unknown reactions.

The most difficult problem is related to detrending. There is no superior method of removing aging and competition trends. Analysis must be carried out carefully to avoid losing parts of the investigated long-term trend signal. No statistical theory explains how well the observed signal in tree-rings estimates the expected signal. In most cases, we must rely on statistical measures of signal strength, frequency domain definitions of signal and noise, and knowledge about sites, stand histories, and tree biology (Cook and Briffa, 1990).

Height growth, especially top height growth, is often used as an indicator of site productivity. For height growth data, the definition of stand height, the measurement method, and the method used for relating height to age has to be consistent. The calculation of volume growth on the basis of height and basal area growth must be done according to the same rules. Long-term deviation of height growth from existing site index curves must not be interpreted as trend when site index curves are based on long-term observations within time periods of changing site conditions.

## 4. Recent changes in forest growth

It is difficult to separate growth trends which reflect long-term changes in site productivity from episodic changes caused by extreme events such as frost, drought, snow and storm damage, fire, insect attack or fungal disease or by a combination of several events which will be followed by a reverse change. The analyses are based on the above mentioned data sources: inventory data, research plot data and tree analysis data. Since existing data have to be used, the quantity and quality of available data varies. In some countries no long-term growth data are available in an appropriate form. Some papers presented in the EFI publication refer to case studies on rather small areas, others - as for example some inventory data - refer to larger regions or nations (see table 1 in Spiecker *et al.*, 1996). The time span of available data varies from a few decades to several centuries. Since the data analyzed were usually not collected for analysis of long-term growth trends, some variation in the applied standards is unavoidable. Because of these shortcomings, the EFI Research Report does not present growth development of forests in Europe in a uniform and statistically representative way. However, the results may still allow some conclusions to be made at the European level. Twenty two papers written by 45 scientists from 12 countries are presented in this report. Most studies were conducted in Northern and Central Europe. Only two studies refer to Southern Europe.

In total 61 scientists independently reviewed the individual contributions and gave valuable comments.

The growth of European forests has changed considerably in recent decades. Although the methods applied varied according to the data available, most studies showed the same general trend: site productivity has increased on many sites (Figure 6). An increasing growth trend has been observed in the southern regions of Northern Europe, in most regions of Central Europe and in some parts of Southern Europe. The results derived from long-term observations on permanent plots and from tree analysis are supported by inventory results which are representative of large areas, but cover generally shorter observation periods. Site productivity in terms of wood volume increased on various sites in recent decades by up to 50 %, in some cases even more. Annual tree height increment increased by up to 5 cm and more, varying with species, site and age of the trees. No clear trend was found in the most northern part of Europe, in rare cases in Central Europe and in some observations in Southern Europe. A decreasing trend was found in exceptional cases where extreme growth conditions such as intense exposure to pollutants or exceptional climatic conditions occurred. Several other publications on changes in forest growth support the findings described (f. e. Schmidt, 1969; Pretzsch, 1985; Spiecker, 1986; Gerecke, 1988; Kenk *et al.*, 1988 and 1991; Becker, 1989 and Becker *et al.*, 1994; Kauppi *et al.*, 1992; Keller, 1992; Kuusela, 1994; Spelsberg, 1994 and Elfving and Tegnhammar, 1996).

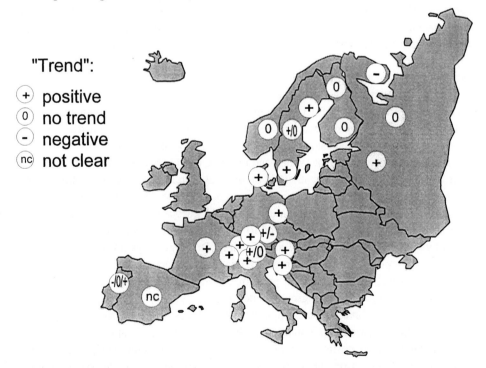

Fig. 6 Growth trends in European Forests, 22 studies from 12 countries. Summarized conclusions by the individual authors (Spiecker *et al.* 1996).

These findings indicate changes in growth conditions. The generally accelerating growth may have been caused by one factor, a factor combination or by regionally changing factors which finally had similar effects on growth.

## 5. Possible causes of accelerated growth

### 5.1 LAND USE HISTORY AND FOREST MANAGEMENT

Possible causes of these changes could be any growth-influencing factors that have changed with time. There is a lack of knowledge about the causes of accelerated growth (Spiecker et al., 1996). In many regions of Europe, the former use of forests, not only for wood production, but also for litter raking and pasturing have had a long-term effect on soil conditions because of large-scale nutrient depletion. Furthermore, former agricultural land was often abandoned to forest because of poor site productivity and marginal return rates, which, in some cases were caused by excessive utilization. Recovery of the soil may have increased site productivity in recent decades. In addition, soil preparation is used in order to create better conditions for forest regeneration.

Intensive harvesting, including removal of forest biomass, may potentially affect productivity. Traditional stemwood harvesting is not considered to lead to impoverishment of the soil because the nutrient content of wood is rather low (Mälkönen, 1976). In field experiments with slash removal Staaf and Olsson (1991) found that soil acidity had increased in plots where slash had been removed and that removal of all slash caused higher acidity than removal of slash without needles. According to Kreutzer (1979), nutrient losses by intensive whole-tree utilization are similar to that of litter utilization and may affect the production conditions on poor sites.

Forest management practices such as regeneration methods, tending, thinning and harvesting regimes have an effect on site productivity. The growth of young stands may be strongly influenced by soil preparation, selection of species and provenances, quality of plant material and weed control. Intensified thinning may have an effect on site productivity by altering nutrient cycling and reducing competition for light, nutrients, and water. Species composition has an effect on the root systems, litter quality, and nutrient storage.

Fertilizers and lime have been applied to parts of European forests for many decades in order to increase site productivity and to overcome some effects of site degradation caused by former land use and soil acidification. Drainage plays an important role in peatland forestry of Northern Europe and is mainly focused on increasing wood production on peat soils. In various parts of Central Europe, flood control and river diversion have had an effect on ground water levels.

The results of national forest inventories show changes in site productivity as well as in species and age distribution. Younger age classes of forests in Central Europe are often represented by a large portion of conifer stands with high volume increments. On the other hand, especially in remote and inaccessible areas or because of the recession of the forest industry, forests became overmature, resulting in growth recession and instability.

In Germany, for example, large forested areas were clear-cut for reparation payments soon after World War II and were often afforested with highly productive even-aged forests. In recent decades the negative impact of these activities on biodiversity of stands encouraged the conversion of stands to so-called close-to-nature forest types.

Wildlife management often has a decisive effect on stand structure and species composition. Selective browsing and other influences caused by roe deer, red deer and other game affect the growth and survival rates of individual trees and tree species. Damage caused by game mainly affects growth and survival of young trees and may result in a shift of species and age distribution possibly delaying stand establishment and tree growth.

## 5.2 CLIMATE, NATURAL DISTURBANCES, CHANGES IN $CO_2$, $O_3$, AND $SO_2$ AND NITROGEN DEPOSITION

Growth responses to climatic influences vary with species, provenance, age, competition status, site conditions, and preconditioning factors like recent extreme climatic events. Air temperature may affect the length of the growing season. Other factors, not only changes in mean conditions, may influence site productivity. Most evident are the effects of extreme events such as late spring frosts, summer droughts, unusual cold and wet or hot, and dry summers as well as extreme and abrupt temperature changes. Such events occurring locally may not be reflected in data on climate and weather conditions generally available. Extreme climatic events not only have a direct effect on trees, but also on insects and microbial pathogens, on disturbances such as fire or windthrow, and on biological as well as chemical and physical processes in the soil. Correlations between these factors and growth have been reported in many parts of Europe. Drought generally slows growth: in Central Europe such a growth recession due to drought occurred in the late 1940s. The drought that attracted the most attention occurred in the mid 1970s (Figure 2) prior to the start of concerns about defoliation and forest dieback. The long-lasting after-effects of climate complicate the detection of possible causes of changes in site productivity.

Natural disturbances play an important role in forest growth. Fire, drought, storms, snow, avalanches, pests, disease, insect attacks, and wildlife may interrupt natural succession, and change factors driving forest growth, such as competition for light, nutrients and water and tree species composition. They may change tree competition and may have an impact on selection processes by changing growth relations and survival rates.

The effect of predicted global warming on tree growth is not yet entirely understood. A positive correlation between an increase in air temperature and plant growth is predicted for temperate and boreal regions when precipitation rate does not decrease. Conversely, there may be some negative effects of global warming that secondarily cause decrease in tree growth, for instance late spring frosts, storms, droughts and insect calamities. Badeau et al. (1996) found in investigations at high elevations that trees growing outside the range of forest management and exposed to relatively low nitrogen deposition showed a clear positive growth trend; this may be due to the increased content of $CO_2$ in the atmosphere or to other climatic conditions. $CO_2$ may stimulate photosynthesis, reduce respiration and relieve water and low-light stress (Hättenschwiler, 1997; Jarvis, 1998). It is likely that the effect of $CO_2$ on tree growth varies in different parts of Europe.

The tropospheric $O_3$ content is believed to have been increased during this century. The problem of the impact of $O_3$ on tree vitality and growth, especially in high elevation forests of Central Europe, is widely discussed. However, no clear evidence of a long-term influence on tree growth under the various natural conditions exists (Sandermann et al., 1997). Long-term effects of $SO_2$ and other atmospheric depositions in low concentrations on tree growth under natural conditions have been investigated in only a few cases. Around major industrial sources the deposition of $SO_2$ on the foliage of trees combined with indirect effects of acidification and high content of heavy metals in the soil are considered to cause forest dieback.

Nitrogen deposition is regarded as a further probable growth promoting factor during recent decades. The positive growth trends observed in many parts of Europe may be at least partially caused by increased nitrogen depositions. The fact that no increased growth trend was detected in some parts of Northern Europe may be explained by low nitrogen deposition.

The significance of each of the above mentioned possible causes possibly varies in space and time. Growth responses to the influencing factors are modified by species, provenance, site and stand conditions. It is difficult to draw general conclusions regarding their effects on forest growth. It is possible that one single factor or a factor combination, or even various factor combinations at different locations, influence growth. It is equally possible that several different factors influence forest growth simultaneously and that their total synergistic effects change the individual monofactorial effects.

## 6. Conclusions

Numerous case studies clearly indicate that productivity of many forest sites in Europe has changed. Site productivity has been increasing for several decades. Some studies showed no trend and in rare cases a decreasing trend was observed. The rate of growth acceleration varies with species and provenances, with site including micro-variation of site, age, management practices, extreme events, diseases, disturbances and other factors. No complete information exists on species, provenance and site specific growth changes on a European level.

The observed growth trends may involve additional risks: trees may be more susceptible to drought, frost, insect attacks as well as to diseases. The risk of windthrow may increase when trees get taller and when standing volume increases. To predict future forest development as well as assess risk, a better understanding of the causal relation between the changing environmental factors and the growth changes is needed. Information about causes and consequences of accelerated growth are still lacking.

The traditional aim of forest growth studies is the development of growth models of trees and stands on different sites under different treatments. Today we have to consider, that sites change on a larger scale. This leads to the conclusion that past growth observations may not reflect actual growth adequately and past experience cannot be used to predict forest growth. New tools for sustainable forest management are needed. Long-term empirical field data and results from laboratory experiments combined with modelling are a good basis for interdisciplinary research to fill these information gaps.

But also our uses of forest ecosystems change with time. We need more information about changes in site properties, about site adaptation of species, about effects of management practices on growth, about wood quality and other properties of trees as well as about changes in the goals of forest management. New adaptive planning tools have to be applied to encompass uncertainty and risk. The amount and complexity of the scientific problems evolving from the observed forest growth trends show that solutions can only be developed by an interdisciplinary cooperation of scientists at an international level. This cooperation will lead to a more comprehensive understanding and will provide a more realistic and reliable information basis for decision support.

## References

Abetz, P.: 1984, *Allgemeine Forstzeitung* **95** (11), 322-325.
Badeau, V., Becker, M., Bert, D., Dupouey, J.-L., Lebourgeois, F. and Picard, J.-F.: 1996, in: Spiecker, H., Mielikäinen, K., Köhl, M. and Skovsgaard, J. P. (eds.) *European Forest Institute Research Report* No. 5. Springer, 167-181.
Becker, M.: 1989, *Canadian Journal of Forest Research* **19**, 1110-1117.
Becker, M., Bert, G. D., Bouchon, J., Picard, J. F. and Ulrich, E.: 1994, *Revue Forestière Francaise* **XLVI** (4), 335-341.
Bert, G.D.: 1992, *Thèse de doctorat de l'Université de Nancy*, 200 p.
Briffa, K.: 1992, in Bartholin, T. S., Berglund, B. E., Eckstein, D. and Schweingruber, F. *Proceedings of the International Dendrochronological Symposium, Ystad, South Sweden, 3-9 Sept. 1990*, 64-71.
Cook, E. and Briffa, K.: 1990, in Cook, E. R. and Kairiukstis, L. A. (eds.) *Applications in the Environmental Sciences*. Dordrecht; Boston; London: Kluwer Academic Publishers, 153-162.
Cook, E., Briffa, K., Shiyatov, S. and Mazepa, V.: 1990, in Cook, E. R. and Kairiukstis, L. A. (eds.). *Applications in the Environmental Sciences*. Dordrecht; Boston; London, Kluwer Academic Publishers, 104-123.
Deutscher Verband Forstlicher Forschungsanstalten, Sektion Ertragskunde: 1988, *Allgemeine Forst- und Jagdzeitung* **159** (7), 115-116.
Dupouey, J.-L., Denis, J.-B. and Becker, M.: 1992, in Bartholin, T. S., Berglund, B. E., Eckstein, D. and Schweingruber, F. (eds.), *Proceedings of the International Dendrochronological Symposium, Ystad, South Sweden, 3-9 Sept. 1990*, 85-88.
Elfving B. and Tegnhammar, L.: 1996, *Scandinavian Journal of Forest Research* 11, 26-37.
Eriksson, H. and Johansson, U.: 1993, *Plant and Soil* **154**, 239-247.
European Commission: 1997, *reports on forestry information and survey systems* Vol. 1 and 2, 1328 p.
Genssler, H.: 1959, *Diss. Hann. Münden*, 191 p.
Gerecke, E.: 1988, *Mitt. d. FVA Rheinland-Pfalz* Nr. 8, 134 p.
Hättenschwiler, S.: 1997, *Diss. Basel,* 103 p.
Holmsgaard, E., Holstener-Jörgensen, H. and Yde-Andersen, A.: 1961, *Det Forstlige Forsoegsvaesen in Danmark*, **XXVII**, 1, 167 p.
Jarvis, P.G. (ed.): 1998, Cambridge, Cambridge University Press, 380 p.
Kahle, H-P.: 1994, *Diss. Freiburg i.Br.*, 184 p.
Kauppi, P. E.; Mielikäinen, K. and Kuusela, K.: 1992, *Science* **256**, 70-74.
Keller, W.: 1992, *DVFFA, Sektion Ertragskunde, Tagungsber.*, 123-129.

Kenk, G.: 1990, *Forstwiss. Centralblatt* **109**, 86-100.
Kenk, G. und Spiecker, H.: 1988, *Kernforschungszentrum Karlsruhe-PEF* 35 (1), 371-381.
Kenk, G. and Fischer, H.: 1988, *Environmental Pollution* **54**, 199-218.
Kenk, G., Spiecker, H. und Diener, G.: 1991, *Kernforschungszentrum Karlsruhe PEF Forschungsbericht* **82**, 59 p.
Krauss, G., Müller, K., Gärtner, G. and Härtel, F. with comments by Schanz, H. and Blanckmeier, H.: 1939, *Tharandt. forstl. Jahrb.* 90. (cit. by Holmsgaard, E., Holstener-Jörgensen, H. and Yde-Andersen, A.: 1961)
Kreutzer, K.: 1979, *Forstw. Cbl.* **98**, 298-308
Kuusela, K.: 1994. Forest Resources in Europe 1950-1990. *European Forest Institute Research Report* 1, 154 p.
LeBlanc, D. C.: 1993, *Canadian Journal of Forest Research* **23**, 1494-1496.
Lorenz, M.: 1987, *Mitteilungen der Bundesanstalt für Forst- und Holzwirtschaft Hamburg* 156, 77 p.
Mälkönen, E.: 1976, *Silva Fennica* **10**, 3, 157-164
Magnussen, S. and Penner, M.: 1996, *Canadian Journal of Forest Research* **26**, 9-22
McLaughlin, S. and Bräker, O. U.: 1985, *Experientia* **41** (3), 310-319.
Moosmayer, H.-U.: 1957, *Mitt. des Vereins f. Forstl. Standortskunde und Forstpflanzenzüchtung* **7**, 3-41
Päivinen, R., Schuck, A. and Lin, L.: 1999 in Karjalainen, T., Laroussinie, O. and Spiecker, H. (eds.), *EFI - Proceeedings* Nancy 1998, in preparation.
Pretzsch, H.: 1985,. *Allgemeine Forstzeitschrift* 42, 1122-1226
Sandermann, H., Wellburn A. R. and Heath, R. L.: 1997. Ecological Studies **127**, Springer, 398 p.
Schadauer, K.: 1996, in: Spiecker, H., Mielikäinen, K., Köhl, M. and Skovsgaard, J. P. (eds.) *European Forest Institute Research Report* No. 5. Springer, 275-289.
Schmidt, A.: 1969, *Forstwissenschaftliches Zentralblatt* **88**, 33-40.
Schneider, O. and Hartmann, P.: 1996, in: Spiecker, H., Mielikäinen, K., Köhl, M. and Skovsgaard, J. P. (eds.). *European Forest Institute Research Report* No. 5. Springer, 183-198.
Schweingruber, F. H., Eckstein, D., Serre-Bachet, F. and Bräker, O. U.: 1990, *Dendrochronologia* **8**, 9-38.
Spelsberg, G.: 1994, *Allgemeine Forst- und Jagdzeitung* **165** (4), 77-80.
Spiecker, H.: 1986, *Allgemeine Forst- und Jagdzeitung* **157** (8), 152-163.
Spiecker, H.: 1987, *Allgemeine Forst- und Jagdzeitung* **158** (4), 70-76.
Spiecker, H.: 1991, *Fertilizer Research* **27**, 87-93.
Spiecker, H.: 1992, in Bartholin, T. S., Berglund, B. E., Eckstein, D. and Schweingruber, F. (eds.), *Proceedings of the International Dendrochronological Symposium, Ystad, South Sweden, 3-9 September 1990*, 308-312.
Spiecker, H., Mielikäinen, K., Köhl, M. and Untheim, H.: 1994, *EFI-Working Paper*, 4, 68 p.
Spiecker, H., Mielikäinen, K., Köhl, M. and Skovsgaard, J. P. (eds.): 1996, *European Forest Institute Research Report* No. 5. Springer, 372 p.
Staaf, H. and Olsson, B.: 1991, *Scandinavian Journal of Forest Research* **6**, 19-29.
Thornthwaite, C. W. and Mather, J. R.: 1955, *Drexel Inst. Lab. of Climatol. Pub. in Climatol.* **8** (1), 5-85.
Untheim, H.: 1996, *Mitteilungen der Forstlichen Versuchs- und Forschungsanstalt Baden-Württemberg, Freiburg i. Br.*, Heft 198, 239 p.
Van Deusen, P.C.: 1987, *Canadian Journal of Forest Research* **17**, 1487-1495.
Van Deusen, P.C.: 1991, *Forest Science* **37** (5), 1365-1375.
Wiedemann, E.: 1923, Kommissionsverlag W. Laux, Tharandt, 180 p.
Zahner, R.: 1988, *Tree-Ring Bulletin* **48**, 13-20.

# CONSEQUENCES OF N DEPOSITION TO FOREST ECOSYSTEMS - RECENT RESULTS AND FUTURE RESEARCH NEEDS

HEINZ RENNENBERG and ARTHUR GESSLER

Albert-Ludwigs-Universität Freiburg, Institut für Forstbotanik und Baumphysiologie,
Professur für Baumphysiologie, Am Flughafen 17, D-79085 Freiburg, Germany
E-mail: here@sun2.ruf.uni-freiburg.de

(Received 30 October 1998; accepted 25 February 1999)

**Abstract.** Wet and dry deposition of atmospheric nitrogen (N) compounds into forest ecosystems and their effect on physical, chemical and microbial processes in the soil has attracted considerable attention for many years. Still the consequences of atmospheric N deposition on N metabolism of trees and its interaction with soil microbial processes has only recently been studied. Atmospheric N deposited to the leaves is thought to enter the general N metabolism of the leaves, but the processes involved, the interaction with different metabolic pathways, and the connection between injury by atmospheric N and its metabolic conversion are largely unknown.

Laboratory and field experiments have shown that N of atmospheric $NO_2$ and $NH_3$, desposited to the leaves of trees, is subject to long-distance transport in the phloem to the roots. This allocation can result in considerable decline of N uptake by the roots. Apparently, the flux of N from the soil into the roots can be down-regulated to an extent that equals N influx into the leaves. This down-regulation is not mediated by generally enhanced amino-N contents, but by elevated levels of particular amino acids. Field experiments confirm these results from laboratory studies: Nitrate ($NO_3^-$) uptake by the roots of trees at a field sites exposed to high loads of atmospheric N is negligible, provided concentrations of Gln in the roots are high.

At the ecosystem level, consequences of reduced N uptake by the roots of trees exposed to high loads of atmospheric N are (1) an increased availability of N for soil microbial processes, (2) enhanced emission of gaseous N-oxides from the soil, and (3) elevated leaching of $NO_3^-$ into the ground water. How recent forest management practices aimed at transforming uniform monocultures to more structured species-rich forests will interact with these processes remains to be seen. Possible implications of these forest management practices on N metabolism in trees and N conversion in the soil are discussed particularly in relation to atmospheric N deposition.

Keywords: nitrate, ammonium, uptake, regulation, nitrogen oxides, ammonia

## 1. Introduction

In pre-industrial times N has been a major growth limiting factor in non-agricultural terrestrial ecosystems (Cole and Rapp, 1981; Dickson, 1989). As a consequence microbial processes in forest soils, metabolic processes in forest trees, and forest ecosystem functioning are considered to be adapted to N limitation rather than N excess (Rennenberg et al., 1998). Due to human activities N inputs to forest ecosystems have increased causing forests to be exposed to a surplus of reactive N from dry and wet deposition (Wellburn, 1990; Pearson and Stewart, 1993). Whilst atmospheric input of reactive N compounds into remote forest ecosystems amounts to less than 10 kg N ha$^{-1}$ y$^{-1}$, values of 60 kg N ha$^{-1}$ y$^{-1}$ or more (Tietema and Beier, 1995) have been measured in heavily polluted regions.

The atmospheric N load of anthropogenic origin consists of oxidised N compounds (mainly NO, $NO_2$, $NO_3^-$) from fossil fuel combustion by automobile traffic and industrial processes (Wellburn, 1990) and of reduced N compounds ($NH_3$ and $NH_4^+$)

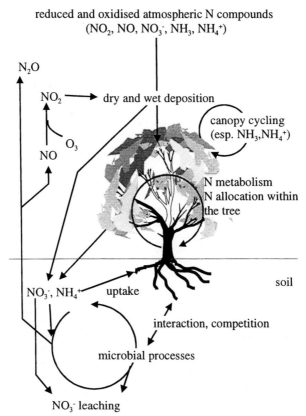

*Fig. 1.* Input of reduced and oxidised N compounds into forest ecosystems, ecosystem internal cycling of N compounds and the possible release of N into the groundwater and/or atmosphere.

mainly emitted from intensive agriculture (Fangmeier *et al.*, 1994). In 1994 c. 85% of $NH_3$ emitted in Germany originated from livestock breeding and slurry application, 10% from application of chemical fertilisers and 5% from industrial combustion, power plants and traffic (Bundesministerium für Umwelt, Naturschutz und Reaktorsicherheit, 1996). The main process leading to $NH_3$ emission is the microbial decomposition of urea, the main N-fraction of livestock urine (Sommer *et al.*, 1993). High atmospheric $NH_3$ concentrations above an annual average value of 20 ppb are only observed in close vicinity of emitters since $NH_3$ is subjected to fast deposition and to rapid conversion into $NH_4^+$ (Asman and Van Jaarsfeld, 1990; Hoek *et al.*, 1996). Dry deposition of $NH_3$ therefore dominates input of reduced N to ecosystems close to emission sources (within the range of meters to a few km). Wet deposition of $NH_4^+$, that has an atmospheric residence time of 4 to 15 days (Böttger *et al.*, 1978), can contribute significantly to atmospheric N load of even remote ecosystems (Geßler and Rennenberg, 1998).

In contrast to reduced N, 50% of oxidised N in Germany was emitted in 1994 from automobile traffic, 22% from power plants, 12.5% from industrial processes and 10% from air traffic (Bundesministerium für Umwelt, Naturschutz und Reaktorsicherheit, 1996). The oxidised N compound preferentially emitted from fossil fuel combustion is

NO that readily reacts with $O_3$ or peroxyl radicals to form $NO_2$ (Crutzen, 1979; Kolar, 1990). Light of a wavelength < 400 nm causes conversion of $NO_2$ to NO during daytime (Kolar, 1990). Both, NO and $NO_2$ are atmospheric trace constituents with high reactivity (Wellburn, 1990) and, hence, are subject to transformation into nitric and/or nitrous acid, respectively, during atmospheric transport processes (Levine et al., 1984).

The abundance of oxidised N compounds is highest in the urban environment were annual average concentration of $NO_2$ can amount to 20 - 60 ppb (Bundesministerium für Umwelt, Naturschutz und Reaktorsicherheit, 1996). Since the residence time of $NO_2$ in the atmosphere is high as compared to $NH_3$, spatial variability of $NO_2$ concentration is much lower. Therefore $NO_2$ produced in the urban environment is also an important contributer to the atmospheric N load of rural ecosystems, at least in central Europe, where population and traffic density are generally high (Stulen et al., 1998).

The present paper focuses on the impacts of the $NO_2$, $NO_2^-$, $NO_3^-$ and $NH_3/NH_4^+$ on trees and forest ecosystems, since these atmospheric constituents contribute most to the N load of terrestrial ecosystems in Europe (United Kingdom Review Group on Impacts of Atmospheric Nitrogen, 1994).

## 2. Input of atmospheric N compounds into forest ecosystems

Reduced and oxidised N compounds in the atmosphere are subjected to dry and wet deposition onto/into above ground parts of the vegetation and into the soil (Figure 1). Whereas the effects of deposition of atmospheric N compounds on physical, chemical and microbial processes in the soil have attracted considerable attention for many years, the consequences of atmospheric N deposition on N metabolism of trees and its interaction with soil microbial processes has only recently been studied.

The input of atmospheric N compounds into the forest soil occurs mainly via wet deposition of $NH_4^+$ and $NO_3^-$ by bulk precipitation and throughfall. $NH_4^+$ in the soil is removed by (1) root uptake of trees and understorey vegetation and (2) nitrification processes. In soils with neutral to basic pH autotrophic dominates over heterotrophic nitrification, whereas in acidic soils heterotrophic nitrification with much lower rates of $NH_4^+$ turnover prevails (Papen et al., 1994). $NO_3^-$ produced by these processes or deposited from the atmosphere is highly mobile within the soil and can be leached into the ground water, depending on N input into the soil and on the soil's capacity to retain $NO_3^-$ (Rennenberg et al., 1998). The two most important processes that remove $NO_3^-$ from the soil and, therefore, can reduce $NO_3^-$ leaching are (1) denitrification under anaerobic conditions and/or in anaerobic microsites, and (2) root uptake by the vegetation.

Both microbial processes, nitrification and denitrification, compete with plant roots for N-substrates i.e. $NH_4^+$ and $NO_3^-$, respectively. The complex processes regulating the interaction between microorganisms and the vegetation are presently not understood in any detail. For example, the release of protons associated with $NH_4^+$ uptake by roots (Marschner et al., 1991) can decrease the pH of the soil solution and may have a negative feedback on autotrophic nitrification. Clear-cut evidence that this interaction takes place in forest soils has, however, not been reported. Also the role of mycorrhizal fungi and root associated microorganisms for N acquisition by trees and, hence, the competition between roots and soil microbial processes for N is not understood. It is generally assumed that mycorrhization can improve N acquisition of trees (Martin and Lorillou, 1997), but information on the significance at different availabilities of particular N compounds ($NH_4^+$, $NO_3^-$, amino acids) and the relation with microbial N acquisition under field conditions is lacking. A better understanding of these interacting processes is crucial, since the amount of $NH_4^+$ available for nitrification, as well as the amount of $NO_3^-$ that can be consumed by denitrifying bacteria, is decisive for the amount of radiatively active trace gases ($N_2O$, NO) emitted as by-products from both processes (Rennenberg et al., 1998). Studies in a forest ecosystem exposed to high loads of N revealed that the NO to $N_2O$ ratio is highly dependent on the tree species, with a preference for NO emission from the soil of a spruce stand and with a preference of $N_2O$ in the beech plot (Rennenberg et al., 1998). At least part of the NO emitted will be converted inside the canopy in the presence of ozone to $NO_2$. The $NO_2$ produced can be taken up by the shoots inside the forest canopy leading to a complex cycling of N within the ecosystem that includes the canopy atmosphere (Figure 1).

Since atmospheric N input contributes to soil acidification the balance of other nutrients is also affected. Leaching of $K^+$, $Ca^{2+}$ and $Mg^{2+}$ (Roelofs et al., 1985; Van Dijk et al., 1989) and slowing down of the P cycling (Carreira et al., 1887) may therefore be consequences of excess $NO_3^-$ and or $NH_4^+$ input into the soil. The reduction of the cation and P availability may cause and/or intensify nutrient imbalances within plants. For Sitka spruce it has been shown that especially P and K deficiencies inhibited N uptake by the roots increasing the amount of $NO_3^-$ leached into the ground water (Stevens et al., 1993; Harrison et al., 1995).

## 3. Uptake of $NO_2$ and $NH_3$ by the foliage and impacts on N metabolism in the leaves

Both $NO_2$ and $NH_3$ can be taken up by above ground tree parts either from dry deposition and/or wet deposition (Wellburn, 1990; Pearson and Stewart, 1993; Brumme et al., 1992; Burkhardt and Eiden, 1994). Generally, uptake and/or emission of gaseous compounds proceeds via the stomata and/or the cuticle. The driving force for this exchange is the concentration gradient between the gas phases inside and outside the leaf. Figure 2 shows the different mechanisms of $NH_3$ exchange between the leaves and the atmosphere. The gas phase concentration of $NH_3$ inside the leaf is in an equilibrium with the concentration of $NH_3$ and $NH_4^+$ in the aqueous phase of the apoplast. This equilibrium concentration inside the leaf is referred to as the $NH_3$ compensation point. When the atmospheric concentration is equal to the equilibrium concentration inside the leaf, net flux of $NH_3$ via the stomata does not occur (Farquhar et al., 1980). $NH_3$ compensation points are known to depend on plant species or cultivar, development stage, temperature and on N nutritional status (Sutton et al., 1995; Husted and Schjoerring, 1996; Husted et al., 1996). Experiments in the field and under controlled conditions showed that high N availability increases $NH_3$ compensation points significantly (Sutton et al., 1995; Husted and Schjoerring, 1996). It may be concluded that this increase is one mechanism by which plants cope with excess N supply (Geßler and Rennenberg, 1998). Since atmospheric concentrations of $NH_3$ are highly variable in space and time, plants and forest ecosystems supplied with excess N, may emit significant amounts of NH$_3$ when atmospheric NH$_3$ concentrations are low, especially following a period of high NH$_3$ deposition.

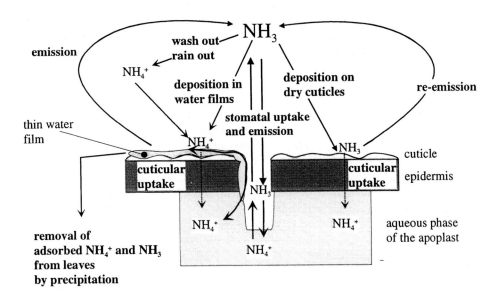

*Fig. 2.* Mechanisms of NH$_3$ gas exchange between the atmosphere and foliage. Under dry conditions the main pathways of either emission or deposition are the stomata. At high air humidities the existence of thin water films on the cuticles has to be considered (cf. Burkhard and Eiden, 1994). Due to its high solubility, NH$_3$ can be deposited into these water films on the leaf surface. The NH$_4^+$ dissolved may pass the stomata by diffusion within the water films that are connected with the aqueous phase of the apoplastic space. NH$_3$ may be re-emitted when the water films are drying and, thus, NH$_4^+$ concentration will increase. In addition NH$_3$ or NH$_4^+$ absorbed can be washed out be precipitation and can contribute to the N input into the soil. Cuticular uptake of NH$_3$ is probably negligible.

For NH$_3$ it is supposed that only stomatal influx contributes significantly to uptake into the aqueous phase of the apoplast, whereas cuticular uptake is negligible (Van Hove et al, 1989). Nevertheless NH$_3$ may be additionally absorbed at the leaf surface, especially under conditions where thin water films are formed on the cuticles. NH$_4^+$ ions dissolved in these water films may pass through the stomata by diffusion within the water films that are thought to be connected with the aqueous phase of the apoplast (Burkhardt and Eiden, 1994). If the equilibrium between NH$_4^+$ in water films on the leaf surface and NH$_3$ in the ambient atmosphere is changed e.g. due to drying of water films and a subsequent increase in NH$_4^+$ concentration and/or a decrease in atmospheric NH$_3$ concentration, re-emission of NH$_3$ may occur. NH$_3$ re-emitted may be recaptured within the canopy e.g. by stomatal uptake depending on the atmospheric NH$_3$ concentration and the NH$_3$ compensation point. Thus, also reduced N compounds may cycle within the forest canopy (Figure 1) (Sutton *et al.*, 1995). NH$_3$ or NH$_4^+$ absorbed to the leaf surface can also be removed by precipitation events and thereby contribute to the N supply of the soil.

Also gas exchange of $NO_2$ between the apoplast and the atmosphere takes place mainly via the stomata (Wellburn, 1990; Neubert et al., 1993) (Figure 3). A whole set of experiments has revealed a linear increase of $NO_2$ deposition rate with increasing atmospheric $NO_2$ concentration indicating the concentration difference between the gaseous phase inside the leaf and the atmosphere as the driving force of $NO_2$ gas exchange (Johansson, 1987; Latus et al., 1990; Thoene et al., 1991; Thoene et al., 1996). In the aqueous phase of the apoplast $NO_2$ taken up is either subject to disproportionation and/or to reaction with apoplastic ascorbate (Ramge et al., 1993). Since disproportionation of $NO_2$ in water is thought to be slow at atmospheric $NO_2$ concentrations (Lee and Schwartz, 1981; Norby et al., 1989; Ramge et al., 1993), the reaction with ascorbate may be of particular significance for maintaining a high flux of $NO_2$ into the leaves and may prevent internal resistances. However the apoplastic ascorbate concentration may vary considerably between tree species (Polle et al., 1990; Polle et al., 1995; Luwe, 1996; Schwanz et al., 1996a) and may depend on environmental factors (Polle and Rennenberg, 1992; Schnug et al., 1995; Polle et al. 1995; Schwanz et al., 1997 a; b; c) and/or the stage of development (Luwe, 1996). Also regeneration of ascorbate from the dehydroascorbate produced in the reaction with $NO_2$ may be an important factor determining the flux of $NO_2$ into leaves. The apoplastic ascorbate system is thought to be kept in a reduced state by the exchange of apoplastic dehydroascorbate for symplastic ascorbate across the plasma membrane by active transport processes (Rautenkranz, 1994; Dietz, 1997; Horemans et al., 1998). Also a reduction of apoplastic dehydroascorbate by transmembrane electron transport involving Cyt b in the plasmalemma has been proposed (Asard et al., 1995). Recently, apoplastic regeneration of ascorbate from dehydroascorbate by an apoplastic dehydroascorbate reductase has been suggested in barley leaves (Vanacker et al., 1998). Since the capacity of apoplastic ascorbate regeneration may limit $NO_2$ conversion (Ramge, 1993; Luwe et al., 1993) and thus $NO_2$ influx, further experiments are required to elucidate species specific differences in the mechanisms and the efficiency of apoplastic ascorbate regeneration. Previous measurements of apoplastic ascorbate integrate over a major part of the apoplast. However, $NO_2$ taken up through the stomata will first react with ascorbate in the apoplast of cells bordering the substomatal cavity as recently also observed for $O_3$ (Jacob and Heber, 1998). Further studies on the interaction between atmospheric $NO_2$ and apoplastic ascorbate should take this spatial heterogeneity into consideration. Jacob and Heber (1998) suggested that antioxidants different from ascorbat, e.g. α-tocopherol, may be involved in the apoplastic defence against $O_3$. These antioxidants may also play a role in the elimination of $NO_2$ in the apoplast.

As observed for $NH_3$, also $NO_2$ compensation points have been determined (Weber and Rennenberg, 1996; Slovik et al., 1996a; b). For different forest tree species $NO_2$ compensation points between 0.6 to 1.6 ppb $NO_2$ have been measured indicating that forest vegetation can be both, source and sink, for $NO_2$ (Rondón and Granat, 1994; Thoene et al. 1996; Geßler, 1998). The chemical source of $NO_2$ within cells however is largely unknown. From in vitro studies with soybeans it appears that NR (nitrate reductase EC 1.6.6.1) may be involved in the reduction of $NO_2^-$ to $NO_2$ (Dean and Harper 1988). Whether $NO_2$ emitted from trees originates from this reaction appears doubtful, since nitrate reductase activity at low atmospheric $NO_2$ concentrations is low

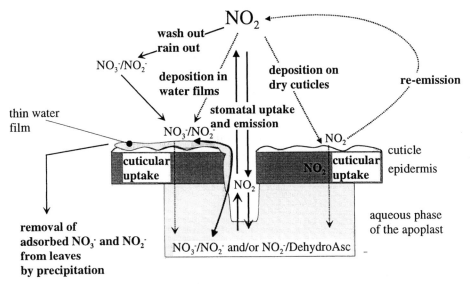

*Fig. 3.* Mechanisms of $NO_2$ exchange between the atmosphere and the foliage. The uptake and/or emission of $NO_2$ occurs mainly via the stomata. Deposition on dry or wet cuticles is of minor significance and $NO_2$ uptake through the cuticles is at least one order of magnitude lower than stomatal uptake. The reaction taking place when $NO_2$ dissolves in the aqueous phase of the apoplast is either the disproportionation reaction yielding equal amounts of $NO_3^-$ and $NO_2^-$ and/or reaction with apoplastic ascorbate.

in many forest tree species that reduce $NO_3^-$ preferentially in the roots (Stewart *et al.*, 1988; Schneider *et al.*, 1996; Geßler *et al.*, 1998b).

Although $NO_2$ exchange via the stomata seems to exceed penetration of the cuticle by one to two orders of magnitude (Wellburn, 1990), $NO_2$ can be absorbed on the cuticle. Besides reversible interaction with the leaf surface $NO_2$ is thought to interact irreversibly with phenolic components of the cuticle (Lendzian and Kerstiens, 1988, Kisser-Priesack *et al.*, 1990) or dissolve into water films on the leaf surface, just like $NH_3$ (Thoene *et al.*, 1996; Weber and Rennenberg, 1996). However, these processes do not seem to play an important role, from a quantitative point of view, in $NO_2$ deposition for the majority of species examined (Wellburn, 1990; Neubert *et al.*, 1993; Geßler, 1998).

If stomatal compensation points for $NH_3$ and $NO_2$ respond to the N status of plants, this may significantly reduce net uptake of these atmospheric N compounds into the apoplast. Further studies are necessary to estimate the variation of stomatal compensation points in the field in relation to $NH_3$ and $NO_2$ concentrations in the atmosphere and pedospheric N availability. If $NO_2$ and $NH_3$ deposited on the leaf surface can be re-emitted, a discrimination between net emission via stomata and cuticular re-emission has to be achieved.

Reactive atmospheric N compounds that have entered the leaves are present as $NH_4^+$ or $NO_2^-$ and/or $NO_3^-$ in the aqueous phase of the apoplast (Figure 4). It has been estimated that the reaction of $NO_2$ in the apoplast generates about three times less $NO_2^-$ than $NO_3^-$ (Amman *et al.*, 1995). From this finding it may be assumed that (1) $NO_2^-$ is converted to $NO_3^-$ in the apoplast and/or (2) $NO_2^-$ uptake into the cytoplasm and its reduction is very efficient and may prevent $NO_2^-$ accumulation (Srivastava, 1998).

$NH_4^+$, $NO_2^-$ and $NO_3^-$ are taken up actively into the cytoplasm of mesophyll cells (Hadwiger-Fangmeier, 1992; Ammann et al., 1995). $NO_3^-$ taken up is thought to be reduced by NR to $NO_2^-$ (Tischner et al., 1988; Thoene et al., 1991) that is transported into the chloroplast in the same way as $NO_2^-$ taken up from the apoplast.

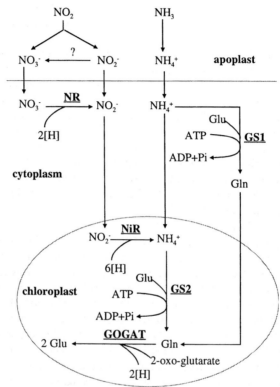

Fig. 4: Main enzymatic processes involved in the assimilation of atmospheric $NH_3$ and $NO_2$ taken up by the leaf. NR: nitrate reductase; NiR: nitrite reductase, GS: glutamine synthetase (GS1: cytoplasmatic isoform; GS2: chloroplastic isoform); GOGAT: glutamate synthase.

Ferredoxin-dependent reduction of $NO_2^-$ to $NH_4^+$ by NiR (nitrite reductase; EC 1.7.7.1) takes place within the plastides (Lam et al., 1996).

For $NH_4^+$ taken up from the apoplast into the cytoplasm assimilation by GS (glutamine synthetase EC 6.3.1.2) has to be assumed. Different studies revealed the presence of two isoforms of GS, one located in the cytoplasm (GS1), the other in the chloroplasts (GS2) (Maeck, 1995; Woodall et al., 1996; Oliveira et al., 1997). Recently two additional isoforms of GS (one cytosolic, one chloroplastic) have been reported (Maeck, 1995; Woodall et al., 1996; Sakakibara et al., 1996). It is uncertain which of these isoform(s) is (are) responsible for the assimilation of $NH_4^+$ taken up from the apoplast into the cytoplasm of the leaves.

$NH_4^+$ may be transported from the cytoplasm into the chloroplasts to serve as substrate for GS2. The role of GS2 in the leaves is supposed to be the assimilation of $NH_4^+$, both that produced from $NO_2^-$ reduction in the chloroplast and that released in photorespiration (Stulen et al., 1998). $NH_4^+$ taken up from apoplast may also be assimilated by cytosolic GS1. This isoform is supposed to assimilate glutamine for transport processes (Schjoerring et al., 1993; Pearson and Ji, 1994), but is also partially involved in assimilation of $NH_4^+$ released in photorespiration. Thus, $NH_3$ taken up by the leaves interacts with the photorespiratory $NH_4^+$ pool. The Gln produced in the cytoplasm either has to be transported into the chloroplast were it can be used as substrate for GOGAT (glutamate synthase; EC 1.4.7.1) in Glu production or may be exported directly by long-distance transport.

The assimilation of inorganic N in the leaves is unusual in many forest trees, since the main site of N assimilation is thought to be the roots (Gojon et al, 1991; 1994; Schneider et al., 1996; Geßler et al., 1998b). Induction of enzymes of N reduction and assimilation is required in response to atmospheric N influx into leaves (Thoene et al., 1991; Pérez-Soba et al., 1994). However, this may cause disturbances in regulatory processes. Since photorespiration is a cycling process the enhancing of the $NH_4^+$ pool in the chloroplasts and the cytoplasm may interact with its regulation. In both, the cytoplasm and the chloroplast, special attention has to be paid to the production of protons during the assimilation of atmospheric $NH_3$ or $NH_4^+$. Assimilation of 1 mole $NH_3$ leads to the production of 0.22 mole $H^+$, assimilation of 1 mole $NH_4^+$ to 1.22 mole $H^+$ (Raven, 1988). Also the uptake and assimilation of $NO_2$ is thought to yield net $H^+$ production, since the formation of protons from the solution of $NO_2$ in the aqueous phase of the apoplast exceeds the production of $OH^-$ from $NO_3^-$ assimilation (Raven, 1988). If proton generation by uptake and assimilation of gaseous N pollutants exceeds the buffer capacity of the cell, disturbance of pH homeostasis has to be expected. In addition $NO_2$ is known to mediate lipid peroxidation (Kunert, 1987; Ramge et al., 1993) and $NO_2^-$ produced from $NO_2$ may be involved in the production of nitroso-derivatives and the destruction of amino acids (Elstner, 1984). At very high concentrations $NH_4^+$ may cause uncoupling of photophosphorylation due to membrane permeation. However, the accumulation of such high $NH_4^+$ concentrations is unlikely, even in most polluted environments, but acute damage to vegetation growing very close to $NH_3$ emitters may be caused by this uncoupling process (Hadwiger-Fangmeier et al., 1992). Another consequence of the uptake of $NH_4^+$ by the leaves may be the excretion of $K^+$, $Mg^{2+}$ and $Ca^{2+}$ (Roelofs et al., 1985; Alenas and Skarby, 1988). In contrast, the exposure of leaves to gaseous $NH_3$ resulted in N uptake, but no cation leaching occurred
(Van der Eerden et al., 1992; Van der Eerden and Pérez-Soba, 1992). In both cases, however, the cation:N ratio decreased indicating nutrient imbalances. These effects may be enhanced, if pedospheric uptake of the nutrients leached from the canopy also decreases due to reduced $K^+$, $Ca^{2+}$ $Mg^{2+}$ and P availability in the soil.

## 4. Interaction between atmospheric and pedospheric N uptake by long-distance transport processes.

Uptake of reduced and oxidised atmospheric N compounds assimilated in the leaves can serve as an additional N supply to support growth, development and storage in trees (Wellburn 1990; Fangmeier et al., 1994).

Fig. 5. Rates of net uptake of $NH_4^+$ by roots of 12-week old beech seedlings after pre-incubation with different amino compounds. Roots were pre-incubated for two days in nutrient solutions containing 50 µM $NH_4^+$. After adding different amino compounds (at 10 mM), the roots were pre-incubated for an additional 2 d. Roots were then washed 3 times with double demineralised water, and transferred into amino-acid-free nutrient solutions with the same composition as the pre-incubation solutions (cf. Geßler et al., 1998a). $NH_4^+$ uptake was determined according to Geßler et al. (1998a). Air and solution temperature were 20°C and the photon flux density amounted to 150 µmol $m^{-2}$ $s^{-1}$. Different letters indicate significant differences at 95% confidence level. The data shown are means (± SD) of five independent experiments.

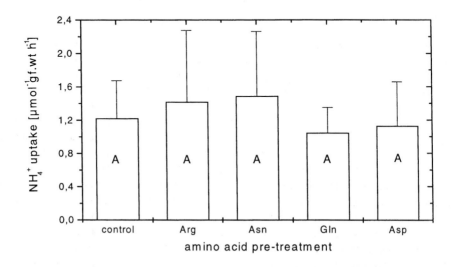

$NO_2$ fumigation of spruce shoots not only stimulates NR activity (Egger et al., 1988; Thoene et al., 1991, Nussbaum et al., 1993), but also results in export of N compounds, originating from $NO_2$ uptake, from the shoot to the roots (Nussbaum et al., 1993): Spruce trees fumigated with [$^{15}$N]$NO_2$ showed $^{15}$N labelling in amino acids in both, the needles and the bark (Gln and Glu). In the roots [$^{15}$N]$NO_2$ fumigation of the leaves resulted in labelling of Ala in spring and of Ala, Gln and Glu in autumn.

Exposure of plant leaves to $NH_3$ is known to stimulate GS (Pérez-Soba et al., 1994) and subsequently increases the contents of amino compounds in tissues exposed to $NH_3$ (Van der Eerden et al., 1990; Geßler et al., 1998c). When branches of c. 80 year-old beech trees were exposed to 100 µg [$^{15}$N]$NH_3$ $m^{-3}$, the leaves, petioles, bark and wood inside the fumigation chambers were enriched with $^{15}$N, and also the bark and the wood of adjacent parts growing outside the fumigation chambers (Geßler et al., 1996). This result indicates long-distance transport of $^{15}$N compounds in the phloem from fumigated parts of the plant to basipetal branch sections not exposed to [$^{15}$N]$NH_3$.

N uptake by roots is thought to be strongly regulated and adapted to the N demand of the plant (Imsande and Touraine, 1994). A pool of amino compounds cycling between the shoot and the roots is considered to serve as signals for the plant internal N status (Cooper and Clarkson, 1989). Since reduced and oxidised atmospheric N compounds, taken up by the leaves, are assimilated and the products are subject to long distance transport within the phloem, interaction with the cycling pool of amino compounds and influence on pedospheric N uptake by the roots can be assumed (Geßler et al., 1998c). Experiments with spruce seedlings have shown that exposure of the shoot to $NO_2$ results in a significant increase in transport of organic N compounds from the shoot to the roots. As a consequence $NO_3^-$ and $NH_4^+$ uptake were reduced to an extent that matched $NO_2$ uptake (Muller et al., 1996). A similar effect has been observed in experiments with beech seedlings exposed to $40\mu g$ $NH_3$ $m^{-3}$ for 3 d. The fumigation caused an increase in the Arg, Glu, Lys and His contents in the leaves and the Glu, Asp, Asn, Gln, Ile, Leu and Phe, but not the Arg contents in the phloem (Geßler et al., 1998c). In the roots of the seedlings total soluble non-protein N (TSNN) contents were not affected by the fumigation treatment, but the Gln contents increased significantly in the fumigated plants. As a consequence, $NO_3^-$ net uptake was reduced to an extent similar to the amount of N taken up as $NH_3$ by the shoot (Geßler et al., 1998c). From these results it appears that it is not the total pool of TSNN, but rather particular amino compounds, e.g. Gln, which serve to regulate root $NO_3^-$ uptake in keeping with plant N demand.

In order to test this hypothesis, the consequences of the enrichment of particular amino compounds in the roots on pedospheric N uptake have been examined. Roots of beech seedlings were directly exposed to different amino compounds, frequently translocated in the phloem of mature beech trees (Schneider et al., 1996; Geßler et al, 1998a; b). In these experiments a whole set of amino compounds accumulated in the fine root tissues but only exposures that enhanced the root Gln and Asp contents resulted in reduced $NO_3^-$ uptake by the roots. Exposures that enhanced Arg, Asn, γ-amino butyric acid, Ala, Ser, Orn and/or Gly contents, but not the Asp or Gln contents, did not affect $NO_3^-$ net uptake (Geßler et al., 1998a). It can, therefore, be concluded that Asp and/or Gln are important phloem-allocated compounds involved in the regulation of $NO_3^-$ uptake in beech roots. A major constraint of the experiment described above is that direct exposure of roots may lead to conditions that differ from the natural situation, in which amino compounds are delivered from the shoot to the roots by phloem transport (Cooper and Clarkson, 1989; Imsande and Touraine, 1994). To avoid artificial experimental conditions, Gln and Asp were directly loaded into the phloem by bark flap feeding (Geßler et al, 1998c). This technique resulted in an enrichment of fine root tissues with the compounds fed, but TSNN contents were not significantly affected by the treatment. Also in these experiments enrichment of roots with Gln or Asp resulted in a significant reduction of $NO_3^-$ uptake, indicating again that these particular amino compounds, rather than TSNN contents, regulate $NO_3^-$ uptake. Especially Gln appears to be a general regulator of $NO_3^-$ uptake and assimilation, since inhibition of

$NO_3^-$ uptake by Gln was also observed for prokaryotes, fungi (Clarkson and Lüttge, 1991), microalgae (Syrett and Peplinska, 1988), maize (Lee et al., 1992) soybean (Muller and Touraine, 1992) and spruce seedlings (Geßler et al., 1998a). In addition Gln is known to repress the activity and/or expression of NR in different species (Martino and Smarrelli, 1989; Deng et al., 1991; Shiraishi et al., 1992; Li et al., 1995). Further studies are required in order to characterise the effects of amino acids on the $NO_3^-$ carrier systems in the roots and the molecular events involved.

The effect of the enrichment of amino compounds was also tested for $NH_4^+$ uptake by the roots of beech seedlings. In these experiments the roots were directly exposed to different amino compounds (Arg, Asn, Gln or Asp) as described by Geßler et al.

Fig. 6. Net rates of uptake of $NO_3^-$ (■) and $NH_4^+$ (□) by fine roots of adult beech trees in the field. Roots dug out from the soil in depth between 0 and 10 cm, but not excised from the tree, were incubated in solutions containing similar concentrations of inorganic anions and cations as the natural soil water of the site. Uptake rates were calculated from the linear decrease in $NH_4^+$ and $NO_3^-$ concentrations during the incubation of the roots as described by Geßler et al (1998a). Depletion of $NO_3^-$ or $NH_4^+$ by microbial processes could be excluded. The data shown are means (± SD) of 15-17 independent experiments. Different letters indicate significant differences at $p < 0.05$. Soil temperature (—○—) was measured in the uppermost layer of the mineral soil (0-5 cm depth).

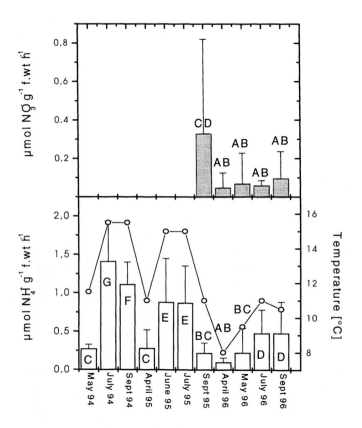

(1998a). As a consequence the amino acids fed were enriched in the fine roots to an extent comparable to that found in the $NO_3^-$ uptake experiments (Geßler et al., 1998a).

In contrast to $NO_3^-$ uptake the enrichment of Gln or Asp (and also that of Arg and Asn) did not reduce $NH_4^+$ uptake by the beech roots (Figure 5). The significance of the regulatory processes observed for beech seedlings was also tested under field conditions. For this purpose, net uptake of $NO_3^-$ and $NH_4^+$ by the roots of 90 year-old beech trees exposed to high N loadings from agriculture was studied (Figure 6) using the depletion technique (Rennenberg et al., 1996; Geßler et al., 1998a).

In 1994 and during spring and summer 1995, $NO_3^-$ net uptake was not observed. Net uptake of $NH_4^+$ amounted to up to c. 1.4 µmol $g^{-1}$ f.wt and showed high correlation with soil temperature. In September 1995, roots of the adult beech trees absorbed $NO_3^-$ for the first time since the onset of the experiments. Subsequently $NO_3^-$ net uptake was observed during the entire vegetation period in 1996. 1995 proved to be a prolific seed year for the beech trees at the field site studied ("Höglwald"). Since the formation of beech nuts requires large amounts of N especially for the formation of the embryos (Matschke, 1992), an increased N demand and, as a consequence, changes in the regulatory status of $NO_3^-$ uptake by the roots - starting in September 1995 - can be assumed. If the hypothesis for the regulation of $NO_3^-$ uptake developed from the results of the laboratory experiments is valid for field conditions, the content of Gln in the fine roots of adult beech trees should reflect the differences in $NO_3^-$ uptake before and after September 1995.

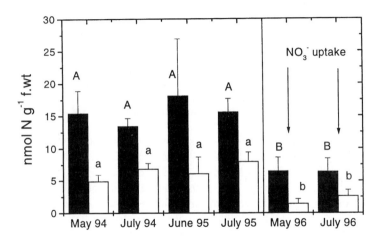

Fig. 7. Content of total soluble non-protein N (TSNN) (■) and of Gln (□) in the fine roots of adult beech trees in the field. Roots were dug out from the mineral soil (0-10 cm depth) and washed with double demineralised water. N compounds were extracted and analysed as described by Geßler et al. (1998b). The data shown are means (± SD) of three to five independent root samples with two replicate analyses, each. Different letters indicate significant differences at p<0.05.

Figure 7 shows the TSNN and Gln contents in beech fine roots in May/June and July of the years 1994 to 1996. In May and July 1996, when $NO_3^-$ uptake was observed, TSNN and Gln contents in the fine roots were significantly lower than those measured at the same time period in 1994 and 1995. From the results of the field study and the results of the laboratory experiments under controlled conditions outlined above, it is concluded that Gln levels in May and July 1994, as well as in June and July 1995, inhibited $NO_3^-$ uptake and that the reduced Gln contents in 1996 released this inhibition. Still, it must be realised that other factors, such as the presence of $NH_4^+$ in the soil solution or soil temperature may have additionally modulated net $NO_3^-$ uptake in this field study (Geßler et al., 1998a). Since the composition of the soil solution and hence the $NO_3^-$:$NH_4^+$ ratio within this solution did not change significantly between 1994 and 1996 (data not shown), the influence of soil solution N can be excluded. Soil temperature was generally higher in 1994 and 1995 as compared to 1996 but never exceeded 15.5°C (Figure 6). Laboratory experiments with beech seedlings revealed that temperatures below 20°C do not influence $NO_3^-$ net uptake significantly (Geßler 1998a). Thus, soil temperature is also unlikely to be a factor responsible for the interannual differences in $NO_3^-$ uptake in this field study with beech.

In support of results from the experiments with beech seedlings the differences in TSNN and Gln contents observed in the fine roots did not affect the $NH_4^+$ uptake of mature beech trees. Since the $NH_4^+$ uptake by the roots of spruce trees is assumed to be inhibited by the accumulation of TSNN (Muller et al., 1996; Rennenberg et al., 1998) it may be concluded that differences exist between deciduous and coniferous species in their mechanisms for adapting total pedospheric N uptake to the N demand. Further studies are required in order (1) to prove if general differences in the regulation of N uptake exist between different tree species and (2) to assess the ecological implications of such differences. In addition, the influence of the availability of other nutrients (e.g. $K^+$, $Ca^{2+}$, $Mg^{2+}$ and $PO_4^{3-}$) on these regulatory processes has to be examined in order to characterise the 'physiological window' (Geßler and Rennenberg, 1998) within which adverse effects of high N input can be prevented.

## 5. Consequences of atmospheric N input on the forest ecosystem level

If trees are able to down-regulate N uptake by the roots to a certain extent in order to avoid excess N nutrition when exposed to atmospheric N, higher amounts of $NO_3^-$ and/or $NH_4^+$ will become available for other processes within the soil (Figure 8). Reduced $NH_4^+$ uptake by roots will both increase the amount of $NH_4^+$ available for nitrification and increase emissions of the radiatively active trace gases, NO and $N_2O$ from soil. The reduction of $NH_4^+$ uptake by the roots also results in decreased release of $H^+$ by the roots. As a consequence the pH of the soil solution may rise and autotrophic nitrification will be favoured over heterotrophic nitrification causing higher rates of $NH_4^+$ turn-over. Since $NO_3^-$ uptake is linked with the efflux of $OH^-$ (Touraine et al., 1994), a reduction of $NO_3^-$ uptake may to result in a decrease of pH and an increase in soil nitrate pool. As a consequence leaching of the highly mobile $NO_3^-$ ion into ground water may rise (Rennenberg et al., 1998). Enhanced $NO_3^-$ availability will also increase both denitrification and the production of NO and $N_2O$ in this process. Enhanced amounts of these radiatively active trace gases may be emitted into the atmosphere thereby stimulating the anthropogenic greenhouse effect (Rennenberg et al., 1998). Despite considerable knowledge on these effects and interactions numerous open questions remain to be solved (e.g. Rennenberg et al., 1998). Quantifying the competition between root uptake and microbial conversion of $NO_3^-$ and $NH_4^+$ has not been possible to date. Likewise, most studies have been made in static, climax forest ecosystems, but recent forest management practices aim to transform uniform conifer monocultures into more structured species-rich forests by selective felling of adult trees and planting of deciduous tree seedlings. These management practices cause dynamic situations that influence rhizosphere processes in so far unknown ways. Felling of adult

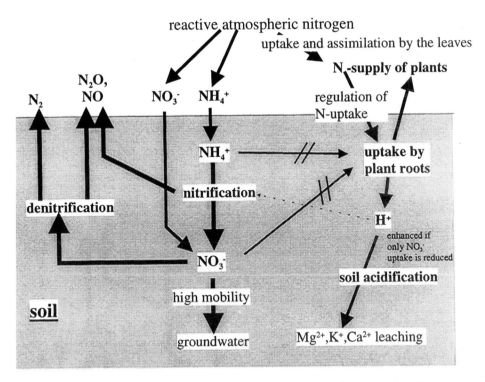

*Fig. 8.* Regulation of $NO_3^-$ and/or $NH_4^+$ uptake of plant roots and its impact on soil nutrient dynamics. The reduction of $NH_4^+$ uptake by the roots the increases the substrate pool for nitrification; a reduction of $NO_3^-$ uptake by the roots increases the substrate pool for denitrification. As a consequence enhanced release of radiatively active trace gases (NO, $N_2O$) into the atmosphere is observed. The increase of $NO_3^-$ contents in the soil solution may also result in leaching of $NO_3^-$ into the ground water. A reduction of $HCO_3^-/OH^-$ release due to reduced $NO_3^-$ uptake may lead to a decrease in soil pH and this favours heterotrophic nitrification over autotrophic.

trees will increase photon flux density and temperature on the forest floor, subsequently stimulating mineralisation and enhancing N availability (Larcher, 1984). This may increase N uptake by the planted seedlings and the remaining adult trees and, as a consequence, increase growth transiently. However, the influences of these changes on the competition and interaction between soil microorganisms and trees are unknown. If adult trees are well supplied with N, e.g. from atmospheric deposition, the additional N in the soil, produced by enhanced mineralisation, may exceed the uptake capacity of the trees and lead to enhanced $NO_3^-$ leaching and/or $NO/N_2O$ emissions. On the other hand, enhanced soil temperature may also lead to a decrease in water availability, due to enhanced evaporation depending on climate and soil properties. This may cause adverse effects on N mineralisation and nutrient uptake of the planted seedlings and understorey vegetation (Orlander *et al*, 1996). Whether these negative effects can compensate or even overcompensate for the positive effect of enhanced soil temperature remains to be established for different soils under different environmental conditions.

Studies on the activity of nitrifying and denitrifying soil microorganisms, the capacity of roots of trees and herbaceous species to take up N and the interaction between microorganisms and plants under these conditions are required for high spatial and temporal variation in N availability in order to assess the effects of forest management practices on N losses into ground water and atmosphere.

## Acknowledgement

Work by the authors was financially supported by the Bundesminister für Bildung, Wissenschaft, Forschung und Technologie (BMBF) under contract No. BEO 51 0339614 and BEO 51 0339615.

## References

Alenas I. and Skarby L. 1988. *Water Air Soil Pollut.* **38**, 223-237.
Amman M., Stalder M., Suter M., Brunold C., Baltensperger U., Jost D.T., Turler A., Gaggeler H.W. 1995. *J. Exp. Bot.* **46**, 1685-1691.
Asard H., Horemans N., Caubergs A.J. 1995. *Protoplasma* 184, 36-41.
Asman W.A.H. and Van Jaarsfeld H.A. 1990. In: *Ammoniak in der Umwelt. Kreisläufe, Wirkungen, Minderung.* Gemeinsames Symposium 10. bis 12. Oktober 1990, FAL, Braunschweig-Völkenrode. Landwirtschaftsverlag, Münster-Hiltrup. Germany pp. 2.1-2.35.
Böttger A., Ehalt P.H., Gravenhorst G. 1978. *Berichte der Kernforschungsanlage* Jülich, FRG, No. 1558.
Brumme R., Leimcke U., Matzner E. 1992. *Plant Soil.* **153**, 155-164.
Bundesministerium für Umwelt, Naturschutz und Reaktorsicherheit (ed.). 1996. *Bericht der Bundesregierung an den Deutschen Bundestag - Sechster Immissionsschutzbericht der Bundesregierung.* Bonn, Germany: Bonner Universitäts-Buchdruckerei.
Burkhardt J. and Eiden R. 1994. *Atmos. Environ.* **28**, 2001-2017.
Carreira J.A., Harrison A.F., Sheppard L.J., Woods C. 1997. *For. Ecol. Manage.* **92**, 153-166.
Clarkson D.T. and Lüttge U. 1991. *Progress in Botany* **52,** 61-83.
Cole D.W. and Rapp M. 1981. In: Reichle D.E., (ed.) *Dynamic Properties of Forest Ecosystems.* Cambridge: Cambridge University Press, pp. 341-409.
Cooper H.D. and Clarkson D.T. 1989. *J. Exp. Bot.* **40,** 753-762.
Crutzen P.J. 1979. *Annu. Rev. Earth Planet. Sci.* **7**, 443-472.
Dean J V. and Harper J E. 1986. *Pl. Physiol.* **82**, 718-723.
Deng M.D., Moureaux T., Cherel I., Boutin J.P., Caboche M., 1991. *Plant Physiol. Biochem.* **29**, 239-247.
Dickson R.E. 1989. In: Dreyer E. (ed.) *Forest Tree Physiology, Annales de Sciences Forestieres.* Paris: Elsevier, INRA **46**, 631-647.
Dietz K-J. 1997. *Progress in Botany* **58**; 221-254.
Egger A., Landolt W., Brunold C. 1988. In: Bucher JB, Bucher-Wallin I. (eds.). *Air pollution and forest decline*, *Proceedings of the 14th International Meeting for Specialists in Air Pollution Effects on Forest Ecosystems*, Interlaken, Switzerland, pp 401-403.
Elstner E.F. 1984 In: Hock B, Elstner EF. (eds.) *Pflanzentoxikologie.* Bibliographisches Institut Mannheim, Gemany. pp 67-94.
Fangmeier A., Hadwiger-Fangmeier A., Van der Eerden L.J.M., Jäger H-J. 1994. *Environ. Pollut.* **86**, 43-82.
Farquhar G.D., Firth P.M., Wetselaar R., Weir B. 1980. *Pl. Physiol.* **66**, 710-714.
Geßler A., Weber P., Schneider S., Rienks M., Rennenberg H. 1996. In: Sutton M.A., Lee D.S., Dollard G.J., Fowler D. (eds.) *Atmospheric Ammonia: Emission, deposition and environmental impacts - poster proceedings.* Institute of Terrestrial Ecology (Edinburgh Research Station), GB. pp. 116-119.
Geßler A. and Rennenberg H. 1998. In: de Kok L.J., Stulen I. (eds.). *Responses of plant metabolism to air pollution.* Backhuys Pub. Leiden, The Netherlands, pp 81-94.
Geßler A., Schneider S., von Sengbusch D., Weber P, Hanemann U., Huber C., Rothe A., Kreutzer K., Rennenberg H. 1998a. *New Phytol.* **138**, 175-285.
Geßler A., Schneider S., Weber P., Hanemann U., Rennenberg H. 1998b. *New Phytol.* **138**, 385-399.
Geßler A., Schultze M., Schremmp S., Rennenberg H. 1998c. *J. Exp. Bot.* **49**, 1529-1537.
Geßler A. 1998. Ph.D. thesis, University of Freiburg, Germany.
Gojon A., Bussi C., Grignon C., Salsac L. 1991. *Physiol. Plant.* **82**, 505-512.

Gojon A., Plassard C., Bussi C. 1994. In: Roy J, Garnier E. eds. *A whole Plant Perspective on Carbon*-N *Interactions*. The Hague: SPB Academic Publishing, pp. 131-147.
Hadwiger-Fangmeier A., Fangmeier A., Jäger H-J. 1992. *Ammoniak in der bodennahen Atmosphäre - Emission, Immission und Auswirkungen auf terrestrische Ökosysteme*. Ministerium für Umwelt, Raumordnung und Landwirtschaft des Landes Nordrhein-Westfalen (MURL). Düsseldorf, Germany.
Harrison A.F., Stevens P.A., Dighton J., Quarmby C., Dickinson A.L., Jones H.E., Howard D.M. 1995. *For. Ecol. Manage.* **76**, 139-148.
Hoek G., Mennen M.G., Allen G.A., Hofschreuder P., Van der Meulen T. 1996. *Atmos. Environ.* **30**, 3141-3150.
Horemans N., Asard H., Caubergs R.J.1998. *Pl. Physiol.* **114**, 1247-1253.
Husted S., Mattson M., Schjoerring J. K. 1996. *Plant Cell Environ.* **19**, 1299-1306.
Husted S. and Schjoerring J.K. 1996. *Pl. Physiol.* **112**, 67-74.
Imsande J. and Touraine B. 1994. *Pl. Physiol.* **105**, 3-7
Jacob B. and Heber U. 1998. *Plant Cell Physiol.* **39**, 313-322.
Johansson C. 1987. *Tellus* **39B**, 426-438.
Kisser-Priesack G.M., Bieniek D., Ziegler H. 1990. *Naturwissenschaften* **77**, 492-493.
Kolar J. 1990. *Stickoxide und Luftreinhaltung*. Springer Verlag Heidelberg.
Kunert K.J. 1987. *Projekt Europäisches Forschungszentrum für Maßnahmen zur Luftreinhaltung* PEF. PEF-Bericht 34. Kahrlsruhe, KfK, Karlsuhe.
Lam H-M, Coschigano KT, Oliveira IC, Melo-Oliveira R, Coruzzi GM. 1996. *Ann. Rev. Pl. Physiol. Pl. Mol. Biol.* **47**, 569-593.
Larcher W. 1984. *Ökologie der Pflanzen*. 4th edition Ulmer, Stuttgart, Gemany.
Latus C., Förstel H., Führ F. 1990. *Naturwissenschaften* **77**, 283-285.
Lee R.B., Purves J.V., Ratcliffe R.G., Saker L.R. 1992. *J. Exp. Bot.* **43**, 1385-1396.
Lee Y-N. and Schwartz S.E. 1981. *J. Geophys. Res.* **86**, 11971-11983.
Lendzian K.J and Kerstiens G. 1988. *Aspects of Applied Biology.* **17**, 97-104.
Levine J.S., Augustson T.R., Anderson I.C., Hoell J.M. Jr. 1984. *Atmos. Environ.* **18**, 1797-1804.
Li X-Z., Larson D.E., Glibetic M., Oaks A. 1995. *Physiol. Plant.* **93**, 740-744.
Luwe M. 1996. *Plant, Cell Environ.* **19**, 321-328.
Maeck G. 1995. *Planta* **196**, 231-238.
Marschner H., Häussling M., George E. 1991. *Trees* 5, 14-21.
Martin F. and Lorillou S. 1997. In: Rennenberg H, Eschrich W, Ziegler H. (eds.). *Trees - Contributions to modern tree Physiology*. Backhuys Publishers, Leiden The Netherlands. pp 423-440.
Martino S.J. and Smarelli J. 1989. *Plant Science* **61**, 61-67
Matschke J. 1992. In: Lyr H., Fiedler H. J., Tranquillini W. (eds.) *Physiologie und Ökologie der Gehölze*. Gustav Fischer Verlag, Jena. pp. 497-539.
Muller B., Touraine B., Rennenberg H. 1996. *Plant, Cell Environ.* 19, 345-355
Muller B. and Touraine B. 1992. *J. Exp. Bot.* **43**: 617-623
Neubert A., Kley D., Wildt J., Segschneider H.J., Förstel H. 1993. *Atmos. Environ.* 27A, 2137-2145.
Norby R.J., Weerasuriya Y., Hanson P. 1989. *Can. J. For. Res.* **19**, 889-896.
Nussbaum S., v. Ballmoos P., Gfeller H., Schlunegger U.P., Fuhrer J., Rhodes D., Brunold C. 1993. *Oecologia* **94**, 408-414.
Oliveira I.C., Lam H-M., Coshigano K., Melo-Oliveira R., Corruzzi G. 1997. *Pl. Physiol. Biochem.* **35**,185-198.
Orlander G., Nilsson U., Hallgren J., and Griffith J. 1996. New Zealand Journal of Forestry Science, **26**, 99-117.
Papen H., Hermann H., Butterbach-Bahl K., Rennenberg H. 1994. *Forschungsbericht KfK-PEF* **127**.
Pearson J. and Ji Y.M. 1994. *Plant Cell Environ.* **17**, 1331-1337.
Pearson J. and Stewart G.R. 1993. *New Phytol* **125**: 283-305.
Pérez-Soba M., Stulen I., Van der Eerden L.J.M. 1994a. *Physiol. Plant.* **90**, 629-639.
Polle A., Chakrabarti K., Schurmann W., Rennenberg H. 1990. *Pl. Physiol.* **94**; 312-319.
Polle A., Wieser G., Havranek W.M. 1995. *Plant Cell Environ.* **18**, 681-688.
Polle A. and Rennenberg H. 1992. *New Phytol.* **121**, 635-642.
Ramge P., Badeck F-W., Ploechl M., Kohlmaier G.H. 1993. *New Phytol.* **125**, 771-785.
Rautenkranz A.A.F, Li L., Mächler F., Martinoia E., Oertli J.J. 1994. *Pl. Physiol.* **106**, 187-193.
Raven J.A. 1988. *New Phytol.* **109**, 1-20.
Rennenberg H., Kreutzer K., Papen H., Weber P. 1998. *New Phytol* **139**, 71-86.

Rennenberg H., Schneider S., Weber P. 1996. *J. Exp. Bot.* **47**, 1491-1498
Roelofs J.G.M., Kempers A.J., Houdijk A.L.F.M., Jansen J. 1985. *Plant Soil.* **84**: 45-56
Rondón A. and Granat L. 1994. *Tellus* **46B**, 339-352.
Sakakibara H., Shimizu H., Hase T., Yamazaki Y., Takao T., Shimonishi Y., Sugiyama T. 1996. *J. Biol. Chem.* **271**, 29561-29568.
Schjoerring J.K., Kyllingsbaek A., Mortensen J.V., Byskov-Nielsen S. 1993. *Plant, Cell Environ.* **16**, 161-167.
Schneider S., Geßler A., Weber P., v. Sengbusch D., Hanemann U., Rennenberg H. 1996. *New Phytol.* **134**, 103-114
Schnug E., Haneklaus S., Borchers A., Polle A., 1995. *Zeitschr. Pflanzenernähr. Bodenk.* **158**, 67-69.
Schwanz P., Haberle K.H., Polle A. 1996b. *J. Plant Physiol.* **148**, 351-355.
Schwanz P., Kimball B.A., Idso S.B., Hendrix D.L., Polle A. 1996c. *J. Exp. Bot.* **47**, 1941-1950.
Schwanz P., Picon C., Vivin P., Dreyer E., Guehl J.M., Polle A. 1996a. *Pl. Physiol.* **110**, 393-402.
Shiraishi N., Sato T., Ogura N., Hakagawa H. 1992. *Plant Cell Physiol.* **33**, 727-731
Slovik S., Balazs A., Siegmund A. 1996a. *Plant Soil.* **178**, 295-310.
Slovik S., Siegmund A., Fuhrer H.W., Heber U. 1996b. *New Phytol.* **132**, 661-676.
Sommer S.G., Kjellerup V., Kristjansen O. 1993. *Soil Plant Sci.* **42**, 146-151.
Srivastava H.S. 1998. In: de Kok L.J., Stulen I. (eds.). *Responses of plant metabolism to air pollution.* Backhuys Pub. Leiden, The Netherlands, pp 65-80.
Stevens P.A., Harrison A.F., Jones H.E., Williams T.G., Hughes S. 1993. *For. Ecol. Managem.* **58**, 233-247.
Stewart G.R., Hegarty E.E., Specht R.L. 1988. *Physiol. Plant.* **74**, 26-33.
Stulen I., Pérez-Soba M., De Kok L.J., Van der Eerden L. 1998. *New Phytol.* **139**, 61-70.
Sutton M.A., Fowler D., Burkhardt J.K., Milford C. 1995. *Water, Air, Soil Pollut.* **85**, 2057-2063.
Syrett P.J. and Peplinska A.M. 1988. *New Phytol.* **109**, 289-296.
Thœne B., Rennenberg H., Weber P. 1996. *New Phytol.* **134**, 257-266.
Thœne B., Schröder P., Papen H., Egger A., Rennenberg H. 1991. *New Phytol.* **117**, 575-585.
Tietema A. and Beier C. 1995. *For. Ecol. Manage.* **71**, 143-152.
Tischner R., Peuke A., Godbold D.L., Feig R., Merg G., Hüttermann A. 1988. *J. Plant Physiol.* **133**, 243-246.
Touraine B., Clarkson D.T., Muller B.. 1994. In: Roy J., Granier E. (eds.) *A whole plant perspective on carbon-Nitrogen interactions.* The Hague: The Netherlands: SPB Academic Publishing pp. 11-30.
United Kingdom Review Group on Impacts of Atmospheric Nitrogen. 1994. (Pitcairn C.E.R., ed.) *Impacts of nitrogen deposition on terrestrial ecosystems. Report of UK review group on impact of atmospheric N.* London, UK Department of the Environment.
Van der Eerden L.J.M., Dueck T.A., Elderson J., Van Dobben H.F., Berdowski J.J.M., Latuhihin M., Prins A.H. 1990. *Project 124/125, phase II, Dutch priority programme on acidification, IPO report R 90/06, RIN report 90/20.*
Van der Eerden L.J.M. and Pérez-Soba M. 1992. *Trees* **6**, 48-53.
Van der Eerden L.J.M., Lekkerkerk L.J.A., Smeulders S.M., Jansen A.E. 1992. *Environ. Pollut.* **76**, 1-9
Van Dijk H.F.G., Creemers R.C.M., Rijniers J.P.L.W.M., Roelofs JGM. 1989. *Environ Pollut.* **62**, 317-336.
Van Hove L.W.A., Adema E.H., Vredenberg W.J., Pieters G.A. 1989. *Atmos. Environ.* **23**, 1479-1486.
Vanacker H., Carver T.L.W., Foyer Ch. 1998 *Pl. Physiol.* **117**, 1103-1114.
Weber P. and Rennenberg H. 1996. *Atmos. Environ.* **30**, 3001-3009.
Wellburn A.R. 1990. *New Phytol.* **115**, 395-429.
Woodall J., Havill D.C., Pearson J. 1996. *Pl. Physiol. Biochem.* **34**, 697-706.

# THE IMPACT OF NITROGEN ON FOREST SOILS AND FEEDBACKS ON TREE GROWTH

B A EMMETT

*Institute of Terrestrial Ecology, Deiniol Rd, Bangor, Gwynedd, LL57 2UP, UK*

(Received 31 March 1999; accepted 9 April 1999)

**Abstract**: The effects of pollutant nitrogen on forest soils and the potential feedbacks on tree growth are discussed using data from recent plot and catchment manipulation studies. Results indicate that N applied to the soil affects both soil N transformations and base cation status in some forest soils. Whilst reductions in tree growth are infrequently reported, a greater understanding of the effect of increased N deposition on soil N transformations, and associated changes in soil acidification and nutritional balance, is essential if sensitive stands are to identified. In particular, the factors controlling soil N accumulation rates and thus the onset of nitrate leaching are not clearly understood with increased nitrogen availability potentially increasing nitrogen accumulation due to abiotic fixation and lignolytic enzyme suppression, whilst the onset of nitrification as the C/N ratio of forest floor material declines, may reduce N retention efficiency. The switch from increased to decreased tree growth rate in response to N additions in some experiments highlights the need for long-term studies as a necessary component of future research.

**Keywords**: Forests, N deposition, soil nitrogen transformations, nitrification, tree growth, acidification, BC/Al ratio, $^{15}N$

## 1. Introduction

In recent decades, a trend towards increasing growth has been identified in many European forests (Spiecker *et al.*, 1996). The deposition of reactive nitrogen (N) to forest ecosystems is a possible contributing factor in some localities. To investigate the effects of N deposition on both tree growth and water quality a series of manipulation experiments have been conducted in Europe (e.g. the NITREX project (Wright and van Breemen 1995)) and the United States (e.g. McNulty *et al.*, 1996; Magill *et al.*, 1996 & 1997). In these studies, both positive and negative effects on tree growth have been recorded. The causes of reductions in tree growth in response to increased N inputs are not clearly understood and may include increased susceptibility to secondary factors such as drought or fungal attack in addition to soil acidification, reduced base cation/Al (BC/Al) ratios in soil water and altered nutritional supply (Bobbink *et al.*, 1992). This paper examines the experimental evidence from recent plot and catchment studies in Europe and the USA concerning the impact of N deposition on soils and tree growth.

## 2. Impact on trees

Various European N fertilisation experiments in forest stands have resulted in either a positive or neutral response of tree growth to N additions (e.g. Farrell, 1985; Malkonen, 1990; Tamm, 1991). However, in a few areas with acid soils and already high N deposition, N fertilisation may result in a decrease in tree growth possibly due to soil acidification and a decline in availability of base cations. For example, Nys (1989) reported a significant decline in wood accumulation rates from 13 to 11.5 $m^3$/ha/yr in a heavily polluted spruce forest, the Croix-Scaille in the French Ardennes, following an addition of 100 kgN/ha/yr as ammonium nitrate

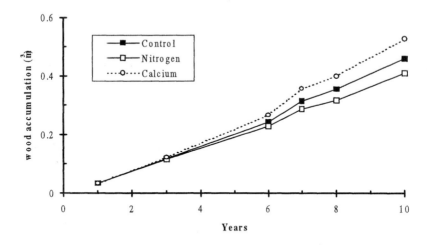

Fig. 1 The effect of two 100 kgN/ha applications of N as ammonium nitrate to a spruce stand in the French Ardennes. Ambient inputs were estimated at *ca.* 50 kgN/ha/yr. (Nys 1989). Comparison with the response to calcium carbonate additions at 2500 kg/ha is also shown.

in 1981 and 1983 (Figure 1). The authors attributed the decline to a decrease in the availability of calcium, due to the acidification of the soil. In the United States, Magill *et al.* (1997) applied N at 150kgN/ha/yr as ammonium nitrate to hardwood and pine stands in Harvard Forest. A decline in tree growth rate was observed in the pine stand, but an increase in the hardwood stand. The authors suggested that the different response to N additions in the two stands was related to land use history and the effect of this on soil N retention capacity, and thus the potential for nitrate and associated base cation leaching. A decline in tree growth rate has also been observed in the Skogaby experiment in Sweden in response to N additions (Nilsson *et al.*, 1998). Initial increases in tree growth in response to N inputs of 100 kgN/ha/yr as ammonium sulphate were observed despite a reduction in BC/Al ratios in soil solution (Binkley and Högberg 1997). However, recent data indicate that production rates decreased after the first six years of N additions and were 15% below production values of the control trees after ten years (Nilsson *et al.*, 1998). These studies indicate that on some sensitive soils, large N inputs alone can result in a decline in BC/Al ratios and a decrease in tree growth. However, the studies involved high doses of N which are unrepresentative of the inputs observed in many parts of Europe and the United States.

Examination of tree growth response to experimentally changed N inputs of a more moderate nature that simulate increased N deposition (< 60 kgN/ha/yr) (Table I), reveals that, in general, no response of tree growth is observed within the experimental period. One exception is a study in the spruce-fir stands of New England in which McNulty *et al.* (1996) reported both a trend towards increased mortality and, interestingly, a change in recruitment of different species following additions of a little as 22 kgN/ha/yr.

TABLE I

Tree growth responses to experimentally changed N inputs (excluding large fertiliser dose experiments)

| Site | Forest Type | Change in N inputs (kgN/ha/yr) | Effect of N on tree growth | Reference |
|---|---|---|---|---|
| Hubbard Brook | Mixed hardwood | 7 to 47 | 0 | Christ et al. (1995) |
| Bearbrook | Mixed hardwood/spruce | 5 to 33 | 0 | Magill et al. (1996) |
| Havard Forest | Hardwood | 8 to 58[1] | 0 | Magill et al. (1997) |
|  | Red pine | 8 to 58[1] | 0 | " |
| Mt. Ascutney | Spruce-fir | 4 to 35 (max)[2] | - | McNulty et al. (1996) |
| 6 European NITREX sites | Spruce or fir | + and -[3] | 0 | Emmett et al. (1998b) |
|  | Scots pine | 54 to < 5 includes reduction of S | - | Boxman et al. (1998) |

[1] Low-N treatment only reported here. A High-N treatment of 150 kgN/ha/yr resulted in a positive response in the hardwood stand and negative in the red pine. No replication in this study.
[2] Range of N application rates from 15.7 to 31.4 kgN/ha/yr. Positive responses observed in early years in the lower application rate plots.
[3] Addition experiments involved application rates of *ca.* 35 kgN/ha/yr. Reduction experiments reduced inputs from > 30 kgN/ha/yr to < 5 kgN/ha/yr.

Furthermore, there was a switch in net tree growth from an initial neutral or positive response relative to control trees, to a decline after several years of N applications (Figure 2), as observed in Skogaby. Another example is a 50% increase in tree growth observed at Ysselsteyn in the Netherlands following a reduction of N and sulphur inputs using an exclusion roof (Boxman *et al.*, 1998) (Figure 3). The authors associated the improvement in tree growth with an increase in the availability of potassium and magnesium, as reflected in foliar K:N and Mg:N ratios. The sensitivity of these particular stands to increased N availability could be due to a variety of factors, including the stand age, site fertility, previous land use practices and species involved. Whilst the decline in tree growth is usually associated with nutrient limitations resulting from soil acidification, the importance of secondary stress factors is poorly understood in mature forests.

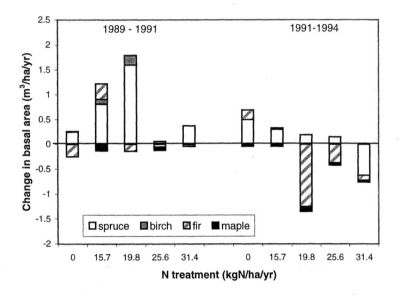

Fig. 2 Changes in basal area in response to a range of N additions in a mixed spruce-fir stand in Mt Ascutney, USA. (Redrawn from McNulty et al., 1996).

## 3. Feedbacks between soil and trees

As various authors have noted, the nitrification rate is the key process that, once stimulated, can result in radical changes in ecosystem functioning. Such changes include acidification, declining BC/Al ratios in sensitive soils and the onset of N leaching (Aber et al., 1989; Gundersen et al., 1998; Berg et al., 1997; Emmett et al., 1998a). A major controlling factor of nitrification is N availability, and thus nitrification would be expected to increase in response to N deposition. Evidence for a direct effect of increased N availability on soil N processes, including nitrification, has been reported in several studies, as summarised in Gundersen et al. (1998) and Aber et al. (1998). No significant changes in nitrification rates were observed in a series of European studies (Gundersen et al., 1998), though some evidence of increases in nitrification was found in studies in the United States (Aber et al., 1998). One factor delaying the response of N transformation rates to chronic N additions may be the slow change in the quality of soil organic matter and the associated biotic community. Strong relationships between N deposition fluxes and nitrification rates across a deposition transect in the United States (McNulty, 1991) suggested a link between N inputs and accelerated nitrification rates if forests have experienced elevated N inputs over a prolonged period. Analysis of European data indicates that nitrification is more closely related to the C/N ratio of the forest floor material than N inputs (Gundersen et al., 1998; Emmett et al., 1998a) and also illustrates the importance of competition from other soil sinks even in the presence of large N inputs in the short term.

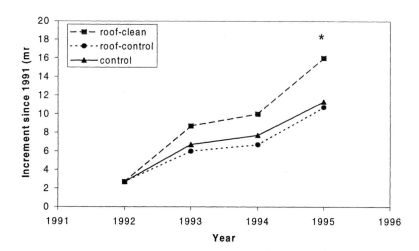

Fig. 3 The change in tree diameter growth of the five dominant trees at Ysselsteyn under a transparent roof which reduced N and S inputs to pre-industrial levels ('roof-clean'). Data for a 'roof-control' are also shown which enabled the effect of the structure to be separated from the reduction in inputs. (Redrawn from Boxman *et al.* 1998). Significant response indicated by *. Increases in Mg:N and K:N ratios in foliage were also observed in the 'roof-clean' treatment.

A link between N additions and a decline in BC/Al ratios in soil solution or on soil exchange sites has been observed in several studies (e.g. Emmett *et al.*, 1998b; Stanturf *et al.*, 1994). This occurs as a result of the acidifying effect of either ammonium uptake or nitrification of the applied ammonium (Figure 4).

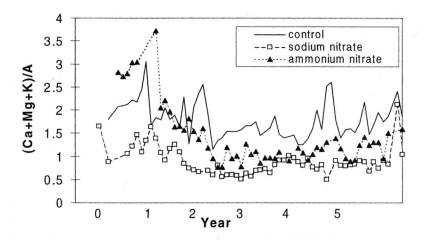

Fig. 4 BC/Al molar ratios in soil solution in response to weekly additions of N as ammonium nitrate or sodium nitrate at 35kgN/ha/yr relative to the control. (Redrawn from Emmett *et al.* 1998).

The effect of N inputs on soil solution will also depend on the pH, with elevated base cation leaching being observed in non-acidified soils and aluminium leaching in acidified soils (Berg et al., 1997). In the long-term, however, a decline in the BC/Al may still occur even in the less acid soils, due to the depletion of base cations on the exchange sites. In addition, Berg et al. (1997) have hypothesised that a reduction in tree growth may accelerate forest die-back because of the acceleration of soil N transformations in response to heat and water flux as the canopy opens up. This in turn would result in further acidification and a decline in soil BC/Al ratios.

Linking these changes in BC/Al ratios to tree growth responses is difficult due to other changes induced following N additions. Some evidence does exist for both the Skogaby site, as discussed previously, and the Mount Ascutney site in the United States. In the latter study site, Ca/Al molar ratios in spruce foliage declined with net spruce growth as the N addition increased from zero to 31.4 kgN/ha/yr (Figure 5).

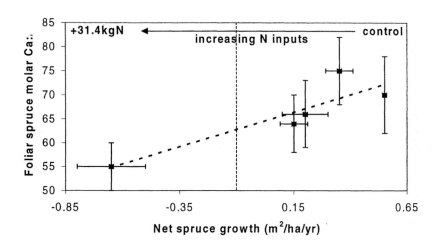

Fig. 5 The relationship between declining Ca/Al ratio in spruce foliage and tree growth at Mt Ascutney (Redrawn from McNulty et al., 1996).

Why this relationship occurred is not clear but dark respiration was greater in trees with low Ca/Al ratio foliage, possibly related to reduced membrane integrity (Schaberg et al., 1997). A decline in phosphorus (P) availability has also been proposed as a contributory factor in declining tree growth following N additions. In one study in Glencorse, Scotland, applications of N in combination with sulphuric acid caused a significant reduction in stem growth. The authors suggested this response may have been due to a N-induced P deficiency in the N treated plots. Using a root bioassay, increased P stress was detected, possibly due to increased P sorption by the soil and decreased P mineralisation of the soil organic pool (Carreira et al., 1997). However, separating out the controlling factors is complicated because of the combination of N with acid inputs and a reduction in the (Ca+Mg)/Al ratio from 9.5 to 0.4, in addition to a decline in P availability.

## 4. Changes in soil N accumulation

Large-scale $^{15}$N studies at the plot or catchment scale have recently been carried out to determine the relative importance of soil and trees as a sink for deposited N. These studies encompassed a range of N loadings in the United States and Europe (Tietema et al., 1998; Nadelhoffer et al., 1999). N was applied at moderate rates to a variety of different forest stands for several years prior to enrichment of the N additions with $^{15}$N for one to three years. The amount of $^{15}$N in trees, soils and, in some cases, drainage waters, was determined at the end of the enrichment period. In general, < 20% of N inputs, as indicated by $^{15}$N retention, were recovered from the trees, similar to the 15% calculated by Kauppi et al. (1995) for Finnish forests. The soil was the major sink for N but in some sites with large N inputs, leaching losses of $^{15}$N also represented a significant proportion of the total. The controls on long-term N accumulation rates in forest soils are not well understood but will ultimately determine the amount of N leached as nitrate and therefore net acidification of the soil (though ammonium may also be leached from more organic soils (Yesmin et al., 1996).

One of the controls on N accumulation is the fate of internally cycled N in litter. Increases in N concentration of litter have been observed across N deposition gradients (Tietema and Beier, 1995). The change in litter quality, in combination with increased N availability in soil solution, is known to accelerate initial stages of decomposition but retard later stages. These effects have been reviewed by Fog (1988), Berg (1996) and most recently Berg and Matzner (1997). Generally, a decrease in microbial activity in response to increased N supply is greatest in litter from low fertility sites or substrates with high C/N ratios (e.g. Bååth et al., 1981). Berg and Matzner (1997) suggest that the retardation of the later stages of litter decomposition is due to a suppression of lignolytic enzymes in white rot fungi. Increase in N accumulation rates with N deposition could therefore result from a combination of enzyme suppression and the formation of chemically stable compounds by the reactions of both ammonium and nitrate with lignin and phenolic compounds (Berg and Matzner, 1997).

Aber et al. (1998) have suggested that the main mechanism for increasing N storage in soils with increasing deposition is mycorrhizal assimilation. Reduced decomposition rates could delay release of internally cycled N but would not account for the large accumulation of externally deposited N in the soil (Nadelhoffer et al., 1999). Rates of abiotic fixation are insufficient to account for the magnitude of retention reported, and carbon (C) limitation of assimilation by soil microbes would prevent microbial immobilisation. Aber et al. (1998) therefore argue that only mycorrhizal fungi have a sufficient C supply to account for the large N accumulation in the soil. This N would be cycled within the soil pool through assimilation and exudation. Continued deposition of N could result in a decline in mycorrhizae, thus reducing competition and increasing availability of N for other soil microbes such as nitrifiers.

Evidence for an increase in N accumulation rates with N deposition may be inferred from a positive relationship between N concentrations in forest floor material and N inputs in wet deposition (Tietema and Beier, 1995; McNulty et al., 1991). However, the stimulation of nitrification in soil with the %N of the forest floor will reduce the efficiency of the soil to store N. Several studies in a variety of different forest ecosystems all observed a stimulation of nitrification rates when the N content of forest floor material increased above 1.4%N (Wilson and Emmett, *In Press*). This onset of net nitrification has been suggested to result from either reduced nitrate immobilisation by soil microbes (Stark and Hart, 1997) or increased gross production (Tietema, 1998). The consequence of an increase in nitrification can be, in many mature forest systems, the onset of N leaching (Gundersen, this volume) due to a lower retention capacity of nitrate relative to ammonium, whether internally generated (Emmett and Quarmby, 1991) or exogenously applied (Emmett *et al.*, 1998b). Following long-term exposure to elevated N deposition, the net effect of the two opposing processes, increased soil N accumulation and the onset of nitrification, is an increase in the amount of N accumulated in the soil (Figure 6) but a decrease in the retention efficiency of deposited N (Figure 7) (Tietema *et al.*, 1998). It is this reduced efficiency that may result in a decline of tree growth on some acidification-sensitive soils.

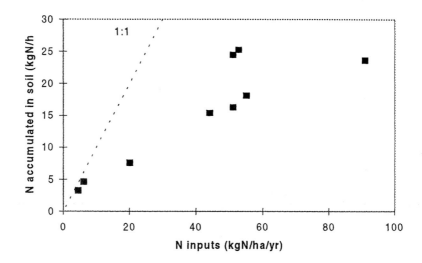

Fig. 6 The relationship between N inputs and the accumulation of $^{15}$N applied in soil. Values calculated from retention of $^{15}$N applied for 1 - 1.5 years in control and N input manipulated plots in various NITREX sites (Data from Tietema *et al.*, 1998)

## 6. Conclusions

There is some, albeit limited, experimental evidence from recent studies to suggest that deposited N may have a negative effect on tree growth on certain sensitive soils. Further work is needed to determine the potential extent of such negative responses to deposited N in the long-term, and to elucidate the relative importance of changes in soil solution chemistry and P supply, together with secondary stress factors in different ecosystems. Long-term experimental studies will also be essential if we are to understand the implications of changes in N accumulation rates, resulting from continued N deposition, on the growth of trees.

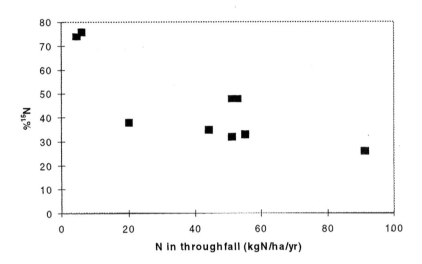

Fig. 7 The decline of soil retention efficiency of $^{15}$N inputs with N inputs in control and N deposition manipulated plots in several NITREX sites (Data from Tietema et al., 1998).

## References

Aber, J.D., Nadelhoffer, K.J., Steudler, P., and Melillo, J.M.: 1989, *Bioscience*, **39**, 378-386.
Aber, J., McDowell, W., Nadelhoffer, K., Magill, A., Berntson, G., Kamakea, M., McNulty, S., Currie, W., Rustad, L. and Fernandez, I.: 1998, *Biosci.* **48**, 921-934.
Bååth, E., Lundgren, B. and Söderström, B., 1981, *Zbl. Bakt. Hyg., I. Abt. Orig., C* **2**, 90-98.
Berg, B., Ekbohm, G., Johansson, M., McClaugherty, C., Rutigliano, F. and Virzo De Santo, A.: 1996, *Can. J. Bot.*, **74**, 659-672.
Berg, B. and Matzner, E.:1997, *Environ. Rev.*, **5**, 1-25.
Berg, M.P., Verhoef, H.A., Bolger, T., Anderson, J.M., Beese, F., Couteaux, M.M., Ineson, P., McCarthy, F., Palka, L., Raubauch, M., Splatt, P. and Willison, T.: 1997, *Biogeochem.* **39**, 295-326.
Binkley, D., and Högberg, P.: 1997, *For Ecol Manage.*, **92**, 119-152.
Bobbink, R., Boxman, D., Fremstad, E., Heil, G., Houdijk, A. and Roelofs, J. 1992, *Critical Loads for Nitrogen - report from a workshop held at Lokeboerg, Sweden.* Nord 1992:**41**. pp111-160.
Boxman, A.W., Van der Ven, P.J.M and Roelofs, J.G.M. 1998, *For. Ecol. Manage.* **101**, 155-164.
Carreira, J. A., Harrison, A. F., Sheppard, L. J., Woods, C.: 1997, *For. Ecol. Manage.* **92**, 153-166.
Christ, M., Zhang, Y., Likens, G.E. and Driscoll, C.T.: 1995, *Ecol. Appl.* **5**, 802-812
Emmett B.A. and Quarmby, C.:1991, *Biogeochem.* **15**,47-63.
Emmett, B.A., Boxman, D., Bredemeier, M., Gundersen, P., Kjønaas, O.J., Moldan, F., Schleppi, P., Tietema, A., and Wright, R.F., :1998a, *Ecosystems* **1**, 352-360.

Emmett, B.A., Reynolds, B., Silgram, M., Sparks, T.H. and Woods, C.:1998b, *For. Ecol. Manage.* **101**, 165-176.
Farrell, E.P.:1985, *Irish For.***42**, 76-91.
Fog, K., 1988, *Biol. Rev.*, **63**, 433-462.
Gundersen, P., Emmett, B.A., Kjønaas, O.J., Koopmans, C., Tietema, A.:1998, *For. Ecol. Manage.*, **101**, 37 - 56.
Gundersen, P.: 1995, *Water Air and Soil Pollut.*, **85**, 1179-1184.
Kauppi, P., Tomppo, E. and Ferm, A.: 1995, *Plant Soil* 168-169, 633-638.
Malkonen, E.; 1990, *Plant Soil* **128**. 75-82
Magill, A.H., Downs, M.R., Nadelhoffer, K.J., Hallett, R.A. and Aber, J.D.: 1966, *For. Ecol. Manage.* **84**, 29-37.
Magill, A.H., Aber, J.D., Hendricks, J.J., Bowden, R.D., Melillo, J.M and Steudler, P.A.:1997, *Ecol. Appl.*, **7**, 402-415.
McNulty, S.G., Aber, J.D. and Boone, R.D. :1991, *Biogeochem.* **14**, 13-29.
McNulty, S.G., Aber, J.D. and Newman, S.D.:1996, *For Ecol and Manage.* **84**, 109-121.
Nadelhoffer, K.J., Emmett, B.A., Gundersen, P., Kjønaas, O.J., Koopmans, C.J., Schleppi, P., Tietema, A. and Wright, R.F.: 1999, *Nature.* **398**, 145-147
Nys, C. :1989, Rev. For.Fr. **41**, 336-347..
Nilsson, L.O., Bergholm, J. and Ostergren, M.: 1998, *Nitrogen, the confer-N-s, Abstract Book*, 23-27 March 1998, Noordwijkerhout, The Netherlands.
Schaberg, P.G., Perkins, T.D. and McNulty, S.G.: 1997, *Can. J. For. Res.* **27**. 1622-1629.
Spiecker, H., Mielikäinen, K., Köhl, M., and Skovsgaard, J.P.: 1996, *Growth Trends in European Forests*, European Forest Institute Research Report No. 5, Springer-Verlag Berlin Heidelberg, pp. 372.
Stanturf, J.A., and Stone, E.L.: 1994, *For. Ecol. Manage.* **65**, 265-277.
Stark, J.M. and Hart, S.C.: 1997, *Nature* **385**, 61-64.
Tamm, C.O., 1991, Nitrogen in terrestrial Ecosystems, *Ecol. Studies* **81**, Springer-Verlag.
Tietema, A.: 1998, *For. Ecol. Manage.* **101**, 29-36.
Tietema, A. and Beier, C.:1995, *For. Ecol. Manage.* **71**, 143-152.
Tietema, A., Emmett, B.A., Gundersen, P., Kjønaas, O.J.and Koopmans, C.J.:1998, The fate of $^{15}$N-labelled nitrogen deposition in coniferous forest ecosystems. *For. Ecol. Manage.* **101**, 19-28.
Van Breemen, N. and van Dijk, H.F.G.:1988, *Environ. Pollut.***75**, 283-308.
Wilson, E.J. and Emmett, B.A., In Press, *The Impact of Nitrogen Deposition on Natural and Semi-natural Ecosystems*, (Eds. Langan and Wilson), Chapman and Hall, London.
Wright, R.F. and Van Breemen, N.: 1995, *For. Ecol. Manage.* **71**, 1-5.
Yesmin, L., Gammack, S.M. and Cresser, M.S.: 1996, *Water Res.* **30**, 2171-2177.

# ATMOSPHERIC INPUT TO DANISH SPRUCE FORESTS AND EFFECTS ON SOIL ACIDIFICATION AND FOREST GROWTH BASED ON 12 YEARS MEASUREMENTS

MADS F. HOVMAND [1] and J. BILLE-HANSEN [2]

[1] National Environmental Research Institute, DK-4000 Roskilde, Denmark. MFH@DMU.DK
[2] Danish Forest and Landscape Research Institute, DK-2870 Hørsholm, Denmark. JBH@FSL.DK

(Received 4 December 1998; accepted 25 February 1999)

**Abstract.** Atmospheric deposition of strong acid and sulphur compounds has been measured at Danish forest sites since 1985. Results from 12 years of measurements are reported from Ulborg experimental forest site situated on the sandy soils of West Jutland. Measurements and model calculated deposition estimates indicate a cumulated acid input from the atmosphere of 40 $kmol_c$ during the last 30 years; far more than the forest soils could neutralise by weathering or by base cation exchange. During the period 1985-1997 an increase in soil water concentration of protons (acid) and aluminium were seen and the molar ratio between calcium and aluminium dropped to a minimum of 0.1. A decrease in forest growth and an increase in needle litter fall were seen in the same period.

**Keywords:** Forest, atmospheric deposition, sulphur dioxide, soil water, aluminium, tree ring.

## 1. Introduction

Beginning several decades ago, acid deposition over Scandinavia was considered to be a problem mainly for the mountain and coastal lakes in Norway (Wright et al., 1975) and Sweden (Hultberg, 1985; Nilsson, 1985). Later on, long term investigations also showed changing acidity in forest soils (Tamm and Hallbäcken, 1988). Due to the regional nature of acidification, different Danish investigations were initiated in 1983, 1985 and 1988 (Bille-Hansen and Hovmand, 1987; Rasmussen, 1988). The forest ecosystem investigation reported here was started in 1985.

Acid deposition to the forest is here defined as the sum of dry deposited sulphur dioxide ($SO_2$) and the wet deposition of strong acid (protons, $H^+$), as these are the most important acidifying components. Dry deposition of gaseous nitric acid ($HNO_3$), hydrochloric acid, sulphuric acid ($H_2SO_4$) or organic acids are not taken into consideration, since these compounds are found at very low concentration levels in the rural area of Denmark. The yearly average deposition of gaseous $HNO_3$ at the Ulborg site was 0.1 $kmol_c$ $ha^{-1}$ $yr^{-1}$ (Andersen, 1997).

The potential acidification from wet and dry deposition of N compounds was described by Breemen et al. (1982). This potential has not been included in the calculation of the cumulated soil acidification, since percolation of nitrate ($NO_3^-$) is a precondition for the activation of the acidification potential from the cumulated atmospheric N input. The present N status of the forest in Ulborg exhibits a very low $NO_3^-$ leaching (less than 0.1 kg $kmol_c$ $ha^{-1}$ $yr^{-1}$).

*Water, Air, and Soil Pollution* **116**: 75-88, 1999.
© 1999 *Kluwer Academic Publishers. Printed in the Netherlands.*

Soil acidification is a natural process caused by soil respiration, production of organic acids and plant uptake of base cations. Soil acidification is also caused by atmospheric deposition of sulphur (S) and nitrogen (N) compounds, originating from human activity. Oxidised S and N compounds form strong acids ($H_2SO_4$ and $HNO_3$), which are able to remove cations adsorbed to soil particles. Base cations, as magnesium ($Mg^{2+}$), calcium ($Ca^{2+}$) and potassium ($K^+$), are washed out of the top soil by $H_2SO_4$ and $HNO_3$, thereby loosing base cations important for the growth of the trees and other vegetation. As the top soil at this experimental plot has no free carbonate and little base cation exchange capability, the input of protons (acid) could lower soil water pH and dissolve aluminium ($Al^{3+}$).

The purpose of this paper is to estimate the acidifying potential of the long term cumulated input of atmospheric deposition of S and protons to Danish spruce forest standing on sandy soils. Further, the accumulated atmospheric input is related to the available base cation content of the soil and the Ca/Al ratio in the soil water. A response in the tree growth vitality to a changing soil water acidity is elucidated.

## 2. Experimental

### 2.1. Site description

The forest ecosystem investigation reported here is from the Danish forest experimental site situated in the Ulborg Forest District, Western Jutland (56.18, 8.26). The Ulborg site is one of three Danish forest research sites operated as integrated monitoring sites since 1985 (one since 1988) (Figure 1). The experimental plots are part of a species trial experiment carried out at 13 locations in Denmark, each site with 12 tree species including Norway spruce (*Picea abies L*) all planted in 1964/65 (Holmsgaard and Bang, 1977).

The three sites are equipped with monitors and samplers for determination of air pollution and chemical composition of precipitation, throughfall (drip from the tree canopy), litterfall and soil water. The investigation also includes chemical analysis of whole trees, humus layer build up and wood increment. The research activities at the sites are based on long-term continuous monitoring of these parameters (Bille-Hansen *et al.*, 1994). The three monitoring sites are a part of the Level-II intensive monitoring network established between "The International Cooperative Programme on the Assessment and Monitoring of Air Pollution Effects on Forest (ICP Forest)" and the European Union.

The site reported here is a typical West Jutland location dominated by spruce plantations on sandy and poor soils, Haplic Podzols (FAO). The clay content is below 5%, and coarse sand increases with depth from 60 to 75 %. The maximum plant available water content is rather low, 96 mm, corresponding to the potential evapotranspiration in one summer month. The content of base cations are low 12 $kmol_c$ $ha^{-1}$ in the mineral soil and 5 $kmol_c$ $ha^{-1}$ are located in the O-horizon. In accordance with the low content of base cations, the base saturation amounts to only 8.6 % for the whole profile, including the O-horizon. Major soil chemical characteristics from soil borings in 1996 are summarised in Table 1.

Fig. 1. Location of the Danish EU-level II forest sites. Results from the Ulborg experimental site are reported here.

The total content of the O-horizon was obtained after wet destruction in Aqua Regia. The exchangeable fractions were obtained after $NH_4NO_3$-extractions for base cations (BCE), whereas the acid cations, $H^+$ and $Al^{3+}$ were obtained by titration on a 1 M KCl extraction (ACE). BS refers to base saturation excluding sodium ($Na^+$).

TABLE I

Important soil characteristics. (ACE) the extraction of acid cations, $H^+$ and $Al^{n+}$ (BCE) and (BS) is base cations and base saturation not including $Na^+$.

|  | Ca | Mg | Na | K | BCE-Na | ACE | BS |
|---|---|---|---|---|---|---|---|
|  | \multicolumn{3}{c}{kg ha$^{-1}$} |  |  | kmol$_c$ ha$^{-1}$ |  | % |
| Total |  |  |  |  |  |  |  |
| O horizon | 162 | 42 |  | 38 | 12.5 |  |  |
|  |  |  |  |  |  |  |  |
| **Exchangeable** |  |  |  |  |  |  |  |
| O horizon | 61 | 24 | 9 | 11 | 5.3 | 3 | 68.0 |
| 0 to 80 cm | 87 | 64 | 127 | 97 | 12.3 | 180 | 6.3 |
| **Sum** | **148** | **88** | **136** | **108** | **17.6** | **182** | **8.6** |

2.2. MONITORING OF ATMOSPHERIC INPUT

Aerosol and gas concentrations in the ambient air was measured on a daily basis by the filter pack method. The filter pack consisted of four filters arranged as a "sandwich". Air is sucked through the filters by a high volume sampler. The first filter collects aerosols and this filter was analysed by wet chemical analysis for $NH_4^+$ and $NO_3^-$, while elements including S were analysed by Proton Induced X-ray Emission (PIXE). $HNO_3$,

$SO_2$ and ammonia ($NH_3$) were collected on impregnated filters following the particle filter (Andersen and Hovmand, 1994; Hovmand and Kemp, 1996).

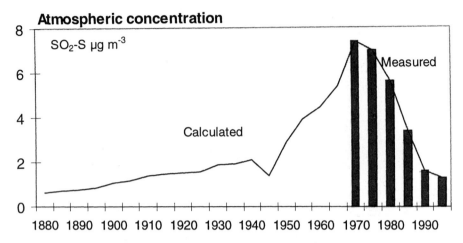

Fig. 2. Measured and calculated sulphur dioxide concentrations in rural Denmark.

Precipitation and throughfall were sampled by 12 funnels in the plot. The funnels were connected to polyethylene flasks, the water samples were collected every half month. The funnels were either the NILU-type (Norwegian Institute of Air Research) with a sampling area of 314 cm² or Danish samplers in black polyethylene with same sampling area. Sampling and analysis were described in more detail by Hovmand *et al.* (1993) and Hovmand and Kemp (1996).

2.3. MONITORING OF BIOGEOCHEMICAL PARAMETERS

Soil water was sampled from 20 tension lysimeters, 10 were installed in 40 cm depth and 10 installed below the root zone, 90 cm depth. Only values from lysimeters installed in 90 cm depth are reported here. The tension lysimeters installed in 1985 were of the ceramic type (P80). A new set of lysimeters of the teflon type were in operation from 1993 and ran in parallel up to 1996, when sampling from the ceramic system was stopped. Only the 1997-sampling from the teflon system are reported here, older values are from the ceramic system. A minimum of 1000 ml soil water were sampled and disregarded from every lysimeter before the first analysis. The following chemical parameters were analysed for: Conductivity, pH, $Cl^-$, $NO_3^-$, $SO_4^{2-}$, $PO_4^{3-}$, DOC, $Na^+$, $K^+$, $Ca^{2+}$, $Mg^{2+}$, $Al^{n+}$, $Mn^{n+}$, $Fe^{3+}$, $NH_4^+$.

In the winter 1997/98 disks were removed from felled 15 trees at breast height and used to measure annual ring width. A microscope attached digital readout system magnified specimen images of annual rings' boundaries down to 0.01mm. Litter was sampled monthly from April 1985 to March 1996 by using 20 littertraps, thereafter using 10 litter traps. The traps, with a diameter of 31 cm, were placed 100 cm above the forest floor, equidistant distributed along one of the plot diagonals. The samples were sorted into needles, twigs/branches, bud scales and cones + seeds and weighed after drying. Further information on methods for sampling and analysis of soil water, litterfall and tree ring analyses are described by Bille-Hansen *et al.* (1994) and Pedersen and Bille-Hansen (1999).

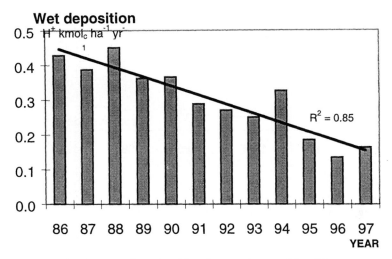

Fig. 3. Measured wet deposition of protons (strong acid) at Ulborg.

## 3. Results and discussion

### 3.1. ATMOSPHERIC $SO_2$ DEPOSITION

Dry deposition of $SO_2$ has been the major acidifying atmospheric component in this century. Since 1978 $SO_2$ has been measured at the station Tange located in the Central Jutland, 60 km east of Ulborg. This makes analysis of a long term trend for $SO_2$ concentrations in the area possible. The first $SO_2$-concentrations measured in the Danish rural area go back to 1971, reported by Jørgensen (1978). Further extrapolations back in time (1965 to 1971) are based on the model calculations of Mylona (1993), incorporating specific regional characteristics for Denmark described by Hovmand (1999).

The $SO_2$ deposition is estimated from modelled (1965-1971) and measured (1971-1997) concentrations values multiplied with an anticipated dry deposition velocity of 0.011 m s$^{-1}$. Using an average deposition velocity is a rough assumption, but the dry deposition fluxes obtained by this method are consistent with the fluxes measured as net-throughfall of S (Hovmand and Kemp, 1996). The measured and calculated yearly mean $SO_2$ concentrations have been decreasing since the beginning of the seventies, as shown in Figure 2. Therefore the increase in accumulated acidity was most pronounced in the beginning of the period (1965-1998), as shown in Figure 4.

### 3.2. WET DEPOSITION OF STRONG ACID (H$^+$) AND TOTAL ACID DEPOSITION

As a consequence of a decrease in atmospheric average $SO_2$ concentrations during the last 30 years, the acidity of the rain has also been falling. Measurements in the period 1985-1997 of wet deposition of protons from Ulborg Forest District (Figure 3)

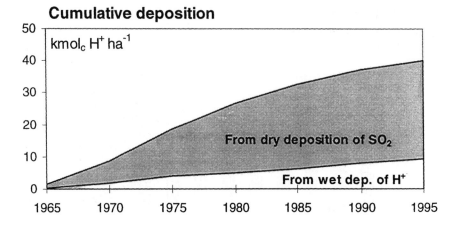

Fig. 4. Total cumulative proton input to spruce forest, estimated from $SO_2$ dry deposition and pH in rain water.

show a clear decreasing trend. Continuous and reliable measurements of pH in rain from the rural background area of Denmark extend back to the beginning of the 1980's (Rasmussen, 1988; Bille-Hansen and Hovmand, 1987). Estimates of pH in rain before 1985 are based on sporadic measurements from southern Scandinavia reviewed by Hovmand (1999). The cumulative total $H^+$ flux, defined as the sum of gaseous $SO_2$ deposition and wet deposition of $H^+$, is shown Figure 4. During the period of tree growth in the experimental plot (1965-1997) the cumulative input of protons has been 40 $kmol_c$ $H^+$ $ha^{-1}$. S deposition was calculated from measured atmospheric concentrations of $SO_2$ and particulate S and measured sulphate in rain in the period 1971-1997, while data from 1965 to 1971 are model derived. Measured and calculated cumulative input of non-marine S was to 878 kg S $ha^{-1}$ (55 $kmol_c$ $ha^{-1}$) in the period 1965 to 1997 (Figure 5). Measurements and calculations are described by Hovmand (1999).

3.3. THROUGHFALL MEASUREMENTS 1985-1997

Throughfall has been sampled under Norway spruce in Ulborg and analysed for $SO_4^{2-}$, $NH_4^+$, pH, $Ca^{2+}$ and $Na^+$. In addition $Mg^{2+}$, $K^+$, $Cl^-$ and $NO_3^-$ have been measured, though the results are not reported here. Throughfall flux of $SO_4^{2-}$ is believed to be close to the total atmospheric S deposition to forest (Hultberg and Grennfelt, 1992). Throughfall fluxes in Figure 6 show a decrease in $SO_4^{2-}$ with time. This is in accordance with the measured and calculated decrease in S deposition at the site.

Also a very large decrease in the $H^+$ concentration in throughfall is seen, probably reflecting the combined effect of a decrease in $SO_2$ deposition and a nearly constant deposition of $NH_3$ over the years. The relatively constant level of NH4$^+$ in throughfall indicates a constant input of $NH_3$ and $NH_4^+$. At the present concentration levels of $SO_2$ a substantial part of the acid produced from the $SO_2$ deposition is neutralised at the

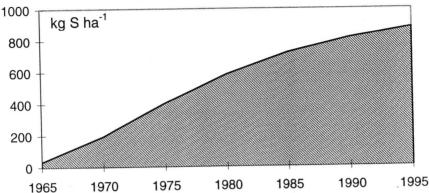

Fig. 5. Total cumulative sulphur input to spruce forest in Ulborg, estimated from measurements in Ulborg from 1985; data before 1985 are estimated from measurements at different sites in western Denmark.

canopy surface by the deposited $NH_3$.

Sodium deposition estimated as throughfall, $NH_4^+$ and acidity, fluctuated during the years (1985-1997) with a median value of 65 kg ha$^{-1}$ yr$^{-1}$. Higher values, up to 100 kg ha$^{-1}$ yr$^{-1}$ were seen in 1985 and 1990 (Figure 7).

3.4. TRENDS IN SOIL BASE SATURATION

The sandy soil of Ulborg is believed to have a very low weathering rate (Raulund-Rasmussen, 1989). Ca released to the soil by weathering is estimated to be 36 mg m$^{-2}$ yr$^{-1}$ (18 mol$_c$ Ca ha$^{-1}$ yr$^{-1}$) compared to an average atmospheric input of 400 mg m$^{-2}$ yr$^{-1}$ (200 mol$_c$ Ca ha$^{-1}$ yr$^{-1}$). The estimated weathering of soil minerals amounts to 100 mol$_c$ ha$^{-1}$ yr$^{-1}$ for K, Mg and Ca (Van der Salm *et al.*, 1999) compared to an average atmospheric input of 1100 mol$_c$ ha$^{-1}$ yr$^{-1}$. The atmospheric deposition of these elements is therefore an important net-input to the ecosystem, relative to weathering of soil minerals.

A decrease in average base saturation from approximately 15% to approximately 5% in the soil column was recognised from 1985 to 1996 (Bille-Hansen, 1998). This indicates a loss of available base cations from the soil, partly by root uptake and bioaccumulation, partly by strong acid input and leaching from the soil. The present level of available base cations ($Mg^{2+}$, $K^+$, $Ca^{2+}$) is 17.6 kmol$_c$ ha$^{-1}$. The mineralization rate of spruce litter has been investigated. Experiments with litterbags placed in the O-horizon showed a 50% release of the Ca from the bags over a period of three years (Vesterdal, 1999) indicating a relatively slow turn over of this element. Cation leaching from the upper 90 cm of the soil column is a permanent ecosystem loss of nutrients, in contrast to the cation accumulation in litter and in the O-horizon.

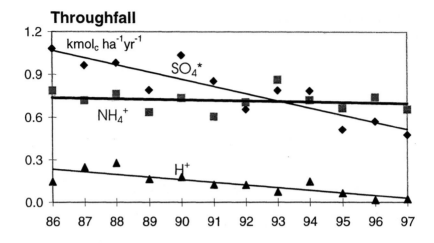

Fig. 6. Throughfall composition of non-marine $SO_4^{2-}$, ammonium and protons in Norway spruce, Ulborg.

3.5. SOIL WATER COMPOSITION, 1986-1997

The soil water concentrations of $Na^+$, $Al^{n+}$ and $Ca^{2+}$, as well as the pH value were quite stable onto 1988. In 1989/90 a drastic decrease in pH took place, ~0.4 units. At the same time the $Na^+$, $Ca^{2+}$ and $Al^{n+}$ concentrations began to increase. A small increase in pH took place in 1992 and lasted to 1995, when pH dropped to the present level, ~0.2 units below the 1985 value (Figure 8). The Na-deposition relative to throughfall precipitation was high in 1989 (Figure 7). As a consequence of the relatively low precipitation amount in 1989, only 194 mm water percolated through the soil that year, which is 65% of the normal percolation. The effect was an increase in soil water $Na^+$ concentration initiating a pronounced decrease in pH and an increase in $Al^{n+}$ concentrations. This incident was reported by Pedersen and Bille-Hansen (1995) with data up to 1993. Similar sea salt induced pH changes, although with less drastic effects, have been reported by Skartveit (1980), Wright et al. (1988) and Chapman et al. (1995). Due to high atmospheric $Na^+$ input in 1990 (Figure 7), $Na^+$ concentrations in soil water maintained high during the following years 1990-1993 (Figure 8). In 1994 a small drop in Na-concentration was caused by the high precipitation that year. After 1994 the $Na^+$ concentration remained high due to below normal water input from the precipitation.

The soil water $Al^{n+}$ concentration tedned to fluctuate (Figure 8), but showed a general upward trend during the period 1989-1997. The decrease in pH and increase in $Al^{n+}$ is a result of a displacement of polyvalent cations, e.g. aluminium hydroxides into soil solution. The released $Al^{n+}$ is for a substantial part leached with $Cl^-$ as the driving anion (Bille-Hansen, 1998).

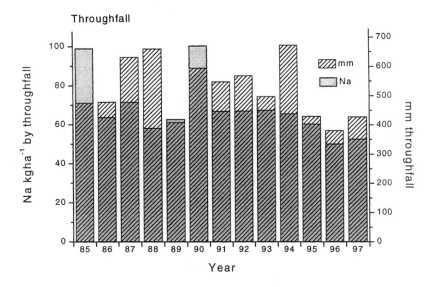

Fig. 7. Throughfall flux of water in mm $yr^{-1}$, and throughfall deposition of Na in kg $ha^{-1}$ $yr^{-1}$. The figures for 1985 covers only from May 1985.

The $Ca^{2+}$-concentration trend (Figure 9) follows the $Na^+$ concentration trend to some extent as sea spray contains $Ca^{2+}$ the addition to high concentrations of $Na^+$, $Mg^{2+}$, $Cl^-$. At poor sandy soil sites, like the one at Ulborg, where the amount of exchangeable bases are low (Table 1), are the ion concentrations in soil water are influenced by several processes besides atmospheric deposition, such as canopy leaching, soil exchange reactions and root uptake. The molar Ca/Al - ratio is often used as an acidification indicator in forest ecosystem analysis (Cronan and Grigal, 1995). According to their review certain thresholds values can be identified, where the likelihood of damage is 50 % or more. For soil water this threshold limit is at a Ca/Al molar ratio of 1. Below a Ca/Al ratio of 1, damage probability rises to above 50 %. At a Ca/Al molar ratio of 0.1 there is a 95% probability of damage to roots and shoots leading to nutrient imbalance in the biomass (Cronan and Grigal, 1995). The capability of the roots particularly to take up P and base cations is impaired at such low Ca/Al ratios (Staaf et al., 1996). The soil water Ca/Al ratio for the Norway spruce stand at Ulborg was ~0.5-0.6 in 1985/86. Considerably below that seen in the last quarter in 1986 and continuing through 1987 (Figure 9) the Ca/Al- ratio had fallen to around 0.1, but thereafter showed a small increase. In 1994 monthly values fell to another low well below 0.1.

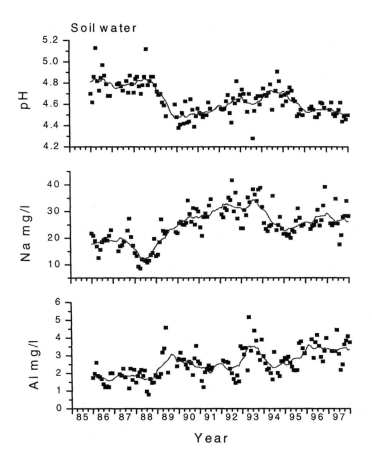

Fig. 8. Trends in soil water chemistry. Each square represents a monthly mean value for 10 tension lysimeters. Discontinuations in the data series are due to either summer drought or frozen soil water.

3.6. TREE GROWTH AND LITTERFALL

Tree growth was estimated by ring analysis for the period 1969 to 1997 shown in Figure 10. Tree ring analysis was taken in addition to normal mensurational characterisation, the ring analysis gives a detailed picture of the year to year radial increment. The trend function, age curve, was fitted according to Pan et al. (1997).

The general trend follows the age curve, the peaks in the beginning of the 1970s, 1980s and in 1990-91 are high increment years, whereas the mid 1970s, 1980s where poor years for increment. Such peaks have been recognised in tree ring analysis from all over the Western part of the country and can be correlated to years with high or low precipitation. However, the general downward trend, with increments below the age curve, from 1992 and onwards coincides with the very low Ca/Al-ratios in the soil water. The main trends in litterfall are illustrated in Figure 11. Prior to 1989/90 litterfall was approximately half or less than half the litterfall in the following years. In the late

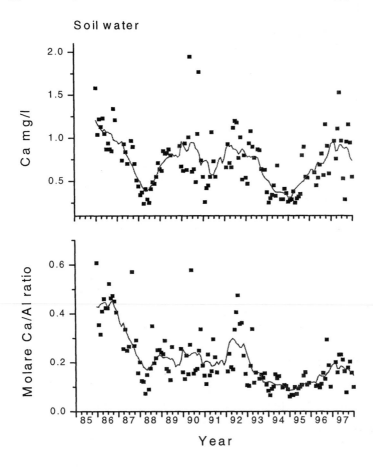

Fig. 9. $Ca^{2+}$ concentration in soil water (upper) and molar Ca/Al ratio (lower). Squares represent monthly mean values for 10 tension lysimeters.

eighties and the early nineties Norway spruce stands suffered from the syndrome called The "Red Picea-phenomenon" a widespread disease in Danish spruce forest causing extensive needle reddening and loss. The eastern part of Denmark was less affected by this disease than the western part of the country. This disease was ascribed to the high sea salt deposition pulse initiating a drop in pH (Pederson, 1993; Pedersen and Bille-Hansen, 1995). Several stands were severely damaged and extensive felling took place especially in middle aged and old stands (above 70 years in Denmark).

The peak in litterfall 1989/90 from Norway spruce is partly ascribed to damage from attacks by the green spruce aphid *(Elatobium abietinum)* known to cause more damage to sitka spruce *(P. sitchensis Bong. Carr.)* than to Norway spruce. Similar peaks in needle loss in 1989/90 has been reported from sitka stands in other countries such as

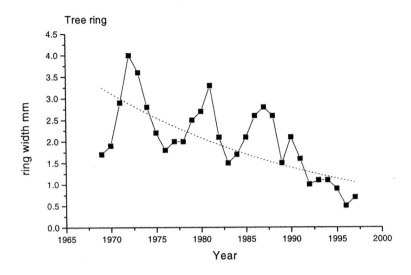

Fig. 10. Tree ring analysis of Norway spruce from 1969 to 1997. The dotted line indicates the trend function for radial increment, age curve.

Scotland and England also caused by aphid attack (Thomas and Miller, 1994; Carter, 1989).The extremely high litterfall in 1996/97 occurred after several years with very low molar Ca/Al ratio in soil water and a warm and dry summer in 1996.

## 4. Conclusion

Measurements at a Norway spruce plot in western Jutland, planted in 1965, indicate a summarised dry and wet deposition of 40 $kmol_c$ protons (strong acid) per ha over a period of 30 years from 1965 to present time. We speculate that this strong acid input, together with plant uptake of K, Mg and Ca, during a period of build up of living and dead biomass in the forest ecosystem, has deprived the sandy soils of most of its sparse pool of available base cations, so its present level of available ($Mg^{2+}$, $K^+$, $Ca^{2+}$) now is only ~18 $kmol_c\ ha^{-1}$.

Continuous soil-water analysis since 1985 to the present time shows an increase in soil-water concentration of protons and Al and a decrease in molar Ca/Al ratio from 0.6 in 1985 to a minimum of 0.1 in 1994-96. Despite the fact that annual S deposition has been going down since the mid 1970's, the effect of accumulated acid deposition over the years, combined with biomass build up in the forest ecosystem, has made the poorly "buffered" soils vulnerable to acidification and lead to long lasting changes in soil-water acidity.

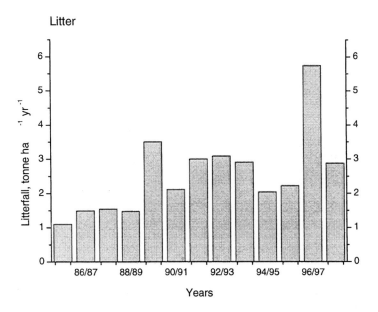

Fig. 11. Needle litterfall in the Norway spruce stand. The sampling year runs from April one year to March the following year.

We consider that the low growth rate and the relatively high litterfall in Norway spruce are partly due to increased soil water acidity in combination with periodically elevated sea salt concentrations in the soil water. Other factors such as natural fluctuations in soil water content and attacks by aphids also contributed to the reduced vitality of the forest.

## Acknowledgements

We thank Dr. Helle Vibeke Andersen for reviewing the manuscript and Dr. Andreas Bergstedt for assistance with tree ring analysis. Sincere thanks are extended to laboratory and field technicians for their skilful work. This work was supported by the Danish Research Councils, the National Forest and Nature Agency and the Forest and Forestry Division of the EEC, DG VI.

## References

Andersen, H.V.: 1997, "Final report, 1992-1996. Sub-programme 1 : Atmosphere and air pollution, *The Danish environmental research programme*. National Environmental Research Institute, DK-4000 Roskilde

Andersen, H. V. and Hovmand, M. F.: 1994, *Atmospheric Environ.* **28**, 3495-3512

Bille-Hansen, J.: 1998, In: Hansen, K.(Ed.) *Monitoring of Forest Damage in the Nordic Countries, Proceedings from a combined SNS Ad hoc group meeting on Monitoring of Forest Damage and the 4th International ECE/EU*. Intercalibration Course for Northern Europe, 15th-17th June 1998, Denmark, Danish Forest and Landscape Research Institute, DK-2970 Hørsholm, p 56-65.

Bille-Hansen, J. and Hovmand, M.F.: 1987, *Miljøstyrelsens Luftforureningslaboratorium, MST LUFT*-**A112** (in Danish). National Environmental Research Institute, DK-4000 Roskilde

Bille-Hansen J., Pedersen L.B, Hovmand M.F., Andersen H.V. Jensen N.O., Hummelshøj P., Ro Poulsen H. and Mikkelsen T.: 1994, Background information for surveillance on the permanent observation plots in Denmark, Level II, EU-report. Danish Forest and Landscape Res. Inst. DK-2970 Hørsholm.

Breemen, N. van, P.A. Burrough, E.J. Velthorst, H.F. van Dobben, T. Dewit, T.B. Ridder and Reinders, H.F.R. : 1982, *Nature* **299**, 548-550.

Carter, C.I., 1989. The 1989 outbreak of the green spruce aphid, Elatobium abietinum. Forestry Commission Occasional Paper No.19, 1 ed. Forestry Commission, Edinburgh.

Chapman, P.J., Reynolds, B., and Wheater, H.S.:1995, *Water, Air, and Soil Pollut*. **85**, 1089-1094.

Cronan, C.S. and Grigal, D.F.:1995, *J.Environ.Qual*.**24**, 209-226.

De Vries, W. and Breeuwsma, A.: 1987, *Water, Air,* and *Soil Pollut*. **35**, 293-310.

Holmsgaard, E. and Bang, C.: 1977, *Statens Forstlige Forsøgsvæsen*, (Copenhagen) **35**,159-196.

Hovmand, M. F.: 1999, *Forest Ecology and Management* **114**, 19-30

Hovmand M.F., Grundahl L., Kemp K. Aistrup W. and Runge E.H.: 1993, *DMU Report no.* **91**, National Environmental Research Institute, DK-4000 Roskilde.

Hovmand, M.F., Andersen, H.V., Bille-Hansen J. and Ro-Poulsen, H.: 1994, *DMU Report no.* **98**, National Environmental Research Institute, DK-4000 Roskilde.

Hovmand, M.F. and Kemp, K.: 1996, *Atmos Environ*. **30,** 2989-2999.

Hultberg, H.: 1985, Budgets of base cations, chloride, nitrogen and sulphur in acid Lake Gårdsjöen catchment, SW Sweden. In: Andersson, F. and Olsson B., "Lake Gårdsjöen" *Ecol. Bull.* (Stockholm) 37

Hultberg, H. and Grennfelt, P.: 1992, *Environ Pollut,* **75,** 215-222.

Nilsson, I. S.: 1985, Why is Lake Gårdsjöen acid? In: Andersson, F. and Olsson B., "Lake Gårdsjöen" Ecol. Bull. (Stockholm) **37**

Jørgensen, V.: 1978, (in Danish with summary in English). *Tidsskrift for Planteavl* (Copenhagen) **28**: 633-656.

Mylona, S.: 1993, *EMEP/MSC-W Report 2/93. Meteorological Synthesising Centre-W*. P.O. Box 43-Blindern, N-0313 Oslo 3, Norway.

Pan, C., Tajchman, S.J., and Kochenderfer, J.N.: 1997, *For.Ecol.Manage*. **98**, 77-87.

Pedersen, L.B. and Bille-Hansen, J.: 1995, *Plant and Soil,* **168-169**, 365-372.

Pedersen, L.B. and Bille-Hansen, J:1999, *For.Ecol.Manage*.**114**, 55-70.

Rasmussen, L.: 1988, Effects of Acid Rain on the Ion Balance and Leaching of Metals and Anions in Danish Spruce Forest Ecosystems in the period 1983-1987. In Danish with English abstract. Laboratory of Environment, Technical University of Denmark, DK-2800 Lyngby.

Raulund-Rasmussen, K.: 1989, *Scand. J. For. Res.* **4**, 417-425

Skartveit, A., 1980. Observed relationships between ionic composition of precipitation and runoff. In: D. Drablös and A. Tollan Ecological impacts of acid precipitation. International conf., ecological impacts of acid precipitation. 1 edn. SNSF Project., Oslo, Norway. 242-244.

Staaf, H., Persson, T. and Bertills, U.: 1996. Skogsmarkskalkning. Resultat och slutsatser från Naturvårdsverkets försöksverksamhet, 1 ed. Naturvårdsverket, Stockholm, pp. 290.

Tamm, C.O. and Hallbäcken, L.: 1988, *Ambio* ,**17**: 56-61.

Thomas, R.C. and Miller, H.G.: 1994, *Forestry*, **67**, 329-341.

Van der Salm,C., De Vries,V., Olsson,M. and Raulund-Rasmussen,K.: 1999. *Water, Air,* and *Soil Pollut.* **109**, 101-135.

Vesterdal, L.: 1999, *Can .J. For. Res.* **29,** 95-105.

Wright, R.F., Dale, T., Gjessing, E.T., Henriksen, A., Johannessen, M. and Muniz, I.P.: 1975, Impact of Acid Precipitation on Freshwater Ecosystems in Norway. Editor SNSF; Finn H. Brække, Norwegian Forest Research Institute, Oslo-Ås.

Wright, R.F.: 1988, Norton, S.A., Brakke, D.F., and Frogner, T., *Nature,* **334**, 422-424.

# THE EFFECT OF N DEPOSITION ON NITROUS OXIDE AND NITRIC OXIDE EMISSIONS FROM TEMPERATE FOREST SOILS

U. SKIBA, L.J. SHEPPARD, C.E.R. PITCAIRN, S. VAN DIJK AND M.J. ROSSALL

*Institute of Terrestrial Ecology, Bush Estate, Penicuik, Midlothian EH26 0QB, UK*
[1] *Institute of Terrestrial Ecology, Merlewood, Grange-over-Sands LA11 6JU, UK*

(Received 30 November 1998; accepted 25 February 1999)

**Abstract:** Long-term and short-term N deposition effects on $N_2O$ and NO emissions from forest soils were compared. Long-term $NH_3$ deposition (> 20 years) from a poultry farm to a downwind woodland (decreasing from 73 to 18 kg N ha$^{-1}$ y$^{-1}$, 30 to 110 m downwind of the farm) resulted in the re-emission of 6% and 14% of $NH_3$-N deposited as $N_2O$-N and NO-N, respectively. However, when in short-term (2-3 years) field experiments the atmospheric N deposition to mature conifer plantations was raised by fumigation with $NH_3$ to 15 kg N ha$^{-1}$ y$^{-1}$ or by acid mist to 48 and 96 kg N ha$^{-1}$ y$^{-1}$ the N deposited was immobilised. In the acid mist experiment more than 2 years of acid mist (48 and 96 kg N ha$^{-1}$ y$^{-1}$) were required to significantly increase $N_2O$ emissions from -0.3 µg $N_2O$-N m$^{-2}$ h$^{-1}$ (control) to 0.5 and 5.7 µg $N_2O$-N m$^{-2}$ h$^{-1}$, respectively. This suggests, that N deposition simulation studies in soil ecosystems, which have previously not been exposed to high rates of N (by deposition or fertilisation), need to be long-term. Also, measurements of $N_2O$ and/or NO may be a non-destructive, quick indicator of the N status of the soil.

## 1. Introduction

The recent interest in increased atmospheric concentrations of the greenhouse gas nitrous oxide ($N_2O$) and the photochemical oxidant $NO_x$ has stimulated a great deal of research into soils as a source of these gases. In soils nitric oxide (NO) and $N_2O$ are produced by nitrifying and denitrifying bacteria. The magnitude of subsequent emissions depends on substrate availability (ammonia ($NH_4$) or nitrate ($NO_3$)), climate and soil properties that either promote nitrification or denitrification (Williams et al., 1992).

NO and $N_2O$ emissions from many temperate climate forest soils are small compared to emissions from agricultural soils. In temperate forests emission rates rarely exceed 20 µg N m$^{-2}$ h$^{-1}$ or 1 to 2 kg N ha$^{-1}$ y$^{-1}$ (Johansson, 1984; Williams et al., 1988; Castro et al., 1993; Rennenberg et al., 1998; Skiba et al., 1998a). Larger emissions, approaching rates typical for agricultural soils, however, have been measured from forests where nitrogen (N) deposition rates are > 20 kg N ha$^{-1}$ y$^{-1}$ or when the soil mineral N concentration is enhanced by other means i.e. N fertilisation, liming, drainage and clear felling.

In particular the effect of N deposition on trace gas fluxes has been studied, usually as part of larger studies investigating the impact of N deposition on forest health and $NO_3$ leaching. Two approaches have been adopted to study the effect of N deposition on forest ecosystems: 1. Comparing similar forest ecosystems in areas receiving high and low rates of atmospheric N deposition (e.g. Rennenberg et al., 1998; Skiba et al., 1998b). 2. Simulating high N inputs in large scale experimental designs (Bowden et al., 1991; Sheppard et al., 1997 & 1998; Skiba et al., 1998b.). Both approaches have been applied to measure the effect of N deposition on soil $N_2O$ and NO emissions. These studies suggest, that at least in the short term, soils with a low N status can show different responses to high N concentrations compared to *in vivo* observations. The importance of soil N status in determining the trace gas flux is developed in this paper, using comparisons of $N_2O$ and NO emissions from four different experiments in areas of high N deposition and in simulation studies.

## 2. Site description and methods

$N_2O$ and NO fluxes were measured from 1) a forest downwind of a poultry farm along a gradient of decreasing ammonia ($NH_3$) deposition rates, 2) a low N input pine forest, 3) a low N input pine forest exposed to a point source of $NH_3$ and 4) a spruce forest exposed to acid mist.

The poultry farm (340,000 birds) is situated in the central belt of Scotland, at the edge of an open mixed woodland of pine, birch, oak, rowan and elder on a brown forest soil (pH in $CaCl_2$ = 4.1) (Pitcairn et al.,1998), and has been operating for approximately 25 years. Atmospheric $NH_3$ concentrations were measured by passive diffusion along a transect of the prevailing wind direction above and below the poultry units over a one year period (1995 to 1996) (Pitcairn et al., 1998). $N_2O$ emissions (3 measurements / position) from the woodland floor were measured 30, 50, 110 and 250 m downwind from the farm on 7 occasions between May and November 1997. NO was measured for 2 short periods in October and November 1997, from duplicate plots at 30, 50 and 250 m downwind of the farm, the 110 m site was not accessible with the mobile laboratory.

Devilla forest is a 40 year old Scots pine forest in Central Scotland growing on a brown forest soil (pH in $CaCl_2$ = 3.5). The soil had a sandy clay loam texture overlain by an organic layer of varying thickness. $N_2O$ was measured at 2 to 3 weekly intervals between May to October 1993 and June 1994 to February 1995 in association with an experiment examining the phenomenon of co-deposition of $NH_3$ and $SO_4$. Up to 15 kg $NH_3$ -N $ha^{-1}$ was released between April and October in 1993 and 1994 (Cape et al., 1995). Chambers (3 at each position) for $N_2O$ measurements were installed 25 m upwind of the $NH_3$ fumigation source and 10 and 25 m downwind of the source. NO fluxes were measured almost continuously from 4 September to 11 September 1997. Four 0.95 $m^2$ frames were installed in the same part of the forest in which the above $NH_3$ experiment was carried out.

Deepsyke Forest, Scottish borders, is a 10 year sitka spruce plantation growing on a drained moorland peat. Groups of 10 trees were treated with acid mist at pH 2.5 ($H_2SO_4$ and $NH_4NO_3$) providing 48 kg N $ha^{-1}$ $y^{-1}$ or 96 kg N $ha^{-1}$ $y^{-1}$ and enhancing precipitation by 10%. The control at pH 5 provided 2 kg N $ha^{-1}$ $y^{-1}$ and annual wet and dry deposition of N contributed a further 6.7 kg N $ha^{-1}$ $y^{-1}$ to all treatments (Sheppard et al., 1999). Treatments were replicated 4 times. $N_2O$ fluxes were measured between April and August and May and October during the first and second year or the third year of spraying, respectively. Average weekly inputs of N for the measurement period were 1.7 (1996) and 1.9 (1997 and 1998) kg N $ha^{-1}$ $week^{-1}$ for the 48 kg N $ha^{-1}$ $y^{-1}$ treatment and 2.8 (1996), 3.7 (1997) and 3.1 (1998) kg N $ha^{-1}$ $week^{-1}$ for the 96 kg N $ha^{-1}$ $y^{-1}$ treatment.

$N_2O$ fluxes were measured by the static chamber method. Small round frames (22 l volume) were inserted into the ground for the duration of the experiment (MacDonald et al., 1997). Chambers were sealed for 1 hour for flux measurement, when gas samples were withdrawn by syringe and were analysed by ECD gas chromatography for $N_2O$.

NO fluxes were measured by the dynamic chamber method. Frames (0.95 $m^2$ * 0.15 m depth) were inserted into the ground onto which a 0.55 m$^3$ perspex chamber was attached for the duration of the flux measurements. Ozone free air was pushed through the chamber at a flowrate of 60 l min$^{-1}$, inlet and outlet air was sampled at intervals of 3 min and was analysed for NO, NO$_2$ by chemiluminescence and O$_3$ using a UV photometric analyser (Skiba *et al.*, 1993).

Soil temperature was measured and soil samples were collected for analysis of available soil NH$_4^+$ and NO$_3^-$ (Grimshaw, 1989) and volumetric soil water content.

## 3. Results and discussion

### 3.1. LONG-TERM HIGH NH$_3$ DEPOSITION RATES

Poultry and other intensive animal farms can be large point sources of atmospheric NH$_3$ which can result in long-term deposition of a decreasing gradient of atmospheric N to the downwind environment (Asman *et al.*, 1998; Pitcairn *et al.*, 1998). Where soil/plant ecosystems downwind of the source are relatively uniform, they provide opportunities to study the long-term effects of elevated N deposition rates. The poultry farm described here provides a large point source of NH$_3$, most of which was deposited within the first 100 m downwind of the source: NH$_3$ deposition rates were 73 kg NH$_3$-N ha$^{-1}$ y$^{-1}$ at the

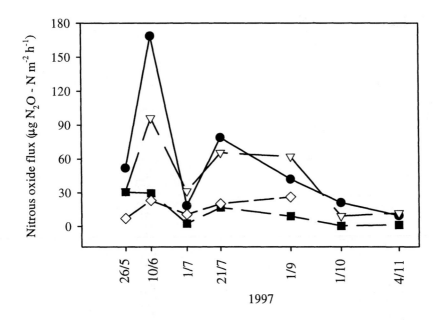

Fig. 1. Nitrous oxide emissions downwind of the poultry farm in Central Scotland. Distance downwind of the farm and NH$_3$ deposition rates were: ● 30 m & 81 kg N ha$^{-1}$ y$^{-1}$, ▽ 50 m & 52 kg N ha$^{-1}$ y$^{-1}$, ◇ 110 m & 25 kg N a$^{-1}$ y$^{-1}$, ■ 250 m & 17 kg N ha$^{-1}$ y$^{-1}$. Each point is a mean of 3 chambers.

woodland edge closest to the building (30 m); reducing to 45, 18 and 10 kg $NH_3$-N $ha^{-1}$ $y^{-1}$ 50, 100 and 250 m downwind, respectively; approaching background levels of 6.7 kg $NH_3$-N $ha^{-1}$ $y^{-1}$ 650 m downwind (Pitcairn et al., 1998). The enhanced N deposition rate close to the poultry houses significantly increased soil available $NH_4$ and $NO_3$ concentrations and emissions of NO and $N_2O$ in a linear fashion ($p < 0.05$ for $N_2O$ and $NH_4$, $p < 0.01$ for $NO_3$) (Skiba et al., 1998b). $N_2O$ fluxes measured between May and November 1997 are shown in Figure 1. Closest to the farm (30 m downwind) $N_2O$ emissions ranged from 10 to 169 µg $N_2O$-N $m^{-2}$ $h^{-1}$, but only 0.5 to 31 µg $N_2O$-N $m^{-2}$ $h^{-1}$ 250 m downwind. In spite of large spatial variabilities between the 3 chambers at all sites, overall differences in $N_2O$ emissions between the closest (30 m) and furthest site (250 m) were significant ($p < 0.01$). Changes in soil temperature and water content for the 7 measurement days explained $> 80\%$ of the variations in $N_2O$ at plots 30, 100 and 250 m downwind, and 55% for plots 50 m downwind of the farm.

Nitric oxide was only measured on 2 occasions in October and November 1997. Fluxes at 30 and 50 m downwind were not significantly different and ranged from 92 to 233 µg NO-N $m^{-2}$ $h^{-1}$; 250 m downwind fluxes were lower and ranged from 54.3 to 110.4 µg NO-N $m^{-2}$ $h^{-1}$ (Table I).

TABLE I

Nitric oxide emissions downwind of the poultry farm

| Distance from poultry farm (m) | Total N deposition (kg N $ha^{-1}$ $y^{-1}$) | NO emission (µg N $m^{-2}$ $h^{-1}$)[1] | |
|---|---|---|---|
| | | 1 October 1998 | 4 November 1998 |
| 30 | 80.1 | 144.7 ± 55.8 | 116.6 ± 3.1 |
| 50 | 52.1 | 223.2 ± 28.8 | 118.3 ± 1.5 |
| 250 | 17.4 | 104.5 ± 6.68 | 55.4 ± 1.5 |

[1] measurements were made from 2 plots for four 6 min periods at 12 min intervals mean ± std

At all sites emissions of NO were significantly lower (45% reduction) on the 4 November compared to the 1October and coincided with a soil temperature decrease from 13 °C (1 Oct) to 8.6 °C (4 Nov) and an increase in soil available water from 20% to 28% (v/v).

The NO and $N_2O$ emissions reported here compare with fluxes measured in other European forests situated in areas of high atmospheric N deposition. For example, in a N saturated forest in South Germany (N deposition $> 30$ kg N $ha^{-1}$ $y^{-1}$) continuous measurements provided mean monthly NO and $N_2O$ emissions which ranged from 29 - 130 µg NO-N $m^{-2}$ $h^{-1}$ and 3.5 - 16.4 µg $N_2O$-N $m^{-2}$ $h^{-1}$. $N_2O$ emissions for three pine forests in

North Germany (N deposition 20 kg N ha$^{-1}$ y$^{-1}$) ranged from 6 - 53 µg N$_2$O-N m$^{-2}$ h$^{-1}$ Butterbach-Bahl et al., 1997).

3.2. NITRIC OXIDE FLUXES FROM A LOW N INPUT PINE FOREST

At Devilla forest, a low N input forest, total N deposition rates were < 15 kg N ha$^{-1}$ y$^{-1}$ and NH$_3$ deposition rates were unlikely to exceed 3 kg N ha$^{-1}$ y$^{-1}$ (INDITE, 1994). During the 1 week measurement period in October 1997 NO was mainly deposited to the soil, with an average flux of - 0.8 ± 1.4 (range - 0.6 to 1.3) µg NO-N m$^{-2}$ h$^{-1}$. Only very short periods of NO emissions were observed from 2 of the 4 plots (Figure 2).

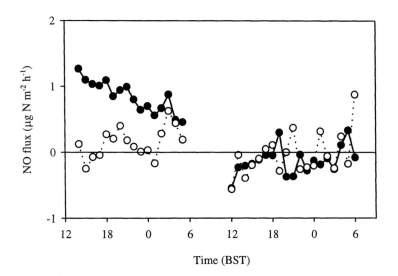

Fig. 2. Nitric oxide fluxes from a Scots pine forest at Devilla, Central Scotland. Mean hourly fluxes from 2 plots (● & ○), 9 - 11 September, 1997.

Average fluxes were 0.8 ± 0.2 µg NO-N m$^{-2}$ h$^{-1}$ and 0.1 ± 0.2 µg NO-N m$^{-2}$ h$^{-1}$, respectively. Whether these fluxes, measured from four, small (1 m$^2$) forest plots were representative for the entire forest or the entire year cannot be determined from this study. However, similar small fluxes were also measured from other forests receiving low rates of atmospheric N deposition. For example, the average NO emission measured from two Swedish pine forests in June, August and September was 1.1 µg NO-N m$^{-2}$ h$^{-1}$ (Johannson, 1984). In Scotland NO emissions measured in June/July and September/October from small plantations of alder (N fixing), sitka spruce, southern beech and birch were 3.0, 1.9, 0.2 and -1.2 µg NO-N m$^{-2}$ h$^{-1}$, respectively (Skiba et al., 1994).

Compared to NO emissions downwind of the poultry farm, however, NO emissions at

Devilla were negligible. Measurements at the poultry farm woodland were made 2 weeks after the Devilla measurements. Climate, soil type and soil pH were similar at both sites, but vegetation cover (mixed deciduous woodland downwind of the poultry farm and a pine forest at Devilla), $NH_3$ deposition rates and therefore mineral N concentrations were different. Differences in tree species have been shown to influence both NO and $N_2O$ emission rates (Butterbach-Bahl et al., 1997). These, however, are unlikely to account for the large flux differences observed between the woodlands at Devilla and the poultry farm. Differences in N deposition between the sites, $\geq 10$ and $< 3$ kg N ha$^{-1}$ y$^{-1}$ (downwind of the poultry farm and the pine forest, respectively) are most likely responsible for the different soil $NO_3$ concentrations (14 and 0.3 µg $NO_3$-N g$^{-1}$, respectively) and mean NO flux measured over the short study periods ($> 50$ and $<1$ µg NO-N m$^{-2}$ h$^{-1}$, respectively). Unfortunately $N_2O$ was not measured during this week. However, frequent measurements between May 1993 and February 1995 suggest that these would be smaller than at the poultry farm and unlikely to exceed 12 µg $N_2O$-N m$^{-2}$ h$^{-1}$ (Figure 3) (Skiba et al., 1998a), and are further discussed below.

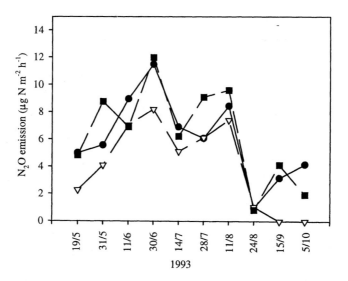

Fig. 3. The effect of $NH_3$ deposition (15 kg N ha$^{-1}$y$^{-1}$) on soil $N_2O$ emissions at Devilla. ● 20 m upwind, ▽ 10 m downwind, ■ 25 m downwind. Each point is the mean of 3 chambers, differences between sites were insignificant.

3.3. EXPERIMENTAL $NH_3$ DEPOSITION AT THE PINE FOREST

At Devilla forest the experimental point source release of $NH_3$, which provided 15 kg N ha$^{-1}$ between May and October in 1993 and 1994 increased $NH_4$ concentration in the throughfall: 38 meq $NH_4$-N m$^{-2}$ 10 m downwind and decreased to 15 meq $NH_4$-N m$^{-2}$ 25 m downwind of the source (Cape et al.,1995). Upwind $NH_4$ concentrations in the throughfall were 7 meq $NH_4$-N m$^{-2}$. This 5 fold enhancement of N inputs, however did not increase soil available $NH_4$ and $NO_3$ concentrations or the emissions of $N_2O$. $N_2O$ fluxes were measured at 2 to 3 weekly intervals between May and October 1993 (Figure 3) and June 1994 and February 1995. For many measurement dates $N_2O$ emissions were lowest at the site 10 m downwind which received the highest $NH_3$ deposition rates, although differences were not significant (Figure 3).

This lack of response implies that the additional N input to the soil was either taken up by the trees or converted to microbial biomass. Average $N_2O$ emissions were $5.3 \pm 3.0$ µg $N_2O$-N m$^{-2}$ h$^{-1}$ and the average soil available $NH_4$ and $NO_3$ concentrations were $4.1 \pm 3.2$ and $6.9 \pm 3.6$ mg N kg dry soil$^{-1}$, respectively (Skiba et al., 1998a). As found for the poultry farm, seasonal variations in $N_2O$ emission were controlled by seasonal changes in soil water and temperature which together accounted for 83% of the variation in flux ($p < 0.01$).

### 3.4. EXPERIMENTAL APPLICATION OF ACID MIST TO A SITKA SPRUCE FOREST

Nitrous oxide fluxes measured before and during the first, second and third acid mist spraying season (May - October) are shown in Figure 4. In all years (1996 - 1998) average and maximum $N_2O$ emission were largest for plots receiving a N input equivalent to 96 kg N ha$^{-1}$ y$^{-1}$, however differences were only significant in the third treatment season ($p < 0.05$) (Table II).

TABLE II

The effect of acid mist on $N_2O$ fluxes (µg N m$^{-2}$ h$^{-1}$) at Deepsyke forest

| Treatment | 1996[1] | 1997[1] | 1998[1] |
|---|---|---|---|
| pH 5 | 2.6 (-15.5 to 22.8) | -1.7 (-8.8 to 1.8) | -0.3 (-0.9 to 0.8) |
| pH 2.5 (48)[2] | 2.4 (-1.1 to 10.2) | 0.4 (-0.9 to 2.0) | 0.5 (-0.3 to 1.3)* |
| pH 2.5 (96)[3] | 6.9 (-16.4 to 28.7) | 3.2 (-4.1 to 14.8) | 5.7 (0.2 to 18.6)* |

[1]Mean and range of 7 (1996), 5 (1997) & 6 (1998) measurements. [2] 48 kg N ha$^{-1}$ y$^{-1}$,
[3] 96 kg N ha$^{-1}$ y$^{-1}$, * fluxes were significantly different from control, $p \leq 0.05$.

On plots receiving a N input equivalent to 96 kg N ha$^{-1}$ y$^{-1}$ compared to the control, increased $N_2O$ emissions were accompanied by significantly larger soil available $NH_4$ concentrations (13.5 and 9.1 µg $NH_4$-N g dry soil$^{-1}$, respectively, n = 5, $p < 0.05$) during the first spraying season. However, the soil $NO_3$ concentrations remained unchanged (0.75 µg $NO_3$-N g dry soil$^{-1}$). For subsequent years soil N data are not available yet. On plots receiving a N input equivalent to 48 kg N ha$^{-1}$ y$^{-1}$ the range and average $N_2O$ emissions were the same as for the control plots in the first treatment season. In the second season a slight increase in the average flux was observed, and only in the third season were emission rates significantly larger than from the control plots ($p < 0.05$) (Table II, Figure 4).

Before the start of the 1996 and 1997 acid mist spraying season the drained peat was a sink for $N_2O$. Maximum uptake rates of -15 µg $N_2O$-N m$^{-2}$ h$^{-1}$ were measured on 15 and

20 May 1996, suggesting that occasionally the forest can be a significant sink for $N_2O$. Occasional smaller rates of $N_2O$ uptake was observed during all spraying seasons, however, there appears to be no common explanation for the sink activity (Figure 4). Again the temporal variations in flux were controlled by soil temperature and moisture changes. During the first year of spraying $N_2O$ emissions from all sites were significantly larger than for equivalent sites in subsequent years. Temperature, rainfall, spraying patterns and sampling time in relation to spraying could not explain this difference.

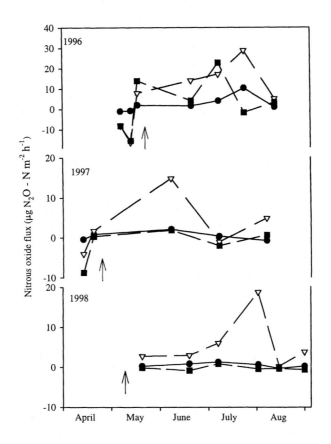

Fig. 4. Effects of acid mist on $N_2O$ fluxes from a mature sitka spruce plantation. ■ Control, pH 5 & 2 kg N $ha^{-1}y^{-1}$, ● pH 2.5 & 48 kg N $ha^{-1}y^{-1}$, ▽ pH 2.5 & 96 kg N $ha^{-1}y^{-1}$. The arrows indicate the start of the spraying season.

NO was not measured at Deepsyke. However, in a mixed coniferous forest on a peaty soil at the summit of Dunslair Heights, Scottish Borders, with an annual N deposition rate of > 24 kg N $ha^{-1}$ $y^{-1}$, NO was deposited and not emitted (Skiba et al., 1994). This suggests, that at Deepsyke, a similarly wet organic soil, NO emissions are unlikely to be of significance, rather the soil may provide a significant sink for NO.

The lack of response to the small elevated inputs of N at Devilla (15 kg N $ha^{-1}$ $y^{-1}$) and to the large elevated inputs of N at Deepsyke (48 and 96 kg N ha$^{-1}$ y$^{-1}$) imply, that for soils with a history of low N deposition rates, raising atmospheric N inputs will, in the short term, have no effect. The experimentally applied N appears to be immobilised, i.e. taken up by vegetation or converted to microbial biomass, making it unavailable for nitrification, denitrification and subsequent release of N$_2$O and NO. At Deepsyke and Devilla there was no evidence of luxury uptake of N by the vegetation in response to elevated N inputs (Cape et al., 1995; Sheppard et al., 1998). This suggests, that microbial immobilisation was the dominant sink for the deposited N. Immobilisation was also the main sink for N in a temperate forest soils in the North-eastern part of the USA, where N fertilisation as high as 150 kg N ha$^{-1}$ y$^{-1}$ only produced small increases in nitrification rates and N$_2$O emission (Bowden et al., 1991). Equally, the application of 150 kg N, as (NH$_4$)$_2$SO$_4$, to a spruce forest, growing on a soil with low N contents in the Black Forest, Germany, enhanced N$_2$O and NO emissions only for a very short time after fertiliser application. The overall rates of emission were low compared to emissions from spruce forests suffering long-term exposure to high N inputs (Rennenberg et al., 1998). These experiments indicate, that for low N forest soils, trace gas emissions may not be affected for several years by increasing N inputs.

On the other hand, in a very similar experiment, with same N rates in acid mist as at Deepsyke, but where sitka spruce was growing on a previously agricultural, base rich, mineral soil, the same rates of N significantly increased N$_2$O and NO emissions during the second season of spraying (Skiba et al., 1994, 1998b). This soil had previously received fertiliser N and retained the ability to respond to N inputs via nitrification and denitrification and causing loss of loose N$_2$O and NO very quickly. This memory effect has also been discussed by Mosier et al. (1996) for a prairie grassland in the USA, where N fertilisation remained apparent in N$_2$O emissions 12 years later.

## 4. Conclusions

Two important points emerge from the above comparisons between long-term and short-term exposure of forest plantations to elevated N deposition rates.

1. The history of N inputs to a site will precondition the response to subsequent N inputs. Experimental applications of elevated atmospheric N to a 'pristine' soil may in the short-term (several years) underestimate the likely impact of N deposition on trace gas emissions. This may also apply to other ecological parameters of interest, especially tree health and NO$_3$ leaching, and should be considered when designing large-scale field experiments.

2. The different response to high N inputs in a soil previously only subjected to low N concentrations compared to soils that have been subjected to long-term high N inputs, suggests, NO and N$_2$O measurements to be useful indicators of the N status of the soil and could complement measurements of C/N ratios carried out for the same purpose (Gundersen et al., 1999). In low N input forest soils N input derived NO and N$_2$O emission are likely to be < 2.5% of the N input, and increasing the N input is unlikely to change this percentage in the short-term. In N saturated forest soils, >2.5% of the N input is likely to be emitted as N$_2$O and NO (Skiba et al., 1998b). The advantage of using N trace gases as indictors of the N status of forests is, that measurements made with manual static chamber

systems are quick and cheap to perform; and most importantly, measurements are non-destructive. The same identical site can be remeasured for years without the problems of the spatial variability of materials collected for destructive sampling. Thus monitoring $N_2O$ and NO trace gas fluxes can provide valuable information on the decline or recovery of a forest ecosystem.

## Acknowledgements

The authors wish to thank NERC for financial support and the landowners for access on their land.

## References

Asman W.A.H. and Sutton, M.A.S.: 1998, *New Phytologist.* **139**, 27 - 48.
Bowden, R.D., Melillo, J.M., Steudler, P.A., and Aber, J.D.: 1991, *J. Geophys. Res.* **96** D5, 9321 - 9328.
Butterbach-Bahl, K., Gasche, R., Breuer, L., and Papen, H.: 1997, *Nutrient Cycl. in Agroecosys.* **48**, 79 - 90.
Cape, J.N., Sheppard, L.J., Binnie, J., Arkle, P. and Woods, C.: 1995, *Water, Air and Soil Pollut.*, 2247 - 2252.
Castro, M.S., Steudler, P.A., Melillo, J.M., Aber, J.D. and Millham, S.: 1993, *Biogeochem.* **18**, 119 - 135.
Grimshaw, H.M.: 1989, *Chemical Analysis of Ecological Materials* (ed. S.E. Allen), 2nd Edition, Blackwell, Oxford, pp7 - 45.
Gundersen, P.: 1999, *Water, Air and Soil Pollut.*, this issue.
INDITE: 1994, *Impacts of Nitrogen Deposition on Terrestrial Ecosystems*, UK Review Group, Department of the Environment, London, pp.100
Johansson, C.J.: 1984, *J. Atmos. Chem.* **1**, 429 - 442.
MacDonald, J.A., Skiba, U., Sheppard, L.J., Ball, B., Roberts, J.D., Smith, K.A. and Fowler, D.: 1997, *Atmos. Environ.*, **31**, 3693 - 3706.
Mosier, A.R., Parton, W.J., Valentine, D.W. and Schimel, D.S.: 1996, *Transactions of the 9$^{th}$ Nitrogen Workshop*, Braunschweig, September, 1996, Technische Universität Braunschweig, 157 - 160.
Pitcairn,C.E.R., Leith, I.D., Sheppard, L.J., Sutton, M.A., Fowler, D., Munro, R.C., Tang, S. and Wilson, D.: 1998, *Environ. Pollut.*,**102**, 41 - 48.
Rennenberg, H., Kreutzer, K., Papen, H., and Weber, P.: 1998, *New Phytol.* **139**, 71-86.
Sheppard, L.J., A. Crossley, J.N. Cape, F. Harvey, J. Parrington and C. White: 1998, *Phyton*, **139**, *in press*
Sheppard L.J., A. Crossley, F. Harvey, D. Wilson and J.N. Cape: 1997, *Environ. Poll.*, **98**, 175 - 184.
Skiba, U., Sheppard, L.J., MacDonald, J. and Fowler, D.: 1998a, *Atmos. Environ.* **32**, 3311 - 3320.
Skiba, U., Sheppard, L.J., Pitcairn, C.E.R., Leith, L., Crossley, A., van Dijk, S., Kennedy, V.H., and Fowler, D.: 1998b, *Environ. Poll*, **102**, 457 - 461.
Skiba, U., Fowler, D. and Smith, K.A.: 1994, *Environ. Monitor. and Assess.*, **31**, 153 - 158.
Skiba, U., Smith, K.A. and Fowler, D.: 1993, *Soil Biol. Biochem.* **25**, 1527 - 1536.
Williams, E.J., Guenther, A., and Fehsenfeld, F.C.: 1992, *J. Geophys. Res.* **97**, 7511 - 7519.
Williams, E.J., Parrish, D.D., Buhr, M.P. and Fehsenfeld, F.C.: 1988, *J. Geophys. Res.* **93**, 9539 - 9546.

# NITROGEN AND ITS EFFECT ON GROWTH, NUTRIENT STATUS AND PARASITE ATTACKS IN BEECH AND NORWAY SPRUCE

WALTER FLÜCKIGER AND SABINE BRAUN
*Institute for Applied Plant Biology, CH-4124 Schönenbuch, Switzerland*

(Received 25 October 1998; accepted 25 February 1999)

**Abstract.** In permanent observation plots across Switzerland, nitrogen (N) concentration in the foliage of mature beech has increased by 15% and phosphorus (P) concentrations in beech and mature Norway spruce decreased by 12 and 13% respectively between 1984 and 1995, leading to increased N:P ratios. Modelled N deposition was correlated with stem increment in both beech and spruce, with indications of P limitation in some beech plots. Experimental application of 0-160 kg N ha$^{-1}$ yr$^{-1}$ over four to five years caused nutrient imbalances in various afforestation plots comparable to those observed in the permanent observation plots. The changes in the trees caused by N treatment led to increased attacks by parasites such as *Apiognomonia errabunda*, *Phomopsis sp.*, *Phyllaphis fagi* in beech and *Botrytis cinerea*, *Sacchiphantes abietis* and *Cinara pilicornis* in Norway spruce. The results suggest current N deposition in Switzerland induces significant changes in the forest ecosystem.

Keywords: Nitrogen, *Fagus sylvatica*, *Picea abies*, nutrition, growth, parasite attack

## 1. Introduction

In the past 40 years forest yield has increased in many parts of Europe (Spiecker *et al.*, 1996). Zingg (1996) reported a significant increase in increment for Norway spruce, silver fir and beech growing predominantly below 1000 m asl. in Switzerland over the last 10-20 years. Climate change and increasing $CO_2$-concentrations, accumulation of nutrients in the forest floor as a result of a decreased exploitation, and increased nitrogen (N) deposition have been suggested as possible causes. Rising N loads since the 1950's have been well documented (Goulding and Blake, 1993). Increasing N concentration in the foliage of various conifer species was reported for recent decades by the Dutch (Van den Burg, 1990b), French (Landmann, 1989) and Germans (Sauter, 1991). Nihlgård (1985) proposed that forests would develop towards nutrient imbalances with rising N deposition. Imbalance of nutrients may cause an increased susceptibility to pathogens and pests (Huber, 1980; Denno and McClure, 1983).

Calculations described in Braun *et al.* (1996) show a nearly constant N deposition of 11-13 kg N ha$^{-1}$ yr$^{-1}$ on Swiss forests in the first half of this century followed by increasing N loads up to 33 kg N ha$^{-1}$ yr$^{-1}$ between mid 1940 and mid 1980. Two thirds of this N deposition are $NH_y$ from agricultural activities. Throughfall measurements in permanent observation plots of Norway spruce across Switzerland revealed N loads of 18-33 kg N ha$^{-1}$ yr$^{-1}$ except in alpine plots >1500m asl where the N loads were 6-11 kg N ha$^{-1}$ yr$^{-1}$ (Flückiger and Braun, 1998). The aim of this study was to investigate the possible long-term effects of N deposition in alpine and non alpine permanent observation plots and afforestations on growth, nutrient status and parasite attacks.

## 2. Materials and Methods

### 2.1. FIELD STUDIES

Between 1984 and 1997 ninety permanent observation plots (size 0.25-0.5 ha) with mature beech and Norway spruce stands were established in the Jura, Northwestern

Switzerland, Central Plateau, Lower Alps, Alps and south side of the Alps. Stand age ranged between 65-175 years for beech and 85-310 years for Norway spruce. 60 adult beech and Norway spruce trees were marked per plot.

Stem increment of beech and Norway spruce was determined every 4 years by measuring the diameter at breast height (DBH) at marked points with a high resolution diameter gauge to the nearest mm. Soil was sampled by horizon and analysed as described in Braun et al. (1996).

Foliar samples were repeatedly taken from eight labelled beech or Norway spruce trees per plot in 1984, 1987, 1991 and 1995. Branches were taken from the top of the crown (beech) and from the $7^{th}$ to $10^{th}$ whorl (Norway spruce) respectively from a helicopter in the end of July after completion of extension growth to have a better indication of nutritional status (van den Driessche 1974). Methods for analysis of N, P, potassium (K) and magnesium (Mg) are described in Braun et al. (1996).

Plot N deposition was calculated by Rihm after the methods outlined in BUWAL (1994). Data were evaluated with multivariate regression analysis using a mixed model (function lme of S-PLUS (S-PLUS, 1997) to take into account the clustered structure of the data. The results are presented graphically as partial correlation plots which consider the multivariate approach and averaged for each plot (see Braun et al., 1999). Because of the multivariate data evaluation, no $r^2$ can be given for single parameters.

2.2. FERTILIZATION EXPERIMENT

At three sites in the Jura and Lower Alps, beech and Norway spruce and at another four sites in the Alps Norway spruce of local provenance were planted in afforestation plots in 1992 (Table I). The trees were treated three times per season, in April, July and October, for six consecutive years with dry $NH_4NO_3$ to give loads of 0, 10, 20, 40, 80 and 160 kg N ha$^{-1}$ yr$^{-1}$. Foliage was sampled every year in August and nutrient concentrations determined in leaves and current year's needles according to analytical methods described in Braun et al. (1996). Trees were investigated for pathogens and pests in April, June and September using isolation techniques in the case of pathogens.

Insects:
*Cinara pilicornis* (Hartig): Aphid with orange-brown or greyish-green apterae, size 2.1-4.7 mm, secreting white wax wool, occurs in spring in small colonies on the abaxial side of previous year's twigs of Norway spruce, moving onto new shoots after budbreak (Blackman and Eastop, 1994), quantified as numbers of infested shoots.
*Phyllaphis fagi* (L.): Aphid with 2.0-3.2 mm large apterae, pale yellowish green, covered with wax wool, living on undersides of mostly young beech leaves, causing them to curl downwards, wither and die prematurely (Blackman and Eastop, 1994). Quantification was done by scoring the extent of the leaf curling.
*Sacchiphantes abietis* (L.): induces a pineapple like gall on *Picea* spp., quantified from the number of galls.

Pathogenic Fungi (all quantified by counting the number of dead shoots per plant):
*Apiognomonia errabunda* (Rob.) Höhn: attacks leaves (necrotic spots in the mid-rib region) and young shoots of beech, prevalent during warm and humid weather in spring (Butin, 1989).
*Phomopsis sp.*: attacks leaves and young shoots of beech which turn brown and die (Uecker, 1988; Farr et al., 1989).

*Botrytis cinerea* Pers.: attacks young shoots of various trees, predominantly conifers (Butin, 1989).

TABLE I

Brief description of sites used in the N fertilization experiments in afforestation plots. The chemical parameters are given as averages over 0-40cm depth except for the minimum base saturation which is the lowest value encountered in the profile. The average 0-40cm includes the O horizon; for calculation layers were weighted by thickness (corrected for stones) and density.

| Plot | Altitude (m asl) | Soil type | Geology | pH ($CaCl_2$) 0-40cm | average base saturation 0-40cm (%) | min. base saturation (%) | Annual rainfall (mm) |
|---|---|---|---|---|---|---|---|
| Axalp | 1700 | Rendzic Leptosol | limestone | 6.8 | 100 | 100 | 1710 |
| Hochwald | 670 | Rendzic Leptosol | limestone | 6.9 | 100 | 100 | 1260 |
| Lurengo | 1600 | Ferralic Cambisol | gneiss | 4.1 | 13 | 3 | 2020 |
| Möhlin | 290 | Haplic Acrisol | gravel | 3.8 | 12 | 4 | 1060 |
| Rötiboden | 1580 | Carbic Podzol | granite | 3.3 | 26 | 6 | 1850 |
| Wengernalp | 1880 | Cambic Podzol | limestone | 3.7 | 23 | 12 | 1970 |
| Zugerberg | 1000 | Dystric Cambisol | till | 4.0 | 12 | 2 | 2010 |

## 3. Results

### 3.1. FIELD STUDIES

Between 1984 and 1995 the nutrient status of beech has changed significantly. N concentration in foliage increased by 15% (from 22.3 ± 0.3 mg $g^{-1}$ d.wt. to 25.5 ± 0.28 mg $g^{-1}$ d.wt.). At the same time, P concentrations decreased by 12% (from 1.26 ± 0.26 to 1.08 ± 0.25 mg $g^{-1}$ d.wt.). This resulted in an increased N:P ratio (Flückiger and Braun, 1998). In 1984, 5% and by 1995 61% of the beech stands had N concentrations above the classification values of 18-25 mg $g^{-1}$ d.wt. recommended by ICP Forests (1996). Furthermore, the proportion of beech stands with low P concentrations (<1.0 mg $g^{-1}$ d.wt.) had increased from 11% in 1984 to 32% in 1995 (Figure 1). There was a significant correlation between the N:P ratio in beech leaves and the modelled NHy-deposition (Figure 2). In Norway spruce, foliar N concentrations did not change but P concentrations decreased between 1984 and 1995 by 13% (from 1.07 ± 0.25 to 0.90 ± 0.23 mg $g^{-1}$ d.wt.), with the consequent increase in N:P ratio.

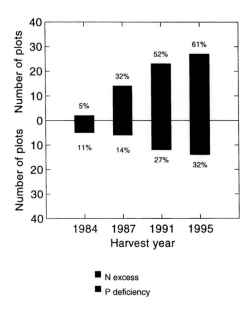

Fig. 1: Proportion of permanent beech observation plots with mean N concentrations >25 mg N g$^{-1}$ d.wt. and P concentrations < 1 mg g$^{-1}$ d.wt. respectively at different harvests between 1984 and 1995 (levels after ICP Forests, 1996, n=44)

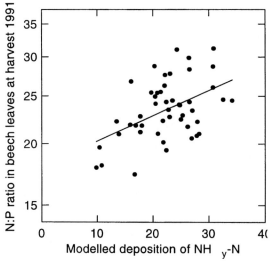

Fig. 2: Correlation between the N:P ratio in beech leaves in 1991 and modelled deposition of reduced N. The correlation is significant at p<0.001 (374 trees in 48 plots). Multivariate analysis (see methods section).

In beech there was a significant correlation between modelled N deposition and stem increment which was modified when the P concentration in the leaves indicated P limitation (Figure 3). In Norway spruce the correlation between stem increment and modelled N deposition was less steep and weaker reflecting the smaller number of plots (Figure 3).

Fig. 3: Correlation of diameter increment between 1991 (beech) or 1992 (spruce) and 1995 with modelled N deposition. Each point represents one plot.

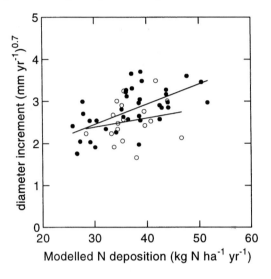

Beech (1539 trees in 55 plots, $p<0.001$). Interaction with foliar levels of P ($p<0.05$): filled symbols: P in leaves $\geq 1$ mg P $g^{-1}$ d.wt., empty symbols: $<1$ mg P $g^{-1}$ d.wt..

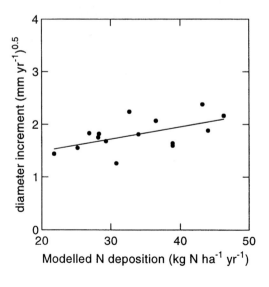

Spruce (517 trees in 15 plots, $p<0.05$). Stem increment was transformed to the power of 0.7 (beech) and 0.5 (spruce) respectively to ensure normal distribution of the residuals. Multivariate analysis (see methods section).

## 3.2. N FERTILIZATION EXPERIMENTS

After five to six years of fertilization with 0-160 kg N ha$^{-1}$ yr$^{-1}$ in afforestation plots, nutrient imbalances were apparent in the young trees which were comparable to the changes observed in permanent observation plots of mature trees (Table II, Figure 4, Figure 5). The largest increases were found in the N:P ratio for acidic as well as calcareous soils, followed by the N:K and N:Mg ratios. These changes mainly reflected decreasing foliar P, K and Mg concentrations.

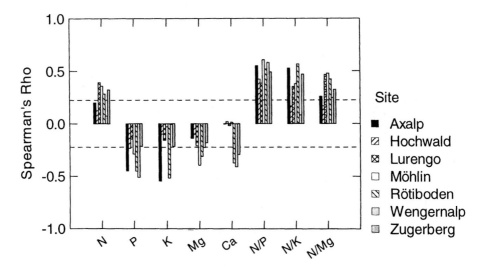

Fig. 4: Nutrient changes in the N fertilization experiment (afforestations) in Norway spruce after four years of treatment. Bars represent the Spearman correlation coefficient for the correlation between N fertilization and the respective nutrient, the dashed lines the significance limit (p<0.05, 6 N-levels with 12 tree replicates in all plots)

TABLE II

Lowest N treatment (kg N ha$^{-1}$ yr$^{-1}$) causing significant changes in nutrient concentrations and ratios in Norway spruce and beech in afforestation plots. Empty fields: no significant differences to control

a) Norway spruce (1996, five years of treatment)

|      | Axalp | Hochwald | Lurengo | Möhlin | Rötiboden | Wengern-alp | Zugerberg |
|------|-------|----------|---------|--------|-----------|-------------|-----------|
| N    |       |          | 40      | 40     | 160       |             | 160       |
| P    | 40    | 160      |         |        | 80        | 80          | 40        |
| K    | 80    |          |         |        | 20        |             | 20        |
| Mg   |       |          | 80      | 160    | 160       |             |           |
| Ca   |       |          |         |        | 160       | 80          | 20        |
| N:P  | 40    | 10       | 160     | 40     | 160       | 20          | 40        |
| N:K  | 80    |          | 160     | 160    | 80        |             | 80        |
| N:Mg | 80    |          | 80      | 80     | 160       | 160         | 160       |

b) Beech (1995, 1996 and 1997, four to six years of treatment)

|  | Hochwald | | | Möhlin | | Zugerberg | | |
| --- | --- | --- | --- | --- | --- | --- | --- | --- |
|  | 1995 | 1996 | 1997 | 1995 | 1996 | 1995 | 1996 | 1997 |
| N |  |  |  |  | 40 |  |  | 20 |
| P |  | 160 |  |  | 80 | 40 | 40 | 160 |
| K |  | 80 | 80 |  |  |  |  | 160 |
| Mg |  |  |  | 160 | 160 |  |  | 10 |
| Ca |  |  |  |  | 160 |  | 40 | 160 |
| N:P |  | 160 |  | 160 | 80 | 40 | 40 | 20 |
| N:K | 80 | 80 | 20 |  |  |  | 160 | 20 |
| N:Mg |  |  |  | 160 | 160 |  |  | 10 |

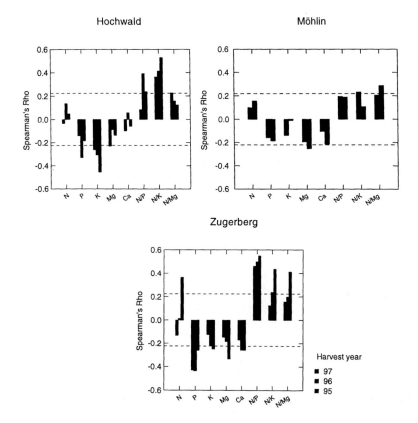

Fig. 4: (n=6x12 in all plots).
Fig. 5: Nutrient changes in the N fertilization experiment (afforestations) in beech after 4-6 years of treatment. For explanation refer to

Parasite attacks on both beech and Norway spruce increased with increasing N fertilization. Beech was significantly more infested by *Apiognomonia errabunda*, *Phomopsis sp.* and *Phyllaphis fagi* when fertilized with N (Table IV, Figure 6). Likewise, fertilized Norway spruce became significantly more infested with *Botrytis cinerea*, *Sacchiphantes abietis* and *Cinara pilicornis* (Table IV). The most important changes were found at the Zugerberg site which had the lowest base saturation of the mineral soils but changes were also observed in Hochwald, on a rendzina soil with low water storage capacity.

TABLE III

Nutrient concentrations (mg g$^{-1}$ d.wt.) in the unfertilized controls of the N experiment (mean ± standard error of mean)

a) Picea abies, current year's needles (sampled in 1996)

| Plot | N | | P | | K | | Mg | | Ca | | N/P | | N/K | | N/Mg | |
|---|---|---|---|---|---|---|---|---|---|---|---|---|---|---|---|---|
| Axalp | 13.0 | ±0.9 | 2.00 | ±0.20 | 4.23 | ±0.15 | 0.71 | ±0.05 | 2.63 | ±0.36 | 6.7 | ±0.3 | 3.13 | ±0.26 | 20.1 | ±2.8 |
| Hochwald | 13.3 | ±0.3 | 1.13 | ±0.06 | 3.87 | ±0.36 | 0.53 | ±0.03 | 4.77 | ±0.23 | 12.0 | ±0.4 | 3.65 | ±0.22 | 26.1 | ±1.8 |
| Lurengo | 13.1 | ±0.9 | 1.92 | ±0.19 | 5.10 | ±0.74 | 0.89 | ±0.06 | 2.43 | ±0.25 | 7.5 | ±0.7 | 3.13 | ±0.42 | 15.2 | ±1.1 |
| Möhlin | 15.3 | ±0.6 | 1.58 | ±0.05 | 3.94 | ±0.10 | 0.61 | ±0.02 | 2.85 | ±0.21 | 9.7 | ±0.3 | 3.88 | ±0.12 | 25.1 | ±1.2 |
| Rötiboden | 16.0 | ±1.3 | 2.39 | ±0.20 | 11.41 | ±1.36 | 0.99 | ±0.05 | 2.78 | ±0.16 | 6.8 | ±0.6 | 1.48 | ±0.14 | 16.2 | ±1.2 |
| Wengernalp | 14.7 | ±0.9 | 1.39 | ±0.08 | 6.12 | ±0.43 | 0.68 | ±0.04 | 3.33 | ±0.32 | 10.6 | ±0.4 | 2.50 | ±0.22 | 22.4 | ±2.0 |
| Zugerberg | 14.7 | ±0.2 | 1.59 | ±0.09 | 4.32 | ±0.20 | 0.55 | ±0.04 | 3.02 | ±0.24 | 9.5 | ±0.4 | 3.47 | ±0.17 | 28.0 | ±1.8 |

b) Fagus sylvatica

| Plot | year | N | | P | | K | | Mg | | Ca | | N/P | | N/K | | N/Mg | |
|---|---|---|---|---|---|---|---|---|---|---|---|---|---|---|---|---|---|
| Hochwald | 1995 | 20.8 | ±0.8 | 1.05 | ±0.09 | 4.88 | ±0.32 | 1.96 | ±0.09 | 11.67 | ±0.83 | 20.9 | ±1.3 | 4.35 | ±0.16 | 10.8 | ±0.5 |
| Hochwald | 1996 | 19.5 | ±0.5 | 0.98 | ±0.07 | 4.18 | ±0.13 | 2.01 | ±0.08 | 12.60 | ±0.43 | 21.0 | ±1.7 | 4.68 | ±0.13 | 9.9 | ±0.6 |
| Hochwald | 1997 | 18.7 | ±0.6 | 0.87 | ±0.08 | 3.68 | ±0.20 | 1.66 | ±0.15 | 13.29 | ±0.76 | 23.4 | ±2.1 | 5.17 | ±0.19 | 12.3 | ±1.2 |
| Möhlin | 1995 | 20.3 | ±0.4 | 1.50 | ±0.04 | 4.58 | ±0.09 | 1.24 | ±0.03 | 6.15 | ±0.40 | 13.6 | ±0.4 | 4.43 | ±0.06 | 16.5 | ±0.7 |
| Möhlin | 1996 | 20.5 | ±0.6 | 1.68 | ±0.07 | 4.64 | ±0.12 | 1.19 | ±0.10 | 6.57 | ±0.44 | 12.4 | ±0.7 | 4.42 | ±0.10 | 18.4 | ±1.3 |
| Zugerberg | 1995 | 21.2 | ±0.4 | 1.57 | ±0.05 | 4.85 | ±0.12 | 1.15 | ±0.10 | 9.51 | ±0.72 | 13.6 | ±0.4 | 4.38 | ±0.10 | 19.9 | ±1.6 |
| Zugerberg | 1996 | 25.6 | ±0.6 | 1.78 | ±0.07 | 5.18 | ±0.17 | 1.69 | ±0.24 | 10.76 | ±0.63 | 14.6 | ±0.6 | 5.00 | ±0.18 | 17.6 | ±1.6 |
| Zugerberg | 1997 | 19.8 | ±0.6 | 1.43 | ±0.06 | 5.58 | ±0.22 | 1.19 | ±0.12 | 8.92 | ±0.61 | 14.0 | ±0.6 | 3.58 | ±0.13 | 21.1 | ±4.4 |

TABLE IV

Infestation of beech and Norway spruce in the N fertilization experiment with parasites. Numbers are mean score classes (score 0-9) ± standard error of mean. Both overall significance (linear trend) and comparison to the control were tested using the linear-by-linear association test (Mehta and Patel, 1992).

| Tree species | Plot | Year | Parasite | Linear trend | N-fertilization (kg N ha$^{-1}$ yr$^{-1}$) | | | | | |
|---|---|---|---|---|---|---|---|---|---|---|
| | | | | | 0 | 10 | 20 | 40 | 80 | 160 |
| Fagus sylvatica | Hochwald | 1997 | Phomopsis sp. | * | 0.25±0.25 | 0.50±0.36 | 1.33±0.36 ** | 0.83±0.37 | 1.33±0.51 * | 1.33±0.51 * |
| Fagus sylvatica | Hochwald | 1998 | Phomopsis sp. | * | 0.08±0.08 | 0.31±0.13 | 2.50±0.69 *** | 4.27±0.57 *** | 3.25±0.54 *** | 1.83±0.42 *** |
| Fagus sylvatica | Mohlin | 1997 | Apiognomonia errabunda | *** | 0.33±0.26 | 0.85±.27 | 0.75±.28 | 1.92±.56 * | 1.50±0.36 | 3.33±0.57 *** |
| Fagus sylvatica | Zugerberg | 1997 | Apiognomonia errabunda | *** | 0.33±0.26 | 2.62±0.24 *** | 3.27±0.38 *** | 4.67±0.36 *** | 4.00±0.44 *** | 5.00±0.49 *** |
| Fagus sylvatica | Zugerberg | 1998 | Apiognomonia errabunda | *** | 0.12±0.12 | 0.23±0.16 | 0.88±0.34 | 1.00±0.34 | 1.00±0.46 | 2.50±0.65 ** |
| Fagus sylvatica | Zugerberg | 1997 | Phllaphis fagi | * | 2.83±0.41 | 2.23±0.38 | 2.55±0.31 | 2.50±0.19 | 2.92±0.38 | 3.82±0.60 |
| Picea abies | Zugerberg | 1996 | Botrytis cinerea | *** | 0.0±0.0 | 0.29±0.17 | 0.49±0.18 | 0.08±0.05 | 2.09±0.53 *** | 4.72±1.15 *** |
| Picea abies | Zugerberg | 1998 | Cinara pilicornis | ** | 5.65±0.63 | 7.20±0.42 | 5.93±0.34 | 6.22±0.70 | 7.45±0.50 * | 7.80±0.52 * |
| Picea abies | Zugerberg | 1998 | Sacchiphantes abietis | ** | 2.62±0.45 | 2.29±0.49 | 3.45±0.52 | 3.45±0.52 | 4.09±0.77 | 4.80±0.68 * |

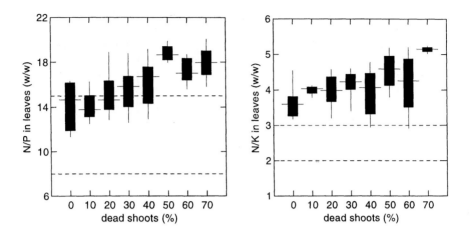

Fig. 6: Nutrient ratios in the leaves of young beech with different extents of shoot dieback caused by *Apiognomonia errabunda* (N afforestation experiment Zugerberg 1997). Both linear trends are significant at $p<0.001$. The box extends from the 25% to the 75% quantile, with the median as horizontal line and non outlier minimum and maximum as whiskers (SYSTAT boxplot, SPSS Inc. Chicago). The dashed lines indicate classification values after various literature data referenced in Van den Burg (1990a).

## 4. Discussion

There are several observations showing that with increasing N loads P status in trees decreases. Houdijk and Roelofs (1993) reported lower P concentrations in Dutch conifer forests in response to increased N loads. Mohren et al. (1986) observed P deficiency in a Dutch stand of *Pseudotsuga menziesii* with increased N deposition. Balsberg-Påhlsson (1992) induced a reduction in foliar P concentrations in mature beech trees in southern Sweden by fertilizing with 204 kg N ha$^{-1}$ yr$^{-1}$ over four years. The similarity of changes in the N:P ratio observed in the field and in the fertilization experiments suggests N deposition as causal factor for the nutrient changes observed in the field. In the forest, N:P ratios exceed the adequate range recommended by various authors in Van den Burg (1990a) and by Hüttl (1990) in many plots. Using ratios overcomes the problem of changing proportions of nonstructural carbohydrates in the foliage (Linder 1995).

The correlations found between N and stem increment in beech and Norway spruce suggest N may be involved as a causal factor in the increased forest growth in Europe. Kreutzer (1993), too, supposed that N deposition plays a predominant role because growth increases first at sites with low N supply and most obviously in the upper part of the stem. This response pattern is comparable to the behaviour observed in N fertilizing experiments. Nevertheless, the significant interaction with P nutrition found in beech suggests that trees in some plots are P limited and are unable to respond to additional N loads.

The observations of increased parasite infestations of both beech and Norway spruce in afforestation plots confirm earlier findings that enhanced N status and nutrient imbalances respectively increase the susceptibility of trees to pathogens and sucking insects. Roelofs et al. (1985) found a relationship between increased N concentration in needles of *Pinus nigra* and infection by the pathogens *Brunchorstia pinea* and *Diplodia pinea*. Lambert (1986) observed the highest infestation by *Dothistroma* needle cast fungus in *Pinus radiata* fertilized with N. However, while N is important, its ratio to other nutrients may be even more influential. De Kam *et al.* (1991) showed that bark necrosis caused by *Sphaeropsis sapinea* was enhanced when *Pinus nigra* was fertilized with N alone. The simultaneous addition of K inhibited lesion development in the same experiment. The length of lesion and shoot dieback of young beech caused by *Nectria ditissima* correlated significantly with enhanced N:K ratios and following N application (Flückiger and Braun, 1998).

Resistance against pathogens is affected by phenolic compounds (Waterman and Male, 1994), which can be decreased by N fertilization (Balsberg-Påhlsson, 1992). Increased concentrations of free amino acids and amides in the phloem due to increased N supply may play a crucial role in the improved performance of aphids (Dixon, 1970; Bolsinger and Flückiger, 1989). The population of *Aphis fabae* increased when the foliar N status of the host plants *Viburnum opulus* and *Phaseolus vulgaris* was increased (Bolsinger and Flückiger, 1987). In this study the most severe attacks by parasites were revealed on poor acid soils, where N fertilization caused the first and most distinct changes of the nutrient status.

## 5. Conclusions

Results from the permanent observation plots and experimental afforestation plots suggest that N deposition may be one cause for the increased stem increment and nutrient imbalances observed in Swiss beech and Norway spruce stands in recent years. Fertilization experiments suggest that trees will become more susceptible to parasite attacks with increasing N loads. Present N loads in Switzerland which exceed the recommended critical load for forests (UN/ECE, 1996) may have an unfavourable long term effect upon forest ecosystems.

## Acknowledgements

This project was supported by the Federal Office of Environment, Forest and Landscape (FOEFL) and by the cantons of Aargau, Basellandschaft, Baselstadt, Bern, Solothurn, Zug and Zürich. We like to thank Beat Achermann, Heinz Balsiger, Dr. Reinhard Eichrodt, Jürg Froelicher, Werner Giss, Dr. Heinz Kasper, Hans Schmid and Dr. Richard Volz for their support and interest in our work. We are also indebted to Brigitte Möcklin and Inês da Costa for plant and nutrient analysis, Heidi Flückiger, Andreas Mebert and Max Fischer for technical assistance, Dr. Erika Hiltbrunner for comments, Dr. Liliane Petrini for identifying the pathogens, Beat Rihm for modelling N deposition and Dr. Ch. Schindler for statistical advice.

## References

Balsberg-Påhlsson, A.: 1992, *Tree Physiol.* **10,** 93-100.
Blackman, R. L. and Eastop, V. F.: 1994, *Aphids on the World's Trees*. CAB International, Wallingford, p. 986
Bolsinger, M. and Flückiger, W.: 1987, *Entomol.exp.appl.* **45,** 237-243.
Bolsinger, M. and Flückiger, W.: 1989, *Environ.Pollut.* **56,** 209-216.

Braun, S., Flückiger, W. and Leonardi, S.: 1996, BUWAL Umweltmaterialien **59** ,1-237
Braun, S., Rihm, B., Schindler, Ch., Flückiger, W.: 1999, *Water Air Soil Pollution* (in press)
Butin, H.: 1989, *Krankheiten der Wald- und Parkbäume*. Georg Thieme-Verlag, Stuttgart, p. 216
BUWAL: 1994, Environmental Series **234** . Bern,1-68
De Kam, M., Versteegen, C. M., Van den Burg, J. and Van der Werf, D. C.: 1991, *Neth.J.Pl.Path.* **97,** 265-274.
Denno, R. F. and McClure, M. S.: 1983, *Variable Plants and Herbivores in Natural and Managed Systems*. Academic Press, p. 717
Dixon, A. F. G.: 1970, *Animal Populations in Relation to their Food Resources*, eds. Watson, A., Blackwell, Oxford, 271-287 .
Farr, D. F., Bills, G. F., Chamuris, G. P. and Rossman, A. Y.: 1989, *Fungi on Plants and Plant Products in the United States.* APS Press, St. Paul, Minnesota, p. 1252
Flückiger, W. and Braun, S.: 1998, *Environ.Pollut.* **102,** 69-76.
Goulding, K. W. T. and Blake, L.: 1993, *Critical Loads: Concepts and Applications. Proceedings of a Conference 12.-14. Feb. 1992, Grange over Sands*, eds. Hornung, M. and Skeffington, R. A., Institute of Terrestrial Ecology, 68-73 .
Houdijk, A. and Roelofs, J. G. M.: 1993, *Environ.Pollut.* **80,** 79-84.
Huber, D. M.: 1980, *Plant Disease*, eds. Horsfall, J. G. and Cowling, E. B., Academic Press, 381-406 .
Hüttl, R. F.: 1990, *Plant and Soil* **128,** 45-58.
ICP Forests: 1996, *Forest condition in Europe, 1996 Report.* UN/ECE, Brussels, Geneva, p. 127
Kreutzer, K.: 1993, *Forest decline in the Atlantic and Pacific region*, eds. Hüttl, R. F. and Müller-Dombois, D., Springer-Verlag, Berlin, 82-96 .
Lambert, M. J.: 1986, *Can.J.For.Res.* **16,** 1055-1062.
Landmann, G.: 1989, *Evolution sur 20 ans de l'alimentation minérale d'un peuplement adulte de sapin pectine et du complexe absorbant dans un site d'altitude du massif Vosgien*. Département de la Santé des Forêts, INRA, Centre de Recherche Forestières, 54280 Champenoux,
Linder, S.: 1995, *Ecological Bulletins* **44,** 178-190.
Mehta, C. and Patel, N.: 1992, *StatXact for SYSTAT. Statistical Software for Exact Nonparametric Inference*. CYTEL Software Corporation, Cambridge MA 01239,
Mohren, G. M. J., Van den Burg, J. and Burger, F. W.: 1986, *Plant Soil* **95,** 191-200.
Nihlgård, B.: 1985, *Ambio* **14,** 1-8.
Roelofs, J. G., Kempers, A. J., Houdijk, A. L. and Jansen, J.: 1985, *Plant Soil* **84,** 45-56.
S-PLUS: 1997, *Guide to Statistics (Version 4)*. Data Analysis Products Division, MathSoft, Seattle.
Sauter, W.: 1991, *Forstwiss.Cbl.* **110,** 13-33.
Spiecker, H., Mielikäinen, R., Köhl, M. and Skorgsgaard, J. P.: 1996, *Growth Trends in European Forests.* Springer-Verlag, New York, p. 372
Uecker, F. A.: 1988, *A World List of Phomopsis Names with Notes on Nomenclature, Morphology and Biology*. J. Cramer, Berlin, Stuttgart, p. 231
UN/ECE: 1996, Convention on Long-range Transboundary Air Pollution . Berlin, Umweltbundesamt,1-142
Van den Burg, J.: 1990a, *Foliar analysis for determination of tree nutrient status - a compilation of literature data.* Rijksinstituut voor Onderzoek in de Bos- en Landschapsbouw "De Dorschkamp", Wageningen, p. 220
Van den Driessche, R.: 1974, *The Bot Rev* 40/3, 347-397
Van den Burg, J.: 1990b, *Forst und Holz* **45,** 597-605.
Waterman, P. G. and Male, S.: 1994, *Analysis of phenolic plant metabolism.* Blackwell Scientific Publications, Oxford, p. 238
Zingg, A.: 1996, *Growth trends in European forests*, eds. Spiecker, H., Mielikäinen, R., Köhl, M. and Skorgsgaard, J. P., Springer-Verlag, New York, 239-265 .

# IMPACTS OF (NH₄)₂SO₄ DEPOSITION ON NORWAY SPRUCE (*PICEA ABIES* [L.] KARST) ROOTS

M. CARNOL[1*,2], P. CUDLÍN[3] AND P. INESON[2]

[1*]*University of Liège, Dept. Botany B22, Microbial Ecology and Radioecology, 4000 Liège, Belgium;* [2]*Institute of Terrestrial Ecology, Soil Ecology Section, Merlewood Research Station, Grange-over-Sands, Cumbria LA11 6JU, United Kingdom;* [3]*Institute of Landscape Ecology, Academy of Sciences of the Czech Republic, Na Sadkach 7, Ceské Budejovice, Czech Republic.* * Address for correspondence*

(Received 25 September 1998; accepted 25 February 1999)

**Abstract.** The effects of enhanced $(NH_4)_2SO_4$ (NS) deposition on Norway spruce (*Picea abies* [L.] Karst) fine root biomass, vitality and chemistry were investigated using root-free in-growth cores reproducing native organic and mineral soil horizons. The cores were covered and watered every 2 weeks with native throughfall or throughfall supplemented with NS to increase deposition by 75 kg ha$^{-1}$ a$^{-1}$ $NH_4^+$-N (86 kg ha$^{-1}$ a$^{-1}$ $SO_4^{2-}$-S). The in-growth cores were sampled after 19 months and assessed for root biomass, necromass, length, tip number, tip vitality and fine root chemistry. Root biomass and fine root aluminium (Al) concentration were negatively correlated, but NS deposition had no effect on root growth or root tip vitality. NS deposition caused increased fine root nitrogen (N) concentrations in the organic horizon and increased Calcium (Ca) concentrations in the mineral horizon. Fine root biomass was higher in the organic horizon, where fine root Al and potassium (K) concentrations were lower and Ca concentrations higher than in the mineral horizon. Results highlighted the importance of soil stratification on fine root growth and chemical composition.

**Keywords**: ammonium sulphate, roots, in-growth cores, *Picea abies*

## 1. Introduction

Increased inorganic nitrogen (N) deposition over the last few decades has become a major concern for the health of European forests. Some forest ecosystems, where N no longer limits primary production, become N saturated and the excess N is thought to contribute to forest decline (Nihlgård, 1985; see review by Ortloff and Schlaepfer, 1996). In particular, N in the form of $NH_4$ in excess of plant and microbial demands can acidify soils if nitrified and leached from the soil, causing loss of base cations or mobilisation of phytotoxic aluminium (Al) (Reuss and Johnson, 1986). Al has long been known to be toxic to plant roots, particularly for cultivated plants (Wheeler *et al.*, 1992). However, in forest soil solution, Al may be organically complexed and, hence, less toxic (Dahlgren *et al.*, 1991).

Decreases in root biomass, mycorrhizal infection and shifts in the mycorrhizal populations have frequently been reported in forests heavily subjected to acid rain (Cudlin and Kropacek, 1990; Jansen and Dighton, 1990). Such observations are often associated with extreme localized pollution. In contrast, the effects of increased N deposition over large areas in Europe are more variable. Decreased (Alexander and Fairley, 1983; van Dijk *et al.*, 1990; Clemensson-Lindell and Persson, 1995), unchanged (Majdi and Rosengren-Brinck, 1994; Seith *et al.*, 1996), and increased (Ahlström *et al.*, 1988) root production have been reported for conifers under high N deposition. This indicates that root responses to increased N deposition may depend strongly on the chemical and physical properties of the soil. Furthermore, root tips might be damaged, and their nutrient uptake capacities impaired, even with unchanged root biomass. Magnesium (Mg) deficiency symptoms in conifers may reflect impaired nutrient uptake through antagonism between NH₄ with Ca and Mg (Jorns and Hecht-Buchholz, 1985; Hecht-Buchholz *et al.*, 1987). Increased N uptake may also lead to relative nutrient deficiencies through increased growth and subsequent limitation of other nutrients, referred to as the 'dilution effect' (Nihlgård 1985; Ortloff and Schlaepfer, 1996). Ca/Al molar ratios < 1 are thought to represent a toxicity threshold for Al in solution culture experiments (Rost-Siebert, 1983; Ulrich *et al.*, 1984). Yet, for forest soils total Al and inorganic monomeric Al might represent better indicators (Joslin and Wolfe, 1988). However, an extensive literature review including laboratory and field studies with a range of species, led Cronan and Grigal (1995) to conclude that the soil solution Ca/Al molar ratio is a satisfactory index by which to assess the ecological risk associated with forest health.

The main objective of this study was to determine the influence of high N deposition on the health of Norway spruce roots, by measuring root parameters in the field subjected to $(NH_4)_2SO_4$, applied to mimic polluted throughfall. As sulphate ($SO_4^{2-}$) behaved as a conservative ion and as major changes in soil solution ionic concentrations at this site were related to nitrate ($NO_3^-$) (Carnol *et al.*, 1997a), the large sulphur (S) input did not seem to interfere with chemical processes and the conclusions of this study.

## 2. Material and Methods

### 2.1. STUDY SITE

The experiment was undertaken in a 42 year old Norway spruce (*Picea abies* [L.] Karst) stand, 3500 stems ha$^{-1}$, located in Grizedale forest, NW United Kingdom (National Grid Ref. SD326915). The stand was located on an acid brown soil with clay texture and slight seasonal water-logging. No ground storey vegetation was present. The site was at 170 m a.s.l. with a mean annual soil temperature (0-5 cm depth) of 7.8 °C and rainfall of 1900 mm (June 1992-1993). The soil chemistry has been described in Carnol *et al.* (1997b) and Raubuch (1992); relevant characteristics are given in Table I.

TABLE I

Total soil element content (µg g$^{-1}$ dw), exchangeable cations (µeq g$^{-1}$ dw) with the percentage (%) of the total exchangeable cations and pH in the organic and mineral horizons (means of seven replicates).

Carnol *et al.* (1997b) Raubuch (1992)

| | Total element content (µg g$^{-1}$) | | | | | Exchangeable cations (µeq g$^{-1}$) | | | | | | | |
|---|---|---|---|---|---|---|---|---|---|---|---|---|---|
| | Al | Ca | Mg | K | P | Al | Ca | Mg | K | C (%) | N (%) | CEC | pH$_{H2O}$ |
| Ol | 3199 | 1462 | 573 | 1008 | 874 | - | - | - | - | 43.9 | 1.5 | - | 4.1 |
| Of | 2372 | 1107 | 560 | 616 | 884 | - | - | - | - | 45.3 | 1.6 | - | 3.9 |
| Oh | 5394 | 298 | 588 | 793 | 903 | - | - | - | - | 34.1 | 1.9 | - | 3.6 |
| mineral (0-5cm) | 19243 | 2963 | 3284 | 1983 | 783 | 113 | 4.3 | 3.6 | 1.2 | 30.7 | 1.9 | 123 | 3.7 |
| (%) | - | - | - | - | - | (91) | (3.6) | (2.9) | (1.0) | - | - | - | - |
| mineral (5-8 cm) | - | - | - | - | - | 112 | 3.7 | 3.2 | 0.9 | 34 | 2.1 | 121 | 3.8 |
| (%) | - | - | - | - | - | (93) | (3.2) | (2.6) | (0.7) | - | - | - | - |

TABLE II

Fine live root tip development stages and vitality classes

(Kocourek and Bystrican, 1989; Chemlikova *et al.*, 1992; Ruess *et al.*, 1996)

| | Stage/Class | Description |
|---|---|---|
| Root tip development | 1 | root initiation |
| | 2 | elongated root tip with root hairs |
| | 3 | club-shaped short roots without hairs, tip slightly swollen, fine roots appear hyaline |
| | 4 | tip considerably swollen or clubbed, with first mycorrhizal structures, the Hartig net, but no fungal mantle |
| | 5 | fully developed mycorrhiza (Hartig net and fungal mantle) |
| Root tip vitality | Turgid | young turgid tip, assumed to be have absorptive function |
| | Shrivelled | older tip, live and functioning vascular tissues, but believed to be not absorptively functioning |
| | Dead | brittle root, dead vascular cylinder |

2.2. SAMPLE COLLECTION AND ANALYSES

Root-free in-growth cores (Persson, 1990) reproduced organic (8 cm) and mineral soil horizons (7 cm) to 15 cm depth. They consisted of a cylindrical nylon mesh bag (Netlon®, mesh size 5 mm), containing the reconstructed Ol, Of, Oh and top mineral horizons, which were inserted into the holes left previously by the removal of the soil cores. The 10 replicates were individually covered with a Perspex® sheet (30*30 cm), 10 cm above the ground. In-growth cores were watered every 2 weeks with throughfall collected in the field, representing the mean throughfall volume over this time. The treatment started 6 months after installation and consisted of $(NH_4)_2SO_4$ added to increase deposition by 75 kg ha$^{-1}$ a$^{-1}$ $NH_4^+$-N (86 kg ha$^{-1}$ a$^{-1}$ $SO_4^{2-}$-S). Native throughfall deposition at this site was 9 kg ha$^{-1}$ a$^{-1}$ $NH_4^+$-N and 34 kg ha$^{-1}$ a$^{-1}$ $SO_4^{2-}$-S (1992-1993). The $(NH_4)_2SO_4$ additions were incorporated into the throughfall at each watering event, to simulate increased deposition by polluted throughfall.

Nineteen months after installation, the in-growth cores were removed from the field and divided into organic and mineral soil horizons. The soil was washed off the roots by aid of a root washing machine, consisting of a 1 mm sieve placed beneath a gently sprinkling water jet. Live and dead roots were then picked manually from amongst the needles and debris from the sieve. Soil remaining on live fine roots was removed by gently cleaning with fine brushes under deionised water. Roots were analysed for biomass, necromass, root length, root tip number (RTN), root tip vitality (defined below) and fine root chemistry. They were divided into coarse (>1 mm) and fine (<1 mm) roots. Distinction between the live and dead root fractions was made on visual criteria; living roots are firm, light coloured and present a good adhesion between the cortex and the periderm (Persson, 1990); dead roots are dark and brittle, and the absence of a live vascular cylinder is verified by teasing the root strand between tweezers. The fine root fraction was sub-sampled for chemical analysis and estimation of vitality, made after fixation in gluteraldehyde. Root lengths were determined with the gridline intersect method (Giovannetti and Mosse, 1980).

Fine root tips were counted under the dissection microscope, and divided into 5 classes of root tip development and 3 vitality classes (Kocourek and Bystrican, 1989;

Chemlikova et al., 1992; Ruess et al., 1996) (Table II). Visual distinction between stages 4 and 5 was possible after 'calibration' of the observer, by examination of thin longitudinal and cross sections under the epifluorescent microscope after staining in a solution of cotton blue in lactoglycerin (Cudlin, 1991). Root dry weights were recorded after drying to constant weight at 85°C.

2.3. CHEMICAL ANALYSES

Sub-samples of fine live roots for chemical analyses were dried at 85°C and wet-digested in a $H_2O_2/H_2SO_4$ digestion mixture containing Se and $LiSO_4$ (Allen, 1989). Al recovery was improved by re-boiling the digestion mixture, which had been diluted with 10 ml distilled water. Al, Ca, Mg and potassium (K) contents were determined using an ICP, and N was determined colorimetrically (indophenol blue) using a Skalar autoanalyser.

2.4. STATISTICAL ANALYSES

Statistical analyses were carried out using the SAS GLM procedure, an ANOVA technique with correction for missing values (Student's T-test). Two-way ANOVA was used to test individually for NS deposition/block and soil layer/block effects. When block was not significant, effects were tested with one-way ANOVA. Percentage data were transformed (arcsine square root; Sokal and Rohlf, 1981) before statistical analyses (SAS Institute Inc., 1989). Relationships between variables were investigated using Pearson's correlation coefficient. Due to the labour intensive nature of the determinations, 8 replicates were analysed for fine root biomass and vitality whereas all 10 replicates were analysed for fine root chemistry and coarse root biomass. Results are expressed per litre of soil volume.

## 3. Results and Discussion

3.1. ROOT BIOMASS

Increased NS deposition did not significantly ($p<0.05$) affect coarse, fine live, fine dead or total fine root biomass (Figure 1), neither did it have any significant effect on the total root tip number (RTN) or root tip density (RTD), defined as the ratio of live root tips (turgid+shrivelled) to live root length (Table III). Similarly, the biomass ratios of dead/live root (31-36%) and total fine root/coarse root (3-9%) were not influenced by the NS deposition treatment. Proportions of coarse, fine live and dead roots relative to total root dry weight varied between 16-32%, 53-63% and 15-20% respectively, with no significant effects of NS deposition or soil horizon. Hence, the addition of 75 kg ha$^{-1}$ a$^{-1}$ $NH_4^+$-N did not alter root biomass production in our experiment. Although, in a previous experiment, increased NS deposition had led to higher soil solution Al concentrations (Carnol et al., 1997a,b), in the range reported to cause a decrease in fine root biomass (Joslin and Wolfe, 1988). Still, the difference in soil solution Al concentrations between controls and NS treatments was relatively small (Carnol et al., 1997b). Furthermore, carbohydrate supply from the tree canopy also controls root

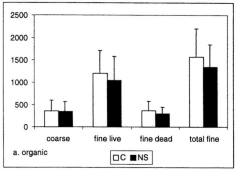 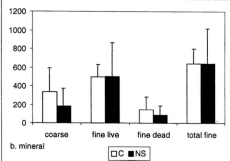

Fig. 1. Coarse, fine live, fine dead, total fine root biomass (mg $l^{-1}$ soil) grown into the organic (a) and mineral (b) soil horizon of in-growth cores (mean and sd) under control (C) and increased $(NH_4)_2SO_4$ deposition (NS). There were no significant NS deposition effects; for significant differences between soil horizons, see text.

TABLE III

Total number of root tips (RTN) and root tip density (RTD, tips $m^{-1}$) of roots grown into in-growth cores (mean and sd); C: control, NS: increased $(NH_4)_2SO_4$ deposition. Different letters denote significant (p<0.01) differences between organic and mineral horizon; there were no significant NS deposition effects.

|  | Organic | | Mineral | |
| --- | --- | --- | --- | --- |
|  | C | NS | C | NS |
| RTN | 8533 (4108)a | 7576 (3785)a | 3521 (1471)b | 3262 (2008)b |
| RTD | 325 (38)a | 331 (61)a | 284 (77)a | 287 (51)a |

biomass production (Marshall and Waring, 1985). In our study, as treated areas were only a small fraction of the whole rooting system, translocation between roots might have moderated treatment effects on root biomass. Whatever the cause, this study has revealed no toxic effects of NS additions (75 kg $ha^{-1}$ $a^{-1}$ $NH_4^+$-N) on root biomass.

Soil horizon however did exert a significant influence on root growth. The dry weight of coarse roots was significantly higher in the organic layers with a mean across both treatments of 350 mg $l^{-1}$ compared to 260 mg $l^{-1}$ in the mineral layer. Dry weights of fine live, fine dead, total fine roots and RTN were also significantly higher in the organic horizon (Figure 1, Table III). The ratio dead/live roots was unaffected. Decreasing root density downward from organic surface horizons to mineral soil horizons is in agreement with general findings from the literature (Schneider et al., 1989; Majdi and Persson, 1993, 1995). This experiment has demonstrated that when the local soil strata is reproduced in in-growth cores, roots re-establish normally by comparison with the use of uniform substrate where the vertical distribution of fine roots tends to be lost (Matzner et al., 1986).

3.2. ROOT TIP VITALITY

No roots were observed in development stage 2 (for description of stages, see Table II). The development stages 1 and 3 showed a similar distribution in both soil horizons irrespective of NS treatment (Table IV), with a mean value across horizons and treatments of 0.4% for stage 1 and 1.1% for stage 3. Development stages 4 and 5 were

TABLE IV

Proportions of turgid, shrivelled and dead root tips and root tips in the development stages 1-5 (%; mean and sd) for roots grown into the organic and mineral horizon of in-growth cores. C: control, NS: increased $(NH_4)_2SO_4$ deposition.

|  | Variable | Organic | | Mineral | |
|---|---|---|---|---|---|
|  |  | C | NS | C | NS |
| Vitality | Turgid | 46.6 (22.6) | 49.8 (13.7) | 61.8 (20.7) | 60.4 (16.6) |
|  | Shrivelled | 39.6 (20.4) | 34.4 (8.8) | 29.6 (15.9) | 26.7 (17.2) |
|  | Dead | 13.8 (10.5) | 15.8 (8.9) | 8.6 (6) | 12.9 (8.3) |
| Development | Stage 5 | 51.5 (21.3) | 54.3 (24.6) | 59.4 (28.9) | 64 (33) |
|  | Stage 4 | 46.5 (21.2) | 44.3 (24.8) | 39.4 (28.4) | 34.6 (32.8) |
|  | Stage 3 | 1.6 (1.6) | 1.1 (1) | 0.8 (0.9) | 0.9 (0.7) |
|  | Stage 1 | 0.4 (0.2) | 0.2 (0.3) | 0.4 (0.3) | 0.4 (0.2) |

TABLE V

Proportions of root tip development and vitality classes (%) on the total number of root tips (mean and sd) for roots grown into the organic and mineral horizon of in-growth cores. C: control, NS: increased $(NH_4)_2SO_4$ deposition.

| Development | Vitality | Organic | | Mineral | |
|---|---|---|---|---|---|
|  |  | C | NS | C | NS |
| 5 | turgid | 30.1 (13.7) | 33.7 (17.1) | 42.5 (29.9) | 41.7 (27.8) |
|  | shrivelled | 16.1 (21.7) | 13.5 (13.1) | 12.5 (13.7) | 16 (20.7) |
|  | dead | 5.3 (4.7) | 7.1 (5.0) | 4.4 (3.7) | 6.3 (5.6) |
| 4 | turgid | 15.5 (10.4) | 15.6 (8.2) | 18.7 (12.6) | 18.2 (19.9) |
|  | shrivelled | 22.7 (14.9) | 20.2 (10.7) | 16.6 (16.3) | 9.9 (8.5) |
|  | dead | 8.3 (8.3) | 8.5 (8.1) | 4.1 (3.5) | 6.6 (7.8) |
| 3 | turgid | 0.9 (1.2) | 0.4 (0.4) | 0.4 (0.7) | 0.4 (0.2) |
|  | shrivelled | 0.6 (0.4) | 0.5 (0.6) | 0.4 (0.4) | 0.4 (0.4) |
|  | dead | 0.1 (0.2) | 0.2 (0.1) | 0 (0) | 0.1 (0.1) |
| 1 | turgid | 0.1 (0.2) | 0.1 (0.1) | 0.1 (0.1) | 0.1 (0.1) |
|  | shrivelled | 0.2 (0.1) | 0.1 (0.1) | 0.2 (0.2) | 0.3 (0.2) |
|  | dead | 0.1 (0.1) | 0 (0.1) | 0.1 (0.1) | 0 (0.1) |

dominant and shared the remaining proportion. There were no significant differences between treatments, and mean values were 42.0% for stage 4 and 56.5% for stage 5, showing that most root tips were mycorrhizal.

Proportions of turgid, shrivelled and dead root tips were 53, 33 and 13% respectively, and there were no significant treatment effects (Table IV). However, a lower proportion of turgid root tips and a higher proportion of shrivelled root tips were apparent in the organic compared to the mineral horizon. This could indicate recent unfavourable conditions, such as summer drought, for root growth in the upper horizon, or a slower ageing processes in the mineral soil (Hendrick and Pregitzer, 1992). In the mineral layer, the proportion of dead root tips was slightly (but not significantly) higher in the NS treatment compared to control.

Individual root tip classes were dominated by dead, shrivelled or turgid root tips in vitality classes 4 and 5 (Table V) and were unaffected by horizon and deposition treatments. However the proportion of 'turgid 5' roots was higher (not significant) in the mineral layers (42%), compared to the organic layers (32%) and was compensated by lower proportions of shrivelled and dead roots of stages 4 and 5.

These results showed that root vitality, as defined by visual criteria was not influenced by the high NS deposition treatment. Wallander and Nylund (1992) likewise observed little change in the fungal biomass viability in *Pinus sylvestris* seedlings treated with high $NH_4$-N concentrations, although the extramatrical mycelium decreased. Arnebrant (1994) found that species composition of the mycorrhiza fungi changed in response to high N supply.

### 3.3. FINE ROOT CHEMICAL COMPOSITION

In the organic horizon, fine root N content was significantly increased from 21.7 to 24.0 mg g$^{-1}$ dw (Figure 2) in NS treated in-growth cores, indicating enhanced N uptake in response to the additional N. Increased root N content following NS application was also observed in spruce roots growing in the organic and top mineral layer, which was associated with a decreased Mg content (Majdi and Rosengren-Brinck, 1994). Boxman *et al.* (1988) reported cation efflux from roots at high $NH_4$ supply. However, in our study, there was no evidence of decreased Mg or Ca root contents due to increased N uptake. Instead, NS deposition led to an increase in fine root Ca contents, increasing the molar Ca/Al ratio in roots of the mineral soil horizon (Figure 2). NS deposition can increase Ca mobilisation in the upper soil horizons by exchange reactions with $NH_4^+$ or due to Ca leaching associated with the mobile $NO_3^-$ anion (Carnol *et al.*, 1997a). This may explain the increased root uptake of Ca in the mineral horizon under NS.

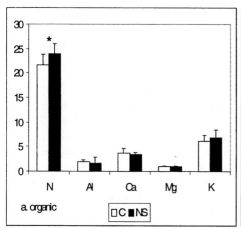
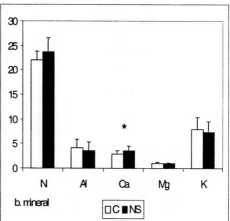

Fig. 2. Chemical content of fine live roots (mg g$^{-1}$ dw) grown into the organic (a) and mineral (b) soil horizon of in-growth cores (mean and sd) under control (C) and increased $(NH_4)_2SO_4$ deposition (NS). * Significant NS deposition effects (p<0.05); for significant soil horizon effects, see text.

Root concentrations of Al, Mg and K were not influenced by NS deposition in either soil horizon. Although soil solution Al concentrations were increased by NS (Carnol et al., 1997b), fine root Al concentrations did not increase. However, soil solution (Carnol et al., 1997b) and root data were not collected simultaneously, and

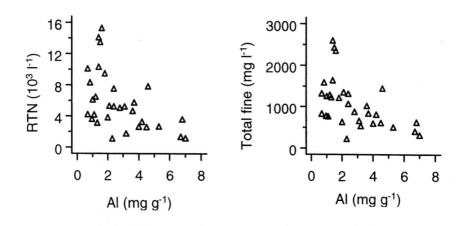

Fig. 3. Relationship between fine root Al content (mg g$^{-1}$ dw) and growth characteristics of Norway spruce roots grown into in-growth cores. RTN: total number of root tips (n=31), total fine: total biomass of fine roots (n=32). Log-linear regression: p<0.001; adj r$^2$: 0.31 (total fine), adj r$^2$: 0.31 (RTN).

year-to-year variation could explain this difference. Furthermore, rhizosphere nutrient availability plays an important role in root uptake (Majdi and Rosengren-Brinck, 1994; Rosengren-Brinck et al., 1995). These authors reported decreased rhizosphere Al concentrations and increased Ca concentrations and Ca/Al ratios following NS treatment. Rhizosphere soil chemistry may thus differ from bulk soil solution chemistry, and this could explain the lack of marked treatment effects.

Fine root Al concentrations were significantly (p<0.05) higher and fine root Ca concentrations significantly lower in the mineral soil horizon, leading to lower Ca/Al, Mg/Al molar ratios compared to the organic horizon, irrespective of treatment. Fine root K concentrations were significantly (p<0.05) higher in the mineral soil horizon (control treatment only). This vertical stratification in fine root chemistry is probably related to a gradient of availability of these elements in the soil profile. Similar increases in root Al concentration and decreases in root Ca concentration with soil horizon have been reported (Dahlgren et al., 1991; Majdi and Rosengren-Brinck, 1994). The lower Ca root concentration in the mineral horizon could be due to competition with Al for adsorption sites (Truman et al., 1986; Hecht-Buchholz et al., 1987), or a lower availability of Ca in this layer. Lower Al uptake in the organic layer has been related to a higher proportion of organically chelated Al in the soil solution (Cronan et al., 1986; Joslin and Wolfe, 1988). Higher fine root K contents in the mineral horizon may be related to a synergism in Al and K adsorption (Truman et al., 1986), or higher K availability in this horizon.

Fine root Al concentration was negatively correlated to the logarithm of fine root biomass (r$^2$: 0.34, adj r$^2$: 0.31; p<0.001) and to the logarithm of RTN (r$^2$: 0.33, adj r$^2$:

0.31; p<0.001) (Figure 3). This significant negative relationship indicated that the reduction in root biomass and RTN in the mineral layer might be due to high concentrations of plant available Al. Root biomass reductions due to Al have been reported in several controlled experiments (Joslin and Wolfe, 1988; Boxman et al., 1991; Oleksyn et al., 1996). Total Al concentrations in soil solution seem to be a good predictor of fine root biomass (Joslin and Wolfe, 1988) and, in this study, total dissolved Al concentrations rose to ca. 0.8 meq l$^{-1}$ (Carnol et al., 1997b), with Ca/Al molar ratios of as low as 0.25. Values of this magnitude have been reported to adversely affect fine root biomass and nutrient uptake (Van Praag et al., 1985; Majdi and Persson, 1993; Cronan and Grigal, 1995).

## 4. Conclusions

After one year, bi-weekly additions of $(NH_4)_2SO_4$ (at 75 kg ha$^{-1}$ a$^{-1}$ $NH_4^+$-N) to local throughfall did not lead to impaired root growth or root tip vitality, as defined by visual criteria. A clear vertical stratification of fine root growth and chemical composition was demonstrated (independent of NS treatment). Lower fine root biomass in the mineral soil horizon was correlated with higher fine root Al concentrations. When reproducing local soil horizons, the in-growth core technique provides a useful method for studying treatment effects on roots of mature trees in the field. The major advantage compared to a sequential coring technique is that measured effects are not confounded by pre-treatment conditions, as all roots sampled have been equally subjected to experimental treatments.

## Acknowledgements

This study was part of the CORE project, funded by the Commission of the European Communities (DGXII). M. Carnol acknowledges funding by a Sectoral grant under the STEP programme (CEC, DGXII) and financial support from the SSTC (Services du Premier Ministre, Services fédéraux des affaires scientifiques, techniques et culturelles) during the preparation of the manuscript. The authors thank staff of the Chemistry Section, ITE Merlewood, Grange-over-Sands, UK, for training in the digestion procedure and performing soil solution and root chemical analyses. We thank the two referees and Dr L.J. Sheppard for their constructive comments on the manuscript.

## References

Ahlström, K., Persson, H. and Börjesson, I.: 1988, *Plant Soil* **106**, 179.
Alexander, I.J. and Fairley, R.I.: 1983, *Plant Soil* **71**, 49.
Allen, S.E. (ed.): 1989, *Chemical Analysis of Ecological Materials*, Blackwell, London p. 368.
Arnebrant, K.: 1994, *Mycorrhiza* **5**, 7.
Boxman, A.W., Krabbendam, H., Bellemakers, M.J.S. and Roelofs, J.G.M.: 1991, *Environ. Pollut.* **73**, 119.
Boxman, A.W., Van Dijk, H.F.G. and Roelofs, J.G.M.: 1988, *Plant Soil* 680.
Carnol, M., Ineson, P., Anderson, J.M., Beese, F., Berg, M.P., Bolger, T., Coûteaux, M.-M., Cudlin, P., Dolan, S., Raubuch, M. and Verhoef, H.A.: 1997a, *Biogeochem.* **38**, 255.
Carnol, M., Ineson, P. and Dickinson, A.L.: 1997b, *Environ. Pollut.* **97**, 1.

Chemlikova, E., Cudlin, P. and Radosta, P.: 1992, in L. Kutschera, E. Hübl, E. Lichtenegger, H. Persson and M. Sobotik (eds.) *Root Ecology and its Practical Application*, Verlag Gumpenstein, Irding.
Clemensson-Lindell, A. and Persson, H.: 1995, *Plant Soil* **168-169**, 167.
Cronan, C.S. and Grigal, D.F.: 1995, *J. Environ. Qual.* **24**, 209.
Cronan, C.S., Walker, W.J. and Bloom, P.R.: 1986, *Nature* **324**, 14.
Cudlin, P.: in 1991, *Methods in Microbiology* Vol. 23, 365, Academic Press.
Cudlin, P. and Kropacek, K.: 1990, in H. Persson (ed.) *Above and Below-Ground Interactions in Forest Trees in Acidified Soils*, p.44, Commission of the European Communities.
Dahlgren, R.A., Vogt, K.A. and Ugolini, F.C.: 1991, *Plant Soil* **133**, 117.
Giovannetti, M. and Mosse, B.: 1980, *New Phytol.* **84**, 489.
Hecht-Buchholz, C., Jorns, C.A. and Keil, P.: 1987, *Journal of Plant Nutrition* **10**, 1103.
Hendrick, R.L. and Pregitzer, K.S.: 1992, *Ecology* **73**, 1094.
Jansen, A.E. and Dighton, J.: 1990, *Effects of Air Pollutants on Ectomycorrhizas*, p.58, Commission of the European Communities.
Jorns, A. and Hecht-Buchholz, C.: 1985, *AFZ* **46**, 1248.
Joslin, J.D. and Wolfe, M.H.: 1988, *Can. J. For. Res.* **18**, 1614.
Kocourek, R. and Bystrican, A.: 1989, *Agr. Ecosyst. Environ.* **28**, 235.
Majdi, H. and Persson, H.: 1993, *Scand. J. For. Res.* **8**, 147.
Majdi, H. and Persson, H.: 1995, *Plant Soil* **168-169**, 151.
Majdi, H. and Rosengren-Brinck, U.: 1994, *Plant Soil* **162**, 7.
Marshall, J.P. and Waring, R.H.: 1985, *Can. J. For. Res.* **15**, 791.
Matzner, E., Murach, D. and Fortmann, H.: 1986, *Water Air Soil Pollut.* **31**, 273.
Nihlgard, B., 1985, *Ambio* **14**, 2.
Oleksyn, J., Karolewski, P., Giertych, M.J., Werner, A., Tjoelker, M.G. and Reich, P.B.: 1996, *Trees* **10**, 135.
Ortloff, W. and Schlaepfer, R.: 1996, *Allg. Forst. Jagdztg.* **167**, 184.
Persson, H.: 1990, in A.F. Harrison, P. Ineson and O.W. Heal (eds.) *Nutrient Cycling in Terrestrial Ecosystems: Field Methods, Application and Interpretation*, 198, Elsevier, Essex.
Raubuch, M.: 1992, *Wechselwirkungen zwischen mikrobiologischen und chemischen Parametern in Waldböden unter Berücksichtigung des atmosphärischen Eintrags*, Göttingen, Berichte des Forschungszentrums Waldökosysteme, Reihe A, Bd. 89.
Reuss, J.O. and Johnson, D.W.: 1986, *Acid Deposition and the Acidification of Soils and Waters*, Springer-Verlag, New York, p. 119.
Rosengren-Brinck, U., Majdi, H., Asp, H. and Widell, S.: 1995, *New Phytol.* **129**, 537.
Rost-Siebert, K.: 1983, *Allg. Forst. Jagdztg.* **38**, 686.
Ruess, L., Sandbach, P., Cudlin, P., Dighton, J. and Crossley, A.: 1996, *Pedobiologia* **40**, 51.
SAS Institute Inc.: 1989, *SAS/STAT User's Guide, Version 6*, SAS Institute Inc., Cary, NC, p. 943.
Schneider, B.U., Meyer, J., Schulze, E.-D. and Zech, W.: 1989, in E.-D. Schulze, O.L. Lange and R. Oren (eds.), *Forest Decline and Air Pollution*. Springer Verlag, p. 370.
Seith, B., George, E., Marschner, H., Wallenda, T., Schaeffer, C., Einig, W., Wingler, A. and Hamp, R.: 1996, *Plant Soil* **184**, 291.
Sokal, R.R. and Rohlf, F.J.: 1981, *Biometry*, Freeman, New York.
Truman, R.A., Humphreys, F.R. and Ryan, P.J.: 1986, *Plant Soil* **96**, 109.
Ulrich, B., Pirouzpanah, D. and Murach, D.: 1984, *Forstarchiv* **55**, 127.
Van Dijk, H.F.G., De Louw, M.H.J., Roelofs, J.G.M. and Verburgh, J.J.: 1990, *Environ. Pollut.* **63**, 41.
Van Praag, H.J., Weissen, F., Sougnez-Remy, S. and Carletti, G.: 1985, *Plant Soil* **83**, 339.
Wallander, H. and Nylund, J.E.: 1992, *New Phytol.* **120**, 495.
Wheeler, D.M., Edmeades, D.C., Christie, R.A. and Gardner, R.: 1992, *Plant Soil* **146**, 61.

# FATE OF NITROGEN COMPOUNDS DEPOSITED TO SPRUCE (*PICEA ABIES* KARST.) AND PINE (*PINUS SILVESTRIS* L.) FORESTS LOCATED IN DIFFERENT AIR POLLUTION AND CLIMATIC CONDITIONS.

T. STASZEWSKI, S. GODZIK, P. KUBIESA AND J. SZDZUJ

*Institute for Ecology of Industrial Areas, Kossutha Str. 6, PL-41-832 Katowice, Poland*

(Received 30 October 1998; accepted 25 February 1999)

**Abstract.** Measurements were made of $NO_3$-N and $NH_4$-N in bulk deposition, throughfall and soil solution on six permanent plots in pine and spruce stands located along a transect from the south to the north of Poland. Location differed both in the level of air pollution level and in climatic parameters. The total N load calculated from throughfall ranged from 12.5 to 34 kg$^{-1}$a$^{-1}$. The load of $NH_4$-N exceeded the $NO_3$-N contribution. Differences in total N load were not reflected in foliar N concentration. Present forest health status of stands determined by defoliation class, and do not appear to be related to their N deposition.

**Keywords:** nitrogen load, pine and spruce stands, wet deposition, ammonium, nitrate, forest health status

## 1. Introduction

The input of nitrogen (N) to forest soil varies considerably across Europe. The load of total inorganic N in throughfall ranges from 1 to 70 kg ha$^{-1}$a$^{-1}$ (Dise *et al.*, 1998). High loads of sulphur (S) and N oversupply have been seen as one possible reason for poor forest health. In Poland, an inventory of forest health has shown no improvement up to the year 1995, despite the previous 30% reduction in sulphur dioxide ($SO_2$) emissions (Wawrzoniak and Małachowska, 1997).

The aim of this study was to determine the amount of $NO_3$-N and $NH_4$-N deposited on spruce and pine forests. Soil solution was also characterised. Tree defoliation, foliar N concentration, phosphorous (P), S, calcium (Ca), magnesium (Mg) and potassium (K) in needles have been taken as criteria for comparison between stands and locations.

## 2. Material and Methods

Investigations were carried out on six permanent plots in spruce (*Picea abies* Karst.) and pine (*Pinus silvestris* L.) stands, located at: Salmopol and Brenna (spruce), Katowice (pine), Puszczykowo - Wielkopolski National Park (pine), and Gać - Słowiński National Park (spruce and pine) (Figure 1). The south – north transect was primarily chosen as representing different air pollution and climates (National Program...1998, Atlas Hydrologiczny Polski, 1997). The first three plots in the south are exposed to air pollutants emitted in the Ostrava region and the Upper Silesia Industrial Region (USIR). The two spruce plots in Brenna and Salmopol differ in elevation above sea level by about 350 m, being located at 650 and 1000 m, respectively. These two plots differ from the Katowice pine in plot both climatic and air pollution parameters. The plot in

Puszczykowo is located close to the city of Poznań. No sources of air pollution occur in the region of Gać. The distance between plots in Brenna and Gać is about 550 km.

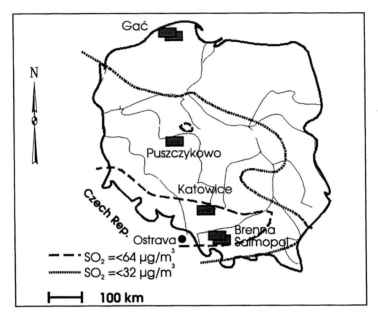

Fig. 1. Location of Sites

Each plot was equipped with 15 randomly distributed throughfall collectors and six pairs of vacuum ceramic cup lysimeters for soil solution sampling (Methods for integrated monitoring..., 1989). Lysimeters were distributed evenly at 25 and 50 cm depth. Bulk deposition was sampled in an open area, close to permanent plots. Polyethylene collectors were used for bulk deposition and throughfall sampling. Soils at locations are podzolic, except for the Gać spruce stand, which is on a very acid deep peat.

In autumn, current and one-year-old needles were sampled from each plot, from the seventh whorl of six trees from the same injury class, and analysed for S, N, K, Mg, Ca, and P. 1 g of unwashed needles, dried and finely ground was mineralised in a MDS 2000 microwave oven with nitric acid. The concentration of K, Mg, Ca was determined by the AAS method (SpectraAA 300 Varian), and P by the ICP (Varian Liberty 220). S concentration was determined (as sulphate ions) after needle combustion in the Schoeninger flask, absorption in 0.3% $H_2O_2$, by ion chromatography (Dionex DX-100). Total N was determined by the Kjeldhal method.

$NH_4^+$ in water samples was determined using the Nessler method and $NO_3^-$ by ion chromatography (Dionex DX100, column AS4A). Water pH was measured using glass electrode and the presented data are volume weighted values. Cations in water samples were determined using the AAS method, and anions by the ion chromatography (Dionex DX 100).

## 3. Results and Discussion

The annual mean pH of bulk deposition collected did not differ markedly between sites, ranging from 4.5 to 5.0. In the southern part of Poland, influenced by pollutants transported from Ostrava and the USIR, the pH of bulk deposition was lower than in other locations (Figure 2). On all plots throughfall pH 0.23 - 0.68 unit more acid than bulk depositions. The smallest decrease was found for the Puszczykowo plot, and the largest decrease was found at the Brenna plot (Figure 2). Soil solution pHs were higher than throughfall (Salmopol, Brenna and Katowice), but lower for the other locations (Puszczykowo, Gać) (Figure 2).

In comparison to 1992/93, the pattern of pH values for both - bulk deposition and throughfall have changed. A 0.5 pH unit increase was found for the bulk deposition at all but one (Brenna), location. In comparison to present data, no pH differences for bulk throughfall were found at the Katowice and Puszczykowo locations in the earlier period (Godzik et al., 1995). These differences most likely reflect improvement of air quality, reduction in emission of $SO_2$ and particulate matter, which took place during this period.

Loads of total N calculated from throughfall measurements significantly exceeded those calculated for bulk deposition data (Figure 3). N loads in spruce stands in Brenna and Salmopol confirm that air quality up to 1200 m does not differ markedly. Lower N loads in the pine stand in Katowice, despite being close to major air pollution sources of USIR, may be explained by about 50% lower rainfall than the Brenna and Salmopol locations. Some differences may be caused by the species specific needle characteristics and tree crown characteristics, e.g. needle surface area (LAI) between these two species. Results from the Gać plots of pine and spruce which are < 200 m from each other, support this explanation (Figure 3). N loads at all sites exceed the critical loads values calculated for these vegetation types (Figure 3) (Mill et al., 1994). However, the N load at all sites was 10% lower than in 1994 (Godzik et al., 1995), and the annual S load over 30% lower (55 compared to 35 kg/ha) on plots in the south of the country: Brenna, Salmopol and Katowice (Staszewski, 1997).

Except for the Salmopol and Brenna locations, where the loads of $NO_3^-$ and $NH_4^+$ are similar, in other locations the $NH_4^+$ ions make up the larger fraction of the bulk and throughfall inputs. These differences were particularly pronounced in pine stands of Puszczykowo and Katowice, and to a lesser extent Gać (Figure 3).

$NO_3^-$ concentrations in soil solution differed between spruce and pine plots (Figure 4). Those in pine stands were below the detection limit of the method used. Such unexpected results may reflect $NO_3^-$ - N uptake by the richer ground flora vegetation found in pine stand compared to spruce stands. Similar results were obtained on unmanaged soils at tree locations in USIR, under dense cover of grass species (Godzik et al. 1997a). The highest concentrations of both N forms were observed in the deep acid peat soil of the spruce stand in Gać. The comparison between $NH_4^+$ concentrations and other nutrients in bulk precipitation and throughfall testifies to their exchange in tree canopies (Table I), and reflects $NH_4^+$ uptake from $(NH_4)_2SO_4$ deposited on the needle surfaces (Roelofs et al., 1985). $K^+$ and $Mg^{2+}$ ions seem to be intensively leached from the needles; this process is the least apparent in the Puszczykowo site. $Ca^{2+}$ leaching was only seen in pine and spruce stands in Gać (Table I).

The comparison of the $NO_3/Cl$ ratios for throughfall and soil solution showed higher values in the former at all sites indicating a low or negligible nitrification rate (Houdijk et al., 1993) (Table II). The values 5 and 10 for $NH_4/Mg$ and $NH_4/K$ ratios, respectively, regarded as critical for forest soil (Houdijk et al., 1993) were slightly exceeded in the Gać and Puszczykowo pine stands only (Table I). The molar ratios of Al/Ca in soil solution were all relatively small when compared to the value of 5, regarded as critical for root damage (Ulrich, 1983) (Table III).

TABLE I

The ratios of ammonium to nutrients of various compartments of ecosystem

| Site | Type of water | $NH_4/Mg$ | $NH_4/K$ | $NH_4/Ca$ |
|---|---|---|---|---|
| Salmopol | bulk | 26.0 | 4.7 | 1.0 |
| Spruce | throughfall | 7.8 | 2.6 | 0.9 |
|  | soil solution 25 cm | 3.8 | 2.1 | 1.2 |
|  | soil solution 50 cm | 0.2 | 0/6 | 2.2 |
| Brenna | bulk | 27.2 | 3.5 | 0.8 |
| Spruce | throughfall | 6.2 | 2.2 | 0.7 |
|  | soil solution 25 cm | 2.9 | 2.2 | 1.0 |
|  | soil solution 50 cm | 3.2 | 3.1 | 1.3 |
| Katowice | bulk | 11.1 | 11.8 | 0.7 |
| Pine | throughfall | 6.6 | 5.4 | 1.3 |
|  | soil solution 25 cm | 1.7 | 1.6 | 0.3 |
|  | soil solution 50 cm | 2.1 | 2.4 | 0.4 |
| Puszczykowo | bulk | 23.7 | 10.6 | 4.5 |
| Pine | throughfall | 25.4 | 7.6 | 5.4 |
|  | soil solution 25 cm | 7.7 | 7.7 | 1.4 |
|  | soil solution 50 cm | 4.9 | 4.1 | 0.9 |
| Gać | bulk | 15.3 | 10.8 | 4.0 |
| pine | throughfall | 5.5 | 3.0 | 2.3 |
|  | soil solution 25 cm | 3.0 | 2.9 | 0.9 |
|  | soil solution 50 cm | 10.5 | 13.1 | 5.1 |
| Gać | bulk | 15.3 | 10.7 | 4.0 |
| spruce | throughfall | 2.7 | 2.8 | 1.2 |
|  | soil solution 25 cm | 1.7 | 1.5 | 0.4 |
|  | soil solution 50 cm | 2.2 | 1.9 | 0.5 |

TABLE II
The comparison of $NO_3/Cl$ concentration in throughfall and soil solution

| Site | Throughfall | Soil solution 25 cm | Soil solution 50 cm |
|---|---|---|---|
| Salmopol spruce | 3.01 | 0.81 | 0.82 |
| Brenna spruce | 3.39 | 1.05 | 1.11 |
| Katowice pine | 0.98 | <0.009 | 0.02 |
| Puszczykowo pine | 1.84 | <0.009 | <0.009 |
| Gać pine | 0.69 | 0.009 | <0.009 |
| Gać spruce | 0.82 | 0.56 | 0.54 |

Nutrients, N concentrations and their ratios in needles were compared with threshold values (Arndt et al., 1987; State Forest Service, 1990). The N concentration ranged from

# FATE OF NITROGEN COMPOUNDS DEPOSITED TO SPRUCE 125

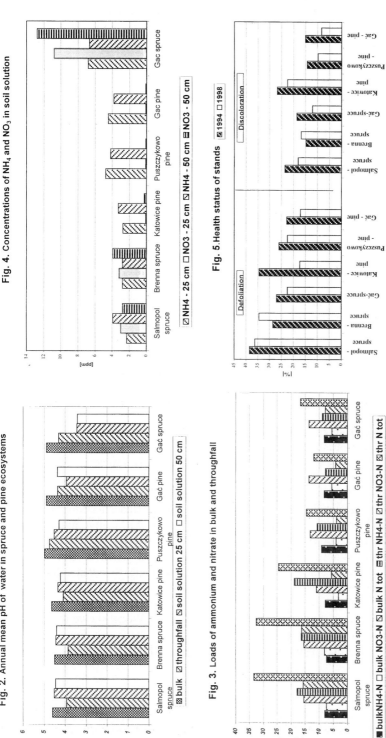

Fig. 2. Annual mean pH of water in spruce and pine ecosystems

Fig. 3. Loads of ammonium and nitrate in bulk and throughfall

Fig. 4. Concentrations of NH$_4$ and NO$_3$ in soil solution

Fig. 5. Health status of stands

0.8 to 1.3% and may be regarded as low. N concentrations in needles do not correlate with the N loads reaching these ecosystems. These values together with those for S are about 20% below those measured in 1995 (Staszewski *et al.*, 1997, Staszewski *et al.* 1998 ). S concentrations ranged from 0.7 to 1.4%. The P concentrations in pine needles were normal, but in spruce, a deficit of P in one-year-old needles was observed (Staszewski *et al.*, 1997; Staszewski *et al.*, 1998 ).

TABLE III
The Al/Ca molar ratios in soil solutions

| Site | Soil solution 25 cm | Soil solution 50 cm |
|---|---|---|
| Salmopol spruce | 0.21 | 0.32 |
| Brenna spruce | o.25 | 0.30 |
| Katowice pine | 0.29 | 0.30 |
| Puszczykowo pine | 0.50 | 0.50 |
| Gać pine | 0.49 | 1.90 |
| Gać spruce | 0.06 | 0.08 |

Cation concentrations had increased in needles by comparison with 1995 values (Staszewski *et al.*, 1997; Staszewski *et al.*, 1998). Ca and K contents ranging from 0.18 to 0.75% and from 0.35 to 0.59% were normal (Staszewski *et al.*, 1997; Staszewski *et al.*, 1998). In 1995 a Ca deficit in needles was observed. In 1997 a deficit of Mg was found in spruce needles from Salmopol and Brenna (0.03 - 0.06%). These sites are characterised by high rainfall (long term mean: 1300 mm), and low soil Mg (Godzik and Szdzuj, 1994).

The load of acidifying compounds was highest at those two sites (Vogt *et al.*, 1994). The medium N concentration in needles was reflected in normal N nutrients ratios. Only for one-year-old spruce needles did the ratios indicate Mg/N deficiency (Staszewski *et al.*, 1997; Staszewski *et al.*, 1998). Despite a reduction in S load, between the years 1992 – 1997, by about 40 %, S remains the most important acidifying element.

These data are reflected in observations of the health of stands based on the level of defoliation and discoloration. Most recent assessments show a relative improvement of health status at all but one (Brenna) stand (Figure 5). The reason for the health status of the stand in Brenna is not clear. Earlier observations have shown insect damage to trees at this site. However, the health status of this stand is still better than that at Salmopol, at a higher elevation. Measurements of air pollutant concentrations at Brenna showed an exceedance of the $SO_2$ critical levels, and AOT40 for ozone (Godzik *et al.* 1997b). The $SO_2$ critical level values are also exceeded at Katowice and Puszczykowo (Staszewski *et al.* 1998). The largest improvement of the forest health status was found in the plot located in Katowice - the Upper Silesia Industrial Region (Figure 5).

## 4. Conclusions

1. The annual N load (total) reaching forest soil ranges from 12.5 kg/ha (sea-side) to 34

kg/ha (mountainous area). The contribution of $NH_4$-N exceeded that for $NO_3$-N on all plots, especially in pine stands.
2. Differences in N load were not reflected in foliar N concentrations. N ratios to other nutrients are in the range described as normal.
3. Forest health status has improved over previous assessments on most plots.
4. The data suggest that the N load is not the main factor adversely affecting forests in Poland. Other factors, e.g. climatic conditions, prevailing high atmospheric concentrations of $SO_2$, and ozone, should be taken into consideration.

## References

Arndt, U., Nobel, W. and Schweizer, B. 1987, *Bioindikatoren. Moglichkeiten, Grenzen und neue Erkenntnisse.* Verlag E. Ulmer. Stuttgart

Atlas Hydrologiczny Polski, IMGW, WG. Warszawa, 1997.

Dise, N.B., Matzner, E. and Gundersen, P.: 1998, *Water, Air, Soil Pollut.* **105**, 143-154

Godzik, S., Kubiesa, P., Staszewski, T. and Szdzuj, J. 1997a. Proc. $4^{th}$ *International Symposium and Exhibition on Environmental Contamination in Central and Eastern Europe*, pp. 32-34.

Godzik, S., Kubiesa, P., Staszewski, T., and Szdzuj, J.: 1997b. *The Beskids Bulletin* **9**. 20-21

Godzik, S., Staszewski, T. and Szdzuj, J.: 1995, *Acid rain research: Do we have enough answers?* Eds. Heij, G.J., Erisman J. W. Elsevier, Amsterdam.

Godzik, S. and Szdzuj, J. Climate and atmospheric deposition studies in forest. *Proc. Intern. Conf.* 1994. Nieborów..

Houdijk, A.L.F.M., Smolders, A.J.P. and Roelofs, J.G.M.: 1993, *Environmental Pollut.* 80,

Methods for integrated monitoring in the Nordic countries.: 1989. *The working group for Environmental monitoring, Nordic Council of Ministers*

Mill, W., Wójcik,A and Rzychoń, D.: 1994, PIOŚ and IETU Report. *Biblioteka Monitoringu Środowiska.* Warszawa.

National Program of Natural Environment Protection till the year 2000. 1998. Draft. Ministry of Environmental Protection and Natural Resources. Warsaw.

Roelofs, J.G.M., Kempers, A.J., Houdijk, A.L.F.M. and Jansen, J.: 1985, *Plant Soil* **84**, 45-46

Staszewski, T., Łukasik, W., Godzik, S., Szdzuj, J., and Uziębło, A.K.: 1998, *Chemosphere*, **36**, 901-905.

Staszewski T.: 1997, *IETU Report (in Polish)*

*State Forest Service: 1990.* Utrecht, The Netherlands, Report no. 1990-11, 1-28

Ulrich, B.: 1983, *Effects of accumulation of air pollutants in forest ecosystems.* Eds. B. Ulrich & J. Pankrath, 127-146. Reidel Publ. Comp. Utrecht

Vogt, R.D., Godzik, S., Kotowski, M., Niklińska, M., Pawłowski, L., Seip, H.M., Sienkiewicz, J., Skotte, G., Staszewski, T., Szarek, G., Tyszka, J., and Aagard, P., 1994. *J. Ecol. Chem.*, **3**, 157-168

Wawrzoniak, J. and Małachowska, J.: 1997, *IBL Report.* Warszawa.

# SIMULATION OF INCREASED NITROGEN DEPOSITION TO A MONTANE FOREST ECOSYSTEM: PARTITIONING OF THE ADDED $^{15}$N

P. SCHLEPPI [1], I. BUCHER-WALLIN [2], R. SIEGWOLF [2], M. SAURER [2], N. MULLER [1] and J.B. BUCHER [1]

[1] *Swiss Federal Institute for Forest, Snow and Landscape Research, CH-8903 Birmensdorf,*
[2] *Paul Scherrer Institute, CH-5232 Villigen, Switzerland*

(Received 25 September 1998; accepted 25 February 1999)

**Abstract.** Nitrogen (N) was added over two years to a spruce-dominated (*Picea abies*) montane forest at Alptal, central Switzerland. A solution of ammonium nitrate ($NH_4NO_2$) was frequently sprinkled on the forest floor (1500 m$^2$) to simulate an additional input of 30 kg N ha$^{-1}$ yr$^{-1}$ over the ambient 12 kg bulk inorganic N deposition. The added nitrogen was labelled with $^{15}NH_4^{15}NO_3$ during the first year. Results are compared to a control plot.
Neither the trees nor the ground vegetation showed any increase in their N content. Only 4.1% of N in the ground vegetation came from the N addition. Current-year needles contained 11 mg N g$^{-1}$ dry weight, of which only 2% was from labelled N; older needles had approximately half as much $^{15}$N. The uptake from the treatment was therefore very small. Redistribution of N also took place in the trunks: 1 to 2-year-old wood contained 0.7% labelled N, tree rings dating back 3 to 14 years contained 0.4%.
Altogether, the above-ground vegetation took up 12% of the labelled N. Most $^{15}$N was recovered in the soil: 13% in litter and roots, 63% in the sieved soil. Nitrate leaching accounted for 10%. Factors thought to be influencing N uptake are discussed in relation to plant use of N and soil conditions.

**Keywords:** nitrogen deposition, montane forest, *Picea abies*, N-15 isotope, nitrate leaching

## 1. Introduction

Deposition of ammonium ($NH_4^+$) and nitrate ($NO_3^-$) in natural or semi-natural ecosystems has increased as a consequence of the emission of inorganic nitrogen (N) by human activities (Galloway, 1995). This supplementary N causes unintentional eutrophication of normally N-limited systems, like most forests in temperate climates. Effects of such emissions on the N cycle of coniferous stands can be studied by experiments which manipulate the deposition rates (e.g. NITREX project: Wright & Rasmussen, 1998). Effects on the vegetation, soil and water depend on the partitioning of deposited N between the different ecosystem compartments. This question has been addressed by supplementing ambient or manipulated throughfall with $^{15}$N (Buchmann *et al.*, 1996; Tietema *et al.*, 1998; Nadelhoffer *et al.*, 1998 a, b). However, these experiments were conducted on podzol-like soils at altitudes below 700 m. Along the northern range of the Alps, coniferous forests often grow on gleyic soils developed on Flysch parent material. This paper reports on effects of simulated increased N deposition to such a montane forest.

## 2. Material and Methods

2.1. SITE AND TREATMENTS

The experimental forest site is located in the valley of Alptal, central Switzerland, at an altitude of 1200 m. The soils are umbric Gleysols with a west aspect and a slope of 20%. The climate is cool and wet: 6°C average temperature and 2300 mm precipitation $yr^{-1}$. The bulk deposition of inorganic N is 12 kg $ha^{-1}$ $yr^{-1}$ equally divided between $NO_3^-$ and $NH_4^+$ (Schleppi et al., 1998 b).

The tree species are predominantly Norway spruce (*Picea abies* (L.) Karst) with 15% silver fir (*Abies alba* Miller). The canopy is relatively open; the leaf area index is 3.8 and the basal area 41 $m^2$ $ha^{-1}$, despite the age of the trees (up to 250 years).

The vegetation and humus types of the Gleysols are closely linked to the microtopography and therefore to the water regime of the site (Schleppi et al., 1998 b). Three different plant associations can be found (Muller, 1997): *Vaccinium* spp. (L.) shrubs grow on acidified mounds, along with the trees; the nutrient richer, wet depressions carry a graminoid community (in canopy gaps) or a *Caltha palustris* (L.) and *Petasites alba* ((L.) Gaertner) community (in the shade).

The experimental plots (approximately 1500 $m^2$ each) were two small catchments from which water runoff was monitored (Schleppi et al., 1998 b). Forty rotating sprinklers (1.5 m above ground) irrigated each plot, one with a $NH_4NO_3$ solution and the other (control) with water (Schleppi et al., 1998 a). The treatment was applied during precipitation events using water collected on a polyethylene sheet (300 $m^2$) spread outside the forest. These frequent small additions (approximately 200 per year) simulated a deposition increase of 30 kg. N $ha^{-1}$ $yr^{-1}$. During the winter, the automatic irrigation was replaced by the occasional application of a concentrated $NH_4NO_3$ solution on the snow with a backpack-sprayer.

During the first year of treatment, between April 1995 and March 1996, the $NH_4NO_3$ was enriched with 1400 ‰ $\delta$ $^{15}NH_4^{15}NO_3$. N isotopes were analysed with an elemental analyser (Carlo Erba, Fisons, Italy) coupled to an isotope-ratio mass-spectrometer (delta S, Finnigan, Germany). The $^{15}N$ enrichments were calculated as the $\delta$ $^{15}N$ values of the treated plot minus those of the control.

2.2. VEGETATION

Twigs were taken from the 7th whorl of five dominant trees per plot in December 1996. They were partitioned into the most recent five age classes. After drying, needles and twigs were separated and ground. The biomass of needles, twigs and branches was estimated using the measured leaf area index and allometric relations verified with a completely dissected branch from the middle of the crown. The $^{15}N$ content of needles older than 5 years was estimated by extrapolation while an interpolation between twigs and trunk was made for the branches.

Wood cores were collected in the spring of 1997 from 8 and 5 spruce trees in the control and treated plots respectively. The following annual rings were separated: $1^{st}+2^{nd}$ (formed during and after the labelling), $3^{rd}$-$6^{th}$, $7^{th}$-$10^{th}$ and $11^{th}$-$14^{th}$. In total, these samples represented about half of the sapwood. Analyses of total N and $^{15}N$ were performed on dried, ground samples. Tree heights and diameters, bark weights, ring widths and core densities were used to estimate bark and wood biomass.

The 10 most frequent species from the ground vegetation were sampled in August 1995 and 1996 for N and $^{15}$N analyses of their aboveground tissues. Weights and soil coverage of five individuals were used to estimate the biomass of herbs (Muller, 1997). Graminoids were harvested from five small quadrats (10 x 10 cm).

2.3. SOIL AND WATER

Soil cores (30 cm deep) were taken in the spring of 1997 on an 8 x 8 m grid. They were separated into their horizons, dried, ground in a mortar and sieved. Organic material > 2 mm was considered as the fraction "litter and roots". Prior to analysis, samples were proportionally bulked within each of four soil types defined according to their morphology and humus form.

Runoff water was sampled proportionally (1 aliquot every 200 l). Weekly bulked samples were analysed for $NO_3^-$ and $NH_4^+$ by ion chromatography and by flow injection, respectively. $^{15}$N was analysed according to Downs et al. (1998): after concentration of quarterly pooled samples over exchange resins, anions and cations were eluted separately and $NO_3^-$ was reduced to $NH_4^+$ with Devarda's alloy. $NH_4^+$ was converted to $NH_3$ and captured in fibreglass filters enclosed in teflon membranes. The filters were analysed by mass spectrometry.

## 3. Results and discussion

3.1. VEGETATION

The $NH_4NO_3$ treatment induced no detectable increase in the N content of the spruce trees or ground vegetation (Figure 1; see also Schleppi et al., 1999). The N content of the needles remained slightly deficient. Twigs and needles were similar in their $^{15}$N content: 2% of the N was labelled in the current-year tissues, while this specific labelling amounted to 1.1% in older twigs and needles. Because of this evident mobility, the applied N was not restricted to growing tissues but also diluted into the larger biomass already present before treatment.

Bark and wood contained on average 0.6% labelled N. $^{15}$N enrichment in the bark was higher than in the wood formed during or after labelling (rings 1+2, Figure 1c). Older wood also contained an appreciable amount of labelled N. Nadelhoffer et al. (1998 b) found a comparable incorporation of $^{15}$N into 3 to 5-year-old wood of mature conifers. Our analysis of even older wood shows that this migration includes, without any significant decline, at least 14 annual rings. A large part of the pre-existing sapwood, if not all, is thus homogeneously labelled.

In the ground vegetation, *Poa trivialis* had a high $^{15}$N enrichment during labelling, but much less a year later (Figure 1d). By contrast, *Vaccinium* plants took up almost no $^{15}$N in 1995 but in 1996 both new and older shoots were labelled. Other species had an intermediate behaviour between *P. trivialis* and *Vaccinium* spp. With 4.1%, the specific labelling of N was higher in the ground vegetation than in the trees.

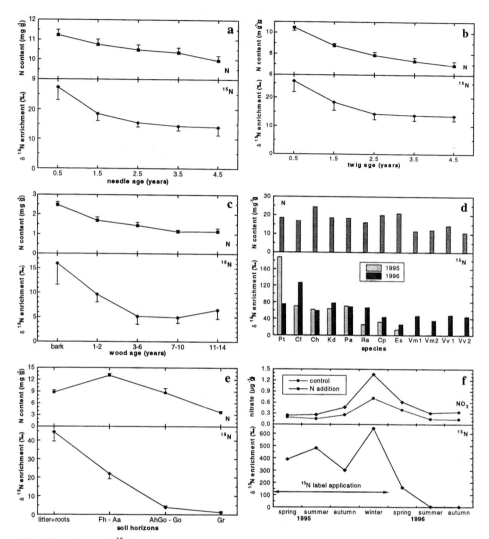

Fig. 1. N contents and $^{15}$N enrichments of different ecosystem pools and fluxes after the addition of 30 kg N ha$^{-1}$ as NH$_4$NO$_3$ with a δ $^{15}$N value of 1400 ‰. a: *Picea abies* needles, b: twigs, c: trunks, d: ground vegetation, e: soil, f: nitrate in runoff water. Bars denote standard errors where available. Species abbreviations: Pt = *Poa trivialis*, Cf = *Carex ferruginea*, Ch = *Chaerophyllum hirsutum*, Kd = *Knautia dipsacifolia*, Pa = *Petasites alba*, Ra = *Ranunculus aconitifolius*, Cp = *Caltha palustris*, Es = *Equisetum sylvaticum*, Vm = *Vaccinium myrtillus*, Vv = *V. vitis-ideae* (1 = current-year, 2 = previous-years tissues)

## 3.2. Soil and water

There were no important differences between soil subtypes, so their results were pooled (Figure 1e). The strongest labelling was found in the "litter and roots" fraction. In the sieved soil, the total N content and especially the $^{15}$N labelling decreased markedly down the soil profile. This corresponds to chemical analyses which showed that the addition did not increase the NH$_4^+$ or NO$_3^-$ concentrations below the 10 cm depth (Hagedorn *et al.*, 1997). Our findings thus contrast with the downward migration of $^{15}$N from $^{15}NH_4$ and especially from $^{15}NO_3$ reported by Buchmann et al. (1996), but this is probably due to the low permeability of the gley horizon at Alptal.

Ammonium in runoff was enriched by about 300 ‰ δ $^{15}N$, but was found only in negligible amounts. Nitrate leaching from the treated plot was almost double that of the control (Figure 1f), whereas it had been equal during the pre-treatment year (Schleppi et al., 1998 b). Highest rates of leaching were observed during the winter of 1996, as snowmelt generated high runoffs together with high concentrations. The $^{15}N$ labelling of $NO_3^-$ was high throughout the year of label application and then declined to a few δ ‰ within 3 months (spring to summer 1996). During the year of tracer application, the amount of labelled N leached as $NO_3^-$ was 2.8 kg. N ha$^{-1}$, close to the increase measured by chemical analyses (+ 3.2 kg N ha$^{-1}$). This indicates that most of the additional $NO_3^-$ leaching was coming directly from the treatment, without isotopic dilution due to exchange processes in the soil. This by-passing is in good agreement with previous conclusions (Schleppi et al., 1998 b) that $NO_3^-$ leaching from a Gleysol, as present at Alptal, is at least partly hydrologically driven (preferential flow through soil macropores) and therefore not sufficient proof that the ecosystem is N-saturated.

### 3.3. $^{15}N$ BUDGET

As measured by the acetylene inhibition method, denitrification increased from 1.7 to 2.9 kg N ha$^{-1}$ yr$^{-1}$ (Mohn, 1999), corresponding to 4% of the N addition not recovered as $^{15}N$. On the other hand, there may be an overestimation of the recovery because the soil samples were taken at the end of winter and therefore included some litter from the already sampled ground vegetation. Considering these possible sources of error and all measurement uncertainties (Table I), the total recovery rate was close to 100%.

Most $^{15}N$ (63%) was retained in the soil, especially in its upper horizon. All pools had a relative labelling within the range observed in comparable European (Buchmann et al., 1996; Tietema et al., 1998) and North-American (Nadelhoffer et al., 1998 a, b) studies on coniferous stands. In our experiment, however, the ratios between vegetation and soil pools (especially trees / upper soil) were unusually low.

TABLE I

Partitioning of the added, $^{15}N$ labelled nitrogen; pool sizes as dry matter, runoff flux as water.

| pool | pool size (kg m$^{-2}$) | N content (g kg$^{-1}$) | N pool (g m$^{-2}$) | $^{15}N$ labelling (%) | |
|---|---|---|---|---|---|
| | | | | specific[1] | relative[2] |
| needles | 1.30 | 8.8 | 11.4 | 1.2 | 5 ± 2 |
| sapwood + bark | 13.7 | 1.4 | 19 | 0.6 | 3 ± 1 |
| ground vegetation | 0.20 | 18.1 | 3.6 | 4.1 | 5 ± 2 |
| litter + roots | 2.03 | 7.6 | 15.4 | 2.5 | 13 ± 2 |
| upper soil | 6.23 | 13.7 | 83 | 1.5 | 42 ± 9 |
| lower soil | 80.1 | 4.9 | 392 | 0.16 | 21 ± 5 |
| runoff | 3600 | 0.0005 | 1.6 | 18 | 10 ± 1 |
| total recovery | | | | | 99 ± 13 |

[1] labelled N / total N content of the given pool or flux
[2] labelled N in the pool or flux / added N amount, ± standard error (based on data for the soil pools; conservative estimates by rules of error propagation for the vegetation and runoff)

Several factors probably explain the rather small N-uptake by the vegetation. All plants analysed were perennial species and therefore rely partly on tissue reserves for their N nutrition. Internal recycling was especially evident for the woody plants (trees, *Vaccinium* shrubs). Ecological factors other than N are probably also limiting plant and root growth at Alptal, especially P and anaerobic soil conditions (Schleppi *et al.*, 1998 b). These factors may hinder the uptake of N, or at least delay it as already observed on trees resampled the second year after labelling (Tietema *et al.*, 1998). In comparison with other $^{15}$N studies, the altitude of our site also reduces the vegetation period, hence the time of active nutrient uptake. Further, the availability of N in the soil solution was increased only in the upper few centimeters of the solum (Hagedorn *et al.*, 1997). As already stated by these authors, this can be explained by (1) a partial denitrification and leaching of the added $NO_3^-$ N and (2) an effective fixation of the added $NH_4^+$ N by the clay-rich soil.

## 4. Conclusion

N added as $NH_4NO_3$ to the throughfall was mainly retained in the soil of the Alptal forest. The amount taken up by the vegetation was relatively low, but could be explained by site characteristics. The ecosystem was leaking part of the deposited N as nitrate in runoff water. Even if this apparent N saturation was probably hydrologically driven, it represents a long-term threat for the water quality which should be considered in calculations and discussions of critical loads of N.

## Acknowledgement

This study, as part of the European project NITREX, was partly financed by the Swiss Federal Office of Science and Education, Bern. It was run in collaboration with the Institute of Terrestrial Ecology of the ETH Zurich (Prof. Flühler). Technical assistance and English corrections were provided by D. Tarjan.

## References

Buchmann, N., Gebauer, G. and Schulze, E. D.: 1996, *Biogeochem.* **33**, 1-23.
Downs, M. R., Michener, R. H., Fry, B. and Nadelhoffer, K. J.: 1999, *Environ. Monit. Assess.*, in press.
Galloway J. N.: 1995, *Water Air Soil Pollut.* **85**, 15-24.
Hagedorn, F., Schleppi, P. and Bucher, J. B.: 1997, *Mitt. Dtsch. Bodenkundl. Ges.* **85**, 241-244.
Mohn, J.: 1999, Thesis, Univ. of Zurich.
Muller, N.: 1997, Thesis 12388, ETH Zurich.
Nadelhoffer, K. J., Downs, M. R., Fry, B., Magill, A. and Aber, J. D.: 1999 a, *Environ. Monit. Assess.*, in press.
Nadelhoffer, K. J., Downs, M. R., Fry, B.: 1999 b, *Ecol. Appl.* **9**, 72-86.
Schleppi, P., Muller, N., Bucher, J. B.: 1998 a, *J. For. Suisse* **149**, 1-15.
Schleppi, P., Muller, N., Feyen, H., Papritz, A., Bucher, J. B., Flühler H.: 1998 b, *For. Ecol. Manage.* **101**, 177-185.
Schleppi, P., Muller, N., Edwards, P. J., Bucher, J. B.: 1999, *Phyton*, in press.
Tietema, A., Emmett, B. A., Gundersen, P., Kjønaas, O. J., Koopmans, C. J.: 1998, *For. Ecol. Manage.* **101**, 19-27.
Wright, R. F., Rasmussen, L.: 1998, *For. Ecol. Manage.* **101**, 1-7.

# SOME EFFECTS OF N ON ECTOMYCORRHIZAL DIVERSITY OF SCOTS PINE (*PINUS SYLVESTRIS* L.) IN NORTHEASTERN GERMANY

J. WÖLLECKE[1], B. MÜNZENBERGER[2], R. F. HÜTTL[1]

1) BTU Cottbus, Chair of Soil Protection and Recultivation, Box 101344, D-03013 Cottbus, Germany. 2) Centre for Agricultural Landscape and Land Use Research Müncheberg, Institute of Microbial Ecology and Soil Biology, Dr.-Zinn-Weg 18, D-16225 Eberswalde, Germany.

(Received 25 September 1998; accepted 25 February 1999)

**Abstract.** In the last three decades high industrial nitrogen (N) emissions have led to eutrophication of a Scots pine stand (*Pinus sylvestris* [L.]) near Schwedt characterized by a broad cover of *Calamagrostis epigejos* [L.] Roth (Poaceae). In comparison to the relatively unimpacted control site (low N site), this high N site showed a remarkably low mycorrhizal frequency (percentage of mycorrhizas on total amount of root tips) with seasonal lows down to 27 %. At the low N site the highest number of mycorrhizal root tips was found in the organic layer. At the high N site the amount of mycorrhizas per soil volume was similar in both organic and mineral soil layers, and also significantly lower when compared to the amount at the low N site. The high N site revealed only nine mycorrhizal morphotypes instead of eighteen found at each sampling date at the control site. 80 % of the coenosis at the high N site were represented by only four morphotypes resulting in a low diversity. The seasonal decrease in the mycorrhizal frequency, the small amount of mycorrhizas and the low diversity suggest that the high N deposition at this site has reduced the ability of the pine trees to withstand natural stresses such as prolonged drought or frost periods. This corresponds well with the 42 % reduction in tree stocking density at the high N site compared to the low N site.

Keywords: mycorrhiza, *Pinus sylvestris*, nitrogen, diversity, mycorrhizal frequency.

## 1. Introduction

During evolution nitrogen (N) was always limiting in terrestrial ecosystems (Raven and Yin 1998). Therefore, forest communities are adapted to low N availability (Ellenberg 1986). Since the beginning of industialization levels of atmospheric N have been increasing. Over the last two decades two Scots pine stands in northeastern Germany have been exposed to different N-deposition regimes (low and high N input). At the high N input site near Schwedt the high N load comes from intensive livestock emissions, fertilizer application and factories producing fertilizer over a longer time period. In the last decade a change in the aboveground vegetation from a *Rubo-Avenello-Cultopinetum sylvestris* to a *Calamagrostio-Cultopinetum sylvestris* could be observed at this site (Hofmann 1996, 1997). The low N site near Eberswalde remains a *Rubo-Avenello-Cultopinetum sylvestris*. As ectomycorrhizal morphotypes are sensitive to altered ecological parameters such as soil chemistry and moisture we investigated the mycorrhizal coenosis (i. e. composition and abundance of morphotypes) as part of the "Forest Ecosystem Research Project Eberswalde". Additionally, the mycorrhizal frequency and the projected surface of mycorrhizas (considering the number of vital mycorrhizal tips and their extent) were quantified.

## 2. Materials and methods

*Study sites.* Both sites represent *Pinus sylvestris* forests located in the continental lowlands of eastern Germany. They were comparable with respect to soil conditions, stocking rate and forest floor vegetation, before one site was exposed to higher N deposition. The low N site (control site) is part of the "Biosphere Reserve Schorfheide - Chorin". The high N site is located directly in the neighbourhood of a large industrial area (PCK Schwedt), 40 km from the control site. Further characteristics of the investigation sites are summarized in Table I.

TABLE I
Characteristics of the investigation sites

| | low N site | high N site |
|---|---|---|
| area status | Biosphere reserve | industrial area - Schwedt |
| association[1] | *Rubo-Avenello-Cultopinetum sylvestris* | *Calamagrostio Cultopinetum sylvestris* |
| soil type | Dystic Cambisol | Dystic Cambisol |
| humus form | raw humus like moder | raw humus like moder |
| pH-($H_2O$) | 4.2 | 4.6 |
| stock rate [trees $ha^{-1}$][2] | 673 | 395 |
| stand age [a][2] | 73 | 76 |
| annual precipitation [mm][2] | 507 - 683 | 487 - 630 |
| bulk-deposition [kg $ha^{-1}$ $a^{-1}$] 1985-1988: N[3] | 10-20 | >35 |
| 1996:[4]    $NO_3$-N | 3.6 | 4.5 |
| $NH_4$-N | 4 | 9 |
| $SO_4$-S | 4 | 8 |
| atmospheric $NH_3$ [$\mu g\ m^{-3}$][4] | 0.5 | 10.3 |
| soil solution $NH_4$-N [mg $l^{-1}$][5] | 0.16 | 0.93 |
| $NO_3$-N [mg $l^{-1}$][5] | 1.28 | 17.95 |
| net-N-mineralisation[6] [mg N $kg^{-1}$ soil d] | 6 | 10 |
| net-nitrification[6] | 0.2 | 1.3 |

1) determination following Hofmann (1997)
2) Müller J. - BFH - Inst. f. Forstökologie u. Walderfassung (pers. comm.)
3) Simon and Westendorff (1991)
4) mean value of 1996-1998 - Zimmerling R. - FAL - Inst. f. Agrarökologie (pers. comm.)
5) Steiner et al. (1998)
6) Papen H. - IFU - Fraunhofer-Inst. f. Atmosphärische Umweltforschung (pers. comm.)

*Root samples.* Twelve randomly distributed soil cores (8 cm int. diam.) were removed from each site, four times a year. These samples were divided into the organic and the mineral soil layer (0 - 10 cm). Mycorrhizas were sorted and counted in water under a stereomicroscope. Characterization of mycorrhizas followed Agerer and Rambold (1996). The projection area of the mycorrhizas (vertical topview on mycorrhizas) was determined using image analysis (Olympus CUE 3). Measurements were made on the mycorrhizas stained in toluidine blue. The mycorrhizal projection areas for the two sites were compared using the *MANN and WHITNEY - U-Test*.

## 3. Results and Discussion

The largest mycorrhizal projection area was found in the organic layer of the low N site (Figure 1). At the high N site the absolute projection area was significantly smaller (p = 0.0009). This difference in mycorrhizal distribution was accompanied by changes in fine root parameters such as reduced fine root biomass and a decrease in the number of root tips per root length at the high N site (Steiner et al., 1996). As mentioned by Wallenda and Kottke (1998), the reduced ramification of fine roots is a well known reaction following N fertilization. The lack of difference in mycorrhizal density between the organic and mineral soil horizon as seen in the presence of high N (p = 0.25) is unusual. Normally, significantly more ectomycorrhizas exist in the organic horizon (Mikola and Laiho, 1962, Fogel and Hunt, 1979, Kottke and Oberwinkler, 1988, Dahlberg, 1990).

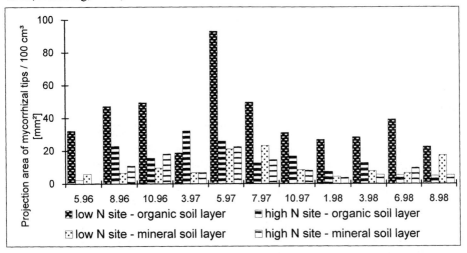

Fig. 1: Projection area of mycorrhizas in the organic and mineral soil layer from both sites.

In general, the mycorrhizal frequency of coniferous stands in Europe is very high with > 90 % of the vital root tips infected (Dahlberg, 1990, Qian et al., 1998). At the low N site the mycorrhizal frequency fluctuated between 76 % and 97 % in the organic soil layer and between 85 % and 96 % in the mineral soil layer (Table II). In contrast, at the high N site the mycorrhizal frequency showed high seasonal fluctuations and was sometimes remarkably low due to seasonal decreases down to 27 %. In experiments with liquid manure a decrease of the mycorrhizal frequency down to 40 % was only observed under very high N inputs of 3000 kg N $ha^{-1}$ $a^{-1}$ (Ritter and Tölle, 1978). On our site levels of mycorrhizal frequency only fell below 40 % following an extended drought (three weeks without rain) resulting in a low soil moisture (Table II). This corresponds well with the results of Nilsen et al. (1998) who found a correlation between drought and mycorrhizal colonization, which was accompanied by differences between the mycorrhizal types. However, our site differences cannot be explained by drought alone as soil moisture between the two sites was not significantly different overall.

The structure of the mycorrhizal coenosis was quiet different between the two sites (Figure 2). At the high N site only 9 mycorrhizal morphotypes were found compared with 18 types at the control site, but all the mycorrhizal morphotypes at the high N site were also found at the control site. There were large differences in the temporal distribution of mycorrhizal morphotypes. At the low N site *Cenococcum geophilum* was present in every sample and formed approximately 5 % of all the mycorrhizas at each sampling. In contrast, at the high N site it formed a more fluctuating proportion between 0 % and 21%. *Russula ochroleuca* was a dominant species at the low N site in 1996 and 1998, but less abundant in 1997.

The other dominant species was the unidentified morphotype 1, probably also a *Russula* species (Figure 2). The three species *Cenococcum geophilum*, *Russula ochroleuca* and *Xerocomus badius* that were most abundant in the coenosis of both sites are known as species with a broad ecological niche (Kottke and Oberwinkler, 1988, Benkert, 1996, Nilsen et al., 1998). No shift in the community structure to morphotypes especially adapted to the high N site conditions was found.

TABLE II

Mycorrhizal frequency (MR = percentage of mycorrhizas on total amount of vital root tips) and soil moisture (SM) of the organic and mineral soil layer.

| Sampling date | low N site | | | | high N site | | | |
|---|---|---|---|---|---|---|---|---|
| | organic soil layer | | mineral soil layer | | organic soil layer | | mineral soil layer | |
| | MR [%] | SM [%] | MR [%] | SM [%] | MR [%] | SM [%] | MR [%] | SM [%] |
| 4.96 | 93.9 | - | - | - | 26.9 | - | - | - |
| 5.96 | 94.3 | 47.6 | 94.6 | 13.4 | 68.7 | 49.0 | 83.6 | 9.5 |
| 8.96 | 81.3 | 56.3 | 91.8 | 11.5 | 54.7 | 61.1 | 62.4 | 7.6 |
| 10.96 | 88.8 | 56.7 | 89.0 | 11.6 | 71.8 | 51.9 | 77.2 | 12.5 |
| 3.97 | 87.5 | 52.1 | 84.9 | 11.4 | 89.8 | 61.2 | 92.1 | 7.4 |
| 5.97 | 97.1 | 61.4 | 95.9 | 11.2 | 70.7 | 60.7 | 87.4 | 10.1 |
| 7.97 | 94.1 | 56.4 | 94.9 | 9.8 | 70.8 | 63.5 | 82.0 | 10.7 |
| 10.97 | 94.9 | 48.3 | 90.4 | 5.8 | 75.6 | 34.6 | 80.0 | 3.1 |
| 2.98 | 92.6 | 66.3 | 90.0 | 11.8 | 78.5 | 62.3 | 87.5 | 8.4 |
| 3.98 | 76.1 | 62.1 | 93.9 | 11.1 | 73.3 | 60.0 | 75.6 | 7.1 |
| 6.98 | 93.7 | 48.4 | 91.0 | 11.5 | 37.9 | 38.3 | 62.2 | 4.5 |
| 8.98 | 91.4 | 37.8 | 94.8 | 4.4 | 32.7 | 35.2 | 59.6 | 4.3 |

From these data it can be hypothesized that the pine trees and their mycorrhizas react less flexibly at the high N site regarding their adaptation potential to natural stress factors such as summer drought or longer frost periods. This corresponds well with the observation of a 42 % higher natural reduction of the stocking density at the high N site compared to the low N site during the last two decades.

Nitrification leads to acidification of the soil. However, at the high N site eutrophication seems to be a more important factor as the soil pH is more alkaline at this site (Table I). As the nitrate concentration is high at this site, alkalization of the rhizosphere may be explained by the high rate of nitrate uptake by fine roots, as demonstrated by Marschner et al. (1986).

Being aware of the difficulties determining critical loads, Wallenda and Kottke (1998) postulated that a possible critical load of 20 - 30 kg N ha$^{-1}$ a$^{-1}$ could still be too high for some ECM communities. However, so far N deposition experiments in the field have only shown minor changes in the below-ground investigation of mycorrhizal root tips. However, their cited data was derived from more or less humid climates. In our investigation drastic changes in mycorrhizal parameters could be observed. The much greater detrimental N effect relates to the combination of high N deposition

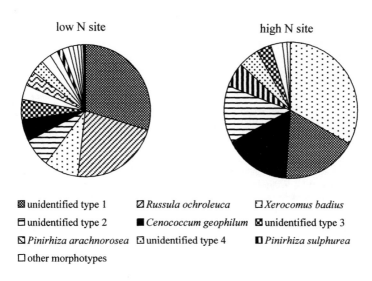

Fig. 2: The structure of the mycorrhizal coenosis of the two sites summarizing three investigation periods.

(> 35 kg N ha$^{-1}$ a$^{-1}$) over an extended period combined with climatic and edaphic dryness at the high N site.

## 4. Conclusion

(i) High N deposition has drastically reduced the mycorrhizal frequency, the total amount of mycorrhizas and the number of mycorrhizal morphotypes.
(ii) The impact of N reducing mycorrhizal infection is exacerbated during drought.

(iii) The interaction between N and drought and its influence on the mycorrhizas of the pine trees may be one cause of the observed reduction in tree density.

**Acknowledgement**

This study was supported by the Federal Ministry of Education, Science, Research and Technology (BMBF), Grant Beo 0339670, Bonn, Germany.

**References**

Agerer, R. and Rambold, G.: 1996, DEEMY v. 1.0 - *a DELTA-based system for characterization and Determination of EctoMycorrhizas*. Institute for Systematic Botany, Section Mycology, University München.
Benkert, D.: 1996, *Verh. Bot. Ver. Berlin Brandenburg* **129**, 151-213.
Dahlberg, A.: 1990, *Scand. J. For. Res.* **5**, 103-112.
Ellenberg, H.: 1986, *AFZ.* **42**, 466-467.
Fogel, R. and Hunt, G.: 1979, *Can. J. Forest Res.* **9**, 245-256.
Hofmann, G. :1996, *Mitt. Bundesforschungsanst. Forst- u. Holzwirtsch.* **185**, 45-72.
Hofmann, G.: 1997, *AFZ/Der Wald - Sonderheft.* 1-85.
Kottke, I. and Oberwinkler, F.: 1988, *KfK-PEF.* **39**.
Marschner, H., Römheld, V., Horst, W.J. and Martin, P.: 1986, *Z. Pflanzenernaehr. Bodenk.* **149**, 441-456.
Mikola, P. and Laiho, O.:1962, *Comm. Inst. For. Fenn.* **55** (18), 1-13.
Nilsen, P., Börja, I., Knutsen, H. and Brean, R.: 1998, *Plant Soil* **198**, 179-184.
Qian, X.M., Kottke, I. and Oberwinkler, F.: 1998, *Plant Soil* **199**, 99-109.
Raven, J.A. and Yin, Z.H.: 1998, *New Phytol.* **139,** 205-219.
Ritter, G. and Tölle, H.: 1978, *Beitr. f. d. Forstwirtsch.* **4**, 162-166.
Simon, K.-H. and Westendorff, K.: 1991, *Beitr. f. d. Forstwirtsch.* **25** (4): 177-180.
Steiner, A., Schneider, B.U., Wöllecke, J., Münzenberger, B., Krakau, U., Bolte, A., Fischer, T. and Hüttl, R.F.: 1996, *Mitt. Bundesforschungsanst. Forst- u. Holzwirtsch.* **185**, 237-239.
Steiner, A. Bolte, A., Schneider, B.U. and Hüttl, R.F.: 1998, *Verh. Ges. Ökol.* **28**, 421-428.
Wallenda, T. and Kottke, I.: 1998, *New Phytol.* **139,** 169-187.

# CHEMICAL COMPOSITION OF AIR, SOIL AND VEGETATION IN FORESTS OF THE SILESIAN BESKID MOUNTAINS, POLAND

A. BYTNEROWICZ[1], S. GODZIK[2], M. POTH[1], I. ANDERSON[3], J. SZDZUJ[2], C. TOBIAS[3], S. MACKO[4], P. KUBIESA[2], T. STASZEWSKI[2] and M. FENN[1]

[1]*USDA Forest Service, Pacific Southwest Research Station, Riverside, California, U.S.A.*
[2]*Institute for Ecology of Industrial Areas, Katowice, Poland.* [3]*The College of William and Mary, Department of Biological Sciences, Gloucester Point, Virginia, U.S.A.* [4]*University of Virginia, Environmental Sciences, Charlottesville, Virginia, U.S.A.*

(Received 16 October 1998; accepted 25 February 1999)

**Abstract.** For the first time concentrations of trace nitrogenous (N) air pollutants, gaseous nitric acid ($HNO_3$), nitrous acid ($HNO_2$), ammonia ($NH_3$), and fine particulate nitrate ($NO_3$) and ammonium ($NH_4$), were measured in the montane forests of southern Poland. Determinations were performed in two forest locations of the Silesian Beskid Mountains in the western range of the Carpathian Mountains, and in an industrial/urban location in Katowice, Poland. The measurements performed in summer 1997 with honeycomb denuder/filter pack systems showed elevated concentrations of the studied pollutants. These findings agree with the low carbon/nitrogen (C/N) ratios and the results of $^{15}N$ analyses of soil and moss samples. High concentrations of N air pollutants help to explain previously determined high levels of $NO_3$ and $NH_4$ deposition to Norway spruce (*Picea abies* Karst.) canopies in these mountains. Ambient concentrations of sulfur dioxide ($SO_2$) and ozone ($O_3$) were elevated and potentially phytotoxic. Deficiencies of phosphorus (P) and magnesium (Mg) in Norway spruce foliage were found while concentrations of other nutrients were normal.

**Keywords**: Norway spruce, ozone, nitrogenous air pollutants, nutrients, forest health

## 1. Introduction

In the 1980s, Poland was one of the most polluted countries in Europe. The annual emissions of 4 million tons of sulfur dioxide ($SO_2$) accounted for about 10% of the total emissions of that pollutant in Europe (Stan Srodowiska w Polsce, 1997). The Katowice principality, which constitutes 2.1% of the territory of Poland, produced as much as 20-25% of Poland's emissions of $SO_2$, dust and nitrogen oxides ($NO_x$). This situation in Poland has dramatically improved since 1989 due to political and economic changes. Initially, in the years 1989-1991, the 25-33% reduction of emissions of the major air pollutants was achieved by decreased industrial production. Since 1992, industrial production has started to increase (by about 10% annually in 1994 and 1995). In 1995, energy use was reduced by 27% compared to1989. As a result of these changes, the 1995 emissions of $SO_2$ and dust compared with 1989 were reduced by 40% and 50%, respectively. However, from 1988 to 1995 the number of motor vehicles sharply increased in Poland – cars by about 70% and trucks by about 40%. Although motor vehicles produced in the 1990s are more economical and have lower emissions than vehicles produced in 1980s, the total emissions of carbon monoxide (CO), $NO_x$ and volatile organic compounds (VOCs) have increased. It is predicted that this trend will continue until Poland reaches a similar level of vehicles per capita as the more developed western European countries. Consequently, it is foreseen that in the near future emissions of $NO_x$ and VOCs in Poland will continue to increase. Because $NO_x$ and VOCs are the precursors of ozone ($O_3$), concentrations of this pollutant will increase as well (Stan Srodowiska w Polsce, 1997). This scenario agrees with predictions that reductions of $NO_x$ emissions for Central and Eastern Europe will not equal those expected for Europe as a whole, where a 20-30 % decrease in $NO_x$ and a 15% decrease in VOCs relative to 1989 are expected (Europe's Environment, 1995).

In forests of Central Europe, especially in Poland, the Czech Republic, Slovakia and Germany, forest decline started to occur in the 1960s and was initially caused by direct effects of industrial air pollutants. In the following decades other stresses were added or intensified in Central European forests, such as soil acidification and N saturation, increasing concentrations of photochemical air pollutants, climatic changes, increased frequency of forest pest attacks, and improper management practices. As a result, the condition of forests in Poland, the Czech Republic, Slovakia, or Romania continued to decline (Europe's Environment, 1995).

The Carpathian Mountains, Europe's second largest mountain range, are of special value due to their protective, economic and cultural importance for Romania, Slovakia, Ukraine, Poland, and the Czech Republic. Within the Carpathian massive, the Silesian Beskid is one of the mountain ranges most exposed to the influences of polluted air masses. Polluted air moves into the Silesian Beskid from the Katowice, Krakow and the Ostrava industrial areas as well as from the infamous "Black Triangle" area shared between Poland, the Czech Republic and Germany. Changes in forests of the Silesian Beskid may be indicative of potential problems in the other parts of the Carpathian Mountains.

The objective of this study was to characterize air pollution exposure and forest nutritional status in selected Silesian Beskid forest sites. Results of $^{15}N$ abundance and chemical characterization of soils and biological material were compared with results of a similar study performed in the Sumava Mountains in the Czech Republic and the previously collected data from the Silesian Beskid range. Results of this study are discussed in the context of the current health status of forests and potential future risks associated with air pollution in the Silesian Beskid and the Carpathian Mountains.

## 2. Methodology

Fig. 1. Location of air pollution monitoring sites in southern Poland.

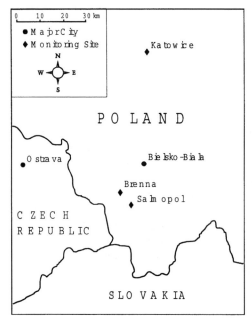

The Brenna and Salmopol research sites are located in southern Poland, in the Silesian Beskid Mountains. The nearest large city, Bielsko Biala is about 15 km NE; the Katowice conurbation, the major industrial area for Poland, is about 60 km N; and the Ostrava industrial region of the Czech Republic about 30 km W (Figure 1). The research area is covered by forests mixed with arable land and pasture. The monitoring stations are located in the vicinity of Norway spruce (*Picea abies* Karst ) stands: the Brenna station at 660 m and the Salmopol station at 1000 m elevation. The climate in the area is typical for the entire Silesian Beskid range with an annual 30 yr mean temperature of 6.6°C and 1300 mm precipitation. For comparison, some of the air measurements were also done in the Katowice conurbation, at the location of the Institute for Ecology of Industrial Areas in Katowice.

$O_3$ concentrations were continuously monitored in Brenna with a Thermo Environmental monitor (model 49C). Sulfur dioxide ($SO_2$) was trapped using impingers filled with 0.3% $H_2O_2$ solution and analyzed as sulfate ($SO_4$) with ion chromatography, and $NO_2$ was absorbed on glass fritts impregnated with sodium metarsenite and determined colorimetrically using the Salzman procedure. Nitric acid ($HNO_3$), nitrous acid ($HNO_2$), ammonia ($NH_3$), $SO_2$ and particulate nitrate ($NO_3$) and ammonium ($NH_4$) were collected with the Ogawa, USA, Inc. honeycomb/filter pack systems (Koutrakis *et al.*, 1993) and determined with ion chromatography and colorimetry. For the Brenna and Salmopol sites air samples were collected during from 0800 to 2000 on August 12–14, 1997, and in Brenna additionally on August 17 and 24, 1997. Samples for the Katowice site the were collected on August 20–23, 1997. Results are presented as 12 h daytime mean concentrations (except $NO_2$ in Katowice, 24 h mean concentrations).

Spruce needle samples for chemical analysis were taken from six trees in two locations in Brenna and two locations in Salmopol. The needles were dried at 70°C, ground and digested in Teflon 'bombs' filled with $HNO_3$ placed in a microwave oven. Concentrations of nutrients in the digest were determined by the ICP method. For N determinations, samples were digested in sulphuric acid with selenium as catalyst, and analysed according to the Kjeldahl procedure.

Moss (*Polytrichum formosum* Hedw. and *Pleurozium schreberi* (Brid.) Mitt.) and soil samples (top 5 cm layer) were collected on August 12 –15, 1997. The samples were dried, ground and digested in a nitric-perchloric solution. The digests were analyzed by atomic absorption spectrophotometry and emission spectrometry for calcium (Ca), magnesium (Mg), sodium (Na), potassium (K), manganese (Mn), zinc (Zn) and phosphorus (P) was determined colorimetrically (Glaubig and Poth, 1993). Total C and N and a natural abundance of $^{15}N$ isotope were determined in ground and homogenized soil and moss sub samples using an elemental analyzer coupled to an OPTIMA stable isotope ratio mass spectrometer (Macko *et al.*, 1987).

## 3. Results and Discussion

### 3.1. OZONE IN AMBIENT AIR

Concentrations of $O_3$ in Brenna during the duration of an intensive study varied between 30 and 90 ppb with the highest values in the afternoon and the lowest at night. Elevated concentrations of the pollutant could be short lived (August 11 – 13) or could extend for most of the day (August 10), or even throughout the night (August 14) with a high peak on the following day (August 15) (Figure 2). Concentrations of the pollutant were the lowest at night, dropping to about 30 - 40 ppb. This is typical for mountain locations where concentrations of nitric oxide (NO), a compound that readily reacts with $O_3$, are too low to completely scavenge $O_3$ from the air (Bytnerowicz et al., 1987). The pollution event on Aug. 14-15 may serve as a good example of photochemical smog build-up during favorable weather conditions with long-lasting high temperatures and stagnant air. The $O_3$ values were similar to the concentrations found in Brenna in 1995 with a mean 24-hour concentration about 42 ppb and the 30 min mean concentrations reaching 90 ppb (Godzik, S. 1997). Similar $O_3$ concentrations were determined in the 1990's in

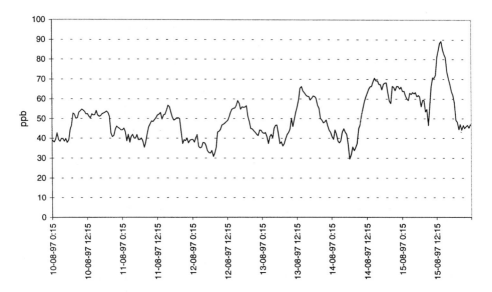

Fig. 2. Ambient concentrations of $O_3$ at Brenna during intensive monitoring in August 1997.

several locations near Krakow, including the Carpathian foothills (Godzik, B., 1997). Our results are also within the range of $O_3$ concentrations occurring in the mountainous locations of the Austrian, Swiss and German Alps (Herman et al., 1998). Much lower $O_3$ concentrations occurred in more distant ranges of the eastern Carpathian in the Ukraine (Blum et al., 1997) or northern parts of Europe (Girgzdiene, 1995). These results indicate an increasing risk to the Carpathian forests resulting from potentially phytotoxic $O_3$ concentrations. As the geographic area of elevated $O_3$ levels increases, other forests of the Carpathian Mountains, including the relatively pristine areas of Slovakia, Ukraine and Romania, may be endangered by $O_3$ phytotoxic effects.

While regional $O_3$ concentrations have been thoroughly investigated in northwestern Europe (Derwent and Jenkin, 1991), eastern and central Europe requires an improved monitoring network and more rigorous investigations. With dynamic economic growth since 1989 and changes in air pollution scenarios this would seem to be one of the top monitoring priorities for that part of Europe.

3.2. SULPHUROUS AND NITROGENOUS POLLUTANTS IN AMBIENT AIR

Average 24-h concentrations of $SO_2$ in Brenna measured with the impingers varied between 8 and 28 µg/m³ and concentrations of $NO_2$ between 0.5 and 6 µg/m³ (Figure 3a). Concentrations of $SO_2$ in Katowice were in a similar range (11-28 µg/m³) while concentrations of $NO_2$ were much higher, in a range of 23 – 48 µg/m³ (Figure 3b). Concentrations of $SO_2$ measured with the honeycomb denuder systems in Brenna and Salmopol were in the range of the values determined with the impinger method (Fig. 4d). The reported $SO_2$ concentrations in the two mountain locations were lower than in the industrial/urban areas of Upper Silesia - the mean annual concentrations in Katowice and Chorzow could be as high as 33 – 47 µg/m³, and the peak values may reach 133 – 182 µg/m³ (Wojewodzka Stacja Sanitarno-Epidemiologiczna, 1998). Measurements of $SO_2$

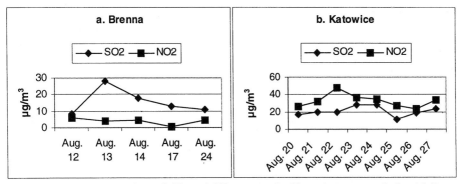

Fig. 3. Ambient concentrations of $SO_2$ and $NO_2$ measured with an impenger method - (a) Brenna monitoring site in the Silesian Beskid Mountains, and (b) Katowice monitoring site in the industrial/urban area of the Upper Silesia.

made in the Silesian Beskid in early 1990s indicated that the annual mean concentrations of the pollutant were about 35-37 µg/m³, exceeding critical levels proposed by the UN-ECE Convention on Long-Range Transboundary Air Pollution (20 µg/m³) and the Polish national standard (32 µg/m³). It may be expected that the $SO_2$ concentrations in the Upper Silesian region will continue to decrease due to the introduction of more efficient energy use and cleaner technologies after 1989.

The annual mean concentration of $NO_2$ in Brenna was 8 µg/m³, well below the UN-ECE proposed levels of 30 µg/m³ and the Polish national standard of 50 µg/m³.

Concentrations of $NO_2$ were higher in the industrial/urban area of Katowice, where $NO_2$ annual mean reached 35 µg/m³ (Godzik et al., 1997). Such values will probably persist in the area since the newly introduced technologies and energy saving measures reduce emissions of $NO_2$ much less than the emissions of $SO_2$ or industrial dust.

Honeycomb/filter pack results indicated that $HNO_3$ concentrations in Brenna and Salmopol (Figure 4a) were slightly lower than the values measured on August 26 and 27 in Katowice (2.0 and 3.2 µg/m³, respectively). Elevated levels of $HNO_3$ in Brenna and Salmopol could result from pollution coming both from the Ostrava industrial region and the Katowice conurbation. The determined $HNO_3$ concentrations were elevated and much higher than levels detected in remote mountains locations in North America (Legge and Krupa, 1989; Miller and Walsh, 1991). The $HNO_3$ levels were similar to values found in the Tyrolian Alps (Puxbaum et al., 1993; Kalina et al., 1998), the Black Forest in Germany (Papenbrock and Stuhl, 1991), and the western Sierra Nevada in California (Bytnerowicz and Riechers, 1995; Bytnerowicz and Fenn, 1996). The $HNO_3$ concentrations in the Silesian Beskid were several fold lower than the concentrations determined in the San Gabriel Mountains (Grosjean and Bytnerowicz, 1993) and San Bernardino Mountains (Bytnerowicz et al., 1997), areas strongly influenced by photochemical smog from the Los Angeles conurbation.

Concentrations of $HNO_2$ (Figure 4b) were several fold higher than the values occurring in remote locations (Legge and Krupa, 1989; Bytnerowicz, unpublished). The concentrations were similar to the highest values determined in various mountain locations in California (Bytnerowicz and Fenn, 1996). The $HNO_2$ values determined in urban Los Angeles were 2-3 fold higher (U.S. EPA, 1993). Elevated concentrations of

HNO$_2$ may be of interest because they can cause increased levels of nitrite (NO$_2^-$), an ion with high phytotoxic potential, in plant cells (Wellburn, 1990).

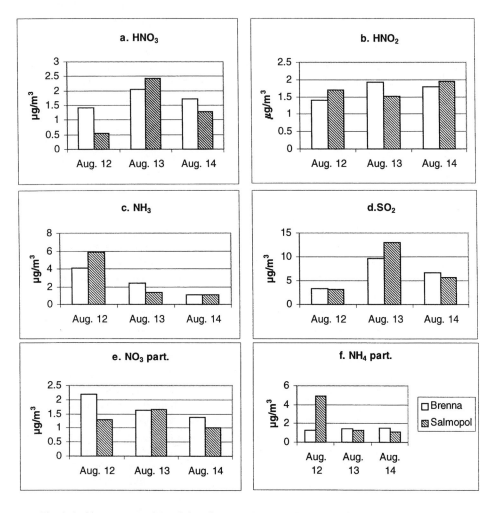

Fig. 4. Ambient concentrations of air pollutants at Brenna and Salmopol during intensive monitoring in August 1997.

Concentrations of NH$_3$ measured in Brenna and Salmopol (Figure 4c) were similar to the values determined on August 26 and 27 in Katowice (4.4 and 3.4 µg/m$^3$, respectively). These values were elevated compared to concentrations found in some remote locations in North America (Roberts et al., 1988). The values were similar to concentrations measured in the western Sierra Nevada (Bytnerowicz and Riechers, 1995) and the San Gabriel Mountains (Grosjean and Bytnerowicz, 1993), but slightly higher than the values measured in the Tyrolian Alps (Kalina et al., 1998). Elevated concentrations of NH$_3$ in the Silesian Beskid may be caused by a combination of agricultural emissions and other activities typical for big conurbations, including municipal waste treatment. The determined values are low compared to values from some urban locations in the Upper Silesia where the annual mean concentrations in Jaworzno and Myslowice can be as high as 15 μg/m$^3$, and in Rudziniec and Tarnowskie Gory 80 – 100 μg/m$^3$ (Wojewodzka Stacja Sanitarno-Epidemiologiczna, 1998). Some areas in the Netherlands or Germany (Gessler and Rennenberg, 1998) have $NH_3$ values comparable to urban locations of the Upper Silesia.

Concentrations of particulate $NO_3$ (Figure 4e) were in the range of values determined in mountain locations close to the photochemical smog source areas in California (Bytnerowicz and Fenn, 1996), about 2 -3 fold higher than values found in the Tyrolian Alps (Kalina et al., 1998), and several fold higher than the values found in the remote mountain locations in North America (Legge and Krupa, 1989; Bytnerowicz, unpublished). Concentrations of particulate $NH_4$ (Figure 4f) were similar to levels typically found in mountain locations in California (Bytnerowicz and Fenn, 1996) and the Tyrolian Alps (Kalina et al., 1998). The $NH_4$ concentrations were several fold higherthan the values determined in the remote mountain locations of North America (Legge and Krupa, 1989; Bytnerowicz, unpublished).

3.3. APPORTIONMENT OF N AIR POLLUTANTS

Apportionment of the N compounds in Salmopol and Brenna (based on the impinger and honeycomb/filter pack determinations) indicated slightly higher proportions of the reduced N pollutants ($NH_3$ and particulate $NH_4$) than the oxidised N pollutants ($NO_2$, $HNO_3$, $HNO_2$ and particulate $NO_3$) (Figure 5). In California, high proportions of the

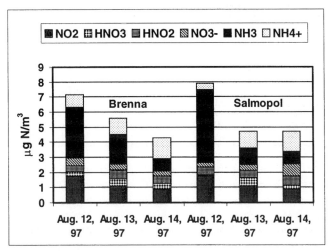

Fig. 5. Apportionment of N air pollutants during intensive monitoring in Brenna and Salmopol.

reduced N compounds were determined near intensive agricultural activities (Bytnerowicz and Riechers, 1995), and a high proportion of the oxidised N compounds near urban areas (Bytnerowicz and Fenn, 1996). In the Tyrolian Alps, the proportion of reduced N compounds was slightly lower than in the Brenna and Salmopol locations (about 46% of total N pollutants - Kalina et al., 1998).

3.4. CONCENTRATIONS OF NUTRIENTS IN PLANTS AND SOILS

Concentrations of N, P, K, S, Ca, M, Mn, Zn and Na in the Norway spruce needles were similar between the two forest locations in Brenna and two locations of the Salmopol sites (Table I). Concentrations of N were within the normal range of this nutrient for Norway spruce needles (1.3 – 1.6 %) (Herman et al., 1998). The S concentrations were elevated compared to needles from clean areas, although significantly lower than the concentrations determined in Brenna and Salmopol in the late 1980s and early 1990s (Godzik et al., 1997). Nutrient ratios for the Norway spruce needles (Table II) were comparable with ranges considered well balanced for this species in Central Europe (Stefan and Furst, 1998). This comparison indicated that the N/K, N/Ca and S/N ratios were within the ranges set for well-balanced trees. However, considering that the well-balanced trees should have an N/P ratio in the range of 7.01 – 10.00 and an N/Mg ratio in a range of 8.01 – 14.00, the values determined at Brenna and Salmopol indicated that

TABLE I
Concentrations of nutrients in Norway spruce needles (means of 6 replicates, one for each tree, % d.wt.).

| Element | Brenna 1 | Brenna 2 | Salmopol 1 | Salmopol 2 |
|---|---|---|---|---|
| N | 1.410 | 1.300 | 1.320 | 1.250 |
| P | 0.118 | 0.092 | 0.113 | 0.078 |
| K | 0.457 | 0.438 | 0.424 | 0.378 |
| S | 0.100 | 0.120 | 0.090 | 0.103 |
| Ca | 0.282 | 0.349 | 0.253 | 0.327 |
| Mg | 0.062 | 0.041 | 0.049 | 0.028 |
| Mn | 0.06 | 0.058 | 0.118 | 0.020 |
| Zn | 0.002 | 0.002 | 0.003 | 0.003 |
| Na | 0.003 | 0.003 | 0.002 | 0.003 |

TABLE II
Nutrient ratios in Norway spruce needles.

| Ratio | Brenna 1 | Brenna 2 | Salmopol 1 | Salmopol 2 |
|---|---|---|---|---|
| N/P | 11.96 | 14.18 | 11.70 | 15.55 |
| N/K | 3.09 | 2.97 | 3.16 | 3.31 |
| N/Ca | 5.00 | 3.72 | 5.22 | 3.82 |
| N/Mg | 22.89 | 32.10 | 26.88 | 44.96 |
| S/N | 0.071 | 0.092 | 0.068 | 0.082 |

deficiencies of P and Mg may be present. Earlier investigations in this area showed deficiencies of Mg and Ca in forest soils (Godzik et al., 1997). Deficiency of P may be expected in some the areas affected by increased N deposition (Fenn et al., 1998). Deficiency of Mg is considered as a key cause of Norway spruce yellowing damage typical for this part of Europe (Herman et al., 1998). The sudden occurrence of this symptom may be attributed to a series of dry years which reduced Mg mineralization and uptake as well as by more intensive (shorter rotation) tree harvesting and leaching of soils by acidic deposition (Europe's Environment, 1995). Elemental concentrations in moss samples (Table III) were similar in the Brenna and Salmopol locations, with the exception of P with much higher concentrations in Salmopol. A tendency for higher Ca, Mn and Zn concentrations in soils at the Brenna sites (Table III) could indicate greater deposition of industrial dust to forests at that lower elevation site. The C/N ratio of soils (about 20) is lower than might be expected for these sites and may reflect the long term deposition of N (Table III). In the Sumava Mountains of the Czech Republic, the C/N

TABLE III

Mineral composition of moss and soil samples presented as means of five replicates (concentrations in % d.wt.)

| Parameter measured | Brenna 1 | | Salmopol 1 | |
|---|---|---|---|---|
| | Moss | Soil | Moss | Soil |
| P | 0.099 | 0.086 | 0.181 | 0.052 |
| K | 0.720 | 0.179 | 0.778 | 0.186 |
| Ca | 0.093 | 0.091 | 0.118 | 0.030 |
| Mg | 0.047 | 0.010 | 0.060 | 0.093 |
| Mn | 0.011 | 0.017 | 0.005 | 0.007 |
| Zn | 0.005 | 0.016 | 0.006 | 0.009 |
| C/N | 25.6 | 19.9 | 20.6 | - |
| $\delta^{15}N$ | -4.6 | -0.9 | -3.9 | - |

ratio values were about 30 in the same period of 1997 (Anderson, unpublished). Evidence of long-term elevated N depositon are also supported by the isotope natural abundance measurements where the soils are less negative than moss (Table III). The observed phenomena could be a consequence of the high deposition of $^{15}N$ depleted N compounds as reflected by the moss. This deposited N may then be nitrified which would strongly fractionate the N and export $^{14}N$-$NO_3$ from the soil. The effect over time would produce less negative natural abundance of N in soil than in moss.

3.5. FOREST HEALTH

Forests in southern Poland, including the Silesian Beskid range, have been weakened due to the long-term effects of anthropogenic disturbances, including air pollution and poor management practices. These forests show high level of crown defoliation which is indicative of the impaired health of trees. In the transnational survey performed in 1992, about 24% of European forests were considered to be damaged (defoliation >25) and *Abies* spp and *Picea* spp were affected the most among the conifers (30 and 26% trees damaged, respectively) (Europe's Environment, 1995). In 1995, only 5.7% of the Polish forests could be considered without damage, while 49.5% of trees displayed slight or moderate damage and 3.2% of trees strong damage or mortality. Among the forest species, Norway spruce was the most affected with 69% of the trees damaged in 1995 (Stan Srodowiska w Polsce, 1997). The most damaged forests are in southern Poland, and include Norway spruce stands in the Sudety Mountains as well as parts of the Silesian Beskid and the Zywiec Beskid, the Carpathian ranges most exposed to the West (Stan Srodowiska w Polsce, 1997). High defoliation (crown transparency) of spruce stands, with 100% trees in injury classes 2 and 3, was determined in Salmopol while the Brenna stands had about 60% of trees injury classes 2 and 3 (Godzik, S., 1997).

4. SUMMARY

This study and the earlier results of air pollution monitoring and evaluation of forest health in Brenna and Salmopol (Godzik, S., 1997) indicate the presence of potentially phytotoxic $O_3$ levels. Although such $O_3$ concentrations alone should not be toxic to Norway spruce (Polle *et al.*, 1995), a potential for $O_3$ phytotoxicity may be enhanced in the presence of the elevated levels of $SO_2$. Species more sensitive to $O_3$ than Norway spruce, such as beech, as well as the understory plants may be affected by such elevated $O_3$ concentrations.

Concentrations of the gaseous and particulate N air pollutants were elevated, and at levels similar to the highest values found in the Alps or strongly polluted forest sites of the western Sierra Nevada. These findings indicate elevated N dry deposition which agrees with high levels of N determined in throughfall of Norway spruce stands. Low C/N ratios in soil and moss samples confirm that long-term elevated deposition of N pollutants have taken place in the Silesian Beskid forests. High levels of N deposition in these forests (28-30 kg N/ha/yr - Godzik et al., 1997) might have led to the observed deficiencies of P and Mg in Norway spruce.

The above findings help to understand why the Norway spruce forests in the Silesian Beskid Mountains are in poor condition. It is reasonable to assume that with increasing concentrations of $O_3$ and continuing high levels of industrial pollution, damage to forests in the Carpathian Mountains will continue and may expand to the east.

## References

Blum, O., Bytnerowicz, A., Manning, W. M. and Popovicheva, L. I.: 1997, *Environ. Pollut.* **98**, 299-304.
Bytnerowicz, A. and Fenn, M.: 1996, *Environ. Pollut.* 92, 127-146.
Bytnerowicz, A., Fenn, M. E., and Arbaugh, M.: 1997, *USDA Forest Service PSW Station, General Technical Report,* **164**, 8 pp.
Bytnerowicz, A., Miller, P.R., Olszyk, D. M., Dawson, P. J., and Fox C. A.: 1987, *Atmos. Environ.* **21**, 1805-1814.
Bytnerowicz, A. and Riechers, G.: 1995, *Atmos. Environ.* **29**, 1369-1377.
Derwent, R. G. and Jenkin, M. E.: 1997, *Atmos. Environ.* **29**, 10-27.
Europe's Environment: 1995, Dobris Report, D. Stanners and P. Bourdeau (eds). *European Environment Agency Task Force (European Commission: DG XI and Phare),* Copenhagen.
Fenn, M.E., Poth, M. A., Aber, J. D., Baron, J. S., Bormann, B. T., Johnson, D. W., Lemley, A. D., McNulty, S. G., Ryan, D. F. and Stottlemyer, R.: 1998, *Ecol. Appl.* **8**, 706-733.
Gessler, A. and Rennenberg, H.: 1998, In: L. J. DeKok and I. Stulen (eds) *Responses of Plant Metabolism to Air Pollution and Global Change,* Backhuys Publishers, Leiden, The Netherlands, 81-94.
Girgzdiene, R.: 1995, *Water, Air Soil Pollut.* **85**, 1509-1514.
Glaubig, R. and M. Poth.: 1993. *Commun. Soil Sci. Plant Anal.* **24**, 2469-2477.
Godzik, B.: 1997, *Eviron. Pollut.* **98**, 273-280.
Godzik, S., Szdzuj, J., Staszewski, T., and Lukasik, W.: 1997, *USDA Forest Service PSW Research Station, General Technical Report,* **164**, 8 pp.
Godzik, S.: 1997, *USDA Forest Service, PSW Research Station, General Technical Report,* **164**, 6 pp.
Grosjean, D. and Bytnerowicz A.: 1993, *Atmos. Environ.* **27A**, 483-492.
Herman, F., Lutz and C., Smidt, S.: 1998 *Environ. Sci. Pollut. Res.* **1**, 2-15.
Kalina, M. F., Zambo, E. and Puxbaum, H.: 1998, *Environ. Sci. Pollut. Res.* **1**, 53-58.
Koutrakis, P., Sloutas, C., Ferguson, S. T., Wolfson, J. M., Mulik, J. D. and Burton, R. M.: 1993, *Environ. Sci. Technol.* **27**, 2497-2501.
Legge, A. H. and Krupa, S. V.: 1989, In: R. K. Olson and A. S. Lefohn (eds). *Transactions of the Symposium on Effects of Air Pollution on Western Forests,* Anaheim, CA, 29-30 June 1989, Air and Waste Management Association, Pittsburgh, PA, USA, pp. 193-206.
Macko, S. A., Estep, M. L. F., Hare, P. E. and Hoering, T. C.: 1987, *Isotope Geosci.* **65**, 79-92.
Miller, D. F. and Walsh, P. A.: 1991 *84th Annual Meeting and Exhibition of the Air and Waste Management Association,* Vancouver, BC, 16-21, June 1991.
Pappenbrock, T. and Stuhl, F.: 1991, *Atmos. Environ.* **25A**, 2223-2228.
Polle, A., Wieser, G. and Havranek, W. M.: 1995, *Plant, Cell Environ.* **18**, 681-688.
Puxbaum, H., Haumer, G., Moser, K. and Ellinger, R.: 1993, *Atmos. Environ.* **27A**, 2445-2447.
Roberts, J. M., Langford, A.O., Goldan, P. D. and Fehsenfeld, F. C.: 1988, *J. Atmos. Environ.* **7**, 137-152.
Stan Srodowiska w Polsce. 1997, *Ministerstwo Ochrony Srodowiska,* Warszawa, Poland.
Stefan, K. and Furst, A.: 1998, *Environmental Science and Pollution Research,* **1**, 63-69
U.S. EPA: 1993, Air Quality for Oxides of Nitrogen, Vol. I., EPA/600/8-91/049aF.
Wellburn, A. R.:1990. *New Phytol.* **115**, 395-429
Wojewodzka Stacja Sanitarno-Epidemiologiczna: 1998, Katowice, Poland.

# FOREST HEALTH IN NORTH AMERICA: SOME PERSPECTIVES ON ACTUAL AND POTENTIAL ROLES OF CLIMATE AND AIR POLLUTION

## S. McLAUGHLIN[1] and K. PERCY[2]

[1]Environmental Sciences Division, Oak Ridge National Laboratory, PO Box 2008, MS 6422, Oak Ridge, TN, 37830, USA; [2]Natural Resources Canada, Canadian Forest Service, Atlantic Forestry Centre, Fredericton, N.B., E3B 5P7, Canada

(Received 25 September 1998; accepted 25 February 1999)

**Abstract.** The perceived health of forest ecosystems over large temporal and spatial scales can be strongly influenced by the frames of reference chosen to evaluate both forest condition and the functional integrity of sustaining forest processes. North American forests are diverse in range, species composition, past disturbance history, and current management practices. Therefore the implications of changes in environmental stress from atmospheric pollution and/or global climate change on health of these forests will vary widely across the landscape. Forest health surveys that focus on the average forest condition may do a credible job of representing the near-term trends in economic value while failing to detect fundamental changes in the processes by which these values are sustained over the longer term. Indications of increased levels of environmental stress on forest growth and nutrient cycles are currently apparent in several forest types in North America. Measurements of forest ecophysiological responses to air pollutants in integrated case studies with four forest types (southern pine, western pine, high elevation red spruce, and northeastern hardwoods) indicate that ambient levels of ozone and/or acidic deposition can alter basic processes of water, carbon, and nutrient allocation by forest trees. These changes then provide a mechanistic basis for pollutant stress to enhance a wider range of natural stresses that also affect and are affected by these resources. Future climatic changes may ameliorate (+ $CO_2$) or exacerbate (+ temperature, + UV-B) these effects. Current projections of forest responses to global climate change do not consider important physiological changes induced by air pollutants that may amplify climatic stresses. These include reduced rooting mass, depth, and function, increased respiration, and reduced water use efficiency. Monitoring and understanding the relative roles of natural and anthropogenic stress in influencing future forest health will require programs that are structured to evaluate responses at appropriate frequencies across gradients in both forest resources and the stresses that influence them. Such programs must also be accompanied by supplemental process-oriented and pattern-oriented investigations that more thoroughly test cause and effect relationships among stresses and responses of both forests and the biogeochemical cycles that sustain them.

**Key Words:** air quality, physiology, growth, biotic and abiotic interactions

## 1. Introduction

The forests of North America are immense in both their geographical range and ecological diversity. The 326 M ha of forested land in the USA (Powell *et al.*, 1992) and 417 M ha in Canada (NRCAN, 1998) are also a highly valuable economic resource for which maintenance of long term productivity is a very high priority. Both actual and perceived potential responses of North American forests to atmospheric pollution during recent decades have figured strongly in policy decisions on air quality regulation that have had significant economic and ecological implications for the region. In the next century, if current increases in atmospheric $CO_2$ continue and projected changes in both temperature and atmospheric deposition of nitrogen (N) occur, changes in forest productivity and function will be of interest not only in terms of shifts in patterns of production, but also from the perspective of the role of forests as sinks or sources for greenhouse gases (Norby, 1998). IPCC (1998) predictions for the effects of Global Climate Change on forest health include both increased growth and range of some forests, but also an increased frequency of declining health in others in response to increased biotic and abiotic stresses associated

---

The authors wish to express their sincere appreciation to Dr. R. Vet, Atmospheric Environment Service, Environment Canada, Downsview, Ontario, and to Dr. T. Dann, Environment Protection Service, Environment Canada, Ottawa, Ontario, for calculation and mapping of $SO_4$, $NO_3H$ ion depositions and cumulative $O_3$ sums.

with climatic warming. Evaluating forest health and forest biogeochemical functions in this more complex future environment will increasingly challenge forest scientists to measure and understand the basic processes underlying forest growth and longer term ecological health.

Perspectives on forest health and current risks posed by chemical and physical climate change differ widely within the forest community (Jenkins, 1997). These differences often stem from varying perspectives of forest values and times scales over which change is evaluated. Kolb *et al.* (1994) differentiated between the utilitarian view of forest health, which stresses timber production as the primary endpoint and the ecosystem perspective, which views forest health in terms of measures of longer term forest function. The latter include properties such as resilience, diversity, and the flow of carbon, water, and nutrient resources required for tree resistance to natural stresses and maintenance of biogeochemical cycles. Assessment of forest health, in fact, has many dimensions (Innes,1993), and at the time of our writing, there exists no universally-accepted definition of a healthy forest. O'Laughlin *et al.* (1994) have defined forest health as "...a condition of forest ecosystems that sustains their complexity while providing for human needs". This definition stresses the utilitarian endpoint, which is defined in anthropocentric terms. We focus here on a more fundamental definition - A capacity to supply and allocate water, nutrients, and energy in ways that increase or maintain productivity while maintaining resistance to biotic and abiotic stresses. We, therefore, define less healthy forests as ones in which trees lose productive capacity and/or become more sensitive to environmental stresses. We should note here that unhealthy trees and stands occur naturally as a part of successional processes by which a balance between forest production, site resources, and climate are attained. In this capacity, for example, grazing insects can serve an important role in establishing a balance in nutrient flow to foliage from nutrient deficient soils (Matson and Addy, 1975). The patterns by which such changes occur however are typically spatially and temporally heterogeneous as are the localized biotic and abiotic factors that influence them.

Ecosystem response to environmental stresses, including air pollutants, is a complex, hierarchical process occurring over time scales ranging from minutes (leaf) to decades (stand). However, air pollution stresses are somewhat distinct from many other forms of environmental stress because of their regional patterns and temporal consistency with changes in air quality. Stress responses begin with sensitive individuals proceeding from branch, to tree, and then stand and ecosystem levels (Hinckley *et al.*, 1992) Underlying these response are process-level changes in uptake and allocation of carbon, water, and nutrient resources. We focus here on linkages between process level changes and ecosystem responses because this approach offers the best chance to understand the individual and interactive effects that will ultimately determine the utilitarian values of forest systems.

This paper summarizes recent advances in understanding of process level effects of regional air pollution on four forest types for which recent syntheses are available. These include the southeastern pine forest, eastern spruce/fir ecosystem, northeastern hardwoods, and western pine forests in Southern California (Figure 1). Our objective is to evaluate how increased understanding of forest process effects can be integrated into regional monitoring and analysis of forest health in relation to both air pollution and its interactions with other biotic and abiotic stresses that affect forests. We have emphasized primarily ozone and acidic deposition (N and S), the pollutants with the greatest potential regional impact on forests in the U.S. and Canada. Measured and potential impacts have been evaluated from the following perspectives:

FOREST HEALTH IN NORTH AMERICA 153

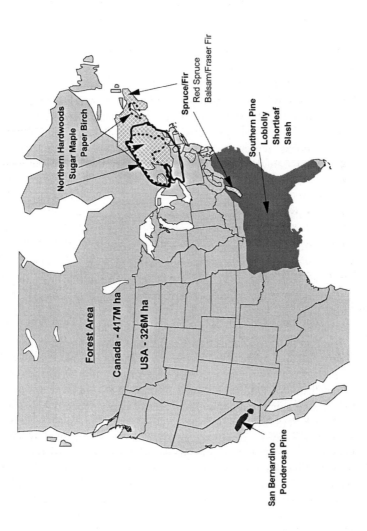

Figure 1. Multidisciplinary research from case studies in southwestern pine, southeastern pine, eastern spruce-fir, and northeastern hardwood forest types was examined to evaluate process level responses to ambient levels of air pollutants in four regional forest types.

(1). Monitoring air quality and forest health in North America.

(2). Evidence of changes in allocation patterns of carbon, water, and nutrients related to ozone or acidic deposition within the four "case-study" forests.

(3). Implications of process-level changes on whole-tree function and biogeochemical processes in current and future climate regimes.

(4). Recommendations for incorporating process-based ecosystem perspectives into regional monitoring of forest health.

## 2. Regional Air Quality in North America

Recent syntheses in both the United States (NAPAP, 1998) and Canada (EC, 1997) demonstrate clear qualitative relationships between spatial patterns of acidic gas emissions and measured air concentrations and wet deposition patterns that were already evident by the 1990 assessments (RMCC,1990). Here, we concentrate on acidic deposition and $O_3$ as they affect our forests regionally, while recognizing that their co-occurrence and interactions with other abiotic (e.g., climate extremes, UV-B, toxics) and biotic stressors ultimately determines the direction and magnitude of forest health responses.

### 2.1. ACIDIC DEPOSITION

Emissions of $SO_2$ in eastern North America have declined from over 20 M t in 1980 to under 17 M t in 1993 (EC, 1997). As a consequence, US air concentrations of $SO_2$ have decreased 37% between 1985 and 1995, particularly in the northeast (NAPAP, 1998). Particulate $SO_4$ concentration reductions have been widespread, except at several mid- to high-elevation, forested areas (NAPAP, 1998). US emissions of anthropogenic $NO_x$ decreased 6.5% from 23.3 M t in 1980 to 21.8 M t in 1995 (NAPAP, 1998). Trends in N species contributing to acidic deposition ($HNO_3$, $NO_3$) were more variable than for $SO_2$. While decreasing $HNO_3$ was reported in the northeastern US (NAPAP, 1998), Canadian data showed and increasing frequency of higher concentration $HNO_3$ events (EC, 1997).

Trends in distribution patterns for wet deposition of $SO_4$ and $NO_3$ are shown in Figures 2 and 3. It should be noted that additional inputs of dry deposition may add 8-37% more S and 15-65% more N to these numbers depending on the region (EC, 1997). Contrasts in wet $SO_4$ and $NO_3$ deposition patterns reveal strong reductions for S in the area enclosed by the 10-30 kg ha$^{-1}$ yr$^{-1}$ isopleths. Also evident was the disappearance of the 30 kg ha$^{-1}$yr$^{-1}$ area in Canada, and its considerable reduction in the US (Figures 2a,b) (EC, 1997). In contrast, $NO_3$ wet deposition did not change dramatically. Areas receiving 5-25 kg $NO_3$ ha$^{-1}$yr$^{-1}$ expanded marginally (Figures 3 a,b) (EC, 1997). When integrated, areas in Canada within the 20 kg ha$^{-1}$yr$^{-1}$ wet $SO_4$ isopleth decreased 44% from 629,000 ha in 1980-83 to 340,000 ha in 1990-93 (EC, 1997).

Hall et al. (1997) recommended a critical threshold of 12 kg $SO_4$ ha$^{-1}$ yr$^{-1}$ for wet deposition to poorly buffered Canadian soils. In the context of forest/lake hydrological process linkages, current wet $SO_4$ deposition exceeds critical loads at all eastern Canadian lake clusters by 7- 12 kg ha$^{-1}$ yr$^{-1}$ and is expected to exceed them by about 6 to 10 kg ha$^{-1}$ yr$^{-1}$ after all emission controls required by the Canada/US Air Quality Agreement are fully implemented (Jeffries, 1997).

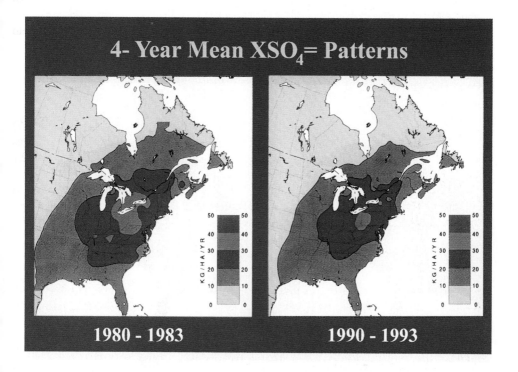

Figure 2. Four-year mean wet deposition patterns of $SO_4^-$ (kg ha$^{-1}$yr$^{-1}$) in eastern North America for (a) 1980-1983 and (b) 1990-1993. Reproduced from EC (1997).

Areas in Canada within the 15 kg wet $NO_3$ ha$^{-1}$yr$^{-1}$ isopleth increased 13% from 373,000 ha in 1980-83 to 420,000 ha in 1990-93. Areas receiving at least 0.4 kg H+ ha$^{-1}$yr$^{-1}$ decreased along the east coast and in southern Ontario, while the area receiving at least 0.2 kg ha$^{-1}$yr$^{-1}$ expanded to the northeast (Newfoundland) and east (Nova Scotia) (Figures 4a,b) (EC, 1997). $NO_3$ and $NH_4$ concentrations showed high variability, and deposition of $NH_4$, generally increased throughout the US (NAPAP, 1998). Acidification potential remains high over large forest areas due to increasing $NO_3$ and decreasing Ca, Mg, K, Na deposition (Likens et al., 1996).

Episodes have a large impact upon temporal changes in annual deposition amounts (Brook, 1995) and may increase annual mean $SO_4$ and H+ deposition by $\leq$ 20% (Beattie and Whelpdale, 1989). In this context, precipitation amount and type are critical in pollutant deposition to forests, particularly those growing at mid to high elevation. Most mountain forests receive more (>50%) precipitation than those at lower elevation; coastal forests are immersed in fog for up to 30% of the growing season. Mountain cloud/fog and coastal fog have been found to have extremely low mean annual weighted pH's, consistently 10-50X more acidic than precipitation (Kimball et al., 1988), with single events recorded as low as

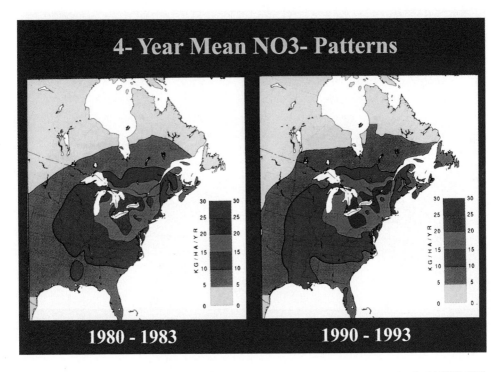

Figure 3. Four-year mean wet deposition patterns of $NO_3^-$ (kg ha$^{-1}$yr$^{-1}$) in eastern North America for (a) 1980-1983 and (b) 1990-1993. Reproduced from EC (1997).

pH 2.8. In the Southern Appalachian Mountains, cloud exposure at high elevation sites has been found to occur on 70% of the days and 30% of the time (Saxena and Lin,1990). In addition 25% of all cloud events over a 4-year period had a pH $\leq$ 3.1 with a minimum pH of 2.2 (DeFelice, 1997).

Recent air quality modeling indicates that eastern regional deposition of $SO_4$ and $NO_3$ is much larger than amounts measured in precipitation (CAPMON, NADP) would indicate (EC, 1997). This raises the real possibility that inputs of acidic deposition to some forests may be substantially higher than estimated.

## 2.2. OZONE

Composite national daily maximum 1 hour $O_3$ concentrations in the US decreased 15% between 1987 and 1996. The highest national composite 1 hour maximum was in 1988. Ozone levels have declined 10% since 1987 at 194 rural monitoring sites (EPA, 1998). In Canada, time series analysis identified a significant declining trend in daily maximum $O_3$ concentrations ranging from -0.05% to -0.08% yr$^{-1}$. Average days per year (1986-1993) exceeding the 82 ppb 1 hour National Ambient Air Quality Objective (NAAQO) ranged

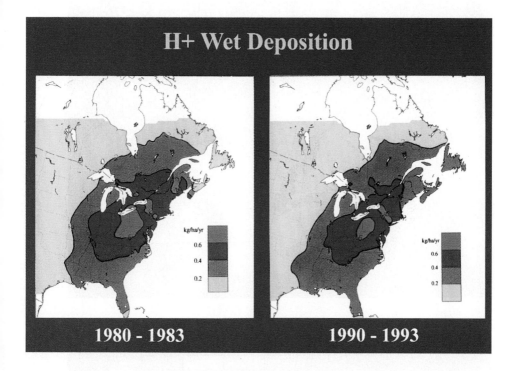

Figure 4. Four-year mean wet deposition patterns of $H^+$ (kg ha$^{-1}$yr$^{-1}$) in eastern North America for (a) 1980-1983 and (b) 1990-1993. Reproduced from EC (1997).

from 18 in southern Ontario, 3 in the Southern Atlantic Region (SAR) and 2 in the Lower Fraser Valley, British Columbia.(Olivotto, 1997). Mapping of percent change in SUM60 between 1985-1987 and 1994-1996 shows a significant decrease of 10-15% in much of the southeast US with decreases of 5-10% or 0-5% in the Lake States and southern Ontario/Quebec. Portions of the New England states and the SAR, however, had increases in SUM60 $O_3$ of between 0-10%. Smaller areas in the midwestern US showed increases of 10-15% SUM60 $O_3$ (Figure 5).

Ozone (photochemical smog) and dry-deposited N remain the most abundant air pollutants in U.S. Pacific Southwest forests (Fujioka *et al.*, 1998). In 1995, 99.3% of $O_3$ exceedances occurred during June to October. The number of adverse episodes in the Los Angeles-Long Beach and Riverside-San Bernardino end of the Southern California Air Basin (SoCAB) declined during 1986-1995. However, the differential between the western (urban) and eastern (urban/forested) ends increased (EPA, 1997). The most abundant reactive compound in the San Bernardino Mountains is still $O_3$, and ambient concentrations there are the highest in North America. Between November 2, 1991 and September 30, 1993, hourly concentrations exceeded the then-existing 120 ppb NAAQS on 235 days (22% of sampling days).

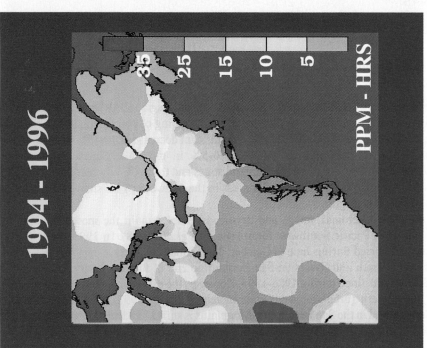
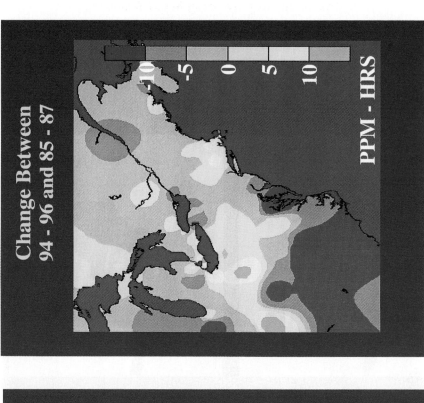

Figure 5. SUM06 (May 1 - July 31; 08:00-19:59 hrs) ozone ppm-hrs calculated using Canadian (NAPS) and US (AIRS) monitoring data for (a) three-year average for 1994-1996 and (b) percent change between the periods 1985-1987 and 1994-1996. Source, T. Dann, Environment Canada-EPS.

## 3. Monitoring North American Forest Health

In the 1980's, the rationale and protocols for the North American monitoring programs changed due to studies that indicated that widespread changes in forest condition were occurring in North America and Europe (Rehfuess, 1981, Ulrich, 1984, Schutt and Cowling, 1985; McLaughlin, 1985; Fraser *et al.* 1985, Prinz,1987). In Table I we have highlighted central features of both US, Canadian, and European forest monitoring networks, and some details of their structure and focus are discussed below.

### 3.1. THE UNITED STATES

The Detection Monitoring component of the US Forest Health Monitoring Program (FHM) (USDA, 1992) was implemented in 1996 on about 40% of approximately 4000 forested plots in the lower 48 states. A spatially and temporally systematic sample has been based on a national hexagonal grid (27 km between grid centres). The plot size and sampling frequency format (Table I) results in all circular 1 ha plots being sampled every 4 years with annual estimates of change being derived from smaller subplots (Smith and Mangold, 1996). This design yields: 1. a base-line condition the first year; 2. a direct measure of 1-year change for sub-plot samples after two years; and, 3. a direct measure of 4-year change after five years. Indicators (Table I) being measured in Detection Monitoring include lichen communities, ozone bioindicator plants, tree growth, crown condition, damage, mortality, regeneration, vegetation structure and plant diversity (Smith and Mangold, 1996).

### 3.2. CANADA

The Acid Rain National Early Warning System (ARNEWS) was initiated in 1984 (D'Eon and Power, 1989). ARNEWS consists of 150 rectangular, 10m X 40m plots with 2m x 2m subplots stratified by forest region. The system currently contains about 11,700 trees (dbh $\geq$ 10cm) comprised of 3,500 hardwoods and 8,200 softwoods (Hall, 1995). Soil samples, foliar samples for nutrient analysis, and radial growth measurements are collected every five years adjacent to the plot (soil) or from numbered off-plot trees (foliage, increment cores). Variables are sampled on all plots at varying frequencies (Table I). Foliar symptoms, hardwood seed crop, pest conditions (type of foliar insect/disease, woody tissue insect/disease and severity) and tree condition (mortality, live crown height, crown damage, needle retention) are assessed seasonally and/or annually. Regeneration, saplings and ground vegetation (mosses, herbs, shrubs) are assessed biennially (D'Eon *et* al., 1994).

### 3.3. JOINT UNITED STATES AND CANADIAN MONITORING

Out of growing concern over sugar maple decline in the early 1980's, the USDA Forest Service and the Canadian Forest Service joined in 1988 with state and provincial agencies to establish the North American Maple Project (NAMP). NAMP consists of a network of 233 sites extending across the natural range of sugar maple (*Acer saccharum* Marsh) (Millers *et al.*, 1991). NAMP sites are five-plot clusters of 20m x 20m plots on which all trees of 10 cm dbh or more are identified. Stands ranged from 50 to 150 years of age in 1988 and had variable levels of damage or crown dieback. Sites were paired where possible, one being tapped and the other untapped (no tapping or management during previous 5

TABLE I
Comparison of the United States Forest Health Monitoring (FHM), the Canadian Acid Rain National Early Warning System (ARNEWS), the North American Maple Project (NAMP) and the European (UN-ECE ICP) forest health monitoring programs.

| | Monitoring Program | | | |
|---|---|---|---|---|
| | FHM | ARNEWS | NAMP | UN-ECE |
| No. plots | 4000 | 150 | 166 | 20,345 |
| Plot strat.[1] | grid 27km$^2$ | forest region | na | grid 16km$^2$ |
| Plot visits | | | | |
|   all | year 1/5 | 1 | 1 | 1 |
|   subset | 1/3 year2 1/3 year 4 1/3 year 5 | | | |
| Associated studies | intensive site | FIGC | none | Level II |
| Tree | | | | |
|   diameter | E | 5 | E | X |
|   height | E | 5 | X | X |
|   dominance | E | 5 | X | X |
|   form | X | 5 | X | X |
|   crown height | X | 5 | X | X |
|   crown closure | X | 5 | X | X |
|   mortality | 1-4 | 1 | 1 | X |
| Crown | | | | |
|   condition | 1-4 | 1 | 1 | 1 |
|   symptoms | 1-4 | 1 | X | 1 |
|   insects/diseases | 1-4 | 1 | 1 | X |
|   foliage | 1-4 | 1 | 1 | X |
|   defoliation | 1-4 | 1 | 1 | 1 |
|   woody | 1-4 | 1 | 1 | X |
|   seed production | X | 1 | X | X |
| Ground | | | | |
|   regeneration | 1-4 | 2 | X | X |
|   vegetation type | 1-4 | 2 | X | X |
| Pests | 1-4 | S | X | X |

TABLE I

*Continued*

| | Monitoring Program | | | |
|---|---|---|---|---|
| | FHM | ARNEWS | NAMP | UN-ECE |
| Nutrition | | | | |
| foliar | X | 5 | X | X |
| organic soil | X | 5 | X | X |
| mineral soil | X | 5 | X | X |
| Metereology/Air Quality | | | | |
| temperature | X | X | X | X |
| precipitation | X | X | X | X |
| $SO_4/NO_3$ | X | M | X | X |
| $O_3$ | B | P | X | X |

FIGC : CFS Forest Indicators of Global Change
Sampling frequency: 5=5 yrs; 4=4yrs; 2=2yrs; 1=yearly; S=seasonally (several times per year); E=on establishment; X=not done
M: modelled; B: bioindicator plants; P: passive monitoring
1. Stratification of plots in network.

years). Variables assessed annually include defoliation, tree vigor, new damage to boles and crown condition (dieback, transparency).

The three forest health monitoring systems were designed to different protocols in order to address unique agency mandates. Each has specific strengths and weaknesses in responding to the monitoring of forest health response to changing air quality. In our view, none, by themselves, will enable forest managers or policy makers to relate effects to causes with the required degree of scientific certainty.

3.4. INSECT AND DISEASE PROBLEMS IN NORTH AMERICA

An important backdrop for consideration of pollution-induced stresses of North American forests is the extent of problems from insects and disease. A summary of the types of diseases, their approximate extent (where available) and the principal regions in which they have been detected, is given for the USA and Canada in Table II. Disease problems to forests are a major consideration in both Canada, where annual forest depletion due to disease during 1982-87 totaled 51 M cubic meters (NRCAN, 1996) and in the U.S.A., where the areal extent of insect problems on forests is 20.4 M hectares (USDA, 1997). With this background we consider the types of physiological effects induced by regional air pollution from two perspectives: (1) the relative magnitude and distribution of effects attributable to air pollution; and (2) the potential for stresses from air pollution to interact with biotic and abiotic stresses.

TABLE II
Estimates of forested areas affected by pests and pathogens in the U.S. and Canada in 1996.
Data are in millions (M) of hectares (where available) by region.

United States[1]

Quantified

Dwarf mistletoe (11.7 M-W), fusiform rust (5.5 M-SE)
Southern pine beetle (3.0 M-SE), spruce budworm (0.3 M-E+W)
Mountain pine beetle (0.1 M-NW), gypsy moth (0.1 M-NE)

Unquantified

Beech bark disease (NE), butternut canker (NE)
Hemlock woody adelgid (NE), dogwood (E)
Anthracenose (NE, SE), European pine shoot beetle (NE, NC)

Total area affected > 20.4 M ha

Canada[2]

Quantified

Aspen 2-leaf tier (3.9 M-W), eastern spruce budworm (1.2 M-E)
Forest tent catepillar (1.2 M-Canada), other insects combined (7.1 M-Canada, W)

Other insects including area of beetle killed trees (4.6 M-Canada)

---
[1]Source: USDA (1997)
[2]Source: NRCAN (1997)

## 4. Regional Air Pollution: North American Case Studies

Evidence of regional threats to forest health from air pollution in the U.S.A. led in the mid-1980s to the National Forest Response Program (Hertel and McKinney-McNeal, 1991), an integrated research program aimed at evaluating evidence and mechanisms of forest responses to acidic deposition and $O_3$. Initial integrated summaries were produced around 1990 in the both the US (NAPAP, 1990) and Canada (Hall and Addison, 1991). The 1990 NAPAP synthesis included both studies of forest health (Barnard et al., 1990) and evaluation of process level changes associated with regional air pollution (Shriner et al., 1990). While very different in context (utilitarian vs ecosystem focus), the conclusions were similar in message - no clear cut evidence of a widespread forest decline, except for high elevation red spruce. Both analyses noted, respectively, uncertainty and evidence for pollution-induced changes in nutrient cycles and biotic and abiotic stress interactions that support longer term forest health. Canadian analyses in the NAPAP era, (Pearson and Percy, 1990) indicated that sugar maple decline was more severe and extensive than historically noted, and that acid deposition was a co-factor in a complex of contributing stressors. White birch decline along the east coast was circumstantially linked to acid fog. Critical loads were unavailable, but 46% of the Canadian land surface mapped was considered highly sensitive to acidification.

These first assessments have now been augmented by more recent syntheses of integrated studies (Eagar and Adams, 1992; Johnson and Lindberg, 1992; Fox and Mickler, 1995; Mickler and Fox, 1997) that were still in progress in 1990. Coupled with more recent results from Canadian studies (Hall, 1995; Hall et al., 1997) and synthesis of the long-standing ecosystem research on oxidant stress effects in the California ponderosa pine forests (Miller and McBride, 1998), additional valuable insights into the nature and extent of regional air pollution stress have been provided that we draw on here.

## 4.1. SOUTHEASTERN PINE FORESTS

The health of southern pine forests has been a subject of great interest and substantial controversy since significant growth declines were detected in the mid 1980's (Sheffield *et al.*, 1985 and Sheffield and Cost, 1987). The evidence of unexplained growth reductions came from analysis of FIA inventory data from unmanaged pine stands by the US Forest Service. With an area of approximately 25 M ha and annual stemwood sales of $4.5 billion, the health of southern pines has been a major economic concern within the region. This concern translated into a large multiagency research effort aimed at quantifying causes of the apparent growth losses of the three predominant pine species in the region. The Southern Commercial Forest Research Cooperative (SCRFC), which was organized to address this task, conducted a series of regionally distributed controlled exposure studies with seedlings and saplings of loblolly pine (*Pinus taeda* L.), shortleaf pine (*P.echinata* Mill) and slash pine (*P. elliota*) to evaluate the potential of $O_3$ and acidic deposition to affect southern pine growth. The summary of forest characteristics, biotic and abiotic stresses, and potential impacts of acidic deposition and ozone for southern pine forests combined information gathered in epidemiological and mechanistic studies over approximately 8 years of research (Fox and Micker, 1995).

A summary of the results of controlled exposure studies with seedlings and saplings of southern pines in open-topped chambers in the SCFRC indicated that significant growth reductions occurred with increased ozone exposure, but no consistent adverse effects of acidic deposition were detected (Teskey, 1995). Most studies reported small reductions in growth at ambient ozone, estimated at 2-5% per year, but these reductions were not typically detectable statistically within the experimental constraints of the experimental systems employed (Teskey, 1995). Contributing to this variability were genetics (multiple families with very different sensitivity levels), treatments imposed (interactions of ozone and acid deposition) and substantial year to year variability in ambient-based treatments imposed.

Additional analyses of the combined results of a wide range of controlled ozone studies with loblolly pine emphasized the role of high genetic variability in sensitivity in growth responses to ozone, particularly at near-ambient levels (Taylor, 1994). Using this much larger data set, Taylor (1994) estimated an ozone response threshold of 33 ppm hrs and a 6% growth loss per 100 ppm hrs of ozone exposure for sensitive families. The threshold for the population mean was similarly estimated at 97 ppm hrs with a 3% growth loss per 100 ppm hrs of cumulative exposure. At typical annual ambient ozone exposure levels of 50-200 ppm hrs (Taylor, 1994), estimated annual growth losses during a single season were 1-10% for sensitive families and 0-3% for the mean population.

Several studies have now indicated that the effects of ozone exposure are cumulative and approximately linear. The concept, which apparently extends to effects occurring over multiple seasons, is supported by studies of both loss of photosynthetic capacity (Richardson *et al.*, 1992) and foliage production and retention (Stow *et. al.*, 1992, Kress *et al.*, 1992). Reduced canopy photosynthesis was noted at a cumulative exposure threshold of 100-150 ppm hrs and reduced foliar retention occurred at 150-230 ppm hrs. An important principal established in these analyses was the carryover of inhibition of photosynthetic potential of ozone impaired foliage into subsequent growing seasons (Sasek *et al.*, 1991).

Studies of responses of more mature loblolly pine trees have included modeling-based estimates developed around changes in gross primary production of branches (Dougherty *et al.*, 1992); regional application of a physiologically-based model with regional ozone exposure data (Hogsett *et al.*1993); and empirically derived measures of tree growth responses in the field (Zahner *et al.*, 1989, and McLaughlin and Downing, 1996). Results of these analyses for pines and selected studies with hardwoods are summarized in Table III and indicate an expected response range of 0-35% annual growth reduction for mature pines, depending on species X year and ozone level. Work by McLaughlin and Downing (1997) and Zahner *et al.* (1989) support the possibility that sensitivity of mature trees to drought may be enhanced by ozone at levels occurring in the field. Using an empirically derived model from weekly growth responses of mature loblolly pine trees to ozone in the field, McLaughlin and Downing (1996) indicated that ozone significantly increased the effects of soil moisture stress on stem growth. This concept is supported by studies with pine seedlings and saplings at ambient or near-ambient ozone levels which document increased transpiration (Lee *et al.*, 1990) or increased water stress (Flagler *et al.*, 1997). In addition several European studies with saplings and mature conifers (Skarby *et al.*, 1987; Wallin and Skarby, 1992; Maier-Maercker and Koch, 1992; Maier-Maercker, 1997), have reported increased transpiration and/or reduced stomatal control of water loss for conifers at ambient ozone levels. Collectively these studies suggest that stomatal closure measured at high ozone levels with seedlings in some controlled exposure studies, may not adequately represent the expected response to ambient exposures in the field, particularly where large trees are involved.

SCFRC studies indicated a low short-term risk to growth from ambient exposure to acidic deposition, and in fact growth was typically stimulated by near-ambient levels (Teskey, 1995). However, acidic deposition exerts it's effects primarily through nutrient mobilization and depletion, and the development of such effects in controlled studies will depend on initial soil nutrient content and soil volume, factors which affect the capacity for nutrient depletion. Of greater relevance from the perspective of evaluating risk to southern pines, is the regional distribution of pine stands on poorly-buffered soils depleted of base cations by previous agricultural practices. Approximately 60% of southeastern pine soils are considered to be susceptible to accelerated cation leaching (Richter and Markewitz, 1995). Binkley *et al.*, (1989) have estimated that 10-15% of commercial southern pine forests may be currently limited by low base cation supply. Long term studies of nutrient cycles at a reference watershed in South Carolina indicate that approximately 80% of the exchangeable calcium (Ca) has been lost from the soil over the past three decades, with 50% of this loss due to uptake by vegetation and 50% due to leaching (Richter *et al.*, 1994). Base saturation declined from $\geq$ 55% in 1962 to 20% in 1972, and to 10% in 1990. This decline was most apparent in shallow soil profiles (< 20 cm) where fine roots are concentrated (Richter and Markewitz, 1995). In considering the potential negative effects of Ca depletion from these soils, it should be noted that many of these same soils are N deficient and are receiving 4.5-14 kg N/ha/y, approximately 15-45% of annual stand requirements, from atmospheric sources (Richter and Markewitz, 1995). Accumulation of N in the upper soil horizons has the potential to stimulate growth at the risk of longer term cation depletion, as we will discuss later.

TABLE III
Some estimates of effects of ozone on growth of forest tree species at ambient ozone levels based on studies with seedlings and mature trees.

**SEEDLING STUDIES**

| Species | Growth Reduction | Conditions | Reference |
|---|---|---|---|
| Multiple | 0-10% | Shoot growth | Chappelka and Samuelson, 1998 |
| Southern pines | 2-5% | Summary estimate - SCFRC field-chamber studies | Teskey, 1995 |
| Loblolly pine | | Synthesis - whole tree biomass | Taylor, 1994 |
| | 0-3% | Mean response and sensitive | |
| | 1-10% | family response to 50-200 ppm*h | |
| Hardwoods | 13% | Response surface derived values | Reich, 1988 |
| Conifers | 3% | at 20 ppm*h | |

**MATURE TREE STUDIES**

| Species | Growth Reduction | Conditions | Reference |
|---|---|---|---|
| Loblolly pine | 2-9% | Whole tree carbon model using branch chamber data (GA) | Dougherty et al., 1992 |
| Loblolly pine | 0-13% | Average annual weekly responses to $O_3$ and interactions of $O_3$ and moisture stress, 5 years (TN) | McLaughlin and Downing, 1996 |
| | 0-5% | Annual $O_3$ effect - no water stress | McLaughlin and Downing, 1996 |
| | 0-30% | Annual $O_3$ effect - moderate water stress | McLaughlin and Downing, 1996 |
| Loblolly pine | 5-12% | Regional forest model simulation based on small tree exposures | Hogsett et al., 1993 |
| Shortleaf Pine | 1% $y^{-1}$ | Empirically derived growth loss 38% cumulative not explained by climate and competition. Natural stands 20-45 yr (GA, SC, NC) | Zahner et al., 1989 |
| Shortleaf Pine | Climatic shift | Dendroclimatic analysis - Increase in non-climatic influences beginning in 1963. (GA)[1] | Grissino-Mayer and Butler, 1993 |
| White Pine | $O_3$ > climate | Dendroclimatic analysis show $O_3$ influences growth more than climate (ME) | Bartholomay et al., 1997 |
| Hardwoods | 2-17% | Regional simulation with canopy-stand model across moisture gradients | Ollinger et al., 1997 |

---

[1] Authors attributed shift to non-climatic physiological factors. We have detected a close relationship between the annual pattern of residual variance (not explained by climate) and annual variations in regional ozone exposure levels.

## 4.2. SOUTHWESTERN PINE FORESTS

In the San Bernardino Forest, ponderosa pine (*Pinus ponderosa* Dougl. Ex Laws), one the most important forest species in California (Olson, 1992), is exposed to the highest ozone concentration in North America by topographical, meteorological and climatic influences which combine to advect the Los Angeles urban pollution plume inland along the western mountain slopes (Miller, 1992). The resultant chronic exposure and ecosystem level changes in the San Bernardino Forest Ecosystem to over 50 years of high levels of $O_3$ and oxidant air pollution has produced perhaps the best example of hierarchical forest response to air pollutants (Miller *et al.*, 1982; Miller *et al.*, 1989; Miller, 1992; Miller *et al.*, 1996). Ozone injury to the mixed conifer forest has been occurring since at least the mid-1950's in the mountains of Southern California (Miller and Millecan, 1971) and since the 1970's in the Sierra Nevada (Pronos *et al.*, 1978). Miller *et al.* (1963) were first to identify the classic $O_3$ symptomology and chlorotic foliar mottle associated with the injury. Species most affected by photochemical oxidants (mainly $O_3$) have been ponderosa and Jeffrey (*Pinus jeffreyi* Grev. and Balf.) pines.

Injury to pine foliage resulted in a decline in the health of sensitive trees; accumulation of weakened trees then resulted in a heavy bark beetle kill. This necessitated extensive sanitation logging to remove declining pine trees. Unable to compete for nutrients, water and light, remaining sensitive individuals declined allowing for enhanced growth of more tolerant species (Miller *et al.*, 1982). Chronic effects of $O_3$ and N compounds continue today. The area has been used as a natural laboratory to study the air pollution effects on ecosystem functioning, including human values, and the 30 year San Bernardino Case Study is now summarized in Miller and McBride (1998).

During early survey years of 1968-1972 $O_3$ concentrations were extremely high (>80 ppb $O_3$ for 1300 h for a five-month year). Ponderosa and Jeffrey pine mortality ranged from 8-10% (2-2.5% per year). Ozone concentrations at high elevation sites remained above 50 ppb at night. An increase in extent of severe foliar injury indicated the beginning of the selective death of ponderosa pine (Miller, 1973). Weakened trees were attacked by bark beetles and elimination of ponderosa pine from the mixed conifer forest was predicted (Cobb and Stark, 1970). This survey led to the establishment of a large, multidisciplinary research team to investigate essential ecological processes underlying mixed conifer forest response to $O_3$. A hierarchical response, parallel to that observed with eastern white pine (*Pinus strobus* L.) (McLaughlin *et al.* 1982), was documented. Average (24 h) concentrations of 50-60 ppb $O_3$: induced foliar injury leading to premature abscission; needle fall resulted in reduced photosynthetic capacity; decreased nutrient availability to stressed trees; reduced production of carbohydrates; and diminishing tree vigor resulting in decreased height and radial growth (Miller *et al.*, 1982). At the ecosystem level, changes in decomposition patterns and altered nutrient cycling altered successional patterns in some species. This cycle of interactions triggered by $O_3$ on sensitive individuals significantly altered ecosystem processes and intra-community relationships. Unlike a catastrophic event such as fire, tree death and associated energy changes within the ecosystem occur gradually. Understanding of such changes in tree vigor, species composition, energy flow, nutrient cycling, and interactions between trophic levels is critical if cause-effect relationships are to be adequately evaluated..

One test of cause-effect linkage is mixed conifer forest response to gradually diminishing $O_3$ stress. Miller *et al.* (1989) reported an improvement (1974-1988) in foliar injury index at all but the most exposed plots in the San Bernardino Mountains in response to declining $O_3$ in 1976-1991(Davidson, 1993). Changes in visual injury were paralleled by improving tree growth patterns. Growth plots in a high-$O_3$ exposure area along the southern rim of the San Bernardino Mountains showed an approximate 40% growth slowdown during the period of highest ozone (1945-72) with a recovery of basal area growth to pre-1945 levels during the most recent period of improved air quality (Arbaugh et al., 1998). The concurrence of long-term reduction in precipitation and high $O_3$ were concluded to be responsible for the growth pattern, and not $O_3$ alone. However, a significant lag in integration of improved canopy condition into improved basal area (BA) growth was identified. BA increases in $O_3$- sensitive ponderosa pine during the same period (Miller et al. 1991) remained less than the BA increase for $O_3$ tolerant species. Only continued ecosystem monitoring will define the new baseline.

Evidence has existed for some time (Hoffer et al. 1981) that polluted air masses accumulating over central California are transported into the Sierra Nevada Mountains (Stohlgren and Parsons, 1987). Surveys recently completed in national parks in the region documented a progression of foliar symptoms on ponderosa pine related to aspect and elevation (Peterson et al. 1991). Standardization of an index to assess crown injury on ponderosa and Jeffrey pines, the Ozone Injury Index (OII), occurred in 1989. The Sierra Cooperative Ozone Impact Study (SCOIAS) and the Forest Ozone Response Study (FOREST) were among the first reports linking crown assessment, including both foliar injury and needle retention, with cumulative ambient $O_3$ indices. This represents a potentially reliable, cost effective approach for monitoring and forecasting $O_3$ injury in remote areas (Arbaugh et al. 1998).

Critical co-factors affecting tree response in SCOIAS/FOREST were site climate, history, tree size and age, N deposition and genetics. The unique role of N deposition in the mixed conifer forests of California has been well established (Bytnerowicz and Fenn, 1996). Unlike other case studies, most deposition occurs in summer in dry N form as acidic vapor, gaseous and particulate species. Deposition may be as high as 25-45 kg N ha$^{-1}$ yr$^{-1}$ and has resulted in localized N saturation in some mixed conifer stands (Bytnerowicz and Fenn 1996).

In summary, the effects of chronic exposure of these forests to acidic deposition and $O_3$ were detected on forest health through intensive process research built upon a historical record of individual tree response (Miller et al. 1996; Bytnerowicz et al. 1998). The multidisciplinary study succeeded in providing a comparative database on mixed conifer forest exposed to moderate and high levels of gas and particle deposition. Acidic pollutants, despite extreme sensitivity of leaf surface wax to acidic hydrolysis by $HNO_3$ (Bytnerowicz et al 1998), were not implicated in a short-term response. The interaction of $O_3$ with periodic droughts (Miller et al. 1996), in the long-term, and summer N pulses (Bytnerowicz et al. 1998), in the very short-term, were judged to be the greatest threats to the health of southwestern pine forests

## 4.3. EASTERN SPRUCE-FIR FORESTS

High levels of mortality of red spruce (*Picea rubens* Sarg.) in the Northern Appalachians (Weeks and Smoronok, 1988) and Fraser fir (*Abies fraseri* (Pursh) Pair.) in the Southern Appalachians (Dull et al., 1988) during the 1970's and 1980s were accompanied by reductions in radial growth of red spruce at many high elevation sites across the region

(Johnson and Siccama, 1983; McLaughlin et al., 1987). The timing and distribution of observed changes in growth suggested a possible linkage to acidic deposition and an integrated series of field and laboratory studies was conducted over the interval 1987-1991 by the Northern Spruce Fir Research Cooperative (NSFRC, Eagar and Adams, 1992). A synthesis of this work concluded that both field and laboratory data support " --that regional scale air pollution has played a significant role in the decline of red spruce in the Eastern United States" (Johnson et al., 1992). This was based on regional and local analysis of the epidemiology and distribution of increased tree mortality and canopy deterioration (Peart et al., 1992); the timing and distribution of radial growth decline of mature trees (Cook and Zedaker, 1992); and mechanistic studies in the field and laboratory which linked changes in carbon allocation (McLaughlin and Kohut, 1992) and winter injury (DeHayes, 1992) to levels of acidic deposition monitored within the region (Mohnen, 1992). These studies represent an excellent model of effective integration of regional monitoring of forest condition with mechanistic studies designed to better understand observed responses. In this case, multiple lines of investigation yielded new mechanistic insights linking acidic deposition as a predisposing agent to increased stress of red spruce forests by abiotic stress. These included both winter injury and nutrient depletion from foliage and soils. A summary of the principal trends in responses observed in field studies from the Northern Appalachians is depicted in Figure 6.

Increased winter injury to current year foliage was an important symptom of damage detected on red spruce in the Northeastern mountains in the late 1950s and early 1960's (Friedland et al., 1984). As this damage was repeated frequently over the following years, clear linkages were found to reduced radial growth of both montane and coastal red spruce forests (Jagels, 1986 and Percy et al., 1993) and increased mortality of montane trees (Johnson et al., 1992). By the mid 1980's increased mortality levels had reduced red spruce as a component of the high elevation canopy by $\geq$ 50% in the Adirondack and Green Mountains and by 25% further North in New Hampshire and Maine. A dendroclimatological analysis of the growth decline for trees from affected areas revealed that the occurrence of increased winter injury was a consequence of increased sensitivity of trees to winter temperatures, not unusually cold winters (Cook and Johnson, 1989). A series of misting experiments with acidic deposition under controlled laboratory conditions (see Dehayes, 1992) coupled with exclusion studies with branches (Vann et al., 1992) and chambered saplings (Dehayes et al., 1991) at high elevation field sites indicated that ambient levels of acidic mist could reduce the cold hardiness of red spruce by 3-10$^0$C. This is an amount sufficient to substantially increase winter injury to this species at high elevation northeastern sites (Johnson et al., 1992).

The growth decline of high elevation red spruce in the Southern Appalachians began typically around 1965, 5 years later than in the North, was confined to elevations above 1700 m, and was not accompanied by the very high red spruce mortality levels of the Northern Appalachians (McLaughlin et al., 1987). However, canopy condition of red spruce declined significantly at both high and low elevation sites during 1985-89 (Peart et al., 1992) and average red spruce mortality was consistently higher (ranging from 3-4% per year) at high elevation sites than intermediate and lower elevations (Nicholas, 1992). Standing dead red spruce comprised 8-17% of the stands at high elevation sites where attrition of dead stems was found to be remarkably high - 33% per year (Nicholas, 1992). While a notable loss in basal area of red spruce has not yet been reported at the southern Appalachian sites (Smith and Nicholas, 1998) mortality has occurred across all size classes

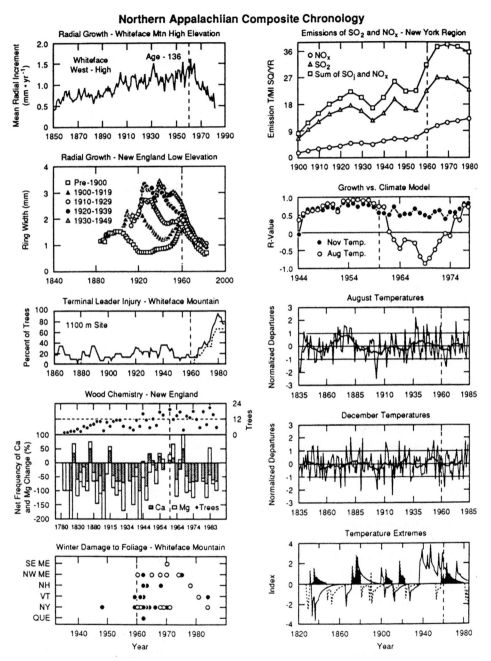

Figure 6. Results of many different analyses revealed a shift in the health of high elevation red spruce around 1960 corresponding to increasing loading from acidic deposition in the Northern Appalachian Mountains (After Johnson et al., 1992). In summary left panels (top to bottom) describe reduced radial growth at high and low elevations, increased terminal leader mortality, shifts in wood cation chemistry, and increased frequency of winter injury to foliage. The timing of these changes coincides with sharp regional increases in emissions of S and N (upper right) and shifts in growth climate relationships that were not attributable to unusual climate (lower right panels). See Johnson et al. (1992) for sources and details.

and available data indicate that recruitment levels are only approximately 50% of mortality levels (Nicholas, 1992).

Physiological studies in the field at both Southern and Northern Appalachian sites have shown that reduced cation availability associated with atmospheric deposition of strong anions reduces net carbohydrate production by foliage (McLaughlin et al. 1990, 1991,Amundson et al., 1992; and Schaberg et al., 1997). Reduced growth, decreased photosynthesis, increased respiration, and foliar cation depletion observed in the field were reproduced with red spruce seedlings under controlled conditions with ambient range (pH 3.0) acidic mist (McLaughlin et al., 1993). Reduced Ca availability induced by both interference with root uptake of Ca from the soil and increased leaching of CA from foliage by acidic mists appear to be an essential feature of observed responses in these studies (McLaughlin and Kohut, 1992). Changes in membrane-associated Ca have now been linked to changes in winter hardiness of red spruce (DeHayes et al., 1997) and subsequent controlled studies have shown that acid mist exposures can alter membrane-associated Ca and reduce winter hardiness of red spruce by $10^0C$ (DeHayes et al., 1999).

Dendroecological studies in the South (McLaughlin et al., 1987 and McLaughlin et al., 1998) in parallel to those in the North (Cook and Johnson, 1989) indicate that reduced growth of mature trees is associated with sensitivity to warmer late summer temperatures. These responses suggest that carbohydrate depletion by increased dark respiration, a physiological symptom associated with Ca deficiency (Bangerth, 1979 and McLaughlin and Wimmer, 1999), may have contributed to observed growth losses. Winter damage to foliage has not occurred to any significant extent in the Southern Appalachians, although a less severe chlorotic stippling, which accumulates over winter with increasing needle age has been detected (Anderson et al., 1991). The absence of severe foliage loss may explain the lower mortality generally observed in southern red spruce stands. On the other hand foliar and soil Ca levels are generally lower (Joslin et al., 1994) and soil solution aluminum levels are substantially higher at southern sites where soil solution Ca:Al levels $\leq 0.50$ occur frequently at some high elevation sites (Johnson et al., 1991).

While most of the research in the spruce fir ecosystem has been directed at red spruce, significant mortality has also occurred in co-dominant and closely-related balsam fir (Abies balsamea (L.) Mill) in the North (Weeks and Smoronok, 1988) and Fraser fir in the South (Dull et al., 1988). Observed declines of these species are regarded by most to be primarily a natural phenomenon linked to cohort senescence (fir waves) and insect damage from the balsam wooly adelgid (Adelges fraserii). However the influence of Ca deficiency on a wide variety of plant processes including both senescence and disease resistance (Pooviah, 1988) suggests that consideration of the role of Ca deficiency in predisposing trees to these stresses is warranted (McLaughlin and Wimmer, 1999). In the South, growth declines of mature fir began about 1965 (Adams et al., 1985) at the same time that adelgid damage was first detected at multiple mountain sites (Eagar, 1984). Physiological studies of Fraser fir indicate that increased dark respiration linked to reduced Ca and increased elevation, parallels symptoms noted for red spruce (McLaughlin et al., 1997). In the North, aggrading stands of balsam fir grew more slowly at high elevations compared to those of the same age at lower elevations after 1960 (Reiners and Lang, 1979), indicating that this species has experienced increasing stress at high elevation sites during the same time interval over which red spruce has been adversely affected by acidic deposition.

## 4.4. NORTHEASTERN HARDWOOD FORESTS

Abnormally high levels of mortality have been recorded for several species of hardwood trees in the northeast since the 1980's. Most concern has centered on sugar maple forests across large areas in Ontario, Quebec, Vermont and New Hampshire.

Extensive crown dieback and mortality of sugar maple was first detected during the late 1970's and early 1980's. By 1982, decline symptoms were evident across 2,000 km$^2$ of sugar maple forest in Quebec alone (Roy *et al.*, 1985). A sustained recovery of these maples has occurred since then with dieback and mortality now reduced to historical levels (Hall *et al.*, 1997). Hall (1995) reported sugar maple had an annual mortality of 0.9% during 1987-1993 with tree condition showing little yearly change. Similarly NAMP data collected during 1985-96 in the US showed the condition of sugar maple trees (n>4,000) in the New England and New York region was good during 1988-1996. Mortality averaged between 0.9% and 1.9% per year (Stoyenhoff *et al.*, 1998). Crown condition improved with transparency declining from over 18% in 1988 to less than 12% in 1996. Dieback declined from circa 7% in 1988 to circa 6% in 1996. No difference was determined for the health of sugar maple growing in sugarbush and non-sugarbush stands. In Canada condition of sugar maple crowns on NAMP (62 sites) plots generally improved from 1988 to 1993, particularly in Quebec which had the highest level of dieback in 1988 (Lachance *et al.*, 1995). Dieback ranged from 5.3% to 8.2% during 1988-1993 while transparency ranged from 9.1% to 21.6%. As in the US, crown condition and mortality did not differ between sugarbush and non-sugarbush stands.

While no relationship was initially found between average NAMP cluster dieback/transparency and average five-year mean wet $SO_4$ or $NO_3$ deposition (Lachance *et al.*, 1995), subsequent analyses based on regional differences in soil buffering provided additional insights. In initial analyses relationships were tested using deposition values from air monitoring stations not collocated with plot clusters and data were averaged regionally into three zones (low, medium, high). In subsequent analyses atmospheric deposition rates were superimposed on soil acidification potential based on criticals load calculation. Exceedances of critical loads were calculated (Arp *et al.*, 1996) for ARNEWS plots (Figure 7), and these analyses indicated that the majority of the plots located north of the limestone region in extreme SW Ontario experienced high exceedances of the calculated critical loads (Hall *et al.*, 1997). Arp *et al.* (1996) also reported using NAMP data for Ontario, that areas of critical load exceedance had consistently higher levels of branch dieback than areas of no exceedance. Hall *et al.* (1997) state that within the Canadian Shield, attainment of a 50% reduction in the greatest critical load exceedance would require a reduction of $SO_4$ deposition rates to 12 kg ha$^{-1}$ yr$^{-1}$ (See Figure 2).

Linkage of tree response to soil sensitivity to acidic deposition is closely tied to soil supply capacity for base cations. This is apparent in the northern Appalachians, where nutrient deficiencies K and Mg were initially associated causally with sugar maple decline (Bernier and Brazeau, 1988a,b; Bernier *et al.* 1989). This was supported by responses to fertilization with K, which improved foliar nutrient status and growth rates in declining maples (Ouimet and Fortin, 1992), and fertilization with Ca and Mg, which led to improved tree vigor (Hendershot, 1991). Studies in mature stands have confirmed that ambient levels of acid deposition have accelerated the loss of base cations from soils (Foster *et al.*, 1992) leading to reduced soil Ca, Mg and K concentrations in sugar maple stands. These effects have been greatest where deposition is highest (Morrison *et al.*, 1996). Similarly, liming

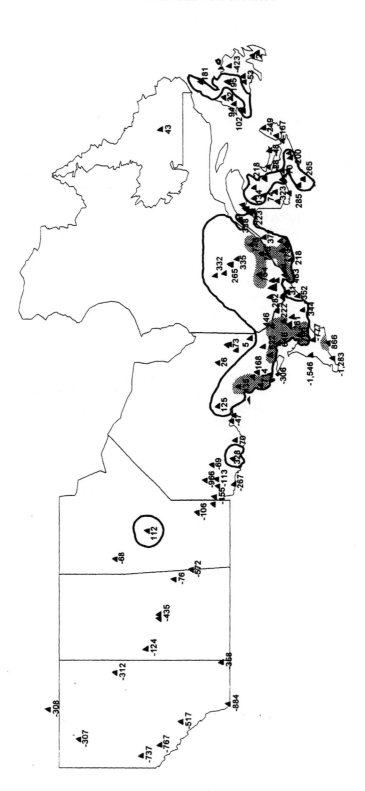

Figure 7. Exceedances of critical acid deposition (eq ha$^{-1}$ yr$^{-1}$) for ARNEWS plots. Dark lines separate regions that exceed levels of $\geq$ 100 eq ha$^{-1}$ yr$^{-1}$ from those below critical thresholds. Shading denotes area of strongest (> 500 eq ha$^{-1}$ yr$^{-1}$) exceedance. Reproduced from Hall et al. (1997).

studies in sugar maple stands in Pennsylvania have demonstrated restoration of more vigorous growth and canopy health (Long et al., 1997). In some sugar maple soil systems efficient in retaining and recycling added nutrients, benefits of cation additions made to soils to offset nutrient imbalances may last several decades (Hendershot and Courchesne, 1994).

Sugar maple is also sensitive to changes in concentration of soil $NO_3$ and increasing N deposition to some N-saturated forests is expected to cause nutrient imbalances, with resultant effects on growth (Yin et al., 1994). In addition despite the scientific and public focus on the role of acidic deposition, other co-occurring stressors of sugar maple forests may be equally important. The role of $O_3$, for instance, has not been adequately investigated. Most of the sugar maple range (NAMP plot distribution) is located in areas of high seasonal ozone exposure and increases in canopy transparency have followed years with higher ozone levels, particularly in 1988 when SUM60 across the sugar maple range increased from 6,600 ppb hrs $O_3$ to over 30,000 ppb hrs $O_3$ (Pearson and Percy, 1997). Transparency also increased during 1988-1993 (Lachance et al., 1995) at those plots where $O_3$ exceedances were highest. Visible foliar injury was reported on three indicator species at 59%, 34% and 39% of FHM plots in the region during 1994-1996 respectively (Stoyenoff et al., 1998).

In addition, winter damage from deep freezing of exposed soil and roots in 1980 in Quebec contributed to sugar maple decline and complicated symptom interpretation. Followup studies on sapflow changes following development of canopy dieback amplified by winter cold treatments of soil (Bertrand et al., 1996 and Robitaille et al., 1995) indicate that the degree of loss of root function may exceed estimates of damage to canopies based on visual symptomology. For example sapflow losses of 39% were found on trees with only 9-13% canopy losses (Robitaille et al., 1995). Such canopy losses have previously been considered normal for sugar maple, but may signal more substantive disruption of belowground processes.

During the early 1980's, the condition of paper birch (*Betula papyrifera* Marsh.) and mountain paper birch (*Betula cordifolia* Regel) growing along the Bay of Fundy coast in Canada also deteriorated significantly. Initial observations of *Betula spp* leaf browning and premature abscission were recorded by the Forest Insect and Disease Survey (FIDS). Subsequent investigation excluded insects and diseases as causal agents. Atmospheric stressors were considered candidates, particularly acidic fog and $O_3$ (Magasi, 1985).

Fogs were collected and found to deposit substantial amounts of acidity (0.1-0.5 keq $ha^{-1}$), which had not previously been detected and which exceeded amounts deposited to the same stands by rain. The fog exposure duration at a nearby site ranged up to 1400 hrs (27% of the time) during the growing season. Significant correlations were determined between $H^+$ ion (P<.001), $NO_3$ (P<.001) concentrations and leaf browning (Figure 8). Continued investigation of the two taxa confirmed the relationship between fog frequency and foliar browning. All trees were affected during 1987-1990 followed by an apparent recovery in tree condition concurrent with historical (1915-1990) lows in fog frequency. Recovery was attributed to the shedding of dead branches and twigs during winter storms and the lack of initiation of new dieback due to low fog frequencies The lower taxon *B. cordifolia* was found to be more prone to browning, slower to recover from dieback and more prone to mortality (Cox et al., 1996).

It has been known for some time that certain mid- to high-elevation species in the northeast like red spruce and balsam fir are adapted to cool, moist environments and are

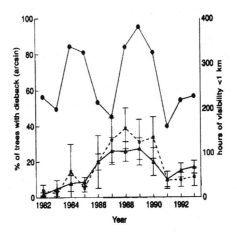

Figure 8. Percentage of Fundy coast *B. cordifolia* (▲) and *B. papyrifera* (■) trees with dieback symptoms in relation to hours of fog (visibility< 1 km) (●) at the St. John airport from 1982 to 1993. Reproduced from Cox et al. (1996).

immersed in clouds/fog for up to 44% of the growing season (Lovett et al., 1982; Schemenauer, 1986). Comparison of northeastern marine-corrected mountain cloud and coastal fog chemistries by Kimball et al. (1988) indicated coastal fogs were up to three times higher in ionic concentrations and were lower in pH. $NO_3$ concentrations were especially high given the higher $O_3$ concentrations measured along the New England coast.

The potential for direct damage to vegetation from acid fog and cloud water is high and Maine coastal fogs have among the highest acidities (pH 2.8) ever recorded (Cape, 1993). When dry deposition is factored into total atmospheric deposition to birch leaf surfaces, very strong, localized acidities are likely (Wisiniewski, 1982), accentuated by meteorologically-driven leaf surface wetting/drying cycles.

## 5. Integrating Effects of Ambient Pollutant Levels on Forest Processes

### 5.1. EFFECTS ON TREE PHYSIOLOGICAL PROCESSES

Despite differences in species, levels and combination of principal and secondary stresses, and primary pathways of effects among the four case study forests, effects of air pollutants on the resources required for growth and defense are often similar. In Figure 9, changes in the primary resources that regulate forest function: carbon production and allocation, water supply, and plant nutrition are depicted. Important features and implications of alterations in these resources are discussed below.

#### 5.1.1. *Effects on Carbon Production and Allocation*

Reduced photosynthetic production in response to chronic pollution stress was found in all four systems examined. This occurred both as a result of changes in photosynthetic rate per unit of foliage mass (a shift in biochemical production potential) as well as from changes in the amount of leaf area produced or retained under chronic pollution stress. The mechanisms of effect, and foliage age classes involved were quite different among the different forests, however. While premature loss of older needle classes has been observed with ozone exposures in both western and southern pine forests, loss of upper canopy foliage of northern hardwoods is most closely associated with nutrient depletion of the outer canopy due to both foliar leaching by acidic deposition, and reduced uptake of Ca and magnesium from poorly buffered soils (Heisey, 1995; Ellsworth and Liu, 1994).

In red spruce significant loss of foliage can be caused by winter injury to current year needles, which are both the most productive and the most sensitive age class. Sensitivity is enhanced by exposure to mists at acidity levels that occur frequently at high elevation sites in the Eastern US, and has been linked to increased frequency of terminal shoot dieback (LeBlanc and Raynal, 1990, and see Figure 6). It is also associated with low Ca retention and is an apparent consequence of foliar leaching of membrane-associated Ca (DeHayes *et al.*, 1997). Losses of foliar Ca by individual rain events may amount to 30% of the foliar Ca content (Joslin *et al.*, 1988). Much slower loss rates have been estimated for sugar maple (10-133 day half-life) suggesting biological non-significance (Scherbatskoy, 1989). However, Ca signaling in leaf physiological processes occurs at levels three to four orders of magnitude lower than those found in cell walls, and even small leaching losses from the small but active metabolic pools in foliage may significantly alter cell physiological functions (McLaughlin and Wimmer, 1999). Both reduced foliar Ca retention and reduced

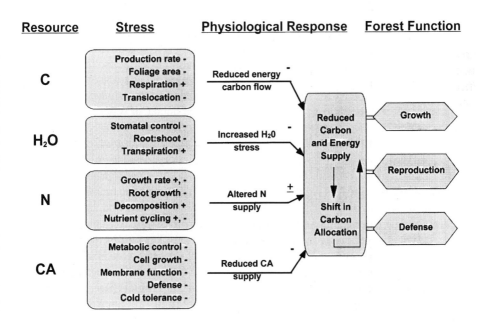

Figure 9. A summary of the physiological changes induced by ambient air pollution in the four case studies indicates multiple pathways of effects on carbon, water, and nutrient resources required for forest health. The net effect of these process-level effects has been multiple indications of reduced carbohydrate production by affected trees. Such changes typically lead to a shift in relative allocation of energy toward growth at the expense of reduced resistance to biotic and abiotic stresses.

Ca supply rates from soil appear to contribute to reduced foliar Ca and increased physiological dysfunction of red spruce foliage (McLaughlin and Kohut, 1992).

Both chronic $O_3$ exposure of pines (Adams et al., 1990, McLaughlin et al., 1982) and exposure of red spruce to acidic deposition in the field (Andersen et al., 1991), reduce basipetal translocation of photosynthetically-fixed carbon from source foliage. Increased retention of photosynthate has also been associated with increased respiratory costs associated with repair of oxidant-stressed foliage and contributes to the reduced availability of carbohydrates for new growth (McLaughlin et al., 1982). Reduced foliar carbon production (reduced Pn rate and leaf area) and reduced allocation away from production centers leads to the reduced root growth frequently associated with ozone exposure (Cooley and Manning, 1987). By contrast the effects of acidic deposition on root growth and survival can occur both from reduced carbon production above ground (winter injury and increased respiration of foliage) as well as from toxicity associated with aluminum mobilization in the soil solution (Cronan and Grigal, 1995). More typically acidic deposition affects root growth by decreasing Ca:Al ratios to levels which interfere with Ca uptake and impair root physiological function (Joslin et al., 1992, and Godbold and Kettner, 1991).

### 5.1.1. Nutritional Effects of Changing N and Ca Availability

Changing nutrient availability induced by atmospheric deposition can also alter plant allocation of carbon in ways that increase sensitivity to other stresses. Such changes include increased deposition and accumulation of N (Aber et al., 1989), reduced deposition of atmospheric sources of Ca (Hedin et al., 1994), and reduced availability of Ca in soils (Likens et al., 1998). Increased absorption of atmospheric sources of N and metabolic utilization of N by forest canopies (Hanson and Lindberg, 1991, Fowler et al., 1998) as well as increasing N availability from chronic accumulation in surface soils has the effect of shifting tree growth and allocation of resources aboveground at a cost of reducing root mass and reducing root:shoot ratios (Persson and Madji, 1995, Schulze, 1989). Such changes can place forests at increasing risks of reduced uptake of water as well as other mineral nutrients.

Ca, on the other hand, has a direct and indirect role in allocation of resources to defense and repair in plants as well as influencing respiratory metabolism. Availability of Ca is particularly important for developing plants because it is not translocated between plant organs and because it is required for a wide variety of metabolic functions associated with growth, structural integrity of cells and membranes, local and interorganismal signaling, and defense and repair processes associated with disease resistance (Hepler and Wayne, 1985). Ca supply is therefore linked to transpiration rate in trees and can be reduced by drought as well as factors which reduce root vitality. Accelerated leaching of Ca from foliage exposed to acidic mists (Joslin et al., 1988) and reduced availability of Ca in soil solutions in poorly-buffered soils exposed to strong anion inputs (Reuss and Johnson, 1986; Robarge and Johnson, 1992) can combine to limit Ca uptake and increase physiological stress in acidic soils.

### 5.1.3. Altered water status

Air pollution stress can increase water stress in forest trees by two principal pathways: alteration of transpiration through control of water flux from leaves at the membrane (substomatal) or stomatal level; or effects on water uptake related to changes in fine root biomass or root function. Much of our thinking regarding ozone and water stress has been based on seedling studies in which water stress has been shown to accelerate stomatal closure in response to high ozone exposures (Tingey and Hogsett, 1985). This assumption led Ollinger et al. (1997) to assume a protective role of water stress in modeling growth responses of mature forests to ozone stress. However, in controlled studies with 3-yr-old beech trees the stomatal closure which occurred when well watered trees were exposed to ozone was impeded under dry conditions (Pearson and Mansfield, 1993) suggesting a more complex regulation of water loss under these conditions.

Larger forest trees typically have much lower ratios of leaf area to root mass than seedlings and have a larger stem volume which can provide an additional hydrologic reservoir during stress episodes (Kozlowski et al., 1991). While stomatal conductances of large trees of some western species can be much lower that for seedlings (Yoder et al. 1994, and Grulke and Miller, 1994), Samuelson and Kelly (1997) have reported both higher conductances and higher ozone uptake within the canopies of large trees than by nearby seedlings in Tennessee. The capacity for ozone to cause stomatal closure as a consequence of water stress may be reduced for larger trees at the moderate water stress levels which limit many physiological processes. Both cell wall growth and protein synthesis, for example, are reduced at stress levels well below those required to cause stomatal closure (Hsaio, 1973).

Studies in the field at ambient levels now provide direct and indirect evidence that moderate ozone levels can increase water stress and reduce growth in larger trees (see McLaughlin and Downing, 1997). Reduced stomatal control of water loss from conifers documented in European studies (Maeir-Maerker,1997) documents one mechanism for this response, however reduced root function provides a potentially additive pathway that has not been adequately evaluated to date. On the other hand in the more arid San Berardino Forest, drought increased resistance of older needles to ozone induced injury. This suppression of $O_3$ injury to older whorls of ponderosa and Jeffrey pines during drought years, was followed by $O_3$ injury and needle abscission upon return of favorable soil moisture (Miller et al., 1996).

## 5.2. EFFECTS ON BIOGEOCHEMICAL CYCLES

The primary effects of current and projected future air pollutant loading on biogeochemical cycles and associated forest health are associated with three processes that operate concurrently in at least some soils in all of the four case studies we have examined. These include depletion of base cations due to increased mobilization and leaching from forest soils (Federer et al., 1989), alteration of nutrient availability in soil solutions through mobilization of acidic cations, AL, Fe, Mn, and H, (Robarge and Johnson, 1992 and Ulrich and Matzner, 1986), and finally the buildup of soil N to levels that saturate surface soils (Aber et al., 1989).

### 5.2.1. Cation Losses

Nutrient budgets for diverse forest stands now document the increased leaching of base cations associated with inputs of the strong anions $SO_4$ and $NO_X$ to forest soils (Johnson et al. 1985; Federer et al., 1989; Johnson and Lindberg, 1992; Richter et al. 1994; Likens et al., 1996; and Likens et al., 1998). In general, acidic deposition has been found to approximately double leaching rates that occur naturally (Joslin et al., 1992 and Johnson and Todd, 1987) Long term depletion of base cations from both foliage and soils in the eastern US has led to accelerated depletion of exchangeable cations from mixed hardwood forested watersheds and predictions of Ca deficiency in coming decades (Federer *et al.*, 1989). Similarly combined effects of forest growth, an important acidifying influence that is a natural part of cation accumulation by aggrading forests (Ulrich and Matzner, 1986), natural leaching and acidic deposition depleted 80% of the base cations from surface soils of a southern pine site in South Carolina over 30 years (Richter *et al.*, 1994). Because of the importance of forest growth in accumulating Ca and other cations, whole tree harvesting, which removes nutrient rich branches as well as boles, can accelerate ecosystem depletion of cation pools (Federer *et al.*, 1989).

### 5.2.2. *Acid Cation Interference*

Two properties of cation exchange in forest soils are pivotal to the capacity of acidic deposition to reduce physiological availability of cations: First, aluminum is mobilized more rapidly than Ca from exchange sites in the soil solutions when strong anions are deposited on poorly buffered soils (Reuss and Johnson, 1986); second strong binding capacity allows the acidic cations Al>Mn>Fe>H to replace Ca and Mg from both physical and biological binding sites. The first property leads to domination of Al in soil solutions in high N and S deposition areas with poorly buffered soils. As Ca:Al ratios fall below 1.0, a wide variety of studies indicate that the risk of adverse effects to forest physiological function increase (Cronan and Grigal, 1995). Analysis of fine root chemistry in both declining red spruce (Shortle and Smith, 1988) and declining sugar maple stands (Adams and Hutchinson, 1992) indicate that low Ca:Al levels in fine roots are associated with reduced growth of trees in poorly buffered soils.

The high replacement potential of acidic cations for Ca, can result in displacement of this essential cation from the cell walls and membranes of fine roots resulting in and loss of membrane integrity and reduced root absorptive capacity (Zhao *et al.*, 1987, Stienen and Bauch 1988 and Schroeder *et al.*, 1988). In addition, Ca binding in cell walls, an important feature influencing the structural integrity of wood, can be reduced by acid cation substitution in xylem walls (Momoshima and Bondietti, 1990) thereby potentially reducing wood structural integrity (McLaughlin and Wimmer, 1999).

### 5.2.3. *Nitrogen saturation*

The long-term inputs of atmospheric sources of N to forest soils have begun to produce measurable changes in forest N cycles in industrialized regions. The phenomenon of N saturation occurs when availability of ammonium and nitrate is soils exceeds the combined plant and microbial demands (Aber *et al.*, 1989; Schulze, 1989). This process can occur naturally where N-fixing species dominate (Cole, 1992), but in other forest types it results from the combined addition of $\leq$ 40 kg ha$^{-1}$y$^{-1}$ of anthropogenic N added to the 20-75 kg/ha/y produced annually by decomposition processes of coniferous forests and 50-175kg/ha/y similarly produced by deciduous forests (Aber *et al.*, 1989). Based on European studies Schulze *et al.* (1989) have suggested critical loading thresholds for N of 3-14 kg ha$^{-1}$ y$^{-1}$ for poorly buffered soils and 3-48 kg ha$^{-1}$ y$^{-1}$ for calcareous soils (note current wet deposition loading patterns for North America in Figure 3).

The net effect of N saturation is accumulation of the acidic nitrate ion in the soil profile, reduced Ca:Al ratios in soil solutions, reduced root mass and depth, and increased N export to streams. Evidences of N saturation are now becoming apparent in many different forest types including high elevation Southern Appalachian spruce-fir (Nodvin *et al.*, 1995), mid-elevation mixed hardwoods in North Carolina (Swank and Vose, 1997),

West Virginia (Gilliam et al., 1996), Northern Hardwoods in both the US (Aber et al.,1989, and Likens et al. 1998) and Canada (Foster et al., 1992) as well as in areas of regionally high N deposition in California (Bytnerowicz and Fenn 1996). Nitrogen saturation has implications for long term stability of forest nutrient cycles because it can lead to breakdown of the cation- rich O horizon, and increased leaching of cations from deeper soil horizons (Shortle and Bondietti, 1992). Evidence suggests that the low soil Ca resulting from this process, affects root function, including uptake of N, thereby potentially contributing further to N accumulation in the soil (McLaughlin and Wimmer, 1999). Under these conditions forests can be expected to become increasingly dependent on recent litter for required Ca levels and roots to colonize shallower soil layers in the forest floor (Persson and Madji, 1995 and Joslin and Wolf, 1992). Another consequence of increasing N can be reduced growth of fine roots (Dougherty et al., 1997).

*5.2.4. Combined Effects on Forest Nutrient Cycling*
So how significant are these combined processes in limiting nutrient availability to regional forests? Both comparative nutrient budgets across diverse forest types (Johnson and Lindberg, 1992) as well as variations in foliar and root nutrient levels across environmental gradients (Friedland et al., 1988; Robarge at al., 1989; and Joslin et al., 1994) indicate that there are significant differences in plant-available nutrients, particularly the base cations, Ca and Mg associated with gradients in acidic deposition to forests. The Integrated Forest Study (IFS, Johnson and Lindberg, 1992) has provided nutrient budgets for at least two representatives for three of the four forest types we have addressed in this review. In Figure 10, we compare for spruce-fir, southern pines, and northern hardwoods, the distribution of Ca pools in vegetation, the forest floor, and soil exchangeable pools (Figure 10a) with the fluxes of Ca, including Ca uptake, leaching, and net increment (Figure 10b). These comparisons emphasize the large differences in nutrient capital both between and within systems, and they also demonstrate why the risks of acidic deposition to nutrient cycles can differ markedly within regions and between forest types. The lower exchangeable Ca pools at high elevation spruce fir forests (in both the Southern (GSM) and Northern (WF) Appalachian Mountains are readily apparent from these contrasts.

Long-term biomass modeling based on critical load exceedance calculations has predicted that forest productivity will decline in some regions with continued $SO_4$ and $NO_3$ deposition (Arp and Oja, 1992). Long-term (30 yrs) ecosystem research at Hubbard Brook, NH reported by Likens et al. (1996) has documented substantial losses of soil base cations due to acidic deposition. Important consequences for forest sustainability are predicted. Long-term research at the watershed level has also been instrumental in documenting $SO_4$ flux through nutrient-poor sugar maple forests indicating that soil buffering capacity of these forests has been depleted. On the other hand, there is also evidence at the watershed scale that reducing S emissions in the Northeastern US is beginning to have measurable effects of nutrient cycles of some forests. For example, at Turkey Lakes Watershed in central Ontario, monitoring of precipitation, nutrient cycling and streamflow chemistries from five basins has detected declining acidic concentrations in precipitation during 1981-1990 and this has resulted in lower concentrations in soil runoff and stream water. Decreases of up to 31% have reduced the flux of $SO_4$ in the soil (Foster and Hazlett, 1991), and will reduce rates of cation loss from these systems.

Foliar and fine root nutrient analyses provide the most important integrators of nutrient availability to forests. Analysis of foliar nutrient data suggest that foliar cation levels principally Ca and Mg are currently in the range of expected deficiency symptoms at wide

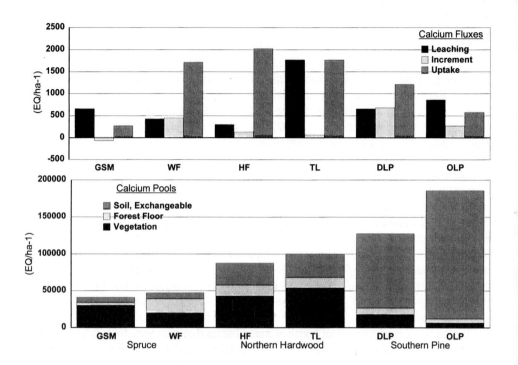

Figure 10. Comparative Ca fluxes (a) and pools (b) for representative spruce, northeastern hardwood, and pine forests are important components of differences in sensitivity of these forests to acidic deposition (after Johnson, 1992). Forest types include: high elevation spruce-fir forests, Great Smoky Mountains (GSM) in TN, and Whiteface Mountain (WF) in New York; northern hardwood forests, Huntington Forest (HF) in NY and Turkey Lakes (TL) in Ontario; and southern loblolly pine forests in Durham, NC (Duke Forest (DF) and Oak Ridge, TN (OLP). Note the low net increment (a) and lower exchangeable soil pools (b) for high elevation spruce and northern hardwood forests compared to managed lowland pine forests

spread locations in the high elevation spruce fir forests (Friedland et al., 1988, Joslin et al., 1994). Reduced base cation levels have also been noted at higher elevations for the northeastern hardwood forest at Hubbard Brook (Likens et al., 1998). Such data do not distinguish the relative roles of innately low soil fertility and any additive effects of acidic deposition on foliar nutrient uptake. However several lines of evidence suggest that acidic deposition had been an important contributing factor to reduced cation availability in areas of high atmospheric deposition: These include (1) accelerated leaching of base cations from foliage by acidic rain and mist events (Joslin et al., 1988), (2) demonstrated physiological significance of relative small changes in the small, active pools of leaf Ca (McLaughlin et al., 1993 and DeHayes et al., 1997) ; (3) close linkages observed between input of $SO_4$ and $NO_3$, elevated levels of Al in soil solutions and depression of Ca:Al ratios in soil solutions to levels (seasonal Ca:Al mean = 0.40, Johnson et al., 1991) at which cation uptake by roots is inhibited; (4) shifting patterns of cation content of annual tree rings that indicate that substitution of acid cations for basic cations has occurred in recent decades in areas receiving high deposition of atmospheric acids (Bondietti et al., 1989, Bondietti et al., 1990, and see Watmough, 1997); and finally, (5) Retrospective analysis of foliar cation levels at a Southern Appalachian spruce/fir site, which indicates that foliar levels of Mg and Ca have decreased by 50% and 33% respectively during the past twenty years (Shepard et al.1995).

## 6. Implications of Changes in Forest Processes for Forest Health

From the utilitarian perspective the direct effects of air pollution on forest health that can be documented at present on North American forests may be considered to be relatively small compared to documentable effects of other types of stresses, such as fire and biotic diseases (see Table II). The most direct linkages can be made to the regional forest types we have chosen as case studies. With red spruce, sugar maple, white birch, and western pines, all predominantly older growth natural communities, the symptoms are reduced growth, weakened trees, increased susceptibility to stresses, and increased mortality. These are the classical indicators of the multicomponent decline complex in which air pollution can be a contributing factor (Manion, 1981). In two of the four systems, northeastern montane red spruce and southwestern pines in the San Bernardino forest, decreasing health had been translated into changes in stand structure and diminished basal area, meeting the utilitarian definition of health loss. With southern pines, a substantial component of which is managed commercial forest, data from controlled chamber studies, dendroecological studies, and models (see Table III) indicate that growth of mature pines is being reduced by current typical ambient ozone levels at annual rates that may vary from 0 to 10% per year. However, at present there are no indications of community-level changes attributable to ozone, and the biological effects of cation depletion of southern pine soils on health of these forests are still conjectural (NAPAP, 1998).

From the ecological perspective, however, we must ask how the process-level indicators of altered forest function from these few case studies relate to responses of these and other forest types to the range of biotic and abiotic stresses that normally regulate forest growth and development. In Figure 9 we summarized relationships among processes affected by ambient levels of air pollutants and changes in water, carbon, and nutrient resources that can be documented at those levels. These are of course the same media through which climate, competition, and disease affect forest processes and forest health. In the regional forest environment it is important to recognize that chronic stress from air pollution can only be expressed through altered sensitivity of forest physiological processes to natural biotic and abiotic stresses. Additionally, internal shifts in resource allocation strategies related to plant allometry (McLaughlin and Shriner, 1980), stage of growth (Mooney and Chu, 1974), nutrition (Clancy et al., 1995) age (Waring, 1987), species type (Waring and Cobb, 1992) and stand level nutrient balance (Mattson and Addy, 1975) significantly influence plant allocation of resources among the competing demands of growth, defense, and reproduction and hence can alter forest sensitivity to insects and diseases.

Evidence from process level studies indicates that regional levels of air pollution stress measured in our case study forests are reducing carbon reserves, increasing water stress, and reducing nutrient availability. Carbon (energy) reserves are critical to plant defenses in many ways, including notably formation of defense chemicals that prevent damage and repair or replacement of damaged tissues after attack. The allocation of carbon to defense in forest trees is a much lower priority than maintaining leaf and root function (Waring and Pittman, 1985), thus chemical and structural defenses may be more significantly reduced at the organ or whole tree level by stresses which reduce carbohydrate production levels. Root diseases of conifers, including littleleaf disease (Phytophtera cinnamomia) of shortleaf pine (Hepting, 1945) and Armillerea root rot (Wargo, 1972) which often develop as secondary pathogens following predisposing stresses (Manion, 1981) are enhanced by reduced carbohydrate supply to roots. The increased sensitivity of ponderosa pine to bark beetle attack in the San Bernardino Mountains following $O_3$ damage is a classical case of reduced defenses of forest trees being related to chronic air pollution stress (Cobb et al., 1968).

Closely tied to reduced carbohydrate reserves, whatever the cause, is reduced root growth and function and ultimately increased water stress. Water stress frequently increases tree sensitivity to insect attack (Clancy et al., 1995) although sensitivity can vary with level of stress (Lorio and Summers, 1986) and species type (Waring and Cobb, 1992). For example, susceptibility of conifers to insect herbivores was consistently increased by water stress (29 out of 31 studies) whereas broadleaves showed almost equal positive and negative responses (Waring and Cobb, 1992).

Significant strides in understanding the interplay of plant physiological responses to stress have come in the area of plant nutrition, specifically significantly expanded understanding of the role of Ca in regulating a wide range of physiological processes. These include intercellular and intraorganismic signaling (McAinsh and Hetherington, 1998 and Trewavas and Malho,1997), chemical defenses (Pooviah, 1988 and Roberts and Harmon, 1992),and physical repair of cellual damage by disease (Dixon and Harrison, 1994). In addition to being essential for cell wall growth (Eklund and Eliasson, 1990), lignin is an important defense chemical against diseases such as damage to firs by the balsam wooley adelgid (Timmel et al., 1986). Poor soil nutrition, including low Ca, P, K, and Mg has been recognized as a predisposing factor is sensitivity of Douglas fir to spruce budworm attack (Kemp and Moody, 1984). Stark (1965) lists 20 studies in which forest fertilization increases resistance of pines to insect attacks. Interestingly in 10 of 12 studies in which Ca was added alone or as a component of fertilization, reduced density or increased mortality of insect larvae was noted. Radiochemical studies suggest that Ca may be toxic to insects (see Stark, 1965). The loss and/ or reduced availability of exchangeable soil Ca from soil reserves in response to long term deposition of S and N, thus has important implications for long term forest growth and forest ecosystem health.

Collectively the physiological effects of air pollutants would be anticipated to either predispose forest trees to other stresses, or amplify their negative effects. Only recently, have patterns of change over time (>10 yrs) at regional and national scales been statistically analyzed with the vigor required to attribute changes in forest health at the species level to individual and/or complexes of biotic and abiotic stressors, including air pollutants (Brandt et al., 1999). For example, in North America the regional patterns of most frequent occurrence of major disease problems documented by forest surveys, are spatially consistent with the patterns of highest levels of ozone and acidic deposition as noted in Figure 11. Such a relationship is compatible with the role of strong anion deposition in depleting available pools of cations, notably Ca, from the relatively low soil pools in some Northeastern forests (see Figure 10). This supports the possibility that both carbohydrates and nutrient resources that influence forest health through altering tree resistance or recovery potential are relatively more affected by pollution stress in the same areas that insects and disease are apparently most problematic. Detailed statistical analyses might reveal a statistical relationship between developmental trends and air quality that is beyond the scope of the more qualitative analyses presented here. While such patterns complicate analyses of cause and effect relationships on a regional basis, they also provide opportunities to examine component stresses from the perspective of process-level interactions involving carbon, water, and nutrient resources. Such analyses would be required to permit more robust tests of cause and effect relationships.

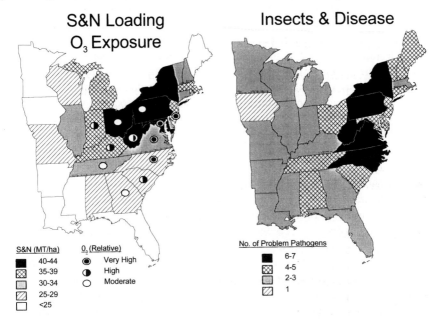

Figure 11. The areas of the eastern US that currently have the greatest frequency of problems from forest insects and disease (USDA, 1997) also receive the highest deposition of S and N, and/or have the highest annual exposure to tropospheric ozone. Deposition data are area - integrated esttimates at a state level from regional isopleths presented by Allen and Gholz (1995) based on summed wet deposition of S and N in kg/ha/y and relative ozone exposure for 1998-1999. The area-integrated ozone exposure index was > 850 for states in the very high category, 775-850 for the high category and 700-775 for the moderate category. By contrast, exposure indices for AL, ME, and IO were 500, 280, and 120, respectively.

## 7. Predicted Changes in Physical Climate and Effects on Forest Physiological Function

The linkages between changing global atmospheric chemistry and climate change have become solidified in recent years (Thomson, 1995, IPCC, 1998). On a global scale, temperatures are predicted to increase by 1-3.5°C by 2100. Spatial and temporal patterns of precipitation are expected to change more variably, leading to increasing frequency of droughts in some regions. Stratospheric ozone depletion since 1980 is not predicted to recover until 2050, even under the most optimistic scenario (Madronich et al., 1995). Resultant increases in biologically-effective ultraviolet radiation (UV-B, 280-315 nm) reaching forests may have important consequences for tree molecular biology (DNA), biochemistry, physiology, growth, regeneration and reproduction (Percy and Gordon, 1998) as well as changes in plant succession and biodiversity (SCOPE, 1992). Ecological systems, including notably forests, are predicted to be significantly impacted by both the magnitude and variability of climatic shifts produced (IPCC, 1998). These responses, which include both positive and negative physiological effects that will influence process level changes induced by regional air pollutants, are reviewed very briefly below. Primary emphasis is on results from the Southeastern US, where an integrated series of studies has recently been completed (Mickler and Fox, 1997).

From a physiological perspective, the effects of increasing $CO_2$ and increasing temperature on net primary production, the effects of temperature on the frequency and severity of drought, and the effects of increasing N on nutrient uptake and allocation to shoots and leaching from soils appear to be the most compelling components of climate change that could significantly influence forest sensitivity to biotic and abiotic stresses. Predicted regional climate models for the Southeastern US associated with a doubling of $CO_2$ levels (Cooter, 1998), include a mean growing season temperature increase of 3.7-6.4°C with more variable and less certain effects on precipitation patterns.

A wide variety of controlled studies now indicate that increasing $CO_2$ will lead to significant increases in leaf photosynthetic rates and increased net primary production of forest trees (Wullschleger et al., 1997; Teskey et al., 1998; and Norby et al., 1999). Although very limited data are available for more mature forest trees, results to date suggest that positive responses are rather robust, in terms of absence of limitations by variations in supply of other resources (Wullschleger et al., 1997). Increasing temperature, on the other hand, affects many essential forest processes and in many different ways. Most notably the carbohydrate economy may be improved by extending the length of the growing season, but adversely impacted by the increased respiratory costs of temperature-driven metabolic processes. A 5°C temperature increase might increase dark respiration by 50% ($Q_{10}$ = 2) and have a major impact on carbohydrate production and allocation. Based on branch studies, Teskey et al. (1998) found that a 2° C temperature increase, while it increased Pn at ambient $CO_2$, reduced the Pn response to increasing $CO_2$. In addition increasing temperature will increase evapotranspiration, thereby increasing potential limitations of water supply on the amount and allocation of growth (Schulze et al., 1987).

Current global models used for predicting the distribution and condition of forests in response to global change (IPCC, 1998), consider increases in $CO_2$ and water availability, on forest growth, but, at present, do not consider potential implications of regional air pollutants on forest growth processes. Modeling results, for a 2X $CO_2$ scenario (Table IV) indicate that with climatic warming the potential forest cover in North America might increase by 24% (without) to 44% (with) $CO_2$ stimulation of growth. However, results from field chamber experiments provide evidence of the potential for $O_3$ and $CO_2$ co-exposure to offset predicted gains from elevated $CO_2$ and to enhance damage from $O_3$ (Kull et al., 1996) and indicate a need for consideration of combined effects of these gases in future assessments.

Much of the simulated increases in Table IV occur as a result of increases in the distribution of temperate forests, which migrate northward, decreasing forest area in the conterminous USA and increasing potential forest cover in Canada and Alaska. By contrast, the modeling projects a significant increase in areas within each forest type in which declining growth would be expected, with 9 to 19% of the area being so affected if the positive effects of $CO_2$ are unconstrained. In the absence of positive $CO_2$ effects, climate change caused predicted productivity to decline over 45% to 89% of the total forested areas of North America.

TABLE IV

Predicted percentages of current forested area in North America that would lose or gain biomass with a doubling of pre-industrial atmospheric $CO_2$ levels. The effects of associated climate change without $CO_2$ potential physiological benefits of 2 x $CO_2$ are shown in the numerator of each response pair.[1]

PERCENTAGE OF FORESTED AREA RESPONDING (-/+ $CO_2$)

|  | Biomass loss (-/+) | Biomass gain (-/+) |
|---|---|---|
| Boreal forest and Taiga tundra | 25/5 | 49/83 |
| Boreal Conifer | 84/16 | 2/54 |
| Temperate evergreen forest | 55/8 | 12/66 |
| Temperate mixed forest | 44/4 | 26/95 |

[1]Source: IPCC (1998), The Regional Impacts of Climate Change. These are considered conservative estimates of expected temperature effects due to delayed temperature equilibration to increasing $CO_2$ (IPCC, 1998). Total forest area predictions for North America are for a 42-44% gain with $CO_2$, and 21-24% gain in the absence of $CO_2$ effects.

The physiological basis of the IPCC (1998) model-predicted declines in forest health in the two times ambient $CO_2$ environment are based upon the effects of high temperatures on forest growth, and associated increases in drought, fire, and pathogens. In reality the spectrum of responses is likely to include both positive and negative responses in forest growth over time and across regions depending on the rate of change of the various climatic components and spatial variability in soils. In regional simulation of responses of loblolly pine stands to climate change in the Southeastern US (McNulty et al., 1996), small increases in temperature ($2^0 C$) were found to have positive effects on Pn unless combined with reduced water supply. Non-linearity of expected responses is particularly likely in the case of N deposition (Aber et al., 1989), which is envisioned to have initial stimulatory effects on forest growth on N-poor soils, such as for southern pines, followed by negative effects as cation depletion continues, leading potentially to N saturation.

Stimulation of growth regionally by N deposition, as has apparently occurred in some areas of Europe in recent decades (Sterba, 1996), also has potential parallel non-linear effects on global climate change. Modeling studies indicate that N-stimulation of growth in a $CO_2$-enriched environment could initially lead to significantly increased carbon and N storage in the soils, but at the risk of forests ultimately behaving as sources of $NO_2$ from microbial release of accumulated N (Norby, 1998 and Goulding et al., 1998). The effects of warming on litter decomposition and release of accumulated nitrate represent an additional risk, particularly to N-saturated soils such as occur in the high elevation spruce-fir ecosystem (Joslin, and Johnson, 1997). Interestingly studies in Germany indicate that liming can drastically reduce $N_2O$ release from forest soils fertilized by N (Brumme and Beese, 1992) a response that suggests important linkages between Ca depletion from soils and the ability of vegetation to utilize/retain added or excess soil N.

## 8. Synthesis and Recommendations

The case studies we have reviewed indicate that stress from regional scale air pollutants is affecting the physiological processes of forest tree species within diverse, widely distributed, and regionally important forests. The processes affected and the stresses they produce are important to the way these forests grow and respond to biotic and abiotic stresses within their regional environments. If one measures forest health strictly in utilitarian terms, the level of effects to the near term economic health of the forests we have examined that can be clearly linked to specific effects of air pollutants may be considered small relative to the total suite of stress responses that affect forest health. However, If one views the potential importance of the processes affected in terms of implications for performance of these species within the framework of biotic and abiotic stresses that affect and are affected by the same resources, the potential ecological effects become potentially much larger, particularly over time. In this regard we consider the potential to be high for ambient ozone and acidic deposition stress to enhance the effects of drought, insects, disease, and high temperature in many situations, but to vary widely across the forest landscape just as these stresses vary. Similar conclusions have been reached in a synthesis of European physiological studies, which highlighted changes in carbohydrate production and allocation as critical to enhanced sensitivity of European forests to secondary stresses (Skarby et al., 1998).

The well documented effects of ambient air pollutants on carbohydrate production and allocation, root growth, water uptake and control, and soil Ca and N levels indicate that regional atmospheric pollution could significantly enhance the adverse effects of global warming, particularly in the United States. While we expect increasing $CO_2$ to ameliorate air pollution effects on tree physiology by enhancing carbohydrate production and allocation to roots, biochemical interactions at the leaf level are less clear (Kull et al., 1996). We expect the effects of increased temperature and increasing drought coupled with the concurrent effects of ozone, and acidic deposition (including N) on root vigor and root depth to increasingly place limitations on $CO_2$ induced enhancement of growth. Under these conditions, the enhancement of stress from drought, disease, and fire associated with climatic warming become more likely and the potential for interactions among the physical and chemical climate would be increased. Also to be considered in this mix are future trends in air pollutants. Reductions in S deposition are already underway and signal a measurable reduction in regional S deposition (Likens et al., 1996). Predictions of future ozone levels depend not only on levels of N emissions, but also on regional climate, which will affect regional air patterns and the extent to which forests serve as sources for NOx and volatile organic carbon. Current predictions for future ozone levels vary widely from likely reduced levels based on expected trends in N emissions (EPA, 1997) to increases of 30% in the next 25 years (NAS, 1992).

8.1. FUTURE MONITORING NEEDS

As Innes (1998) points out, despite the fact that most countries do not attribute damage to air pollutants, considerable care must be taken as absence of record may not mean that the particular cause of injury (e.g. air pollution) was not present. Development of better diagnostic techniques applicable to regional/national/multinational inventories of crown transparency and discoloration remain to be developed (Ferretti, 1998; Innes, 1998). The very nature of effects of chronic air pollution on forest ecological processes assures that attribution of causal relationships between air pollution and forest health will be very difficult without at least annual scale data on changes in system function. Because the carbon, water, and nutrient resources affected by air pollution are affected by many other stresses, it has required multidisciplinary team approaches to couple focused process level studies aimed at documenting mechanisms of response with the patterns of change detected by survey approaches. The case studies we have presented indicate the value of designing surveys that consider both ecological and pollution gradients and make comparisons across ecologically analogous sites that tie in to hypothesized mechanisms of action. Such a gradient approach was developed by Legge et al. (1981) around a point source. It is now implemented in the "CFS Forest Indicators of Global Change Project" using sugar maple, white pine and red spruce ARNEWS/NAMP plots, with on-site process research, across gradients (2000 km) in acid deposition exceedance, $O_3$ exceedance and climate (temp. precip.) in eastern Canada.

Because of the temporal covariances among stressors in the environment, the development of a time series of multiple component environmental data provides an analytical framework within which to identify and model dominant independent variable and their interactions in affecting observed responses. As an example, the development of dendroecology as a method to evaluate annual variability in forest growth responses to physical and chemical climate has provided important tools to identify shifts in climate-growth relationships indicative of the timing and magnitude of increased external stresses, such as regional air pollution (McLaughlin et al., 1987, Zahner et al., 1989; Cook and Johnson, 1989, Petersen et al., 1995). The development of additional analytical approaches such as multispectral scanning, which can be deployed with remote sensing technology, offers important new tools to obtain repeated measures of changes in canopy condition (Royle and Lathrop, 1997), canopy chemical changes such as lignin content indicative of soil nitrification rates (Wessman et al., 1989), as well as providing indicators of more subtle shifts in leaf physiology, such as changes in photosynthesis and stomatal conductance (Carter, 1994) in response to environmental change. The statistical power provided by repeated measures of forest responses at interannual to annual scales can help provide important insights into the roles of both primary and secondary variables in controlling forest growth responses in a changing chemical and physical climate. This is substantiated by the retrospective statistical analysis of ARNEWS tree mortality (1985-1996) in relation to acidic deposition and other abiotic and biotic factors (Brandt et al., 1999).

While we have focused primarily on physiological changes in this review, the development of quantitative indicators of tree condition also has great value in delineation of areas with outward evidences of changing forest health and mapping changes in those areas over time. Merging of monitoring and research expertise at the working levels will give impetus to the development, and implementation of more sensitive, cause-attributable diagnostic indicators. Successful examples of this concept in practice can now be found in both Europe (Swiss Federal Institute for Forest, Snow, and Landscape Research, Birmensdorf), and North America (Canadian Forest Service, Forest Health Network). In the past, monitoring and process-level research programs have tended to operate in parallel However, monitoring data must be fully integrated with process studies if we are to more fully understand the influence of air pollutants and climate change on our forest health and long-term sustainability. The San Bernardino case study (Miller and McBride, 1998) remains the model of such a systematic approach to forest health.

In conclusion, our analyses indicate that changes in depth and vigor of root systems, shifts in pool sizes and allocation patterns of carbon, and changes in supply rates of N and Ca represent important shifts in ecological function that are currently occurring in diverse forest types across North America. The influence of these process level changes on future health of North American forests could be substantially increased in coming decades if projected changes in climate become a reality. Under such conditions, evaluating the diverse influences of multiple stresses, operating in multiple combinations, will require monitoring programs that combine economic and ecological indicators of forest health (Olsen and Schreuder, 1997). Future monitoring programs should address multiple interactive hypotheses for causality with analytical tools that can adequately evaluate consistency, responsiveness, and mechanistic linkages of cause effect relationships across gradients in resource supply, both natural and manipulated. Such efforts should help forest ecologists to manage resources of North American forests in the future in ways that enhance both ecological and economic health.

## References

Aber, J. D., Nadelhoffer, K. J., Steudler, P., Melillo, J. M.: 1989, *BioScience* **39**, 378.
Adams, C. M.. and Hutchinson, T. C.: 1992, *Can. J. For. Res.* **22,** 1489.
Adams, H. S., Stevenson, S. L., Blasing, T. J., Duvick, D. N.: 1985, *Env. Exp. Bot.* **25,** 315.
Adams, M. B., Edwards, N. T., Taylor, G. E., Jr., Skaggs, B. L.: 1990, *Can. J. For. Res.* **20**, 152.
Allen, E.R. and Gholz, H.L.:1995, in S. Fox and R.A. Mickler (eds), Impacts of Air Pollutants on Southern Pine Forests, Springer, New York, p 83.
Amundson, R. G., Hadley, J. L., Fincher, J. F., Fellows, S., Alscher, R. G.: 1992, *Can. J. For. Res.* **22,** 1605.
Andersen, C. P., McLaughlin, S. B., Roy, W. K.: 1991, *Can. J. For. Res.* **21,** 455-461.
Arbaugh, M. J., Miller, P. R., Carroll, J. J.: 1998, *Envir. Pollut.* **101**, 291.
Arp, P.A. and Oja, T.: 1992, CEA No. 841-726, Joint CEA-CPPA-FC-EC Project Tech. Rep. Part Xvl.
Arp, P.A., Oja, T., March, M.: 1996, *Can. J. For. Res.* **26,** 696.
Bangerth, F.: 1979, *Ann. Rev. Phytopathol.***17**, 97.
Barnard, J., Lucier, A. A., Johnson, A. H., Brooks, R. T., Karnosky, D. F., Dunn, P. H.: 1990, NAPAP State of Science and Technology, National Precipitation Assessment Program.
Bartholomay, G. A., Eckert, R. T, Smith, K. T.: 1997, *Can. J. For. Res.* **27**, 361.
Beattie, B. L. and Whelpdale, D. M.: 1989, *Water Air Soil Pollut.* **46**, 45.
Bernier, B. And Brazeau, M.: 1988a, *Can. J. For. Res.*, **18**, 754.
Bernier, B., and Brazeau, M.: 1988b, *Can. J..For. Res.* **18**, 762.
Bernier, M., Pare, D., Brazeau, M.: 1989, *Water, Air Soil Pollut.* **48**, 239.
Bertrand, A., Robitaille, G., Boutin, R., Nadeau, P.: 1996, *Tree Physiol.* **15**, 775.
Bertrand, A., Robitaille, G., Nadeau, P.,Boutin, R.: 1994, *Tree Physiol.* **14**, 413.
Binkley, D., Driscoll, C. T., Allen, H. L., Schoenenberger, P. and McAvoy, D.: 1989, *Acid deposition and forest soils: context and case studies in the Southeastern United States*, Springer-Verlag Ecological Studies 77.
Bondietti, E. A., Baes, C. F. III, McLaughlin, S. B.: 1989, *Can. J. For. Res.* **19**, 586.
Bondietti, E. A., Momoshima, N., Shortle, W. C., Smith, K. T.: 1990, *Can. J. For. Res.* **20**, 1850.
Brandt, J. P., Tomm, B. D., Hurley, J. E.: 1999, *Water Air Soil Pollut.* (submitted).
Brook, J. R.: 1995, *Atmos. Environ.* **29**, 1795.
Brumme, R. and Beese, F.: 1992, *J. Geophysical Res.* **97**, 12,851.
Bytnerowicz, A., and Fenn, M.: 1996, *Environ. Pollut*, **92**, 127.
Bytnerowicz, A., Percy, K., Riechers, Padgett, P., Krywult, M.: 1998, *Chemos.* **36**: 697.

Cape, J.N.: 1993, *Environ. Pollut.* **82**, 167.
Carter, G. A.: 1994, *Int. J. Remote Sensing* **15**, 697.
Chappelka, A. H. and Samuelson, L. J.: 1998, *New Phytol.* **139**, 49.
Clancy, K. M., Wagner, M. R., Reich, P. B.: 1995, pp. 118-180 in W. K. Smith and T. M. Hinckley (eds.), Ecophysiology of Coniferous Forests, Academic Press.
Clarkson, D. T. and Hanson, J. B.: 1980, *Ann. Rev. Plant Physiol.* **31**, 239.
Cobb, F. W., and Stark, R. W.: 1970, *J. For.* **68**, 147.
Cobb, F. W., Wood, D. L., Stark, R. W., Parameter, J. R.: 1968, *Hilgardia* **39**, 141.
Cole, D. W.: 1992, Atmospheric Deposition and Forest Nutrient Cycling, Springer-Verlag. pp. 150-151.
Cook, E. R. and Johnson, A. H.: 1989, *Water, Air, Soil Pollut.* **48**, 127.
Cook, E. R. and Zedaker, S. M.: 1992, pp. 192-231 in C. Eagar and M. B. Adams, Ecology and Decline of Red Spruce in the Eastern United States, Springer-Verlag.
Cooley, D. R. and Manning, W. J.: 1987, *Envir. Pollut.* **47**, 95.
Cooter, E. J.: 1998, pp. 15-54 in R. A. Mickler and S. Fox, *The Productivity and Sustainability of Southern Forest Ecosystems in a Changing Environment*. Springer-Verlag.
Cox, R.M., Lemieux, G., Lodin, M.: 1996, *Can. J. For. Res.*, **26**, 682.
Cronan, C. S. and Grigal, D. F.: 1995, *J. Envir. Qual.* **24**, 209.
Davidson, A.: 1993, *Air Waste* **43**, 226.
DeFelice, T. T.: 1997, *Atm. Res.* **43**, 325.
DeHayes, D. H., Thornton, F. C., Waite, C. E., Ingle, M. A.: 1991, *Can. J. For. Res.* **21**, 1292.
DeHayes, D. H.: 1992, pp. 295-337 in C. Eagar and M. B. Adams, Ecology and decline of red spruce in the eastern United States. Springer-Verlag.
DeHayes, D. H., Schaberg, P. G., Hawley, G. J., Borer, C. H., Cumming, J. R., Strimbeck, G. R.: 1997, *Tree Physiol.* **17**, 687.
DeHayes, D. H., Schaberg, P. G., Hawley, G. J., Strimbeck, G. R.: 1999, *Bioscience* (in review).
D'Eon, S. P., and Power, M.: 1989, *The Acid Rain National Early Warning System (ARNEWS) Plot Network*, Forestry Canada Inf. Report PI-X-91, Petawawa.
D'Eon, S. P., Magasi, L. P., Lachance, D., DesRochers, P.: 1994, Canada's National Forest Health Monitoring Plot Network: Manual on Plot Establishment and Monitoring (Revised), Can. For. Serv. Inf. Rep. PI-X-117, Petawawa..
Dixon, R. A. and Harrison, M. J.: 1994, *Annu. Rev. Phytopathol.* **32**, 479.
Dougherty, P. M., Allen, H. L., Kress, L. W., Murthy, R., Maier, C. A., Albaugh, T. J., Sampson, D. A.: 1997, pp. 149-168 in R. A. Mickler and S. Fox, *The Productivity and Sustainability of Southern Forest Ecosystems in a Changing Environment*. Springer-Verlag.
Dougherty, P. M., Teskey, R. O., Jarvis, P. G.: 1992, pp. 303-312 in R. B. Flagler (ed.), *Transactions: The Response of Southern Commercial Forests to Air Pollution*. Air and Waste Management Association.
Dull, C. W., Ward, J. D., Brown, H. D., Ryan, G. W., Clerke, W. H., Uhler, R. J.: 1988, Evaluation of spruce and fir mortality in the southern Appalachian mountains. USDA Forest Service Southern Region R8-PR 13. 92 pp.
Eagar, C.: 1984, *The Southern Appalachian Spruce-Fir Ecosystem: Its Biology and Threats*, National Park Service, Research/Resources Management Report SER-71.

Eagar, C. and Adams, M. B.: 1992, *Ecology and Decline of Red Spruce in the Eastern United States*. Springer-Verlag. pp. 295.
Eklund, L., and Eliasson, L.: 1990, *J. Exp. Bot.* **41**, 863.
Ellsworth, D. S. and Liu, X.: 1994, *Can. J. For. Res.* **24,** 2118.
Environment Canada (EC):1997. Canadian Acid Rain Assessment Vol. 2: Atmospheric Science Assessement Report, Ottowa.
EPA: 1996, *National Air Quality and Emissions Trends Report*, 1995. EPA 454/R-96-005.
EPA: 1997, National Air Pollutant Emission Trends, 1900-1906, U.S. Environmental Protection Agency EPA-454/R-97-011.
EPA: 1998, *National Air Quality and Emissions Trends Report*, 1996. EPA 454/R-97-013.
Federer, C., Hornbeck, J. W., Tritton, L. M., Martin, C. W., Pierce, R. S., Smith, C. T.: 1989, *Environ. Manage.* **13,** 000.
Ferretti, M.: 1998, *Chemos.* **36**, 1031.
Flagler, R. B., Brissette J. C., Barnett, J. P.: 1997, pp. 73-92 in R. A. Mickler and S. Fox, The Productivity and Sustainability of Southern Forest Ecosystems in a Changing Environment. Springer-Verlag.
Foster, N. W. and Hazlett, P.W.: 1991, pp. 179-191 In Acid Deposition. Origins, Impacts and Abatement Strategies, Springer-Verlag.
Foster, N. W., Morrison, I. K., Yin, X. and Arp, P. A.: 1992, *Can. J. For. Res.* **22**, 1753.
Fowler, D., Flechard, C., Skiba, U., Coyle, M. and Cape, J. N.: 1998, *New Phytol.* **139**, 11.
Fox, S. and Mickler, R. A.: 1995, *Impact of Air Pollutants on Southern Pine Forests*. Springer-Verlag.
Fraser, G. A., Phillips, W. E., Lamble, G. L., Hogan, G. D., Teskey, A. G.: 1985, Can. For. Ser, Rep. E-X-36. 43 p.
Friedland, A. J., Gregory, R. A., Karenlampi, L., Johnson, A. H.: 1984, *Can. J. For. Res.*
Friedland, A. J., Hawley, G. J., Gregory, R. A.: 1988, *Plant Soil* **105**, 189.
Fujioka, F. M., Roads, J. O., Chen, S. C.: 1998, Climatology, in Miller, P. R. and McBride, J. (eds.), *Oxidant Air Pollution Impacts in the Montane Forests of Southern California: The San Bernardino Case Study,* Springer-Verlag, Heidelberg (in press).
Gilliam, F. S., Adams, M. B., Yurish, B. M.: 1996, *Can. J. For. Res.* **26**, 196.
Godbold, D. L. and Kettner, C.: 1991, *J. Plant Physiol.* **138**, 231.
Goulding, K.W.T., Bailey, N. J., Bradbury, N. J., Hargreaves, P., Howe, M., Murphy, D. V., Poulton, P. R., Willison, T. W.: 1998, *New Phytol.* **139**, 49.
Grissino-Mayer, H. D. and Butler, D. R.: 1993, *Southeast. Geogr.* **1**, 65.
Grulke, N. E. and Miller, P. R.: 1994, *Tree Physiol.* **14**, 659.
Hall, J. P.: 1995, *For. Chron.* **71**, 607.
Hall, J. P. and Addison, P. A.: 1991, *Response to air pollution: ARNEWS assesses the health of Canada's forests*, For. Can. Sci. Dir. Inf. Rep. DPC-X-34, Ottawa, 42 pp.
Hall, J. P., Bowers, W., Hirvonen, H., Hogan, G., Foster, N., Morrison, I., Percy, K., Cox, R., Arp, P.: 1997, 1997 Canadian Acid Rain Assessment: Vol. 4 The Effects on Canada's Forests, Natural Resources Canada, 47 pp.
Hanson, P. J. and Lindberg, S. E.: 1991, *Atmos. Envir.* **25**, 1615.
Hendershot, W. W.: 1991, *Fert. Res.* **27**, 63.
Hendershot, W. W. and Courchesne, F.: 1994, *Can. J. For. Res.*, **24**, 609.
Hedin, L. O., Granat, L., Likens, G. E., Bulshand, T. A., Galloway, J. N., Butler, T. J., Rodhe, H.: 1994, *Nature* **367,** 351.
Heisey, R. M.: 1995, *Water Air Soil Poll.* **82**, 675.

Hepler, P, K., and Wayne, R. O.: 1985, *Ann. Rev. Plant Physiol.* **36**, 397.
Hepting, G. H.: 1945, *Phytopath.* **35**, 106.
Hertel, G. D. and McKinney-McNeal: 1991, *USDA For. Serv. Agric. Inf. Bull.* **622**, 19.
Hinckley, T., Ford, D., Segura, G., Sprugel, D.: 1992, Key processes from stand to tree level in Wall, G. (ed.), *Implications of climate change for Pacific Northwest Forest Management.* University of Waterloo, Department of Geography paper 15, Waterloo, pp. 33-43.
Hoffer, T. E., Miller, D. J., Farber, R. J.: 1981, *Atmos. Environ.* **15**, 1935.
Hogsett, W. E., Herstrom, A. A., Laurence, J. A., Lee, E. H., Weber, J. E., Tingey, D. T.: 1993, Proceedings of the 4$^{th}$ US/Dutch International Symposium: Comparative Risk Analysis and Priority Setting for Air Pollution Issues. Air and Waste Management Association.
Hsaio, T. C.: 1973, *Ann. Rev. Plant Physiol.* **24**, 519.
Innes, J. L.: 1993, Forest Health: Its Assessment and Status, CAB International, 677 pp.
Innes, J. L.: 1998, *Chemos.* **36**, 1025.
Intergovernmental Panel on Climate Change (IPCC): 1998, *The Regional Impacts of Climate Change. An Assessment of Vulnerability.* pp. 267-279 in R. T. Watson, M. C. Zinyowooa, R. H. Moss, and D. J. Dokken (eds.), Cambridge University Press.
Jagels, R.: 1986, *I.A.W.A. Bull.*, 299.
Jeffries, D. S. (ed.): 1997, *1997 Canadian Acid Rain Assessment: Vol. 3 Aquatic Effects.* Environment Canada, Burlington.
Jenkins, A. F.: 1997, *J. For.*
Johnson, A. H. and Siccama, T. G.: 1983, *Envir. Sci. Technol.* **17**, 294A.
Johnson, A. H., Mclaughlin, S. B., Adams, M. B., Cook, E. R., DeHayes, D. H., Eager, C., Fernandez, I. J., Johnson, D. W., Kohut, R. J., Mohnen, V. A., Nicholas, N. S., Peart, D. R., Schier, G. A., White, P. S.: 1992, pp. 385-411 in C. Eagar and M. B. Adams, Ecology and Decline of Red Spruce in the Eastern United States. Springer-Verlag.
Johnson, D. W. and Todd, D. E.: 1987, *Plant Soil* **102**, 99.
Johnson, D. W. and Lindberg, S. E.: 1992, *Atmospheric deposition and forest nutrient cycling.* Springer-Verlag.
Johnson, D. W., Richter, D. D., Lovett, G. M., Lindberg, S. E.: 1985, *Can. J. For. Res.* **15**, 773.
Johnson, D. W., Van Miegroet, H., Lindberg, S. E., Todd, D. E., Harrison, R. B.: 1991, *Can. J. For. Res.* **21**, 769.
Joslin, J. D., McDuffie, C. M., Brewer, P. F.: 1988, *Water, Air, and Soil Pollut.* **39**, 355.
Joslin, J. D. and Wolfe, M. H.: 1992, *Can. J. For. Res.* **22**, 893.
Joslin, J. D. and Wolfe, M. H.: 1994, *Soil Sci. Soc. Am. J.* **58**, 1572.
Joslin, J. D., Kelly, J. M., Van Miegroet, H.: 1992, *J. Env. Qual.* **21**, 12.
Joslin, J. D., and Johnson, J. W.: 1997, pp. 571-590 in R. A. Mickler and S. Fox, *The Productivity and Sustainability of Southern Forest Ecosystems in a Changing Environment.* Springer-Verlag.
Joslin, J. D., Wolfe, M. H., Garten, C. T.: 1998, pp. 557-570 in R. A. Mickler and S. Fox (eds.), *The Productivity and Sustainability of Southern Forest Ecosystems in a Changing Environment.* Springer-Verlag.
Kemp, W. P. and Moody, U. L.: 1984, *Environ. Entomol.* **13**, 1291.
Kimball, K. D., Jagels, R., Gordon, G. A., Weathers, K. C., Carlsile, J.: 1988, *Water Air Soil Pollut.* **39**, 383.

Kolb, W. E. et al.: 1994, *J. For.* **92**, 10.
Kozlowski, T. T., Kramer, P. J., Pallardy. S. G.: 1991, *The Physiological Ecology of Woody Plants*, Academic Press.
Kress, L. W., Allen, H. L., Mudano, J. E., Stow, T. K.: 1992, *Environ. Toxicol. Chem.* **11**, 1115.
Kull, O., Sober, A., Coleman, M. D., Dickson, R. E., Isebrands, J. G., Gagnon, Z., Karnosky, D. F.: 1996, *Can. J. For. Res.* **26**, 639.
Lachance, D., Hopkin, A., Pendrel, B., Hall, J. P.: 1995, Health of Sugar Maple in Canada: results from the North American Maple Project, 1988-1993, Canadian Forest Service Report ST-X-10, 27 pp.
LeBlanc, D. C. and Raynal, D. J.: 1990, *Can. J. For. Res.* **20**, 1415.
Lee, W. S., Chevone, B. I., Seiler, J. R.: 1990, *Water Air Soil Pollut.* **51**, 105.
Legge, A. H., Jaques, D. R., Harvey, G. W., Krouse, H. R., Brown, H. M., Rhodes, E. C., Nosal, M., Schelhase, H. U., Amundson, R. G., Walker, R. B.: 1981, *Water Air Soil Pollut.* **15**, 77.
Likens, G. E., Driscoll, C. T., Buso, D. C.: 1996, *Science* **272**, 244.
Likens, G. E., Driscoll, C. T., Buso, D. C., Siccama, T. G., Johnson, C. E., Lovett, G. M., Fahey, T. J., Reiners, W. A., Ryan, D. F., Martin, C. W., Bailey, S. W.: 1998, *Biogeochem.* **41**, 89.
Long, R. P., Horsley, S. B., and Lija.: 1997, *Can. J. For. Res.* **27**, 1560.
Lorio, P. L., Jr. and Sommers, R. A.: 1986, *Tree Physiol.* **2**, 301.
Lovett, G. M., Reiners, W. A., Olson, R. K.: 1982, *Science* **218**, 1303.
Madronich, S., McKenzie, R. L., Caldwell, M., Bjorn, L. O.: 1995, *Ambio* **24**, 143.
Magasi, L. P.: 1985, *Forest pest conditions in the Maritimes in 1985*, Can. For. Sev. Inf. Rep. M-X-159, Fredericton.
Maier-Maercker, U.: 1997, *Trees* **11**, 229.
Maier-Maercker, U. and Koch W.: 1992, *Trees* **7**, 12.
Manion, P. D.: 1981, *Tree Disease Concepts*, Englewood Cliffs, NY, Prentice Hall.
Mattson, W. J. and Addy, N. D.: 1975, *Science* **190**, 515.
McAinsh, M. R. and Hetherington, A. M.: 1998, *Perspectives* **3**, 22.
McLaughlin, S. B.: 1985, *J. Air Poll. Control Assoc.* **35**, 516.
McLaughlin, S. B., McConathy, R. K., Duvick, D., Mann, L. K.: 1982, *For. Sci.* **28**, 60.
McLaughlin, S. B. and Shriner, D. S.: 1980, *Plant Disease Vol. 5*. Academic Press. pp. 407-431.
McLaughlin, S. B. and Downing, D. J.: 1996, *Can. J. For. Res.* **26**, 670.
McLaughlin, S. B. and Downing, D. J.: 1997, pp. 207-230 in R. A. Mickler and S. Fox (eds.), *The Productivity and Sustainability of Southern Forest Ecosystems in a Changing Environment*. Springer-Verlag.
McLaughlin, S. B., Downing, D.J., Blasing, T. J., Cook, E. R., Adams, H. S.: 1987, *Oecologia* **72**, 487.
McLaughlin, S. B., Andersen, C. P., Edwards, N. T., Roy, W. K., Layton, P. A.: 1990, *Can. J. For. Res.* **20**, 485.
McLaughlin, S. B., Andersen, C. P., Hanson, P. J., Tjoelker, M. G., Roy, W. K.: 1991, *Can. J. For. Res.* **21**, 1234.
McLaughlin, S. B., Tjoelker, M. G., Roy, W. K.: 1993, *Can. J. For. Res.* **23**, 380.
McLaughlin, S. B. and Kohut, R.: 1992, pp. 338-384 in C. Eagar and M. B. Adams (eds.), *Ecology and Decline of Red Spruce in the Eastern United States*. Springer-Verlag.

McLaughlin, S. B., Joslin, J. D., Robarge, W., Stone, A., Wimmer, R., Wullschleger, S. D.: 1998, pp. 255-277 in R. A. Mickler and S. Fox (eds.), *The Productivity and Sustainability of Southern Forest Ecosystems in a Changing Environment.* Springer-Verlag.

McLaughlin, S. B. and Wimmer, R.: 1999, *New Phytologist* (in press).

McNulty, S. G., Vose, J. M., Swank, W. T.: 1996, *Ambio* **25,** 449.

Mickler, R. and Fox, S.: 1997, *The Productivity and Sustainability of Southern Forest Ecosystems in a Changing Environment.* Springer-Verlag.

Miller, P. R.: 1992, Mixed conifer forests of the San Bernardino Mountains in Olson, R. K., Binkley, D., Bohm, M. (eds.), *The Response of Western Forests to Air Pollution,* Springer-Verlag, New York. Pp. 461-497.

Miller, P. R.: 1973, pp. 101-117 in J. A. Naegele, *Air Pollution Damage to Vegetation.* Advances in Chemistry Series 112. Am. Chem. Soc., Washington.

Miller, P. R., Chow, J., Watson, J. G.: 1996, Assessment of Acidic Deposition and Ozone Effects on Conifer Forests in the San Bernardino Mountains, California Air Resources Board Report.

Miller, I., Lachance, D., Burkmann, W. G., Allen, D.C.: 1991, North American maple decline project: organization and field methods, USDA Forest service Gen. Tech. Rep. NE-154.

Miller, P. R. and McBride, J. (eds.): 1998, *Oxidant Air Pollution Impacts in the Montane Forests of Southern California: The San Bernardino Case Study,* Springer-Verlag, Heidelberg.

Miller, P. R., McBride, J. R., Schilling, S. L., Gomez, A. P.: 1989, pp. 309-324 in R. K. Olson and A. S. Lefohn, *Effects of Air Pollution on Western Forests.* Air and Waste Manage. Assoc., Pittsburgh.

Miller, P.R., and Millecan, A.A.: 1971, *Plant Disease Rep.* **55,** 555.

Miller, P. R., Parmeter, J. R., Taylor, O. C., Cardiff, E. A.: 1963, *Phytopath* **53,** 1072.

Miller, P. R., Chow, J., Watson, J. (eds.): 1996, *Assessment of Acidic Deposition and Ozone Effects on Conifer Forests in the San Bernardino Mountains,* California Air Resources Board Report NTIS 97115612, Sacremento.

Miller, P. R., Taylor, O. C., Wilhour, R. G.: 1982, *Oxidant air pollution effects on a western coniferous forest ecosystem.* US EPA Report No. EPA-600/D-82-276.

Millers, I., Lachance, D., Burkman, W. G., Allen, D. C.: 1991, *North American Sugar Maple Decline Project: Organization and Field Methods.* U.S. Dep. Agric. For. Serv., Radnor.

Mohnen, V. A.: 1992, pp. 64-124 in C. Eagar and M. B. Adams (eds.), *Ecology and Decline of Red Spruce in the Eastern United States.* Springer-Verlag.

Momoshima, N. and Bondietti, E. A.: 1990, *Can. J. For. Res.* **20,** 1840.

Mooney, H. A. and Chu, C.: 1974, *Oecologia* **14,** 295.

Morrison, I.K., Fournier, R.E., Hopkin, A.A.: 1996, *Proceedings 16th IUFRO International Meeting for Specialists in Air Pollutant Effects on Forest Ecosystems,* Fredericton, 402 pp.

NAPAP: 1990, *National Acid Precipitation Assessment Program,* State of Science/Technology. Report Series.

NAPAP: 1998, *National Acid Precipitation Assessment Program Biennial Report to Congress: An Integrated Assessment,* Silver Spring, MD.

NAS: 1992, *Rethinking the Ozone Problem in Urban and Regional Air Pollution*. National Academy of Sciences. National Research Council. National Academy Press. 500 pp.

Natural Resources Canada (NRCAN).: 1996, *1995 Compendium of Canadian Forestry Statistics*. Canadian Council of Forest Ministers, Ottawa. 202 pp.

Natural Resources Canada (NRCAN).: 1997, *1996 Compendium of Canadian Forestry Statistics*. Canadian Council of Forest Ministers, Ottawa. 234 pp.

Natural Resources Canada (NRCAN).: 1998, The State of Canada's Forests, 109 pp.

Nicholas, N. S.: 1992, *Stand Structure, Growth, and Mortality in Southern Appalachian Spruce-Fir*. Ph.D. Dissertation, Virginia Polytechnic Institute and State University, 176 pp.

Nodvin, S. C., Van Miegroet, H., Lindberg, S. E., Nicholas, N. S., Johnson, D. W.: 1995, *Water, Air, and Soil Pollut.* **85**, 1647.

Norby, R. J.: 1998, *New Phytol.* **139**, 189.

Norby, R. J., Wullschleger, S. D., Gunderson, C. A., Johnson, D. W., Ceulemans, R.: 1999, *Plant Soil Envir.* **22**, (in press).

O'Laughlin, J., Livingston, R. L., Thier, R., Thornton, J., Toweill, D. E., Morelan, L.: 1994, *J. Sust. For.* **2**, 65.

Olivotto, C. (ed.): 1997, *Canadian 1996 Nox/VOC Science Assessment: Summary for Policy Makers*, Environment Canada, Toronto.

Ollinger, S. V., Aber, J. D., Reich, P. B.: 1997, *Ecol. Appl.* **7**, 1521.

Olsen, A. R. and Schreuder, H. T.: 1997, *Envir. Ecol. Stat.* **4**, 167.

Olson, R. K.: 1992, Physiography of forest types in Olson, R. K., Binkley, D. and Bohm, M. (eds.), *The Responses of Western Forests to Air Pollution*, Springer-Verlag, New York. pp. 7-40.

Ouimet, R. and Fortin, J.-M.: 1992, *Can. J..For. Res.* **22**, 699.

Peart, D. R., Nicholas, N. S., Zedaker, S. M., Miller-Weeks, M., Siccama, T. G.: 1992, pp. 125-191 in C. Eagar and M. B. Adams (eds.), *Ecology and Decline of Red Spruce in the Eastern United States*, Springer-Verlag.

Pearson, M. and Mansfield, T. A.: 1993, *New Phytologist* **123**, 351.

Pearson, R. G. and Percy, K. E.: 1990, *The 1990 Canadian Long-range Transport of Air Pollutants and Acid Deposition Assessment Report*: part 5 Terrestrial Effects, Federal/Provincial Research and Monitoring Coordinating Committee (RMCC), 105 pp.

Pearson, R. G. and Percy, K. E.: 1997, Canadian 1996 $NO_x$/VOC Science Assessment: Report of the Vegetation Objective Working Group. Canadian Council of Ministers of Environment (CCME) Multistakeholder $NO_x$/VOC Science Plan, pp.

Percy, K. E. and Gordon, D. C.: 1998, UV-B and Canada's Forests: State of Science and Risk. Pp 75-85 in Legge, A. H. and L.L. Jones (ed.), *Proceedings Air and Waste Management Association Speciality Conference on Emerging Air Issues for the 21$^{st}$ Century*, Calgary

Percy, K. E., Jagels, R., Marden, S., McLaughlin, C. K., Carlisle, J.: 1993,. *Can. J. For. Res.* **23**, 1472.

Persson, H. and Majdi, H.: 1995, *Water, Air, and Soil Pollut.* **85**, 1287.

Peterson, D. L., Silsbee, D. G., Poth, M., Arbaugh, J., Biles, F. E.: 1995, *J. Air Waste Manage. Assoc.* **45**, 36.

Peterson, D. L., Arbaugh, M. J., Robinson, L. J.: 1991, *Holoc.* **1**, 50.

Pooviah, B. W.: 1988, *Senescence and Aging in Plants*, Academic Press, Inc.

Powell, D. S., Faulkner, J. L., Darr, D. R., Zhu, Z., MacCleery, D. W.: 1992, *Forest Resources of the United States*. USDA Forest Service Gen. Tech. Report RM-234, 132 pp.
Prinz, B.: 1987, *Environment* **29**, 10.
Pronos, J., Vogler, D. R., Smith, R. S.: 1978, An evaluation of ozone injury to pines in the *southern Sierra Nevada*. USDA Forest Service Forest Pest Management Report 78-1, San Francisco.
Rehfuess, K. E.: 1981, *Forstwiss. Centralbl.* **100**, 363.
Reiners, W. A. and Lang, G. E.: 1979, *Ecol.* **60**, 403.
Rennenberg, H., Kreutzer, K., Papen, H., Weber, P.: 1998, *New Phytol.* **139**, 71.
Reuss, J. O. and Johnson, D. W.: 1986, *Acid Deposition and the Acidification of Streams and Waters*. Springer-Verlag.
Richardson, C. T., Sasek, T. W., Fendick, E. A.: 1992, *Environ. Toxicol. Chem.* **11**, 1105.
Richter, D. D. and Markewitz, D.: 1995, pp. 315-336 in S. Fox and R. A. Mickler (eds.), *Impacts of Air Pollutants on Southern Pine*. Springer-Verlag.
Richter, D. D., Markewitz, D., Wells, C. G., Allen, H. L., April, R., Heine, P. R., Urrego, B.: 1994, *Ecology* **75**, 1463.
RMCC: 1990, *The 1990 Canadian long-range transport of air pollutants and acid deposition assessment report: Part 1 Executive Summary*. Federal-Provincial Research and Monitoring Coordinating Committee, Ottawa.
Robarge, W. P., Pye, J. M., Bruck, R. I.: 1989, *Plant and Soil* **114**, 19.
Robarge, W. P. and Johnson, D. W.: 1992, *Advances in Agronomy* **47**, 1.
Roberts, D. M. and Harmon, A. C.: 1992, *Annu. Rev. Plant Physiol. Plant Mol. Biol.* **43**, 375.
Robitaille, G., Boutin, R., Lachance, D.: 1995, *Can. J. For. Res.*, **25**, 577.
Roy, G., Robitaille, G., Gagnon, G.: 1985, *Phytoprotection* **66**, 91.
Royle, D. D. and Lathrop R. G.: 1997, *For. Sci.* **43**, 327.
Samuelson, L. J. and Kelly, J. M.: 1997, *New Phytologist* **136**, 255.
Sasek, T. W. and Flagler, R. B.: 1995, Impacts or Air Pollutants on Southern Pine Forests. Springer-Verlag. pp. 424-466.
Sasek, T. W., Richardson, C. J., Fendick, E. A., Bevington, S. R., Kress, L. W.: 1991, *For. Sci.* **37**, 1078.
Saxena, V. K., and Lin, N. H.: 1990, *Atmos. Envir.* **24A**, 329.
Schaberg, P. G., Perkins, T. D., McNulty, S. G.: 1997, *Can. J. For. Res.* **27**, 1622.
Schemenauer, R. S.: 1986, *Atmos. Ocean* **24**, 303.
Scherbatskoy, T. de W.: 1989, *Ionic Relations of Leaf Cuticles*, Ph.D. Thesis, University of Vermont.
Schroeder, W. H., Bauch, J., Endeward, R.: 1988, *Trees* **2**, 96.
Schulze, E. D.: 1989, *Science* **244**, 776.
Schulze, E. D., De Vries, W. Hauhs, M., Rosen, K., Rasmussen, L., Tamm, C. O., Nilsson J.: 1989, *Water, Air, and Soil Pollut.* **48**, 451.
Schulze, E. D., Robichaux, R. H., Grace, J., Rundel, P. W., Ehleringer, J. R.: 1987, *BioScience* **37**, 30.
Schutt, P. and Cowling, E. B.: 1985, *The Amer. Phytopathological Soc.* **69**,
SCOPE: 1992, *Effects of increased ultraviolet radiation on biological systems*. Scientific Committee on Problems in the Environment, Paris.
Sheffield, R. M. and Cost, N. D.: 1987, *J. For.* **87**, 29.

Sheffield, R. M. et al.: 1985, USDA For. Serv. Res. Bull. SE-83.
Shepard, M. R., Lee, C. E., Woosley, R. S., Butcher, D. J.: 1995, *Microchemical J.* **52**, 118.
Shortle, W. C. and Bondietti, E. A.: 1992, *Water, Air, and Soil Pollut.* **61**, 253.
Shortle, W. C., Smith, K. T., Minocha, R., Alexeyev, V. A.: 1995, *J. Biogeography* **22**, 467.
Shortle, W. C. and Smith, K. T.: 1988, *Science* **240**, 1017.
Shriner, D. S., Heck, W. W., McLaughlin, S. B., Johnson, D. W., Peterson, C. E.: 1990, *NAPAP State of Science and Technology*, National Acidic Precipitation Assessment Program.
Skarby, L., Troeng, E., Bostrom, C. A.: 1987, *For. Sci.* **33**, 801.
Skarby, L., Ro-Poulsen, H., Wellburn, F.A.M., Sheppard, L. J.: 1998, *New Phytol.* **139**, 109.
Smith, W., and Mangold, R.: 1996, *Estimating status and change in forest health*, U.S. Dep. Agric. Forest Service Fact Sheet.
Smith, G. F., and Nicholas, N. S.: 1999, *Castanea* (in press).
Stark, R. W.: 1965, *Annu. Rev. Entomol.* **10**, 303.
Sterba, H.: 1996, *Forest decline and growth trends in Central Europe - a review*. Berlin: Springer-Verlag.
Stienen, H. and Bauch, J.: 1988, *Plant and Soil* **106**, 231.
Stohlgren, T. J. and Parsons, D. J.: 1987, *Atmos. Environ.* **21**, 1969.
Stow, T. K., Allen, H. L., Kress, L. W.: 1992, *For. Sci.* **38**, 102.
Stoyenhoff, J., Witter, J., Leutscher, B.: 1998, *Forest Health in the New England States and New York*, University of Michigan, 31 pp.
Swank, W. T. and Vose, J. M.: 1997, *Global Biogeochemical Cycles* **11**, 657.
Taylor, G. E.: 1994, *J. Environ. Qual.* **23**, 63.
Teskey, R. O.: 1995, pp. 467-490 in S. Fox and R. A. Mickler (eds.), *Impacts of Air Pollutants on Southern Pine*. Springer-Verlag.
Teskey, R. O., Dougherty, P. M., Mickler, R. A.: 1998, pp. 279-290 in R. A. Mickler and S. Fox (eds.), *The Productivity and Sustainability of Southern Forest Ecosystems in a Changing Environment.* Springer-Verlag.
Thomson, D. J.: 1995, *Science* **268**, 59.
Timmel, T.E. 1996. pp 1907-1969 in *Compression Wood* Vol 3. Springer-Verlag.
Tingey, D. T. and Hogsett, W. E.: 1985, *Plant Physiol.* **77**, 944.
Trewavas, A. J. and Malho, R.: 1997, *The Plant Cell* **9**, 1181.
Ulrich, B.: 1984, *Atmos. Env.* **18**, 621.
Ulrich, B. and Matzner, E.: 1986, *Experientia* **42**, 344.
USDA.: 1997, *Forest Insect and Disease Conditions in the United States 1996.* Forest Service Health Production, Washington, D.C., 87 p.
USDA Forest Service: 1996, *Sampling and plot design*, Forest health monitoring fact sheet 1909.14, USDA Forest Service, Washington, DC.
Vann, D. R., Strimback, G. R., Johnson, A. H.: 1992, *For. Ecol. Manag.* **51(1-3)**, 69.
Wallin, G. and Skarby, L.: 1992, *Trees* **6**, 128.
Wargo, P. M.: 1972, *Phytopathology* **62**, 1278.
Waring, G. L. and Cobb, N. S.: 1992, Insect-Plant Interactions. pp. 167-226.
Waring, R. H.: 1987, *BioScience* **37**, 569.
Waring, R. H. and Pitman: 1985, *Ecol.* **66**, 889.
Watmough, S. A.: 1997, *Envir. Rev.* **5**, 181.

Watson, R. T., Zinyowera, M. C., Moss, R. H., Dokken, D. J.: 1998, The Regional Impacts of Climate Change, An Assessment of Vulnerability. Cambridge University Press.
Weeks, M. W., and Smoronok, D.: 1993, *Aerial assessment of red spruce and balsam fir conditions in the Adirondack region of New York, etc.* USDA-USFS Technical Report NA-TP-16-93, 141 p.
Wesselink, L. G., Meiwes, K. J., Matzner, E., Stein, A.: 1995, *Environ. Sci. Technol.* **29**, 51.
Wessman, C. A., Aber, J. D., Peterson, D. L.: 1989, *Int. J. Remote Sensing* **10**, 1293.
Wisiniewski, J.: 1982, *Water, Air, Soil Pollut.*, **17**, 361.
Witkowski, Z., Madziara-Borusiewicz, K., Plonka, P., Zurek, Z.: 1987, *Ekologia Polska* **35**, 465.
Wullschleger, S. D., Norby, R. J., Gunderson C. A.: 1997, pp. 79-100 in L. H. Allen, *Advances in Carbon Dioxide Effects Research.* ASA Special Publication No. ffi, X., Foster, N. W. and Arp, P. A.: 1993, *Can. J. For. Res.*, **23**, 617.
Yin, X., Foster, N. W., Morrison, I. K., Arp, P.A.: 1994, *Can. J. For. Res.*, **24**, 1567.
Yoder, B. J., Ryan, M. G., Waring, R. H., Schoettle, A. W., Kaufmann, M. R.: 1994, *For. Sci.* **40**, 513.
Zahner, R., Saucier, J. R. and Myers, R. K.: 1989, *Can. J. For. Res.* **19**:612.
Zhao, X. J., Sucoff, E., Stadelmann, E. J.: 1987, *Plant Physiol.* **83**, 159.
Zoettl, H. W. and Huettl, R. F.: 1986, *Water, Air, and Soil Pollut.* **31**, 449.

## Acknowledgements

The authors wish to express their sincere appreciation to Dr. R. Vet, Atmospheric Environment Service, Environment Canada, Downsview, Ontario and to Dr. T. Dann, Environmental Protection Service, Environment Canada, Ottawa, Ontario for calculation and mapping of $SO_4$, $NO_3$ H ion depositions and cumulative $O_3$ sums. The senior author thanks the U.S. Forest Service for support of some aspects of his basic research reported herein and the Department of Energy, Bioenergy Feedstock Development Program at Oak Ridge National Laboratory for partial support of synthesis activities. This manuscript is Publication No. 4892 of the Environmental Sciences Division, Oak Ridge National Laboratory. ORNL is operated by Lockheed Martin Energy Research for the U.S. Department of Energy under contract No. DE-ACO5-96OR22464.

# OZONE – A RISK FACTOR FOR TREES AND FORESTS IN EUROPE?

R. MATYSSEK[1] and J.L. INNES[2]

[1]*Lehrstuhl für Forstbotanik, Universität München, Am Hochanger 13, D-85354 Freising, Germany,* [2]*Swiss Federal Institute for Forest, Snow and Landscape Research, Zürcherstrasse 111, CH-8903 Birmensdorf, Switzerland*

(Received 25 September 1998; accepted 25 February 1999)

**Abstract:** Tropospheric ozone ($O_3$) may adversely affect tree growth, with critical levels for $O_3$ being exceeded in many parts of Europe. However, unequivocal evidence for $O_3$-induced foliar injury on woody species under field conditions has only been found in a few places. Visible $O_3$ injury appears to occur mainly in the Mediterranean Basin, which is also the area where the least amount of information is available on $O_3$ exposure as well as the sensitivity of individual species. Overall, the quantitative risk assessment of $O_3$ impacts on mature trees and forests is vague at the European scale, as most knowledge is derived from controlled $O_3$ fumigations of young trees, grown in isolation in exposure chambers. Research suggests that risks exist, but these need to be validated for stand conditions. $O_3$-induced changes in resource allocation rather than productivity appear to be crucial as they affect competitiveness and predisposition to parasite attack and may eventually lead to the loss of genetic diversity. 'Free-air' $O_3$ fumigations in forest canopies may reveal processes that are susceptible to $O_3$ stress under field conditions and provide a scientific basis towards quantitative risk assessment and realistic definitions of critical levels for $O_3$ in forest ecosystems.

**Keywords:** Tropospheric ozone, Forests, Ecological risk analysis, Critical levels, Free-air fumigation, Visible injury, European Alps

## 1. Introduction

Of the various different types of air pollutants currently impacting forests in Europe, the greatest concern surrounds ozone ($O_3$). In contrast to pollutants associated with heavy industry, such as sulphur dioxide ($SO_2$), $O_3$ concentrations do not appear to have declined in recent years, and concentrations in some areas may even be increasing (Staehelin *et al.*, 1994; Dollard *et al.*, 1995). Current levels are clearly greater than pre-industrial concentrations (Volz and Kley, 1988; Anfossi *et al.*, 1991), but current trends in $O_3$ concentrations are uncertain, as systematic monitoring in Europe was only started in the late 1980s. In addition, many measurement stations are primarily aimed at assessing $O_3$ levels in relation to human health risks and many monitoring stations are therefore located in urban and suburban areas. These stations have little relevance to $O_3$ concentrations in rural situations because of scavenging by nitrogen oxides ($NO_x$).

The following account concentrates on rural environments as these are of relevance for examining the risks that $O_3$ may impose to forest trees and forest ecosystems. After providing an overview of $O_3$ regimes in the major regions in Europe, visible injury in woody plants is examined as an indicator of $O_3$ impact and in relation to current definitions of 'Critical Levels for Ozone'. Evidence for $O_3$ impacts at the ecosystem level is examined, as is the information required for judging the level of risk to ecosystems. With respect to trees, risks inferred from young plants under controlled exposure conditions must be the basis for scenarios of $O_3$ effects on forest stands. The importance of tree ontogeny is highlighted and linked to potential mechanisms of $O_3$

impact on mature trees and forest ecosystems. Finally, approaches towards a mechanistic understanding of $O_3$ impact under stand conditions are examined and, through this, possible methods for a quantitative risk assessment of forest development are proposed.

In contrast to the recent, excellent review by Skärby et al. (1998), we concentrate on risk assessment and its uncertainties for mature trees and forests as related to the European $O_3$ scenarios. Mechanisms of $O_3$ impact and their interactions with other factors in young plants, as well as the 'Critical Level' concept, are addressed only as a 'reference' for approaching risk assessment in the field. However, whole-plant allocation is viewed as a determinant of $O_3$ impacts on competitiveness and the predisposition of trees to multiple stress and thus as a key factor in determining resource fluxes and the responsiveness of forest ecosystems to $O_3$ stress.

## 2. Characteristics of $O_3$ regimes in Europe

Knowledge of the characteristics of $O_3$ regimes over the European region is at best patchy. In the alpine region (Switzerland, Austria, and southern Germany), the majority of stations show no clear trends (Smidt, 1998). For example, in Switzerland, annual average concentrations, April-September 1000-1700 concentrations, peak hourly concentrations and the numbers of hours above 60 ppb all show no clear trends over the period 1982-1996 (BUWAL, 1997).This contrasts with longer-term data obtained for the upper troposphere, where a definite increase in $O_3$ has been recorded in the period 1969-1989 (Staehelin and Schmid, 1991) and with data from other countries, such as the United Kingdom (Dollard et al., 1995), where increasing concentrations have also been documented.

In central and northern Europe, $O_3$ episodes primarily occur downwind of industrialised areas (Proyou et al., 1991; Grennfelt and Beck, 1994). Episodes are typically associated with periods of high pressure, when atmospheric conditions are relatively stable. $O_3$ episodes can also occur in spring (e.g. Davies and Schuepbach, 1994; Staehelin et al., 1994; Laurila and Tuovinen, 1996; Skärby and Karlsson, 1996), although the causes are uncertain, with incursions of stratospheric $O_3$ being a likely contributing factor (UKPORG, 1987; Ebel et al., 1991; Davies and Schuepbach, 1994). Tropospheric $O_3$ formation is dependent on sufficient concentrations of $NO_x$ and reactive volatile organic compounds (VOCs). In urban areas, concentrations of VOCs tend to limit the formation of $O_3$, whereas in rural areas, $O_3$ production tends to be limited by $NO_x$ concentrations (Chameides et al., 1992). The known emissions of various pollutants, combined with the available monitoring data, has enabled maps to be prepared of the distribution of $O_3$ in Europe (Hettelingh et al., 1996, 1997). One such map, prepared in relation to the accumulated hourly $O_3$ exposure above 40 ppb (AOT40) for forests, is shown in Figure 1. This map is based on 50 km x 50 km grid squares, and therefore the variation in $O_3$ concentrations that occurs with altitude (e.g. Reiter et al., 1987; Proyou et al., 1991; Loibl and Smidt, 1996) is inadequately addressed.

Reliable estimates of $O_3$ concentrations are extremely difficult to obtain from eastern Europe, although concentrations in areas as far east as the Ukraine are known to be sufficient to cause visible injury to Bel-W3 tobacco, and $O_3$-like symptoms have been recorded on sensitive species such as red-berried elder (*Sambucus racemosa*) and wild hop (*Humulus lupulus*) (Blum et al., 1997). Potentially phytotoxic concentrations have also been reported in Poland (Bytnerowicz et al., 1993). Other information is very limited, however modelling studies suggest that AOT40 thresholds are regularly exceeded in the region (Dollard et al., 1995).

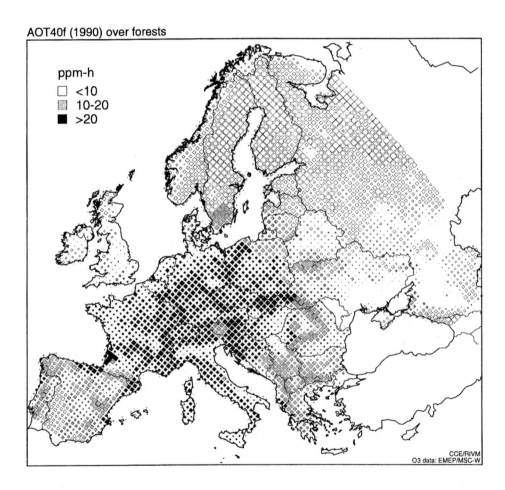

Fig. 1. AOT40 values for forests in Europe in 1990. Areas with <10 ppm.h fall below the critical level identified as potentially causing growth losses in forests. Map prepared by M. Posch of the UN-ECE Coordination Center for Effects (Working Group on Effects, Convention on Long-Range Transboundary Air Pollution) using ozone data from EMEP/MSC-W.

Special conditions surround the occurrence of $O_3$ in the Mediterranean region. The region is dominated by two major weather systems: the Azores anticyclone in the west and a low-pressure system in the east which extends from the Middle East to southwest Asia (Sanz and Millán, 1998). The boundary between the two is located over Italy, and changes in the relative importance of the two systems have a marked impact on air quality in the central Mediterranean. This large-scale pattern is modified by local conditions. Much of the Mediterranean is surrounded by mountains 1500 m or more in altitude. The orography of the region results in site-specific meteorological conditions influenced by sea breezes and up-slope winds, which can have marked impacts on patterns of local air quality (e.g. Lalas *et al.*, 1983). Consequently, in the Mediterranean region, the maps produced by EMEP (Figure 1) appear to be of less value than elsewhere in Europe, as meso-scale processes result in significant diurnal fluctuations that are not taken into account in the EMEP model (Sanz and Millán, 1998).

In the western Mediterranean, the build-up of $O_3$ concentrations occurs through the enhancement of sea breezes by up-slope winds in the coastal region (where there are substantial $NO_x$ sources) and the subsequent seaward return of these air masses at high altitudes. This causes the formation of stacked layers, 2-3 km deep and more than 300 km wide, with the oldest air masses near the sea (Millán *et al.*, 1996, 1997). Such circulation cells result in the build-up of high $O_3$ concentrations, with the usual model of increasing concentrations downwind of $NO_x$ being sources being inapplicable. Similar processes probably occur over the Italian Peninsula, but detailed information is lacking (Millán *et al.*, 1997).

$O_3$ concentrations are extremely difficult to characterise in a way that is meaningful for biological impact studies. For example, there is still considerable debate over the relative importance of short-term peak concentrations versus longer-term accumulated concentrations (Lefohn 1992). For forests, the latter approach has been deemed the most suitable, and the AOT40 index has been devised as a critical level. This is based on the accumulated hourly exposure to $O_3$ above 40 ppb during daylight hours (>50 W m$^{-2}$) between April and September (Kärenlampi and Skärby, 1996). The critical threshold of 10 ppm.h relates to an estimated 10% reduction in the growth of young (0-6 year old) beech (*Fagus sylvatica*) trees exposed to $O_3$ for between one and three seasons, based on five studies of above-ground growth and three studies of above-ground biomass increment (see Fuhrer *et al.* (1997) for further details of these studies). As these figures relate to above-ground responses only, the AOT40 index fails to take into account the substantial effects that are expected to occur below-ground as indicated from experiments with young trees grown under controlled conditions (Matyssek *et al.*, 1995a).

The AOT40 studies suggest that substantial areas of forest may be at risk and, although the objective of the UN-ECE Level I $O_3$ standards is simply to indicate where adverse effects of $O_3$ might occur (Skärby *et al.*, 1998), an $O_3$-induced loss in forest productivity of 10% has been estimated for Europe (Broadmeadow, 1998). There seems to be very little basis for such a statement, which is similar to the scenarios developed in the 1980s for growth losses induced by acid rain and later shown to be without foundation. AOT40 values are widely exceeded in the European Alps (Figure 2), where the presence of suitable precursors combines with altitudinal effects to result in widespread exceedances (Smidt, 1998). At a European scale, a substantial proportion of the forested area appears to be at risk from $O_3$, as judged by the AOT40 standard (Figure 3). Consequently, attention on the impacts of gaseous pollution on forests has increasingly concentrated on $O_3$ as the most important risk facing European forests today.

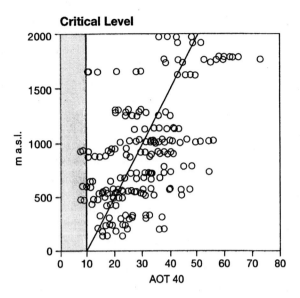

Fig. 2. Measured AOT40 values (1991-1996) for a variety of sites located near forests in the Austrian, German and Swiss Alps. From Smidt (1998), with permission.

## 3. Evidence of visible $O_3$ injury in woody species

In the USA, $O_3$ is widely recognized as a pollutant that can induce visual injury of foliage. Research in the San Bernardino Mountains of California has shown that $O_3$ is responsible for the development of visible injury in ponderosa pine (*Pinus ponderosa*) and Jeffrey pine (*Pinus jeffreyi*) (Miller et al., 1963; Miller and Millecan, 1971; Miller and McBride 1999). Subsequent work has indicated visible injury to many species growing in forest environments, not only in California, but also in other parts of the USA (e.g., Davis and Skelly, 1992; Neufeld et al., 1992; Simini et al., 1992; Miller et al., 1994; Chappelka et al., 1997; Skelly et al., 1997; Chappelka and Samuelson, 1998). Typical symptoms described in the literature include red, brown, purple or black stipple on the upper sides of sun-exposed leaves of broad-leaved trees (Brantley et al., 1994; Skelly et al., 1987; Flagler 1998), and chlorotic mottling of older needles in conifers.

Considering the widespread evidence of $O_3$ injury for American species, the scarcity of published information of confirmed visible symptoms of ambient $O_3$ injury to European species is rather surprising (see Davison and Barnes, 1998). Several studies have reported injury under experimental conditions (e.g. Günthardt-Goerg et al., 1996, 1997), but confirmed records of $O_3$ injury under ambient field conditions are extremely limited. In a study reported by Skelly et al. (1998a, 1998b), several species showing symptoms typical of $O_3$ injury were grown in ambient and filtered (ca. 50% $O_3$) air in southern Switzerland. It was possible to reproduce under experimental conditions the symptoms recorded in the field, and reducing the ambient level of $O_3$ by ca. 50% delayed the onset of symptom development. $O_3$ injury has been confirmed on a number of native Swiss species, including ash (*Fraxinus excelsior*), beech (*Fagus sylvatica*), red-berried elder (*Sambucus racemosa*), silver birch (*Betula pendula*), wayfaring tree (*Viburnum lantana*), black mulberry (*Morus nigra*), buckthorn (*Rhamnus catharticus*) and osier (*Salix viminalis*) (Skelly et al., 1998b). $O_3$-induced visible injury has not been documented on any coniferous species in southern Switzerland, but elsewhere in southern Europe there have been a number of records of $O_3$-like symptoms on Aleppo pine (*Pinus halepensis*), Arolla pine (*P. cembra*), Austrian pine (*P. nigra*) and Greek fir (*Abies cephalonica*) (e.g. Gimeno et al., 1992; Velissariou et al., 1992, 1996; Dalstein et al., 1997). In at least one of these species (Aleppo pine), the visible injury has been confirmed in experimental studies as being caused by $O_3$ (Gimeno et al., 1992; Anttonen et al., 1998).

The severe nature of visible injury to plants in southern Switzerland is particularly interesting. A combination of factors probably influences the occurrence of such injury. Firstly, $O_3$ concentrations are high, and six-month, daylight AOT40 values are often in excess of 30 ppm.h (BUWAL, 1997; Staffelbach et al., 1997). These cumulative values are accompanied by hourly peak concentrations of between 150 and 200 ppb. Secondly, southern Switzerland is unusual in having high amounts of summer rainfall, a characteristic of its insubric climate (Frei and Schär, 1998) which extends along the southern edge of the Alps, from Mont Blanc in the west to the Slovenian border in the east. Thirdly, although this remains to be confirmed, radiation intensity is high in southern Switzerland in comparison to the northern side of the Alps. In most other parts of Europe, this particular combination of conditions is not apparent. For example, in many parts of the Mediterranean region, the highest concentrations of $O_3$ occur in mid- to late summer, when most plants are suffering severe drought and might therefore be expected to experience lower $O_3$ doses than is at first apparent. The response of trees to drought stress in the Mediterranean region varies, with some species [e.g. Holm oak (*Quercus ilex*), Downy oak (*Q. pubescens*), Manna ash (*Fraxinus ornus*), and Eastern hornbeam (*Carpinus orientalis*)] exhibiting reductions in stomatal conductance and photosynthesis during the driest months, while other species (e.g. Phillyrea (*Phillyrea latifolia*) and Turpentine tree (*Pistacia terebinthus*)) show no clear responses (Tretiach, 1993; Damesin and Rambal, 1995; Radoglou, 1996). The effects are likely to vary with local water availability (see Rico et al., 1996), and this will make any risk presented by $O_3$ in the Mediterranean difficult.

## 4. Discrepancies between observed patterns in forests and the AOT40 critical threshold

The AOT40 critical threshold is intended to provide "an environmental standard of threshold to minimise the effects of $O_3$ on sensitive receptors, but does not seek to quantify the impacts of exceeding the critical level under field conditions" (Fuhrer et al., 1997). The maps of AOT40 exceedances (e.g. Figure 1) and risk assessments that have been undertaken to date (e.g. Figure 3) suggest that substantial effects on European forests should be apparent. While visible injury seems not to be widespread, it would be logical to expect to find evidence of growth reductions amongst European trees. This has not been the case, and the available data actually suggest that growth rates of forests in several parts of central Europe are currently increasing (Spiecker et al., 1996). Several possible reasons for this have been proposed, including increased nitrogen deposition, fertilisation by $CO_2$, a more favourable climate, better forest management and an increasing length of the vegetation period.

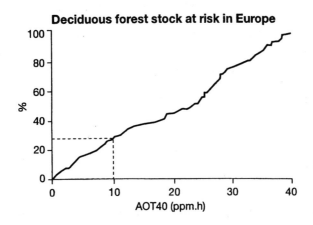

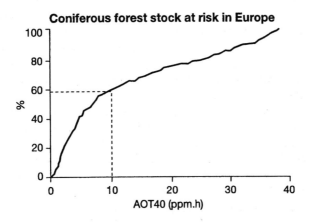

Fig. 3. Deciduous and coniferous forests at risk as judged by exceedance of the AOT40 critical level in 1990. The dotted lines indicate the proportion of forest area that lies below the AOT40 exposure threshold. The proportion of deciduous forest at risk (72%) is much greater than the proportion of coniferous forest (41%). From Hettelingh et al. (1996).

Why should there be discrepancies between the AOT40 risk analyses and field observations? Several points can be raised about the threshold and the way in which it has been defined. It is based on observed growth reductions under experimental conditions. Leaving aside the substantial problems associated with extrapolating results from young trees growing in an experimental environment to mature trees growing in a forest situation (e.g. Fredericksen et al., 1995, 1996a,b,c; Kolb et al., 1997), the approach makes the implicit assumption that timber supply is the only forest product of interest. Today, many foresters would disagree with this, although economic constraints usually mean that timber production remains the primary aim of most forestry enterprises. Moreover, the threshold of 10 ppm.h has very little scientific basis. It is too high to protect the most sensitive species from adverse impacts (Skelly et al., 1998b), and appears to be too low to provide information on species which are less sensitive to $O_3$ (e.g. Norway spruce (*Picea abies*)). There may also be a number of internal and external factors that must provide some form of $O_3$ buffering capacity for trees growing in forests.

Another issue concerns the threshold level of 40 ppb. The choice of this level was based on empirical information from a small number of studies and it remains very artificial. However, the level has not been adopted in the USA (where 60 ppb is used), and in one Scandinavian study, the best fit between response parameters and cumulative exposures was obtained when a threshold of 30 ppb was adopted (Pääkkönnen et al., 1996). Skärby (1994) has also suggested a 30 ppb threshold to be more applicable in relation to Norway spruce (*Picea abies*). As additional studies of AOT thresholds become available (e.g. Mortensen, 1998), it should be possible to evaluate the usefulness of the current 40 ppb threshold.

There are also uncertainties about both the seasonal period (April – September) and the daylight hours used in the calculations of AOT40. For example, under high mid-day radiation conditions, typical of the Mediterranean region, the stomata of many species may narrow (Tretiach, 1993) or even close (Epron et al., 1992; Inclán et al., 1998). This throws serious doubts on the use of mid-day $O_3$ concentration values in the cumulative exposure indices that are calculated for some areas.

Further uncertainties surround the response parameters used in AOT40 calculations. While above-ground growth seems under certain circumstances to be correlated quite well with exposure as expressed as AOT40, other equally valid indices, such as visible injury, do not correlate well. This may be because the relationship between visible injury and growth is very variable, and may not exist for some species (Somers et al., 1998). Consequently, the Level 1 AOT40 threshold must be viewed as a first approximation and there should be increased emphasis on the identification of Level 2 thresholds in future.

## 5. Potential impacts of $O_3$ on forest ecosystems in Europe

The AOT40 value was designed as a means of assessing the potential risk of $O_3$ to forest growth. It is already known that the value will not protect sensitive individuals from visible injury (Skelly et al., 1998a). Effects on the forest ecosystem as a whole are much more difficult to assess. The effects of $O_3$ on wild plants have recently been addressed by Davison and Barnes (1998), whereas effects on animals are covered in Anon. (1996).

The majority of $O_3$ impact studies have concentrated on species of economic importance (commercially important trees and crops), and on response indicators of economic value (growth and yield). However, a picture is slowly beginning to emerge about the possible ecosystem-level effects of $O_3$. $O_3$ can significantly inhibit the growth and reproduction of the most sensitive individuals within a population (Karnosky, 1981; Berrang et al., 1991). This may lead in the long-term to the loss of the most sensitive genotypes from the population (Karnosky, 1981; Berrang et al., 1989, 1991). At a time when the value of genetic diversity is being increasingly realised, these losses may be important. For example, the genetic diversity of a population plays a major role in its resistance to a broad range of other stresses (Müller-Starck and Ziehe, 1991).

One widely used indicator of forest condition is crown defoliation, although this particular indicator has many problems associated with its assessment (Innes, 1990; Innes et al., 1993) and interpretation (Innes, 1993; Ferretti, 1998). Crown condition varies substantially between years and in most cases correlates poorly, if at all, with $O_3$ exposure (Innes et al., 1997; Klap et al., 1997). Until now, $O_3$ has been shown to be an important determinant of tree health only in very special circumstances (e.g. Skelly et al., 1998b) and has not been demonstrated to be a widespread cause of forest decline (Skelly and Innes, 1994; Kandler and Innes, 1995). A major difficulty in the interpretation of such studies is the extent of the information on the spatial and temporal distributions of $O_3$. Normally, interpolated values are used, but the reliability of such values is questionable. Even if they are accurate, they may have very little relevance to the actual dose received by trees. In addition, the use of crown defoliation as a response variable is questionable because it is influenced by so many other factors. Growth is equally difficult to interpret because it too is affected by a range of different factors. Consequently, any effects of $O_3$ may be masked by other causes.

## 6. Discrepancies between available and required knowledge on risks presented by $O_3$

The major difficulty associated with the analysis of the physiological status of mature forest trees is the multi-factorial impact that may bias any response to $O_3$ (Arndt and Seufert, 1990). Despite this, there is no doubt that $O_3$ poses a risk to woody vegetation (Reich, 1987; Pye, 1988), not only through acute peak concentrations, but also through chronic exposures to enhanced mean $O_3$ levels. The impacts of chronic exposures over decades are likely to have effects, but the available information is extremely limited. With the exception of a few experimental $O_3$ exposures of branches or crowns of tall trees (see 11.), most of our knowledge on the risks of $O_3$ derives from young trees exposed to controlled $O_3$ treatments (Matyssek et al., 1995a). Typically, seedlings or saplings have been grown in pots in various kinds of fumigation chamber, often under non-limiting conditions except for $O_3$. They have then been compared - under the exclusion of competition and parasites - with corresponding control plants exposed to low levels of $O_3$ or $O_3$-free air. Using such techniques, even minor $O_3$ effects are detectable, even more so if clonal plants are used for minimizing the statistical variability in individual plant response.

Although such approaches have been valuable in revealing the principles of $O_3$ action and in showing that $O_3$ is an injurious pollutant, the findings are, *a priori*, not representative for forest sites as they are more or less biased by the nature of the exposure (Chappelka and Chevone, 1992). Given the lack of validation for forest sites and the uncertainties about the long-term acclimation and sensitivity of mature trees to

$O_3$, any expert opinion on the risks of $O_3$ must still largely rely on findings from young trees under controlled chamber exposures.

## 7. Risk assessments based on observations of young trees under
## 8. controlled chamber exposures

To the extent that boundary layers enable $O_3$ to reach the leaf surface, the diffusive influx of $O_3$ into the plant is mainly determined by the stomatal number and aperture (Reich, 1987; Laisk et al., 1989). Both stomatal density (which develops during leaf growth) and aperture are modified by chronic $O_3$ impact (Keller and Häsler, 1984; Frey et al., 1996), although the available results are inconsistent in terms of the extent and direction of the response (Matyssek et al., 1995a, 1998). Partial closure of stomata may occur, but often in such cases, and even in the absence of a stomatal response, leaf 'water-use efficiency' declines (i.e. photosynthetic $CO_2$ uptake versus transpiration rate, WUE; Sasek and Richardson, 1989; Schweizer and Arndt, 1990; Matyssek et al., 1991; Lippert et al., 1996; Maurer et al., 1997). This effect on WUE is mainly caused by injury to mesophyll cells. The primary site of the impact by $O_3$ or its oxidative degradation products is on the walls and plasmalemma of these cells. This impact, though being 'buffered' by antioxidative defence systems in the apoplast (Polle et al. 1995; 1998), may induce structural changes in the cell walls (Günthardt-Goerg et. al., 1993, 1997; Maier-Maercker, 1998), although being 'buffered' by antioxidative defence systems in the apoplast (Polle et al., 1995, 1998). The strain on the plasmalemma may eventually lead through unknown biochemical signal transduction to chloroplast degradation and 'programmed cell death' similar to defence reactions induced by pathogenic infections (Sandermann, 1996; Kangasjärvi et al., 1994). Mesophyll cell collapse and, eventually, premature leaf loss are the result (Matyssek et al., 1991). In conifers, needle loss (e.g. through reduced frost tolerance) may be delayed until the growing season succeeding the $O_3$ stress event, and induced changes in the secondary metabolism may persist (Sandermann 1996). Remarkably, all other changes in the plant can be traced back to the primary site of $O_3$ action, i.e. the apoplast and plasmalemma.

Injury to mesophyll cells not only lowers the carbon gain of the leaves, but may also impede, together with the disturbance of phloem loading and transport, assimilate translocation to other tree organs (Günthardt-Goerg et al., 1993; Rennenberg et al., 1996; Polle et al., 1998). The growth limitation of non-photosynthetic organs may be exacerbated by the enhanced assimilate demand of $O_3$-stressed foliage for detoxification and repair, as indicated by raised respiration rate and altered primary metabolism (Maurer et al., 1997; Landolt et al., 1997; Einig et al., 1997; Saurer et al., 1995), or for compensatory new-leaf formation (Mooney and Winner, 1991; Tjoelker and Luxmoore, 1991). Often, the organ that is most limited is the root (Andersen et al., 1991; Coleman et al., 1996), although conclusions from short-term experiments on the long-term uptake capacity of roots for water and nutrients have not yet been clarified (Matyssek et al., 1993, 1995a). C partitioning between root diameter classes may remain stable (Maurer and Matyssek, 1997), but increased fine-root production has been found in seedlings under $O_3$ stress (Kelly et al., 1995). In addition, the above-ground allocation can be affected by $O_3$: Often radial rather than longitudinal stem growth (with unknown consequences for long-term mechanical stability) as well as lateral branching and the weight and size of individual leaves are reduced (Schier et al., 1990; Matyssek et al., 1992, 1993; Samuelson et al., 1996). Whenever crown and root architecture are modified, plant competitiveness may be affected, because the efficiency between space occupation and resource acquisition is influenced (Küppers, 1994; Tremmel and Bazzaz, 1995). The significance for plant competition and, potentially, biodiversity, has been demonstrated in semi-natural plant communities (Nebel and Fuhrer, 1994; Barbo et al., 1998), but not for woody species. The risks caused by $O_3$ through changing resource allocation should not be overlooked relative to potential limitations in plant productivity.

## 8. Action of $O_3$ in concert with other biotic and abiotic influences

The risks highlighted above may be moderated, with simultaneous increases in uncertainty, by factors acting in concert with $O_3$ - and still, we must largely rely on findings from young plants under controlled conditions. One factor often overlooked is the interaction between mycorrhizal fungi and trees. Mycorrhizae represent a considerable sink for carbon, and both mycorrhizal and rhizospheric activities may be inhibited under $O_3$ stress by reduced carbon allocation to roots (Andersen and Rygiewicz, 1991). Nevertheless, increased fungal and bacterial biomass in the rhizosphere as well as raised root respiration have been found in *Pinus ponderosa* seedlings under environmentally relevant $O_3$ stress (Andersen and Scagel, 1997). It is uncertain if the raised metabolic activity of roots and mycorrhizae indicate enhanced nutrient uptake for assuring the maintenance of $O_3$-stressed foliage, or if mycorrhizae become a 'burden' to trees under $O_3$ stress (cf. Mahoney et al., 1985), as the plant-internal competition for assimilates required in foliage maintenance may be exacerbated. Such 'conflicts' appear to have relevance in herbaceous plants (Miller et al., 1997). The observation that $O_3$ exposure of trees may cause morphological changes in their mycorrhizae while stimulating the fungal and presumably non-mycorrhizal biomass in the rhizosphere may have implications for pathogenic root infections (Duckmanton and Widden, 1994). Decline in mycorrhization under $O_3$ stress (Ericsson et al., 1996) may lead to a lowered metabolic defence capacity (Gehring et al., 1997) and - as a consequence - favour infections by root pathogens or even enhance the attraction to phytophagous insects (Langebartels et al., 1997; Bonello et al., 1993).

Interrelationships between $O_3$ impact, mycorrhization and parasite attack are apparent. Since at the biochemical level, plant response to $O_3$ impact resembles defense against pathogenic infection (Sandermann, 1996), $O_3$ may initially strengthen the trees' resistance to parasite attack, but in the long-term it may weaken it by 'over-straining' the defensive mechanisms. This pathway of $O_3$ action predisposing trees to biotic stress and eventually increasing the risk of lethal injury requires further attention.

The attraction of plants to parasite attack is basically driven by their nutritional status (Herms and Mattson, 1992) and the gradually increasing $CO_2$ concentration of the atmosphere and, in some regions, dramatically enhanced nitrogen deposition may therefore distinctly influence host-parasite relationships. The extent that $O_3$ interacts with elevated $CO_2$ and variable nutrient supply, and by this, determines tree predisposition to parasite attack, is however unclear (cf. Manning and von Tiedemann, 1995). $O_3$ effects on nutrition have been found to be inconsistent and may be confounded by processes of senescence and retranslocation or depend on the specific nutrient under consideration (Matyssek et al., 1995a). However, both $O_3$ and $CO_2$ may alter C/N (Maurer and Matyssek, 1997; Hättenschwiler et al., 1996) and, as a result, the balance between growth and competitiveness *versus* the capacity in parasite defence (Loomis, 1953; Herms and Mattson, 1992) - this balance being a major determinant of tree 'fitness'. In more general terms, both elevated $CO_2$ and $O_3$ levels have the potential for modifying the crown architecture and, *via* changing the efficiency in light use, may also affect competitiveness (Reekie and Bazzaz, 1989; Matyssek et al., 1992, 1993; Hättenschwiler et al., 1996). Findings from controlled experiments about high $CO_2$ supply compensating for adverse effects by $O_3$ on photosynthesis and growth (Volin et al., 1998; Kellomäki and Wang, 1997; Olszyk and Wise, 1997; McKee et al., 1997) should be cautioned in terms of ecological significance (Saxe et al., 1998).

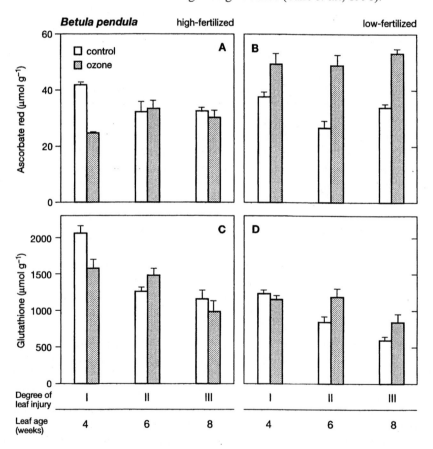

Fig. 4. Antioxidant responses (reduced ascorbate, A, B; glutathione, C, D) of birch cuttings (*Betula pendula*) grown at high or low nutrient supply (fertilization see Maurer et al., 1997) to $O_3$-free air (control) or ozonated air (90/40 nl l$^{-1}$ day/night). Leaves were sampled by the degree of progressive $O_3$ injury, degree I corresponding to 4-week-old, degree II to 6-week-old, and degree III to 8-week-old leaves (redrawn from Polle et al. 1998).

Nutrition does affect the sensitivity in carbon allocation to $O_3$ and may explain some of the apparent 'contradictions' found in the literature about the responses of trees to $O_3$ (Polle et al., 1998). Remarkably, high fertilization levels are not necessarily a prerequisite for enhanced $O_3$ tolerance (Maurer and Matyssek, 1997; Matyssek et al., 1997b). In young birch plants, low nutrition appeared to enhance the antioxidative defence capacity against $O_3$ (Figure 4) and delayed premature leaf loss (Polle et al., 1998; Maurer et al., 1997), although the 'price to be paid' is a re-directed carbon allocation, limiting instead of stimulating root growth at low nutrient supply (Figure 5; Maurer and Matyssek, 1997; Mooney and Winner, 1991). Nevertheless, low nutrition does not necessarily prevent $O_3$ injury, contrasting with the hypotheses of Weinstein et al. (1991). The extent to which nutrition may drive the turn-over of leaves and, through new leaf formation, may limit the decline of productivity under $O_3$ stress, needs to be examined with respect to the seasonal dynamics of the different patterns in shoot growth (determinate vs. indeterminate; Tjoelker and Luxmoore, 1991; Laurence et al., 1994; Matyssek et al., 1997b). Varying nitrogen availability may differ in its effects on $O_3$ sensitivity from responses to variable but balanced supply between the relevant macro and micro-nutrients (cf. Greitner and Winner, 1989).

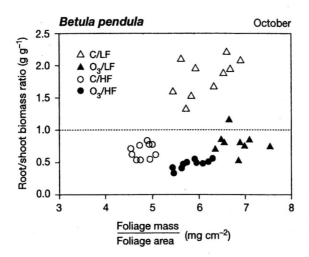

Fig. 5. Root/shoot biomass ratio in cuttings of *Betula pendula* at the end of the growing season as related to the ratio of foliage mass *versus* foliage area (one-sided); same experiment as in Fig. 6, 'C' representing control, '$O_3$' ozonation, 'LF' low-fertilization and 'HF' high-fertilization treatments (redrawn from Maurer and Matyssek 1997).

Nutrition relates to the water status of trees, as a low nutrient supply in particular tends to promote high sensitivity in stomatal regulation (Schulze, 1994). The capacity of chronic $O_3$ exposure to lower the stomatal conductance was found in birch to be most pronounced at low nutrient supply (Maurer et al., 1997). Partial stomatal closure caused by low air humidity or soil drought may limit $O_3$ uptake so that during sunny, dry days with high $O_3$ levels the influx of $O_3$ may be less than that under lower $O_3$ regimes on overcast, humid days (Wieser and Havranek, 1993). However, $O_3$ can influence the sensitivity of the stomata and, through this, of the whole plant to drought, as chronic $O_3$ exposure may lead to sluggishness in stomatal movements and to an inefficient control of the transpiration rate (Barnes et al. 1990a,b; Pearson and Mansfield, 1993; Karlsson et al., 1995; Götz, 1996). This change in stomatal behaviour appears to be related to altered 'mechanics' in the stomatal apparatus, possibly associated with reduced cell wall lignification, or to disturbed osmotic control of the guard cells because of membrane injury (Maier-Maercker, 1998; Heath and Taylor, 1997). Drought/$O_3$ interactions deserve greater attention in global-change research.

Another important factor determining $O_3$ sensitivity is irradiance, as not only high but also low light conditions can promote $O_3$ injury (Volin et al., 1993; Oren and Schoeneberger in Matyssek et al., 1995a). Low-light conditions are relevant in the shade and at dawn and dusk, and stomata may remain open at night, facilitating $O_3$ uptake (Tobiessen, 1982; Matyssek et al., 1995b). Although stomata may be less open under limiting light conditions, the lower rate of $O_3$ uptake can result in greater effects on production and carbon allocation to the stem and root (Figure 6) than higher rates of $O_3$ influx associated with non-limiting irradiance (Matyssek et al., 1995a, 1995b). The lack of light seems to limit the detoxification capacity (Menser, 1964; Foyer et al., 1991). This may have implications for the emphasis given in AOT40 calculations to daylight conditions. High $O_3$ sensitivity of shade leaves (Tjoelker et al., 1995) may relate also to their higher ratio of mesophyll surface area versus projected leaf area (Matyssek et al., 1995a). Implications for light gradients in canopies or scenarios of high $O_3$ regimes at night are imaginable. The state of knowledge on temperature/$O_3$ interaction, in particular the influence of $O_3$ on frost hardiness, has been dealt with comprehensively by Skärby et al. (1998).

Given the multi-factorial impacts at field sites and their capacity to exacerbate, weaken or even mask the action of $O_3$, it is clear that we are far from being able to quantify the risks imposed by $O_3$ on forest stands. This is emphasized when potential interactions between $O_3$ stress and the global-change scenarios are considered as well as the resulting consequences for nutrient availability and predispositions to biotic stress. These gaps in understanding are aggravated by the fact that current knowledge about the action of $O_3$ on trees is derived mainly from exposures of young plants under artificial conditions and are, therefore, not necessarily meaningful for mature individuals under stand conditions.

## 9. Ontogeny and the sensitivity of trees to $O_3$

Although ample information exists on the responses of woody plants to $O_3$ (see 7. and 8.), most of this knowledge derives from young trees, which are uncertain surrogates for mature trees (Kelly et al., 1995). One cause of different sensitivities to $O_3$ between ontogenetic stages and between species may be related to contrasting allocation patterns (Laurence et al., 1994). Young trees typically possess a high proportion of foliage to total biomass, enabling high annual increments in stem growth, but also representing a large surface area vulnerable to $O_3$ attack. In contrast, the respiratory demand for maintaining the high proportion of non-green biomass may be high in mature trees

(Waring and Schlesinger, 1985) and may be increased by $O_3$ defense demands so that reductions in net primary production and relative growth rate may be exacerbated. As plant defense against $O_3$ must

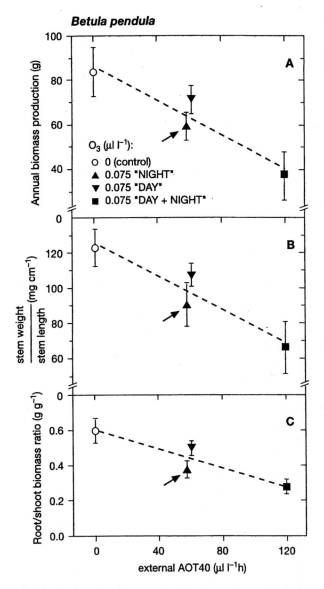

Fig 6. The effect of $O_3$ exposure on the annual whole-plant biomass production (A) and allometric relationships (B, ratio of stem weight vs. stem length; C, root/shoot biomass ratio) in cuttings of *Betula pendula* by the end of the growing season. Plants were exposed to $O_3$-free (control) or ozonated air (75 nl l$^{-1}$), the ozonation being applied each day either during daylight hours (7:00 hrs through 19:00 hrs, 'DAY'), or during the night including dawn and dusk (19:00 hrs through 7:00 hrs, 'NIGHT'), or throughout the entire daily course ('DAY + NIGHT'; arrows highlighting the 'NIGHT' treatment; redrawn from Matyssek et al. 1995b).

be fuelled from the internal resource pool, age-dependent 'conflicts' during ontogeny together with the demand in allocation for substrate and energy by other plant processes may affect $O_3$ tolerance (Herms and Mattson, 1992; Sandermann, 1996). Also, in mature trees rather than unshaded young individuals in experiments, large proportions of the crown can consist of shade leaves that - given the uncertainty about their $O_3$ sensitivity - may substantially contribute to the carbon gain, in particular, of shade-tolerant species (Schulze, 1970). Moreover, the control of carbon allocation between primary and secondary metabolism (which is responsible, for example, for $O_3$ defense) is affected by the resource availabilities governing the competitive interactions between plants as well as host-parasite relationships. There is increasing awareness (particularly in relation to tree responses to air pollutants or global-change scenarios) that both competition for resources and ontogenetic stage may determine the susceptibility of woody plants to environmental stress (Manning and von Tiedemann, 1995; Arnone, 1996; Groninger et al., 1996; Skelly et al., 1997).

If a high sink strength for carbon in mature trees is caused by the maintenance respiration requirement of non-green biomass, then a feedback to increasing photosynthetic capacity of the leaves is imaginable (Stitt, 1991). A comparison between seedlings and mature individuals of Red oak (*Quercus rubra*) revealed higher photosynthetic rates in older trees (Hanson et al., 1994; Figure 7). However, photosynthesis of the mature trees was also more susceptible to $O_3$ stress. Interestingly, the photosynthetic decline was most pronounced in the lower tree canopy (Samuelson and Edwards, 1993), consistent with the high photosynthetic sensitivity to $O_3$ of shade leaves of Sugar maple trees (*Acer saccharum*; Tjoelker et al., 1995). The $O_3$-induced decline in photosynthesis of oak trees may relate to the concomitantly higher stomatal conductance as compared with the seedlings and, because of higher $O_3$ uptake through the stomata, a resulting higher risk of injury (cf. Reich, 1987). There are observations that trees with high metabolic activity (as reflected, for example, in high photosynthetic capacity) are more susceptible to $O_3$ stress (Laurence et al., 1994). However, the photosynthetic limitation by $O_3$ in *Quercus rubra* trees had no measurable impact on growth performance, although allocation was affected (Kelly et al., 1995): Mature oak trees retained carbon in leaves and branches, while root growth was limited (as elsewhere found in seedlings: Matyssek et al., 1992; Polle et al., 1998). In the oak seedlings, however, exposure to $O_3$ led to stimulated fine-root production (which contrasts with many observations of seedlings under $O_3$ exposure, cf. Matyssek et al., 1995a). Differences in the micro-climate between the forest floor and the canopy - although influencing growth - were excluded as key factors determining the different $O_3$ sensitivities in oak trees and seedlings (Samuelson, 1994).

Given the limited knowledge about the responses of mature forest trees to $O_3$, the findings on red oak should not be extrapolated to other species. For example, an investigation on *Sequoia gigantea* revealed higher $O_3$ sensitivity in seedlings than in mature trees (Grulke and Miller, 1994). In this species, it is assumed that light interception overrules age-dependent shifts in sensitivity to $O_3$, as low irradiance in particular may lower $O_3$ tolerance. Using grafts of *Picea rubens*, a rather $O_3$-tolerant species, Rebbeck and Jensen (1993) found a lower $O_3$ sensitivity in scions from mature than from juvenile trees, the latter having higher stomatal conductance and, therefore, higher $O_3$ uptake capacity. When comparing trees of different ages beyond the early juvenile stage in *Pinus contorta* and *P. ponderosa,* Yoder et al. (1994) found that photosynthetic performance and stomatal conductance decreased with tree age, and even questioned maintenance respiration in mature trees as a limiting factor of growth. An implication may be that old trees are less endangered by $O_3$ through reduced $O_3$ uptake. It is hypothesised that the lower stomatal conductance of mature trees is related to increases in hydraulic resistance as stem and crown extend in size with age. Several observations support this hypothesis (Kline et al., 1976; Mattson-Djos, 1989; Sperry et al., 1993; Waring and Silvester, 1993). Given such relationships, $O_3$ uptake (and photosynthesis) may be controlled by water rather than the carbon relations in mature trees (Waring and Silvester, 1993; Yoder et al., 1994; Farquhar et al., 1989). The impact of hydraulic architecture, extensive soil water consumption and the gravitational constraints of large, mature trees on the internal water status may favour hormonal signals that close the stomata (Meinzer et al., 1991). However, little is known - except for the role of ethylene in promoting $O_3$ injury (Sandermann, 1996) - on the hormonal status in trees under $O_3$ stress (Lucas and Wolfenden, 1996). This gap in knowledge exists not only in relation to growth hormones, given the differential $O_3$-induced shifts in the meristematic activities between above- and below-ground organs, but also to the stomatal sensitivity to abscisic acid (ABA) under $O_3$ exposure. Apart from potential changes in membrane and cell wall properties of the stomatal apparatus under $O_3$ stress or shifts in the nutritional status that may affect stomatal sensitivity to ABA (Heath and Taylor, 1997; Maier-Mercker, 1998; cf. Schulze, 1994), it is imaginable that $O_3$-induced dysfunction in phloem transport may result in enhanced ABA levels in the foliage (Neales and McLeod, 1992) that contribute to stomatal closure.

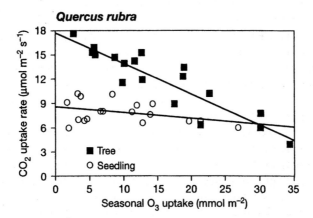

Fig. 7. Photosynthetic $CO_2$ uptake rate as related to the seasonal $O_3$ uptake of leaves in seedlings and mature trees of *Quercus rubra* (redrawn from Hanson et al. 1994).

The extent to which ontogenetic stages differ in $O_3$ tolerance cannot be resolved at present, but age-dependent differences in $O_3$ uptake, anatomy and detoxification, as well as in injury repair, appear to be of paramount importance. These properties are under genetic control and are modified by prevailing site conditions. Thus, general rules about the $O_3$-sensitivity of given ontogenetic stages may not exist, but perhaps stages of high rather than low metabolic activity are prone to limitations by $O_3$, as the interference of $O_3$ with the whole-plant carbon allocation may be most pronounced during stages of high metabolic activity (Laurence et al., 1994). A similar principle has been proposed for plant interactions with biotic stress (Herms and Mattson, 1992). In contrast, low metabolic activity may enhance $O_3$ tolerance, but possibly only in the short term, as the capacity for defense may be exceeded by the chronic $O_3$ exposure (Langebartels et al., 1997). There is a need to examine whether young *versus* mature trees, and species of different foliage type, flushing dynamics and allocation pattern, can be classified in terms of their $O_3$ tolerance by the above principles (Kelly et al., 1995). The need to clarify 'tree fitness' (cf. 8.) under field conditions is apparent, since trees, irrespective of their age, are physiologically 'challenged' whenever $O_3$ is taken up through open stomata.

## 10. Unknown risks by $O_3$ in mature trees and forests

A major problem for the assessment of the risks imposed by $O_3$ to mature trees under stand conditions is the uncertainty about diurnal and seasonal $O_3$ dynamics inside canopies and the significance of potentially enhanced foliage sensitivity to $O_3$ under limiting light conditions (Tjoelker et al., 1995; Matyssek et al., 1995b; Samuelson and Edwards, 1993). During the first half of the day, $O_3$ concentrations underneath the canopy of a spruce stand were found to stay almost as high as above, whereas during dusk and at night, the concentrations were distinctly lower underneath the canopy (Matyssek et al., 1997b). The factors determining the variability of $O_3$ flux from the free atmosphere into the different layers of canopies and the deposition of $O_3$ onto leaves need to be analysed (Fuentes et al., 1994). Pleijel et al. (1996) found the $O_3$ concentration at night to be higher inside than outside a spruce stand, when the wind velocity was low. Apart from the $O_3$ impact on shade foliage under limiting light, the overall constraint by $O_3$ on canopies at night appears to be important, as the nightly $O_3$ impact may increase in the proximity to urban centers (Lefohn and Jones, 1986) and with altitude (Sandroni et al., 1994; Beyrich et al., 1996). Advection from urban areas to the rural surroundings and the lack of nitrogen oxides may result in high $O_3$ concentrations in forests at night, and incursions of $O_3$ from high atmospheric layers may be an important factor at higher altitudes (Zaveri et al., 1995).

The extent to which stomata may be open under light limitation is debatable, particularly at night. However, stomata can open without light stimulation (Tobiessen, 1982), and such a behavior may be favored by warm and humid conditions (Lösch, 1979; Schulze and Hall, 1982). In addition, stomata may stay open in darkness as a result of the sluggishness in stomatal regulation associated with leaf senescence (Field, 1987; Benecke et al., 1981) or as a consequence of chronic $O_3$ exposure (Barnes et al. 1990a,b; Keller and Häsler, 1984; Götz 1996). Stomata that were open at night have been found in conifers and angiosperm trees (Wieser and Havranek, 1993; Matyssek et al., 1995b), as well as in herbaceous plants (Goknur and Tibbitts, 1984; Aben et al., 1989). Winner et al. (1989) found that night-time exposure to $O_3$ reduced production in turnip plants, and similar findings have been reported for birch (Matyssek et al. 1995b).

However, the capacity for $O_3$ uptake must be related to the boundary layers inside canopies and around leaves, as aerodynamic resistances can strongly impede the $O_3$ influx into plants, even though the stomata may be open (Pleijel et al., 1996). $O_3$ flux into the different canopy layers and its significance for shade crown sensitivity, as well as $O_3$ impact during night-time, may depend strongly on the extent of canopy closure and the crown architecture of single trees under particular stand conditions. So again generalisations are not possible about $O_3$ dynamics and the related risks at the canopy level. Case studies for the specific structural features of individual forests are required, especially for stands with unevenly aged and tall trees or mixed forests with tree species contrasting in crown shape or foliage type. Such stand structures are being increasingly favored as forestry moves towards a more ecologically-based management paradigm (Bode, 1997).

In addition to the uncertainties about spatial and temporal variations of $O_3$ regimes in canopies and the consequences for the $O_3$ sensitivity of different foliage types, it is not known whether chronic $O_3$ exposure induces a decline in stomatal aperture in mature trees as observed in young trees, and thus, whether effects on the water balance of whole stands and the 'water-use efficiency' of the stand production are to be expected. Although mature trees may not be limited by chronic $O_3$ exposure to a higher extent than by other site factors – given the increased productivity of forests presently observed in many parts of Europe – $O_3$ can act on the carbon allocation in the trees, as defense costs against $O_3$ need to be covered by the assimilate pool. The resulting differential shifts in growth patterns between the organs, particularly if exacerbated by disturbed assimilate transport, rather than effects on the overall production, appear to be critical for long-term stand development, because competitiveness is affected whenever crown or root architecture become altered. Competition between plants is exerted via the occupation of space, exclusion of neighbours and efficient sequestration of resources. Therefore, changes in crown and root allometry and in the associated cost/benefit balances between the investment of resources in, and gains from, organ structure may be the 'pathway' through which $O_3$ most effectively acts on the stability and biodiversity of forests (cf. Küppers, 1994; Tremmel and Bazzaz, 1995). $O_3$-induced changes in organ differentiation and turnover (e.g. of roots and leaves), associated with raised respiration and its consequences for the net primary production (especially in mature trees), may be most significant under chronic $O_3$ stress. Moreover, changed allocation may promote susceptibility to shortages in water and nutrient supply (through potential long-term limitations on root development) and the predisposition to parasite attack (Langebartels et al., 1997). $O_3$ effects on fruiting and natural recruitment of seedlings also need to be clarified (Chappelka and Chevone, 1992).

Although many forests in Europe are not 'natural' and are subjected to management, the impact of $O_3$ on resource fluxes in the ecosystem, through modifying competitive interactions, may be a determinant of the stability of managed mixed-species plantations and forests. Knowledge about the long-term mechanisms of $O_3$ impact on forests appears to be a pre-requisite for silvicultural measures that may compensate - to some extent at least - for the stress caused by $O_3$. However, such information does not yet exist, and information from controlled $O_3$ exposures in chambers with young trees still requires validation under specific $O_3$ regimes in 'real' forest stands.

## 11. Requirements for a quantitative risk assessment of the $O_3$ impact on forests

Risk assessment of $O_3$ impacts on mature trees has made use of cuvettes or bags for exposing branches in tree crowns to defined $O_3$ regimes (Houpis et al., 1991; Havranek and Wieser, 1994; Musselman and Hale, 1997). This approach has provided valuable insights into the mechanisms underlying $O_3$ uptake under conditions of air or soil moisture deficit in mature trees (Wieser and Havranek, 1995, 1996). Tolerance to chronic $O_3$ exposure and biochemical acclimation was indicated in *Larix decidua* and *Picea abies*, especially under the persistently elevated $O_3$ levels encountered at higher altitudes (Polle et al., 1995; Sandroni et al., 1994). The results of such approaches are subject to the micro-climatic bias introduced by branch enclosure or, perhaps, biochemical interaction with non-fumigated branch and crown parts (Musselman and Hale, 1997; Matyssek et al., 1997a), even though 'net carbon fluxes' may largely indicate carbon autonomy of branches (Stitt and Schulze, 1994; Dickson and Isebrands, 1991; Matyssek and Schulze, 1988). Large 'open-top' chambers are also influenced by micro-climatic bias, even though whole mature trees can be exposed to $O_3$ fumigation (e.g. Hanson et al., 1994).

Such micro-climatic bias can be overcome by using 'free-air' fumigation systems that avoid branch enclosure. Such a system releasing $O_3$ to leaf clusters on several sun and shade branches was constructed within in the canopy of an *Acer saccharum* stand (Tjoelker et al., 1994, 1995). The study revealed enhanced $O_3$ sensitivity in photosynthesis of the shade foliage. Employing 'free-air' fumigations to entire crowns can assure the experimental $O_3$ regime to induce whole-tree responses. With the exception of a number of 'free-air' $O_3$ fumigation experiments in agricultural crop plants (Manning and Krupa, 1992; Musselman and Hale, 1997), the few 'free-air' $O_3$ fumigations on trees have been restricted mainly to young plants, grown in experimental plots, with small crown dimensions (Wulff et al., 1992; McLeod et al., 1992). $O_3$ responses of these plants have, to some extent, resembled findings from similarly-aged individuals under chamber conditions (Pääkkönen et al., 1993; Wulff et al., 1992) and highlighted the importance of genetic variation in $O_3$ sensitivity. In possibly the most extensive study conducted to date (the Liphook experiment) few significant effects of $O_3$ were found on *Picea abies* (Wellburn et al., 1997). So far, 'free-air' fumigations of large, ageing trees have been restricted to other air pollutants such as $SO_2$ (Hendrey et al., 1992) or used to clarify impacts of rising $CO_2$ levels (Lewin et al., 1994).

At present, a 'free-air' $O_3$ fumigation study is being set up by scientists from the University of Munich in a 40 to 60-year-old mixed beech/spruce forest (*Fagus sylvatica/Picea abies*; Matyssek et al., 1997b). In this case study, $O_3$ is released into the canopy through a tubing system, with exposures controlled by feedbacks from on-line monitors. In addition, the weekly $O_3$ exposure in the different crown parts is assessed by passive samplers (Werner, 1992). The canopies of the about 27 m tall trees are accessible *via* scaffolding in sun and shade crowns. Groups of five beech and spruce trees each are exposed to '2 x ambient' $O_3$ levels up to a maximum of 150 nl $O_3$ $l^{-1}$, tracking diurnal and seasonal courses in ambient $O_3$ concentrations above the canopy. Corresponding groups of spruce and beech trees in ambient air serve as 'controls'.

Impacts of O3 are based on a broad spectrum of biochemical, (eco)-physiological and structural analyses in this interdisciplinary project, providing data for biophysical and ecophysiological models required for 'Level 2' critical levels development (cf. Ollinger et al., 1997). 'Free-air' fumigations appear to be the 'ultimate choice' for approaching a quantitative risk assessment of the chronic $O_3$ exposure of mature trees and forest stands and for scaling $O_3$ effects from the tree to the stand level.

## 12. Conclusions

The $O_3$ levels currently prevailing in Europe do not seem to pose an acute danger to mature trees and forests and are unlikely to lead to widespread dieback and break-down in the short-term. However, regional exposure conditions, such as those in southern Europe, may be conducive to $O_3$ injury. At a European level, judgement of the effects of chronic exposure to the persistently elevated $O_3$ levels is difficult, because the database available to achieve a long-term and quantitative risk assessment is still incomplete. Although partly reliable maps showing the distribution of $O_3$ at a regional scale are available, meso-scale information about diurnal and seasonal $O_3$ regimes in stands is scarce. Most importantly, knowledge about the physiological effects of $O_3$ on trees is not representative for the environmental conditions at forested sites, as most experiments deal with young trees, often grown for short periods of time in isolation inside fumigation cabinets. This biases exposure conditions and minimises the complex biotic and abiotic interactions which are typical of the field. Although such experiments have been valuable for revealing the 'principles' of $O_3$ action on woody plants and for indicating the broad spectrum of risks presented by $O_3$, these findings lack ecological significance as long as validation at forest sites is missing.

Due attention must be given to the three atmospheric components that alone or in combination represent the greatest risks to the long-term stability of forests, namely elevated $O_3$ and $CO_2$ concentrations and high nitrogen deposition. Impacts of these three factors relate to changes in allocation patterns, with subsequent and ecologically meaningful effects on 'tree fitness', i.e. competitiveness and predisposition to different kinds of stress. 'Free-air' fumigations of canopies with $O_3$ can approach a quantitative risk assessment in forest stands. Of course, such approaches can only be undertaken locally and are valid primarily for the case study under consideration. However, conducting studies of this kind for a transect of representative forest types throughout Europe would provide a 'realistic' database for urgently needed mechanistic modelling approaches towards a quantitative $O_3$ risk assessment (cf. Ollinger et al., 1997) and an integration of $O_3$ impact across the different geographical and ecological scales. A broad array of plant parameters would contribute to adequate definitions of Level 2 'Critical Levels for Ozone'. Quantitative risk assessment is necessary for achieving an environmental legislation that optimizes and reconciles, on a sound scientific basis, ecological *versus* economic demands.

Even though an ecological risk assessment for $O_3$ impacts on forests still needs to be accomplished, in many regions of Europe $O_3$ regimes prevail today that are known from controlled fumigation experiments to have the potential to adversely affect tree growth. For chronic $O_3$ stress, it seems that the $O_3$ impact on the plant-internal resource allocation rather than on the extent of productivity may be significant for the trees' fitness in the long term. Such a pathway of $O_3$ action may eventually lead to losses in the genetic diversity. It needs to be examined to what extent catastrophic break-down of stands caused by disease may arise from predispositions relating to chronic $O_3$ exposure, and also changes in competitiveness are relevant to managed forests. Knowledge about such 'pathways' of $O_3$ impact may be important for 'buffering' $O_3$-induced constraints by silvicultural measures, and such a view appears to be relevant as no prognosis exists today that would predict, for the decades to come, a decline in the $O_3$ regimes to pre-industrial levels. Despite the requirements for increased knowledge of the actual risks of $O_3$ to mature trees and forests, the information available today about the action of $O_3$ on trees does provide arguments for a responsible, precautionary policy that may prevent adverse effects of $O_3$ on the forests of the 21$^{st}$ century.

## Acknowledgements

We are grateful to Doris Pichler and Verena Fataar for producing the diagrams, and to Michèle Kaennel for her help with the literature search.

## References

Aben J., Thiel, F., Boekestein, A.: 1989, *Ultramicroscopy* **31**, 457.
Andersen, C.P., Hogsett, W.E., Wessling, R., Plocher, M.: 1991, *Can. J. For. Res.* **21**, 1288–1291.
Andersen, C.P. and Rygiewicz, P.T.: 1991, *Environ. Pollut.* **73**, 217–244.
Andersen, C.P. and Scagel, C.F.: 1997, *Tree Physiol.* **17**, 377–387.
Anfossi, D., Sandroni, S., Viarengo, S.: 1991, *J. Geophys. Res.* **96**, 17349–17352.
Anon.: 1996, *Air Quality Criteria for Ozone and Related Photochemical Oxidants. Volume 3*. Report no. EPA/600/P-93/004cF. Office of Research and Development, United States Environmental Protection Agency, Research Triangle Park, NC, 529 pages.
Anttonen, S., Kittilä, M., Kärenlampi, L.: 1998, *Chemosphere* **36**, 663–668.
Arndt, U. and Seufert, G.: 1990, *Environ. Pollut.* **68**, 195–203.
Arnone, J.A. III: 1996, 'Predicting responses of tropical plant communities to elevated $CO_2$: lessons from experiments with model ecosystems', in Körner, C., Bazzaz, F.A. (eds.), *Carbon Dioxide, Populations, and Communities*, Academic Press, San Diego, pp. 101–121.
Barbo, D.N., Chappelka, A.H., Somers, G.L., Miller-Goodman, M.S., Stolte, K.: 1998, *New Phytol.* **138**, 653–662.
Barnes, J.D., Eamus, D., Brown, K.A., 1990a, *New Phytol.* **114**: 713–720.
Barnes, J.D., Eamus, D., Davison, A.W., Ro-Paulsen, H., Mortensen, L.: 1990b, *Environ. Pollut.* **63**, 345–363.
Benecke, U., Schulze, E.D., Matyssek, R., Havranek, W.M.: 1981, *Oecologia* **50**, 54–61.
Berrang, P., Karnosky, D.F., Bennett, J.P.: 1989, *Can. J. For. Res.* **19**, 519–522.
Berrang, P., Karnosky, D.F., Bennett, J.P.: 1991, *Can. J. For. Res.* **21**, 1091–1097.
Beyrich, F., Acker, K., Kalass, D., Klemm, O., Möller, D., Schaller, E., Werhahn, J., Weisensee, U.: 1996, *Atmos. Environ.* **30**, 1271–1281.
Blum, O., Bytnerowicz, A., Manning, W., Popovicheva, L.: 1997, *Environ. Poll.* **98**, 299–304.
Bode, W. (ed.): 1997, *Naturnahe Waldwirtschaft. Prozeßschutz oder biologische Nachhaltigkeit?* Deukalion, Holm, 396 pages.
Bonello, P., Heller, W., Sandermann, H.: 1993, *New Phytol.* **124**, 653–663.
Brantley, E.A., Anderson, R.L., Smith, G.: 1994, *How to Identify Ozone Injury on Eastern Forest Bioindicator Plants*, Protection Report R8-PR 25, U.S. Department of Agriculture Forest Service, Southern Region and Northeastern Area, Asheville NC and Durham NH, 2 pages.
Broadmeadow, M.: 1998, *New Phytol.* **139**, 123–125.

Bundesamt für Umwelt, Wald und Landschaft (BUWAL): 1997, *NABEL Luftbelastung 1996*. Schriftenreihe Umwelt 286. Bundesamt für Umwelt, Wald und Landschaft, Bern, 193 pages.
Bytnerowicz, A., Manning, W.J., Grosjean, D., Chmielewski, W., Dmuchowski, W., Grodzinska, K., Godzik, B.: 1993, *Environ. Poll.* **80**, 301–305.
Chameides, W.L., Fehsenfeld, F., Rodgers, M.O., Cardelino, C., Martinez, J., Parrish, D., Lonneman, W., Lawson, D.R., Rasmussen, R.A., Zimmerman, P., Greenberg, J., Middleton, P., Wang, T.: 1992, *J. Geophys. Res.* **97**, 6037–6055.
Chappelka, A.H. and Chevone, B.I.: 1992, 'Tree response to ozone', in Lefohn, A.S. (ed.), *Surface Level Ozone Exposures and Their Effects on Vegetation*, Lewis, Chelsea, MI, pp. 271–324.
Chappelka, A.H., Renfro, J., Somers, G., Nash, B.: 1997, *Environ. Pollut.* **95**, 13–18.
Chappelka, A.H. and Samuelson, L.J.: 1998, *New Phytol.* **139**, 91–108.
Coleman, M.D., Dickson, R.E., Isebrands, J.G., Karnosky, D.F.: 1996, *Tree Physiol.* **16**, 145–152.
Dalstein, L., Garrec, J.-P., Bonneau, M.: 1997, *Rev. For. Franç.* **49**, 41–48.
Damesin, C. and Rambal, S.: 1995, *New Phytol.* **131**, 159–167.
Davies, T.D. and Schuepbach, E.: 1994, *Atmos. Environ.* **28**, 53–68.
Davis, D.D. and Skelly, J.M.: 1992, *Water, Air, Soil Pollut.* **62**, 269–277.
Davison, A.W. and Barnes, J.D.: 1998, *New Phytol.* **139**, 135–151.
Dickson, R.E. and Isebrands, J.G.: 1991, 'Leaves as regulators of stress response', in Mooney, H.A., Winner, W.E., Pell, E.J. (eds.), *Response of Plants to Multiple Stresses*, Academic Press, San Diego, pp. 4–34.
Dollard, G., Fowler, D., Smith, R.I., Hjellbrekke, A.-G., Uhse, K., Wallasch, M.: 1995, *Water, Air, Soil Pollut.* **85**, 1949–1954.
Duckmanton, L. and Widden, P.: 1994, *Mycologia* **86**, 181–186.
Ebel, A., Hass, H., Jakobs, H.J., Laube, M., Memmesheimer, M., Oberreuter, A., Geiss, H., Kuo, Y.-H.: 1991, *Atmos. Environ.* **25A**, 2131–2144.
Einig, W., Lauxmann, U., Hauch, B., Hampp, R., Landolt, W., Maurer, S., Matyssek, R.: 1997, *New Phytol.* **137**, 673–680.
Epron, D., Dreyer, E., Bréda, N.: 1992, *Plant, Cell, Environm.* **15**, 809–820.
Ericsson, T., Rytter, L., Vapaavuori, E.: 1996, *Biomass Bioenergy* **11**, 115–127.
Farquhar, G.D., Ehleringer, J.R., Hubick, K.T.: 1989, *Ann. Rev. Plant Physiol. Plant Mol. Biol.* **40**, 503–507.
Ferretti, M.: 1998, *Chemosphere* **36**, 1031–1036.
Field, C.B.: 1987, 'Leaf-age effects on stomatal conductance', in Zeiger, E., Farquhar, G.D., Cowan, I.R. (eds.), *Stomatal Function*, Stanford University Press, Stanford, pp. 367–384.
Flagler, R.B. (ed.) : 1998, *Recognition of Air Pollution Injury to Vegetation: A Pictorial Atlas*. A&WMA, Pittsburgh, 160 pages.
Foyer, C.H., Lelandais, M., Edwards, E.A., Mullineaux, P.M.: 1991, 'The role of ascorbate in plants, interactions with photosynthesis, and regulatory significance', in Pell, E., Steffen, K. (eds.), *Active Oxygen/Oxidative Stress and Plant Metabolism*, American Society of Plant Physiologists, pp. 131–144.
Fredericksen, T.S., Joyce, B.J., Skelly, J.M., Steiner, K.C., Kolb, T.E., Kouterick, K.B., Savage, J.E., Snyder, K.R.: 1995, *Environ. Pollut.* **89**, 273–283.
Fredericksen, T.S., Skelly, J.M., Steiner, K.C., Kolb, T.E., Kouterick, K.B.: 1996a, *Environ. Pollut.* **91**, 53–63.
Fredericksen, T.S., Skelly, J.M., Snyder, K.R., Steiner, K.C., Kolb, T.E.: 1996b, *J. Air Waste Management Assoc.* **46**, 464–469.
Fredericksen, T.S., Kolb, T.E., Skelly, J.M., Steiner, K.C., Joyce, B.J., Savage, J.E.: 1996c, *Tree Physiol.* **16**, 485–490.
Frei, C. and Schär, C.: 1998, *Internat. J. Climatol.* **18**, 873–900.
Frey, B., Scheidegger, C., Günthardt-Goerg, M.S., Matyssek, R.: 1996, *New Phytol.* **132**, 135–143.
Fuentes, J.D., Gillespie, T.J., Bunce, N.J.: 1994, *Water Air Soil Pollut.* **74**, 189–210.
Fuhrer, J., Skärby, L., Ashmore, M.R.: 1997, *Environ. Pollut.* **97**, 91–106.
Gehring, C.A., Cobb, N.S., Whitman, T.G.: 1997, *Amer. Nat.* **149**, 824–841.
Gimeno, B.S., Velissariou, D., Barnes, J.D., Inclan, R., Peña, J.M., Davison, A.W.: 1992, *Ecología* **6**, 131–134.
Goknur, A.B. and Tibbitts, T.W.: 1984, *HortScience* **19**, 548.
Götz, B.: 1996, *Ozon und Trockenstreß, Wirkungen auf den Gaswechsel von Fichte*, Libri Botanici Bd. 16, IHW-Verlag Eching, 149 pages.
Greitner, C.S. and Winner, W.E.: 1989, 'Nutrient effects on responses of willow and alder to ozone', in Olson, R.K., Lefohn, A.S. (eds.), *Transaction: Effects of Air Pollution on Western Forests*, Air & Waste Management Association, Anaheim, California, pp. 493–511.

Grennfelt, P. and Beck, J.P.: 1994, 'Ozone concentrations in Europe in relation to different concepts of the critical levels', in Fuhrer, J., Achermann, B. (eds.), *Critical Levels for Ozone: a UN-ECE Workshop Report*, Schriftenreihe der FAC Liebefeld, Swiss Federal Research Station for Agricultural Chemistry, Liebefeld-Bern, pp. 184–194.
Groninger, J.W., Seiler, J.R., Zedaker, S.M., Berrang, P.C.: 1996, *Funct. Ecol.* **10**, 708–716.
Grulke, N.E. and Miller, P.R.: 1994, *Tree Physiol.* **14**, 659–668.
Günthardt-Goerg, M.S., Landolt, W., Bucher, J.B.: 1996, *Wald und Holz* **77**, 16–19.
Günthardt-Goerg, M.S., Matyssek, R., Scheidegger, C., Keller, T.: 1993, *Trees* **7**, 104–114.
Günthardt-Goerg, M.S., McQuattie, C.J., Scheidegger, C., Rhiner, C., Matyssek, R.: 1997, *Can. J. For. Res.* **27**, 453–463.
Hanson, P.J., Samuelson, L.J., Wullschleger, S.D., Tabberer, T.A., Edwards, G.S.: 1994, *Tree Physiol.* **14**, 1351–1366.
Hättenschwiler, S., Schweingruber, F.H., Körner, C.: 1996, *Plant, Cell Environ.* **19**, 1369–1378.
Havranek, W.M. and Wieser, G.: 1994, *Proc. Roy. Soc. Edinburgh Sect. B* **102**, 541–546.
Heath, R.L. and Taylor, G.E.: 1997, 'Physiological processes and plant responses to ozone exposure', in Sandermann, H., Wellburn, A.R., Heath, R.L. (eds.), *Forest Decline and Ozone, a Comparison of Controlled Chamber and Field Experiments*, Ecological Studies 127, Springer-Verlag, Berlin, pp. 317–368.
Hendrey, G.R., Lewin, K.F., Kolber, Z., Evans, L.S.: 1992, *J. Air Waste Manage. Assoc.* **42**, 1324–1327.
Herms, D.A. and Mattson, W.J.: 1992, *Quart. Rev. Biology* **67**, 283–335.
Hettelingh, J.-P., Posch, M., de Smet, P.A.M.: 1996, 'Mapping critical thresholds and stock at risk', in Kärenlampi, L., Skärby, L. (eds.), *Critical Levels for Ozone in Europe: Testing and Finalizing the Concepts*, University of Kuopio, Kuopio, pp. 125–137.
Hettelingh, J.-P., Posch, M., de Smet, P.A.M.: 1997, 'Analysis of European maps', in: Posch, M., Hettelingh, J.-P., de Smet, P.A.M., Downing, R.J. (eds.), *Calculation and Mapping of Critical Thresholds in Europe: Status Report 1997*, RIVM Report No. 259101007, National Institute of Public Health and the Environment, Bilthoven, pp. 3–18.
Houpis, J.L., Costella, M.P., Cowles, S.: 1991, *J. Environ. Qual.* **20**, 467–474.
Inclán, R., Alonso, R., Pujadas, M., Terés, J., Gimeno, B.S.: 1998, *Chemosphere* **36**, 685–690.
Innes, J.L.: 1990, 'Some problems with the interpretation of international assessments of forest damage', in *Proc XIX World Congress IUFRO, Montréal, 5–11 August 1990*, IUFRO, Montrèal, pp 380–387.
Innes, J.L.: 1993, *Forest health: Its Assessment and Status*. CAB International, Wallingford, 677 pages.
Innes, J.L., Ghosh, S., Dobbertin, M., Rebetez, M., Zimmermann, S.: 1997, 'Kritische Belastungen und die Sanasilva-Inventur', in *Forum für Wissen 1997*, Eidg. Forschungsanstalt Wald, Schnee Landschaft, Birmensdorf, pp. 73–83.
Innes, J.L., Landmann, G., Mettendorf, B.: 1993, *Environ. Monit. Assessm.* **25**, 29–40.
Kärenlampi, L. and Skärby, L. (eds.): 1996, *Critical Levels for Ozone in Europe: Testing and Finalizing the Concepts*. University of Kuopio, Kuopio, 363 pages
Kandler, O. and Innes, J.L.: 1995, *Environ. Pollut.* **90**, 171–180.
Kangasjärvi, J., Talvinen, J., Utriainen, M., Karjalainen, R.: 1994, *Plant Cell Environ.* **17**, 783–794.
Karlsson, P.E., Medin, E.-L., Wickström, H., Selldén, G., Wallin, G., Ottoson, S., Skärby, L.: 1995, *Water, Air, Soil Pollut.* **85**, 1325–1330.
Karnosky, D.F.: 1981, *Mittl. Forstl. Bundesversuchsanstalt Wien* **137**, 41–45.
Keller, T. and Häsler, R.: 1984, *Oecologia* **64**, 284–286.
Kellomäki, S. and Wang, K.-Y.: 1997, *Environ. Pollut.* **97**, 17–27.
Kelly, J.M., Samuelson, L., Edwards, G., Hanson, P., Kelting, D., Mays, A., Wullschleger, S.: 1995, *Water, Air, Soil Pollut.* **85**, 1317–1324.
Klap, J., Voshaar, J.O., De Vries, W., Erisman, J.W.: 1997, 'Relationships between crown condition and stress factors', in Müller-Edzards, C., De Vries, W., Erisman, J.W. (eds.), *Ten Years of Monitoring Forest Condition in Europe*, European Commission, Brussels and United Nations Economic Council for Europe, Geneva, pp. 277–302.
Kline, J.R., Reed, K.L., Waring R.H., Stewart, M.L.: 1976, *J. Appl. Ecol.* **13**, 273–282.
Kolb, T.E., Frederiksen, T.S., Steiner, K.C., Skelly, J.M.: 1997, *Environ. Pollut.* **98**, 195–208.
Küppers, M.: 1994, 'Canopy gaps: competitive light interception and economic space filling – a matter of whole-plant allocation', in Caldwell, M.M. and Pearcy, R.W. (eds.), *Exploitation of Environmental Heterogeneity by Plants – Ecophysiological Processes Above and Below-Ground*, Academic Press, San Diego, pp. 111–144.
Laisk, A., Kull, O., Moldau, H.: 1989, *Plant Physiol.* **90**, 1163–1167.

Lalas, D.P., Asimakopoulos, D.N., Deligiorgi, D.G., Helmis, C.G.: 1983, *Atmos. Environ.* **17**, 1621–1632.
Landolt, W., Günthardt-Goerg, M.S., Pfenninger, I., Einig, W., Hampp, R., Maurer, S., Matyssek, R.: 1997, *New Phytol.* **137**, 389–397.
Langebartels, C., Ernst, D., Heller, W., Lütz, C., Payer, H.-D., Sandermann, H.: 1997, 'Ozone responses of trees: results from controlled chamber exposures at the GSF phytotrones', in Sandermann, H., Wellburn, A.R., Heath, R.L. (eds.), *Forest Decline and Ozone, a Comparison of Controlled Chamber and Field Experiments*, Ecological Studies 127, Springer-Verlag Berlin, pp. 163–200.
Laurence, J.A., Amundson, R.G., Friend, A.L., Pell, E.J., Temple, P.J.: 1994, *J. Environ. Qual.* **23**, 412–417.
Laurila, T. and Tuovinen, J.-P.: 1996, 'Monitored data in relation to exceedances of AOT40', in Fuhrer, J. and Achermann, B. (eds.), *Critical Levels for Ozone: a UN-ECE Workshop Report*, Schriftenreihe der FAC Liebefeld, Swiss Federal Research Station for Agricultural Chemistry, Liebefeld-Bern, pp. 115–124.
Lefohn, A.S.: 1992, *Surface Level Ozone Exposures and Their Effects on Vegetation*. Lewis Publishers, Chelsea, MI, 366 pages.
Lefohn, A.S. and Jones, C.K.: 1986, *J. Air Pollut. Control Assoc.* **36**, 1123–1129.
Lewin, K.F., Hendrey, G.R., Nagy, J., McMorte, R.L.: 1994, *Agric. For. Meteorol.* **70**, 15–29.
Lippert, M., Steiner, K., Payer, H.-D., Simons, S., Langebartels, C., Sandermann, H. Jr.: 1996, *Trees* **10**, 268–275.
Loibl, W. and Smidt, S.: 1996, *Environ. Sci. Pollut. Res.* **3**, 213–217.
Loomis, W.E.: 1953, 'Growth and differentiation – and introduction and summary', in Loomis, W.E. (ed.), *Growth and Differentiation in Plants*, Iowa State College Press, Ames, pp. 1–17.
Lösch, R.: 1979, 'Stomatal responses to changes in air humidity', in Sen, D.N., Chawan, D.D., Bansal, R.P. (eds.), *Structure, Function and Ecology of Stomata*, Dehra Dun, pp. 189–216.
Lucas, P.W. and Wolfenden, J.: 1996, *Phyton* **36**, 51–56.
Mahoney, M.J., Chevone, B.I., Skelly, J.M., Moore, L.D.: 1985, *Phytopathology* **75**, 679–682.
Maier-Maercker, U.: 1998, *Trees* **12**, 181–185.
Manning, W.J. and Krupa, S.V.: 1992, 'Experimental methodology for studying the effects of ozone on crops and trees', in: Lefohn, A.S. (ed.), *Surface Level Ozone Exposures and Their Effects on Vegetation*, Lewis Publishers, Chelsea, USA, pp. 93–156.
Manning, W.J. and v. Tiedemann, A.: 1995, *Environ. Pollut.* **88**, 219–245.
Mattson-Djos, E.: 1989, 'The use of pressure bomb and porometer for describing plant water stress in tree seedlings', in Puttunen, P. (ed.), *Proc Nordic Symposium on Vitality and Quality of Nursery Stock*, University of Helsinki, Helsinki, pp. 45–57.
Matyssek, R. and Schulze, E.-D.: 1988, *Trees* **2**, 233–241.
Matyssek, R., Günthardt-Goerg, M.S., Keller, T., Scheidegger, C.: 1991, *Trees* **5**, 5–13.
Matyssek, R., Günthardt-Goerg, M.S., Saurer, M., Keller, T.: 1992, *Trees* **6**, 69–76.
Matyssek, R., Keller, T., Koike, T.: 1993, *Environ. Pollut.* **79**, 1–7.
Matyssek, R., Reich, P.B., Oren, R., Winner, W.E.: 1995a, 'Response mechanisms of conifers to air pollutants', in Smith, W.K. and Hinckley, T.H. (eds.), *Physiological Ecology of Coniferous Forests*, Academic Press, New York, pp. 255–308.
Matyssek, R., Günthardt-Goerg, M.S., Maurer, S., Keller, T.: 1995b, *Tree Physiol.* **15**, 159–165.
Matyssek, R., Havranek, W.M., Wieser, G., Innes, J.L.: 1997a, 'Ozone and the forests in Austria and Switzerland', in Sandermann, H. Jr., Wellburn, A.R., Heath, R.L. (eds.), *Forest Decline and Ozone: a Comparison of Controlled Chamber and Field Experiments*, Ecological Studies 127, Springer-Verlag, Berlin, pp. 95–134.
Matyssek, R., Maurer, S., Günthardt-Goerg, M.S., Landolt, W., Saurer, M., Polle, A.: 1997b, *Phyton* **37**, 157–167.
Matyssek, R., Günthardt-Goerg, M.S., Schmutz, P., Saurer, M., Landolt, W., Bucher, J.B.: 1998, *J. Sustainable Forestry* **6**, 3–22.
Maurer, S, Matyssek, R., Günthardt-Goerg, M.S., Landolt, W., Einig, W.: 1997, *Trees* **12**, 1–10.
Maurer, S. and Matyssek, R.: 1997, *Trees* **12**, 11–20.
McKee, I.F., Bullimore, J.F., Long, S.P.: 1997, *Plant, Cell Environ.* **20**, 77–84.
McLeod, A.R., Shaw, P.J.A., Holland, M.R.: 1992, *For. Ecol. Manage.* **51**, 121–127.
Meinzer, F.C., Grantz, D.A., Smit, B.: 1991, *Aust. J. Plant Physiol.* **18**, 329–338.
Menser, H.A.: 1964, *Plant Physiol.* **39**, 564–567.
Millán, M.M., Salvador, R., Mantilla, E., Artíñano, B.: 1996, *Atmos. Environ.* **30**, 1909–1924.
Millán, M.M., Salvador, R., Mantilla, E.: 1997, *J. Geophys. Res.* **102**, 8811–8823.
Miller, J.E., Shafer, S.R., Schoenenberger, M.M., Pursley, W.A., Horton, S.J., Davey, C.B.: 1997, *Water,*

*Air, Soil Pollut.* **96**, 233–248.
Miller, P.R., Grulke, N.E., Stolte, K.W.: 1994, 'Effects of air pollution on giant sequoia ecosystems', in Aune, P.S. (Technical Coordinator), *Proceedings of the Symposium on Giant Sequoias: Their Place in the Ecosystem and Society*, USDA Forest Service, Pacific Southwest Research Station, Albany, California, General Technical Report PSW-151, pp. 90–98.
Miller, P.R. and McBride, J.R. (eds.): 1999, *Oxidant Air Pollution Impacts in the Montane Forests of Southern California. A Case Study of the San Bernardino Mountains*. Springer Verlag, New York, 424 pages.
Miller, P.R. and Millecan, A.A.: 1971, *Plant Disease Reporter* **55**, 555–559.
Miller, P.R., Parmeter, J.R., Taylor, O.C., Cardiff, E.A.: 1963, *Phytopathology* **53**, 1072–1076.
Mooney, H.A. and Winner, W.E.: 1991, 'Partitioning response of plants to stress', in Mooney, H.A., Winner, W.E., Pell, E.J. (eds.), *Response of Plants to Multiple Stresses*, Academic Press, San Diego, pp 129–141.
Mortensen, L.M.: 1998, *Scand. J. For. Res.* **13**, 189–196.
Müller-Starck, G. and Ziehe, M.: 1991, *Genetic Variation in European Populations of Forest Trees*. J.D. Sauerländer's Verlag, Frankfurt, 271 pages.
Musselman, R.C. and Hale, B.A.: 1997, 'Methods for controlled and field ozone exposures of forest tree species in North America', in Sandermann, H. Jr., Wellburn, A.R., Heath, R.L. (eds.), *Forest Decline and Ozone: a Comparison of Controlled Chamber and Field Experiments*, Ecological Studies 127, Springer-Verlag, Berlin, pp. 277–315.
Neales, T.F. and McLeod, A.L.: 1992, *Plant, Cell, Environ.* **14**, 979–986.
Nebel, B. and Fuhrer, J.: 1994, *Angew. Bot.* **68**, 116–121.
Neufeld, H.S., Renfro, J.R., Hacker, W.D., Silsbee, D.: 1992, 'Ozone in Great Smoky Mountains National Park: dynamics and effects on plants', in Berglund, R.L. (ed.), *Transactions: Tropospheric Ozone and the Environment II*, Air & Waste Management Assoc., Pittsburgh, PA, USA, pp. 594–617.
Ollinger, S.V., Aber, J.D., Reich, P.B.: 1997, *Ecol. Applic.* **7**, 1237–1251.
Olszyk, D.M. and Wise, C.: 1997, *Agric. Ecosys. Environ.* **66**, 1–10.
Pääkkönen, E., Holopainen, T., Kärenlampi, L.: 1996, 'Relationships between open-field ozone exposures and growth and senescence of birch (*Betula pendula* and *Betula pubescens*)', in Skärby, L., Pleijel, H. (eds.), *Critical Levels for Ozone – Experiments with Crops, Wild Plants and Forest Tree Species in the Nordic Countries*, TemaNord 1996: 582, pp. 39–48.
Pääkkönen, E., Paasisalo, S., Holopainen, T., Kärenlampi, L.: 1993, *New Phytol.* **125**, 615–623.
Pearson, M. and Mansfield, T.A.: 1993, *New Phytol.* **123**, 351–358.
Pleijel, H., Wallin, G., Karlsson, P.E., Skärby, L.: 1996, *Atmos. Environ.* **30**, 4077–4084.
Polle, A., Wieser, G., Havranek, W.M.: 1995, *Plant, Cell Environ.* **18**, 681–688.
Polle, A., Matyssek, R., Günthardt-Goerg, M.S., Maurer, S.: 1998, 'Defence strategies against ozone in trees: the role of nutrition' in Agrawal, S., Agrawal, M., Krizek, D.T. (eds.), *Environmental Pollution and Plant Responses*, CRC, Lewis Publishers, New York, in press.
Proyou, A.G., Toupance, G., Perros, P.E.: 1991, *Atmos. Environ.* **25A**, 2145–2153.
Pye, J.M.: 1988, *J. Environ. Qual.* **17**, 347–360.
Radoglou, K.: 1996, *Ann. Sci. For.* **53**, 269–278.
Rebbeck, J. and Jensen, K.F.: 1993, *Can. J. For. Res.* **23**, 450–456.
Reekie, E.G. and Bazzaz, F.A.: 1989, *Oecologia* **79**, 212–222.
Reich, P.B.: 1987, *Tree Physiol.* **3**, 63–91.
Reiter, R., Sladkovic, R., Kanter, H.J.: 1987, *Meteorol.. Atmos. Phys.* **37**, 27–47.
Rennenberg, H., Herschbach, C., Polle, A.: 1996, *J. Plant Physiol.* **148**, 296–301.
Rico, M., Gallego, H.A., Moreno, G., Santa Regina, I.: 1996, *Ann. Sci. For.* **53**, 221–234.
Samuelson, L.J.: 1994, *New Phytol.* **128**, 235–241.
Samuelson, L.J. and Edwards, G.S.: 1993, *New Phytol.* **125**, 373–379.
Samuelson, L.J., Kelly, J.M., Mays, P.A., Edwards, G.S.: 1996, *Environ Pollut.* **91**, 317–323.
Sandermann, H. Jr.: 1996, *Ann. Rev. Phytopathol.* **34**, 347–366.
Sandroni, S., Bacci, P., Botta, G., Pellegrini, U., Ventura, A.: 1994, *Sci. Total Environ.* **156**, 169–182.
Sanz, M.J. and Millán, M.M.: 1998, *Chemosphere* **36**, 1089–1094.
Sasek, T.W. and Richardson, C.J.: 1989, *For. Sci.* **35**, 745–755.
Saurer, M., Maurer, S., Matyssek, R., Landolt, W., Günthardt-Goerg, M.S., Siegenthaler, U.: 1995, *Oecologia* **103**, 397–406.
Saxe, H., Ellsworth, D.S., Heath, J.: 1998, *New Phytol.* **139**, 395–436.
Schier, G.A., McQuattie, C.J., Jensen, K.F.: 1990, *Can. J. For. Res.* **20**, 1714–1719.

Schulze, E.-D.: 1970, *Flora* **159**, 177–232.
Schulze, E.-D. and Hall, A.E.: 1982, 'Stomatal responses, water loss, and nutrient relations in contrasting environments', in Lange, O.L., Nobel, P.S., Osmond, C.B., Ziegler, H. (eds.), *Encyclopedia of Plant Ecology 12B, Physiological Plant Ecology II*, Springer, Berlin, Heidelberg, New York, pp. 182–230.
Schulze, E.-D.: 1994, 'The regulation of plant transpiration: interactions of feedforward, feedback, and futile cycles', in Schulze, E.-D. (ed.), *Flux Control in Biological Systems*, Academic Press, New York, pp. 203–235.
Schweizer, B., Arndt, U.: 1990, *Environ. Pollut.* **68**, 275–292.
Simini, M., Skelly, J.M., Davis, D.D., Savage, J.E.: 1992, *Can. J. For. Res.* **22**, 1789–1799.
Skärby, L.: 1994, 'Critical levels for ozone to protect forest trees', in Fuhrer, J., Achermann, B. (eds.), *Critical Levels for Ozone: a UN-ECE Workshop Report*, Schriftenreihe der FAC Liebefeld, Swiss Federal Research Station for Agricultural Chemistry, Liebefeld-Bern, pp. 74–87.
Skärby, L. and Karlsson, P.E.: 1996, 'Critical levels for ozone to protect forest trees – best available knowledge from the Nordic countries and the rest of Europe', in Fuhrer, J., Achermann, B. (eds.), *Critical Levels for Ozone: a UN-ECE Workshop Report*, Schriftenreihe der FAC Liebefeld, Swiss Federal Research Station for Agricultural Chemistry, Liebefeld-Bern, pp. 72–85.
Skärby, L., Ro-Poulsen, H., Wellburn, F.A.M., Sheppard, L.J.: 1998, *New Phytol.* **139**, 109–122.
Skelly, J.M., Chappelka, A.H., Laurence, J.A., Fredericksen, T.S.: 1997, 'Ozone and its known and potential effects on forests in eastern United States', in Sandermann, H, Wellburn, A.R., Heath, R.L. (eds.), *Forest Decline and Ozone, a Comparison of Controlled Chamber and Field Experiments*, Ecological Studies 127, Springer-Verlag Berlin, pp. 69–93.
Skelly, J.M., Davis, D.D., Merrill, W., Cameron, E.A., Brown, H.D., Drummond, D.B., Dochinger, L.S. (eds.): 1987, *Diagnosing Injury to Eastern Forest Trees: a Manual for Identifying Damage Caused by Air Pollution, Pathogens, Insects, and Abiotic Stresses*, National Acidic Precipitation Program, Forest Response Program, Vegetation Survey Research Cooperative. University Park, PA: Agricultural Information Services, College of Agriculture, Department of Plant Pathology, Pennsylvania State University, 122 pages.
Skelly, J.M. and Innes, J.L.: 1994, *Plant Disease* **78**, 1021–1032.
Skelly, J.M., Innes, J.L., Snyder, K.R., Savage, J.E., Hug, C., Landolt, W., Bleuler, P.: 1998a, *Chemosphere* **36**, 995–1000.
Skelly, J.M., Savage, J.E., Snyder, K.R., VanderHayden, D., Zhang, J., Innes, J.L., Sanz, M.J.: 1998b, This volume.
Smidt, S.: 1998, *Environ. Sci. Poll. Res.*, Spec. Issue **1**, 25–31.
Somers, G.L., Chappelka, A.H., Rosseau, P., Renfro, J.R.: 1998, *For. Ecol. Managem.* **104**, 129–137.
Sperry, J.S., Alder, N.N., Eastlack, S.E.: 1993, *J. Exper. Bot.* **44**, 1075–1082.
Spiecker, H., Mielikäinen, K., Köhl, M., Skovsgaard, J.: 1996, *Growth Trends in European Forests*, Springer Verlag, Berlin, Heidelberg, 372 pages.
Staehelin, J. and Schmid, W.: 1991, *Atmos. Environ.* **25A**, 1739–1749.
Staehelin, J., Thudium, J., Buehler, R., Volz-Thomas, A, Graber, W.: 1994, *Atmos. Environ.* **28**, 75–87.
Staffelbach, T., Neftel, A., Blattner, A., Gut, A., Fahrni, M., Stähelin, J., Prévôt, A., Hering, A., Lehning, M., Neininger, B., Bäumle, M., Kok, G.L., Dommen, J., Hutterli, M., Anklin, M.: 1997, *J. Geophys. Res.* **102**, 23345–23362.
Stitt, M.: 1991, *Plant, Cell Environ.* **14**, 741–762.
Stitt, M. and Schulze, E.-D.: 1994, 'Plant growth, storage, and resource allocation: from flux control in a metabolic chain to the whole-plant level', in Schulze E-D (ed.), *Flux Control in Biological Systems*, Academic Press, San Diego, pp. 57–118.
Tjoelker, M.G. and Luxmoore, R.J.: 1991, *New Phytol.* **119**, 69–81.
Tjoelker, M.G., Volin, J.C., Oleksyn, J., Reich, P.B.: 1994, *Plant, Cell Environ.* **17**, 211–218.
Tjoelker, M.G., Volin, J.C., Oleksyn, J., Reich, P.B.: 1995, *Plant, Cell Environ.* **18**, 895–905.
Tobiessen, P.: 1982, *Oecologia* (Berl.) **52**, 356–359.
Tremmel, D.C. and Bazzaz, F.A.: 1995, *Ecology* **76**, 262–271.
Tretiach, M.: 1993, *Acta Œcologia* **14**, 341–360.
United Kingdom Photochemical Oxidants Review Group (UKPORG): 1987, *Ozone in the United Kingdom*, Department of the Environment, London, 112 pages.
Velissariou, D., Davison, A.W., Barnes, J.D., Pfirrmann, T., Maclean, D.C., Holevas, C.D. 1992, *Atmos. Environ.* **26**, 373–380.
Velissariou, D., Gimeno, B.S., Badiani, M., Fumagalli, I., Davison, A.W.: 1996, 'Records of $O_3$ visible injury in the ECE Mediterranean region', in Kärenlampi, L. and Skärby, L. (eds.), *Critical Levels for*

*Ozone in Europe: Testing and Finalizing the Concepts*, University of Kuopio, Kuopio, pp. 343–350.
Volin, J.C., Tjoelker, M.G., Oleksyn, J., Reich, P.B.: 1993, *New Phytol.* **124**, 637–646.
Volin, J.C., Reich, P.B. and Givnish, T.: 1998, *New Phytol.* **138**, 315–325.
Volz, A. and Kley, D.: 1988, *Nature* **332**, 240–242.
Waring, R.H. and Schlesinger, W.H.: 1985, *Forest Ecosystems, Concepts and Management*, Academic Press, Orlando, San Diego, New York, 340 pages.
Waring, R.H. and Silvester, W.B.: 1993, *Tree Physiol.* **14**, 1203–1213.
Weinstein, D.A., Beloin, R.M., Yanai, R.D.: 1991, *Tree Physiol.* **9**, 127–146.
Wellburn, A.R., Barnes, J.D., Lucas, P.W., McLeod, A.R., Mansfield, T.A.: 1997, 'Controlled $O_3$ exposures and field observations of $O_3$ effects in the UK', in Sandermann, H., Wellburn, A.R., Heath, R.L. (eds.), *Forest Decline and Ozone: a Comparison of Controlled Chamber and Field Experiments*, Ecological Studies 127, Springer-Verlag, Berlin, Heidelberg, pp. 201–236.
Werner, H.: 1992, *Das Indigopapier, sensitives Element zum Aufbau von Passivsammlern zur Messung von Ozonimmissionen*, Forstl. Forschungsberichte München, 122.
Wieser, G. and Havranek, W.M.: 1993, *Trees* **7**, 227–232.
Wieser, G. and Havranek, W.M.: 1995, *Tree Physiol.* **15**, 253–258.
Wieser, G. and Havranek, W.M.: 1996, *J. Plant Physiol.* **148**, 189–194.
Winner, W.E., Lefohn, A.S., Cotter, I.S., Greitner, C.S., Nellessen, J., McEvoy, L.R. jr., Olson, R.L., Atkinson, C.J., Moore, L.D.: 1989, *Proc. Natl. Acad. Sci. USA, Ecology* **86**, 8828–8832.
Wulff, A., Hänninen, O., Tuomainen, A., Kärenlampi, L.: 1992, *Ann. Bot. Fennici* **29**, 253–262.
Yoder, B.J., Ryan, M.G., Waring, R.H., Schoettle, A.W., Kaufmann, M.R.: 1994, *For. Sci.* **40**, 513–527.
Zaveri, R.A., Saylor, R.D., Peters, L.K., McNider, R., Song, A.: 1995, *Atmos. Environ.* **29**, 1043–1065.

# OBSERVATION AND CONFIRMATION OF FOLIAR OZONE SYMPTOMS OF NATIVE PLANT SPECIES OF SWITZERLAND AND SOUTHERN SPAIN

J. M. SKELLY[1], J. L. INNES[4], J. E. SAVAGE[2], K. R. SNYDER[2], D. VANDERHEYDEN[2], J. ZHANG[3] and M. J. SANZ[5]

[1]*Department of Plant Pathology,* [2]*Environmental Resources Research Institute, and* [3] *School of Forest Resources respectively, Pennsylvania State University, University Park PA, USA, 16802;* [4] *Swiss Federal Institute for Forest, Snow, and Landscape Research, Birmensdorf, Switzerland and* [5] *Centro de Estudios Ambientales del Mediterraneo, Valencia, Spain.*

(Received 16 October 1998; accepted 26 February 1999)

**Abstract.** Tropospheric ozone is considered as the major pollutant of concern to the health and productivity of forests in the eastern United States and has more recently become of increasing concern within the forests of southern Europe. Recent observations have clearly demonstrated foliar injury symptoms to be occurring on many tree and native plant species within remote forested areas. Several plant genera (and a few species within genera) found in both the forests of Switzerland and the southern coastal region of Spain exhibit field symptoms typical of ambient ozone exposures. Ozone exposures for many species have been conducted under controlled CSTR conditions and within open-top chambers within the study areas. Results have confirmed that the $O_3$-like foliar symptoms as observed under natural forest and open grown conditions for many native tree, shrub, and herbaceous species in Spain and Switzerland are caused by exposures to ambient $O_3$.

**Keywords:** ozone, pollution, Switzerland, Spain, native species, symptoms

## 1. Introduction

The widespread presence of mountains bordering the Mediterranean Sea creates atmospheric circulation systems which favor the build-up of primary pollutant emissions leading to high ozone concentrations over the coastal and proximal mountainous terrain. In southern Europe, recent studies by Millán *et al.* (1996, 1997) have shown high ozone exposures in southern Spain to be a result of recurring summer atmospheric circulation patterns of ozone and its photochemical precursors, from industrial coastal areas inland to the central plateau and then flowing in upper air layers south over the Mediterranean Sea and returning inland on a 2-3 day cycle. Ozone exposures 2-3 times the EC directives for damage to vegetation (65 $\mu g\ m^{-3}$ 24 h avg.) have been reported. Millán *et al.* (1996,1997) also suggested that similar patterns of ozone occurrence could be expected in other southern Mediterranean regions.

Bacci *et al.* (1990) and Wunderli and Gehrig (1990) previously described patterns of ozone exposures across the sub-Alpine regions of southern Switzerland and northern Italy with the upper Po plain industrial area and the city of Milan, Italy as major contributors of the pollutant precursors. Staffelbach *et al.* (1997 a, b) further described high ozone exposures in the south-facing sub-alpine regions of Sotto Ceneri of Canton Ticino in southern Switzerland. They described situations of air parcels passing over the metropolitan and industrialized Milan area and then arriving at their monitoring site a few hours later with peak ozone hourly concentrations in excess of 100 ppb (200 $\mu g\ m^{-3}$). The Ticino region situated to the south of the Swiss Alps generally experiences a

Mediterranean climate, with hot but relatively moist summer seasons. The Alps form an arc around the plain of the Po valley, acting as a barrier against central European weather; the high Alps also trap northerly-moving air masses from the Po plain of northern Italy directly influencing ozone concentrations during hot and sunny days throughout the southern regions of Switzerland.

Regional scale tropospheric ozone air pollution has become of concern within the forests of central and southern Europe (Derwent, 1990; Schenone and Lorenzini, 1990; Grennfelt and Beck, 1994; Fuhrer *et al.*, 1997; Hettelingh *et al.*, 1996; Skärby *et al.*, 1998; Matyssek and Innes, 1999). Forests within the more remote and mountainous terrain of Switzerland and the coastal areas of southern Spain are exposed to elevated ambient ozone concentrations on a seasonal basis with short periods of more severe ozone exposure episodes associated with stagnant high pressure systems. Several recent reports of ozone-induced foliar injuries to sensitive plant species within the more sunny and warmer climes of the Mediterranean regions are available. (Schenone and Lorenzini, 1990; Innes *et al.*, 1995, 1996; Mea *et al.*, 1998; Ferretti *et al.*, 1998; Bussotti *et al.*, 1998, Skelly *et al.*, 1998).

Sanz *et al.* (unpublished data) and Sanz and Millán (1998) have described patterns of injuries to forest species within the forested ecosystems of southern Spain. They determined that during a three-year period (1994-96) visual ozone injury of *Pinus halepensis* was well correlated with ozone-laden sea breeze penetrations into the coastal valleys of the Castellan region.

Symptoms of ozone induced foliar injury appear as mild chlorosis, adaxial leaf surface pigmentation (stippling) and reddening and premature leaf senescence; symptoms are most prominent on older leaves following season-long ozone exposures. Stippling has been described as the most definitive symptom of ozone injury on broadleaf species (Skelly *et al.*, 1987). The color of the stippling is usually characteristic for a species but can vary with environmental or physiological conditions; dark red, reddish purple and darker brown/black colors are the most common. On young leaves, symptoms tend to develop at the leaf margins and tips; on older leaves, symptoms develop first toward the base. On sensitive species, foliar ozone injury is most apparent during the latter part of the growing season; in late summer and early fall, the entire surface of older leaves may exhibit symptoms. However, in some circumstances, foliage injured early in the growing season may be shed, resulting in reduction in the amount of visible foliar injury on a particular plant (Ghosh *et al.*, 1998).

This report presents findings of field and forest surveys for typical ozone-induced foliar injuries of native and introduced plant species growing under natural conditions in Switzerland and southern Spain; numerous species cultivated within a forest nursery in Canton Ticino, Switzerland were also surveyed during the period. For many species, the observed foliar symptoms were confirmed as being caused by ozone using seedling response to controlled ozone delivery within CSTR (Continuously Stirred Tank Reactor) chambers and/or as planted seedlings of each species within open-top chambers at the nursery site in southern Switzerland or within an open-top chamber facility near Valencia, Spain.

## 2. Materials and Methods

Surveys for foliar ozone injuries to native plant species were conducted during the late summer seasons of 1995 through 1998 in southern Switzerland and within the coastal valleys of southern Spain (1996,1997). Surveys were designed to look for symptoms typical of ozone-induced foliar injuries visibly expressed as adaxial stipple of the older leaves with veins not affected and abaxial leaf surfaces essentially free of symptoms; injuries which appeared on both leaf surfaces were not considered typical of ozone-induced injuries. Not all areas were surveyed every year.

### 2.1. SWITZERLAND

The main part of the surveys in Switzerland was conducted within and around a forest nursery in Canton Ticino (Vivaio Forestale Cantonale Lattecaldo); this site was the most frequently visited during each of the years. The forest nursery was an ideal site for such a survey since it is situated at a relatively high elevation (600 m asl), has high ozone exposures (Deserti, 1998), has many species representative of forested areas of southern Switzerland present each year, and has plants maintained under conditions of good soil moisture due to standard nursery operations and the occurrence of frequent advective storms during the summer season. Surveys comprised several walking tours of the nursery species towards the latter part of each growing season. Plants were examined along the sun-exposed edges of the nursery beds and typical ozone symptoms of adaxial stipple and reddening were recorded for each species; no estimates of the percent of plants showing symptoms were made. Other sites visited during the late summer seasons of 1995-1998 in Switzerland included 14 of the NABEL air quality monitoring stations throughout the whole of Switzerland, and a number of forest sites (NABEL, 1995; Skelly et al., 1998). Native plant species within a few hundred metres of each site were examined for typical ozone-induced foliar symptoms.

### 2.2. SOUTHERN SPAIN

A survey was conducted during the period August 20-22, 1996 within several valley and low plateau areas of southern coastal Spain in the Provinces of Catalonia and Valencia; observations were also made within the Font Roja National Park. Additional species had been observed exhibiting ozone-like symptoms in more casual observations in 1995, 1997, and 1998. Due to the dryness of the general environs, surveys were concentrated on areas with obviously higher soil moisture, e.g. valley bottoms, stream beds and areas receiving artificial irrigation. Surveys and observation methods for ozone-like symptoms were similar to those employed within the Swiss surveys.

### 2.3. CSTR EXPOSURES

If present, mature seeds were collected from symptomatic plants during each of the respective surveys. Seeds were cleaned, dried, and prepared for import through the USDA Plant Introduction Station, Hoboken, NJ. Seeds were delivered to The Pennsylvania State University where they were cold stratified in preparation for germination and seedling growth. During the spring and summer seasons of 1996 and 1997, seedlings were exposed within CSTR chambers (Heck et al., 1978) to 30, 60, 90, and 120 ppb $O_3$ for 7h per day over 5 days each week, for varying numbers of weeks depending upon symptom initiation and progressive symptom development. Ozone concentrations were monitored every 40 minutes in the CSTR chambers with a TECO Model 49 $O_3$ analyzer; soil moisture, temperatures and relative humidity were maintained to promote uptake of the pollutant.

2.4. OPEN-TOP CHAMBERS

At the forest nursery site in Canton Ticino, four replicate treatments of charcoal filtered (50 % ambient) and non-filtered (96 % ambient) open-top chambers (Heagle et al., 1973) and open plots (no-chambers) were established on a former nursery bed site in the early spring 1996. Seedlings of several species were planted within each chamber and repeatedly observed for development of symptoms confirming ozone sensitivity within the open plots and non-filtered air supplied chambers. A TECO Model 49 $O_3$ Analyzer was used to determine ozone exposures within each of the respective treatment chambers and an open plot (ambient). A similar open-top chamber facility had been constructed to the southeast of Valencia, Spain. Several species were evaluated within this facility.

## 3. Results and Discussion

The results of our surveys of native species in areas of southern Spain and across broad areas of Switzerland strongly suggest that numerous native plant species are exhibiting sensitivity expressed as typical foliar symptoms following exposure to ambient ozone in southern Europe (Tables I, II). The most commonly observed symptom for all species was adaxial stipple of older leaves with associated leaf reddening, yellowing and/or premature leaf senescence of the symptomatic leaves. The dimensions and color of the stipple varied widely with very fine light brown stipple on some species to larger areas of red, brown and purple pigmentation and general leaf reddening. Symptoms were not associated with veins and commonly a gradation of increasing severity was noticeable with increasing age of leaves for most species. It is important to note that our field surveys were by no means comprehensive nor were they repetitive within identical areas across years; a fortuitous exception occurred with the observation of many symptomatic species grown over the four-year study period within the forest nursery site in Canton Ticino of southern Switzerland. With few exceptions, symptoms observed in the field were confirmed as being induced by ambient ozone exposures for those species with seed collection, successful propagation and subsequent seedling exposures to the three higher ozone exposures within the CSTR facilities at the Pennsylvania State University; no symptoms of consequence developed within the 30 ppb $O_3$ (control) exposed plants. Similarly, species grown within the open-top chamber facilities at the forest nursery in southern Switzerland and near Valencia, Spain clearly demonstrated typical ozone injury on plants within open plots and non-filtered chambers, with asymptomatic plants of each respective species being present even much later in the season within filtered air chambers (Tables I, II). Several species (i.e. *Liriodendron tulipifera, Liquidambar styraciflua,* and *Platanus occidentalis)* observed in both Switzerland and southern Spain are known to be sensitive to ozone (Skelly et al., 1987). Ozone symptoms were identified within several common genera and for a few symptomatic species common to both countries (Tables I, II).

Open-top chamber investigations conducted at the Ticino nursery site during the growing seasons of 1997 and 1998 have confirmed ozone-induced stipple and leaf reddening on 14 of 16 species investigated. Based upon observations within the open-plots and non-filtered open-top chambers, the ranking of species sensitivities from most to least symptomatic were *P. serotina, M. nigra, S. viminalis, R. cathartica, S. racemosa, V. lantana, R. obtusifolius, B. pendula, P. avium, C. avellana, F. excelsior, R. frangula, A. viridis, F. sylvatica,* with *A. pseudoplatanus* and *C. sanguinea* being asymptomatic. Details of rankings and injury thresholds using AOT 40 and related ozone exposures are being prepared (VanderHeyden, 1999).

TABLE I.

Native plant species of Switzerland observed with ozone-like foliar symptoms during surveys of natural habitats and forest nursery conditions in the late summer seasons, 1995-1998.
[a] Confirmed ozone-induced symptoms via open-top chamber investigations, Canton Ticino, CH
[b] Confirmed ozone-induced symptoms via CSTR chamber investigations, Penn State University

| **Shrubs** | **Trees** | **Forbs** |
|---|---|---|
| *Alnus viridis* [a] | *Acer platanoides* | *Alchemilla* spp. |
| *Berberis* spp. | *Acer pseudoplatanus* [a] | *Artemesia vulgaris* |
| *Corylopsis pauciflora* | *Acer saccharinum* | *Calamentha grandifolia* |
| *Euonymous europeaus* | *Ailanthus altissima* | *Calystegia* spp. |
| *Forsythia* spp. | *Alnus incana* | *Centaurea paniculata* |
| *Ligustrum ovalifolium* | *Betula pendula* [a,b] | *Chenopodium* spp. |
| *Ligustrum vulgaris* | *Carpinus betulus* | *Clematis* spp. |
| *Lilac* spp. | *Cornus alba* [a,b] | *Convolvulus arvensis* |
| *Prunus spinosa* | *Cornus capitata* | *Epilobium angustifolium* [a,b] |
| *Ribes alpinum* [b] | *Cornus mas* | *Epilobium hirsutum* |
| *Rosa canina* | *Cornus sanguinea* [a,b] | *Impatiens parviflora* |
| *Salix pentrandra* | *Cornus stolonifera* [b] | *Lamium galeobdolon* |
| *Salix purpurea* | *Corylus avellana* [a] | *Lonicera caprifolium* |
| *Salix viminalis* [a] | *Crateagus monogyma* | *Lapsana communis* |
| *Sambucus racemosa* [a,b] | *Fagus sylvatica* [a] | *Malva* spp. [b] |
| *Spirea* spp. [b] | *Frangula alnus* | *Oenothera* spp. [b] |
| *Viburnum lantana* [a,b] | *Fraxinus excelsior* [a,b] | *Parthenocissus quinquifolia* |
| *Viburnum opulus* | *Juglans nigra* | *Plantago major* [b] |
| *Viburnum plicatum* | *Liquidambar styraciflua* | *Polygonum* spp. |
| | *Liriodendron tulipifera* | *Reynoutria japonica* |
| | *Morus alba* | *Rubus fruticosus* |
| | *Morus nigra* [a,b] | *Rudbeckia lacinata* |
| | *Malus sylvestris* | *Rumex obtusifolius* [a,b] |
| | *Platanus occidentalis* | *Solidago canadensis* |
| | *Populus nigra* | *Stachys officinalis* |
| | *Prunus avium* [a] | *Succisa pratensis* |
| | *Prunus serotina* [a,b] | *Vitis* spp. |
| | *Rhamnus cathartica* [a,b] | |
| | *Rhamnus frangula* [a] | |
| | *Robinia pseudoacacia* [b] | |
| | *Salix alba* | |
| | *Tilia cordata* [b] | |
| | *Tilia platyphyllos* | |
| | *Ulmus glabra* | |
| | *Ulmus montana* | |

Tropospheric ozone pollution is of major importance within the Mediterranean Basin (Butkovic *et al.*, 1990; Millán and Artinano, 1992; Lorenzini *et al.*, 1994). Although precise information is scarce, many studies have suggested that ozone may reach phytotoxic concentrations in southern Europe. Maps of the cumulative exposure index AOT 40 for Switzerland indicate that exceedances occur over very large areas, particularly

## TABLE II.

Native plant species of southern Spain observed with ozone-like foliar symptoms during surveys of natural habitats and forest nursery conditions in the late summer seasons, 1995-1998.
[a] Confirmed symptomatic via open-top chamber studies, Valencia, Spain 1997, 1998
[b] Confirmed ozone-induced symptoms via CSTR chamber investigations, Penn State University.

| Shrubs | Trees | Forbs |
|---|---|---|
| *Arbutus unedo* [a] | *Ailanthus altissima* [a,b] | *Agrimonia eupatoria* |
| *Cystus salvifolius* [a] | *Cornus sanguinea* [b] | *Albutilon theophrasti* |
| *Lagersteroemia indica* | *Corylus avellana* | *Anthylis cytisoides* [a] |
| *Myrtus communis* [a] | *Fraxinus ornus* | *Calystegia sepium* |
| *Pistacia lentiscus* [a] | *Juglans regia* | *Campanula spp.* |
| *Pistacia terebintus* [a] | *Populus alba* | *Chenopodium album* |
| *Prunus spinosa* | *Populus nigra* | *Colutea arborescens* |
| *Ricinus communis* [b] | *Prunus amylgladus* [b] | *Cystisus patens* [a] |
| *Rosa canina* | | *Epilobium angustifolium* [b] |
| *Sambucus nigrum* | | *Epilobium collium* [b] |
| *Viburnum tinus* | | *Inula viscosa* |
| | | *Ipomea sagitatta* |
| | | *Lagersteroemia indica* |
| | | *Lonicera etrusca* |
| | | *Lonicera implexa* [a] |
| | | *Oenothera rosea* [b] |
| | | *Pathenocissus quinquefolia* |
| | | *Plantago lanceolata* |
| | | *Rubia peregrina* [a,] |
| | | *Rubus ulmifolius* |
| | | *Rumex pulcher* [b] |
| | | *Verbascum sinuatum* [b] |
| | | *Vinca difformis* [a] |

over the important forested zones at ca. 1000 m asl. The threshold AOT 40 at 10 ppm.h for the April-September growing season has been exceeded every year at most Swiss monitoring stations with the highest exceedances reported at a sub-alpine site on Rigi (NABEL 1995). Other southern European regions show exceedances of the AOT 40 values for forested areas. For example, most of the rural stations along the southeast coast of Spain up to 100 km inland of the Mediterranean exceed this threshold, as do most of the mountainous stations above 1000 m asl (Sanz *et al.*, unpublished data). Similarly, Klap *et al.* (1997) have reported that the AOT 40 threshold value was widely exceeded in central and southern Europe in 1990. Thus, there is every expectation that widespread foliar injuries could be occurring in forest and native plant species due to ambient ozone exposures across this vast region (Fuhrer, 1996); the results of our survey tend to support this expectation.

It is interesting to note that many plant species in a variety of habitats exhibited typical symptoms of phytotoxic ozone exposures. Injury thresholds expressed as AOT 40 values for each of the fumigated species are being determined *via* open-top chamber investigations (VanderHeyden, 1999).

## 4. Conclusions

Ambient ozone exposures have been found to be sufficiently high to induce significant amounts of typical foliar injury on native tree, shrub and herb species within areas of southern Europe. Many of these species have been shown to be ozone sensitive following controlled ozone exposures and ozone exclusion studies verifying the appearance of typical ozone-induced foliar injuries. Several plant species and species within genera appear to be ozone sensitive across broad regions of southern Switzerland and southern Spain. Several species may well serve as bioindicators of ozone exposure on a regional basis (e.g. *S. viminalis, R. cathartica, S. racemosa,* and *V. lantana*) but more detailed studies of exposure/response should first be completed. Confirmation of ozone sensitivity (i.e. induced foliar symptoms) remains to be completed for numerous species observed as symptomatic within our surveys.

## References

Bacci, P., Sandroni, S. and Ventura, A.: 1990, *Sci. Total Environ.* **96**, 297-312.
Bussotti, F., Soda, C. and Grossini,P.: 1998, in *Vegetali come bioindicatori di inguinamento atmospherica in citta Italiane: Rusultati, problemi, prospettive*. Ferretti, M. and Pellegrini, M. (eds.), pp.75-80, Pubblicato congiuntamente da:Commune di Bologna, META spa Modena, Provincia di Firenze, Italy.
Butkovic, V., Cvitas,T. and Kasling, L.: 1990, *Sci. Total Environ.* **99**, 145-151.
Derwent, R.G.: 1990, *Environ. Pollut.* **63**, 299-318.
Deserti, M.: 1998, in *Vegetali come bioindicatori di inguinamento atmospherica in citta Italiane:Rusultati, problemi, prospettive*. Ferretti, M. and Pellegrini, M. (eds.), pp.85-92, Pubblicato congiuntamente da:Commune di Bologna, META spa Modena, Provincia di Firenze, Italy.
Ferretti, M., Gavilli,G., Grossini, D., Lorenzini, M. and Pellegrini, M.: 1998, in *Vegetali come bioindicatori di inguinamento atmospherica in citta Italiane:Rusultati, problemi, prospettive*. Ferretti, M. and Pellegrini, M. (eds.), pp.21-34, Pubblicato congiuntamente da:Commune di Bologna, META spa Modena, Provincia di Firenze, Italy.
Fuhrer, J.: 1996, *Key elements in ozone risk analysis* in *Exceedance of critical loads and levels*. Report of a workshop held in Vienna, Austria under the Convention on Long Range Transboundary Air Pollution, 22-24 November, 1995.
Fuhrer, J., Skärby, L. and Ashmore, M.R.: 1997, *Environ. Pollut.* **97**, 91-106.
Ghosh, R., Skelly, J. M., Innes, J. L. and Skelly, L.: 1998, *Environ. Pollut.* **102**, 287-300.
Grennfelt, P. and Beck, J. P.: 1994, in *Critical levels for ozone: a UN-ECE workshop report*. Fuhrer, J. and Achermann, B. (eds.), pp.184-194. Schriftenreihe der FAC Liebefeld, Swiss Fed. Res. Sta. for Agric. Chem., Liebefeld-Bern.
Heagle, A. S., Body, D. E. and Heck, W. E.: 1973, *Environ. Qual.* **2**, 365-368.
Heck, W. W., Philbeck, R. B. and Dunning, J. A.: 1978, *A Continuous Stirred Tank Reactor (CSTR) System for Exposing Plants to Gaseous Contaminants: Principles, Specifications, Construction and Operation*. USDA-ARS Pub. No. ARS-S-181, New Orleans, LA.
Hettelingh, J-P., Posch, M. and de Smet, P.: 1996, in *Critical levels for ozone in Europe: Testing and finalizing the concepts*. Kärenlampi, L. and Skärby, L. (eds.), pp.125-137, University of Kuopio, Kuopio.
Innes, J. L. and Skelly, J. M.: 1995, *Effects of ambient concentrations of ozone on visible symptom development of Prunus serotina in southern Switzerland*, Project. Rept., Swiss Fed. Inst. Forest, Snow, Landscape Res. Birmensdorf.
Innes, J. L., Skelly, J., Landolt, W., Hug, C., Snyder, K. R. and Savage, J. E.: 1996, in *Exceedance of critical loads and levels*. Knoflacher, M., Schneider,J. and Soja, G. (eds.), pp.146-154, Report of a workshop held in Vienna, Austria under the Convention on Long Range Transboundary Air Pollution, 22-24 November 1995
Klap, J. M., deVries, W., Erisman, J. W. and van Leeuwen, E. P.: 1997. *Relationships between forest condition and natural and anthropogenic stress factors on the European scale; a pilot study*. SC Report 150, RIVM Report 722108022. DLO Winand Staring Centre, Wageningen.
Lorenzini, G., Nali, C. and Panicucci, A.: 1994, *Atmos. Environ.* **38**, 51-59.
Matyssek, R. and Innes, J. L.: 1999, *Water, Air, and Soil Pollut.* (this volume)

Mea, M. D, D'Erchia, V., Cesaroni, D. and Bagni, N.: 1998, in *Vegetali come bioindicatori di inguinamento atmospherica in citta Italiane: Rusultati, problemi, prospettive*. Ferretti, M. and Pellegrini, M. (eds.), pp.11-20, Pubblicato congiuntamente da:Commune di Bologna, META spa Modena, Provincia di Firenze, Italy.

Millán, M. M. and Artinano, B.: 1992, *Meso meteorological cycles of air pollution in the Iberian Peninsula (MECAPIP)*, Final report to the European Communities Commission EV4V-0097-E(A).

Millán, M., Salvador, R., Mantilla, E. and Artinano, B.: 1996, *Atmos. Environ.* **30**, 1909-1924.

Millán, M., Salvador, R., Mantilla, E. and Kallos, G.: 1997, *J. Geophys. Res.* **102**, 881-8823.

NABEL: 1995, *Luftbelastung 1994*, Schriftenreihe Umwelt 244, Bern: Bundesamt fur Umwelt, Wald und Landschaft.

Sanz, M. J. and Millán, M.: 1998, *Chemosphere* **36**, 1089-1094.

Schenone, G. and Lorenzini, G.: 1990, *Rivista di Patologia Veg.* **26**, 85-105.

Skärby, L., Ro-Poulsen, H., Wellburn, F. A. M. and Sheppard, L. J.: 1998, *New Phytol.* **139**, 109-122.

Skelly, J. M., Davis, D. D., Merrill, W., Cameron, E. A., Brown, H. D., Drummond, D. B. and Dochinger, L. S.: 1987, *Diagnosing Injury to Eastern Forest Trees*. USDA-For. Ser. Veg. Survey Res. Coop. and Penn State Univ., Univ.Park, PA.

Skelly, J.M., Innes, J. L., Snyder, K. R., Savage, J. E., Hug, C., Landolt, W. and Bleuler, P.: 1998, *Chemosphere* **36**, 995-1000.

Staffelbach, T., Neftel, A., Blatter, A., Gut, A., Fahrni, M., Stahelin, J., Prevot, A., Hering, A., Lehning, M., Neininger, B., Baumle, M., Kok, G. L., Dommen, J., Hutterli, M. and Anklin, M.: 1997a, *J. Geophys. Res.* **102**, 23, 345-23, 362.

Staffelbach, T., Neftel, A. and Horowitz, L. W.: 1997b, *J. Geophys. Res.* **102**, 23, 363-23, 273.

VanderHeyden, D. J.: 1999. *Tropospheric ozone effects on native plants of southern Switzerland: Determining exposure/response relationships*. Master of Science Thesis. Pennsylvania State Univ., Univ. Park, PA, USA. 16802.

Wunderli, S. and Gehrig, R.: 1990, *Atmos. Environ.* **24A**, 2641-2646.

# DECIDUOUS CONIFERS: HIGH N DEPOSITION AND $O_3$ EXPOSURE EFFECTS ON GROWTH AND BIOMASS ALLOCATION IN PONDEROSA PINE.

N.E. GRULKE* and L. BALDUMAN

*Pacific Southwest Research Station, Riverside Fire Laboratory, 4955 Canyon Crest Drive, Riverside, CA 92507 U.S.A.*

(Received 25 September 1998; accepted 25 February 1999)

**Abstract.** Ponderosa pines (*Pinus ponderosa* Dougl. ex. Laws) 21 to 60 yr old were used to assess the relative importance of environmental stressors ($O_3$, drought) versus an enhancer (N deposition) on foliar retention, components of aboveground growth, and whole tree biomass allocation. Sites were chosen across a well-described gradient in ozone exposure (40 to 80 ppb per h, 24 h basis, 6 month growing season) and nitrogen deposition (5 to 40 kg ha$^{-1}$ yr$^{-1}$) in the San Bernardino Mountains east of Los Angeles, California. A high level of chlorotic mottle indicated high $O_3$ injury at sites closest to the pollution source, despite potential for the mitigating effects of N deposition. At the least polluted site, foliar biomass was evenly distributed across three of the five needle-age classes retained. At the most polluted site, 95% of the foliar biomass was found in the current year's growth. High N deposition and $O_3$ exposure combined to shift biomass allocation in pine to that of a deciduous tree with one overwintering needle age class. Based on whole tree harvests, root biomass was lowest at sites with the highest pollution exposure, confirming previous chamber exposure and field studies. Aboveground growth responses in the high-pollution sites were opposite to those expected for $O_3$ injury. Needle and lateral branch elongation growth, and measures of wood production increased with increasing proximity to the pollution source. An enhancement of these growth attributes suggested that N deposition dominated the ponderosa pine response despite high $O_3$ exposure.

**Keywords:** ponderosa pine, ozone exposure, nitrogen deposition, drought stress, biomass allocation

## 1. Introduction

Ponderosa pine (*Pinus ponderosa* Dougl. ex Laws.) is the most sensitive western North American conifer to oxidant air pollution (Miller *et al*, 1983). The effects of ozone ($O_3$) exposure since the 1940's on bole growth of this species and others of the Sierran mixed conifer zone have been documented from dendrological studies (McBride *et al*, 1975; Peterson *et al*, 1991). Long-term changes in canopy health (Miller *et al*, 1989), and forest stand structure (McBride and Miller 1987; McBride *et al*, 1975) have been clearly correlated with exposure to oxidant pollution in the San Bernardino Mountains, east of Los Angeles, California.

Three sites were chosen across a gradient in pollution load in the San Bernardino Mountains (Miller *et al*, 1989). Tree attributes were also measured in an atmospherically clean site in northern California to provide an example of "typical" ponderosa pine. At these sites, $O_3$ exposure ranged from near global backround levels to some of the highest reported for forested lands in North America (Table I). At the sites nearest Los Angeles,

---

* Author to whom all correspondence should be addressed. Nancy E. Grulke, Pacific Southwest Research Station, Riverside Fire Laboratory, 4955 Canyon Crest Drive, Riverside, CA 92507, U.S.A. E-mail: ngrulke@deltanet.com

high nitrogen (N) deposition accompanied high $O_3$ exposure, but decreased from west to east in the San Bernardino Mountains (Bytnerowicz and Fenn, 1996; Fenn et al., 1996; Kiefer and Fenn, 1997; Fenn and Poth, 1999). Long term inputs of high N deposition on the western end of the pollution gradient are reflected in elevated soil N (Grulke et al, 1998) as well as isotopic composition of nitrate ($NO_3^-$) and ammonium ($NH_4^+$) in soil extracts (Fenn and Poth,1999). Drought stress, while an important variable, varied more between years than between sites within a year (Grulke et al., 1998). Growing season length and cumulative degree days increased from the least to the most polluted site (Table I).

Low soil temperature (Gholz and Cropper, 1991), and nutrient (Haynes and Gower, 1995) and moisture availability (Hermann and Petersen, 1969) alter growth rates and resource allocation between above- and below-ground tissues in pine. These field responses as well as responses to experimentally modified variables in chamber exposure studies aid in interpreting the growth and biomass allocation patterns observed in this study. Qualitative comparisons were made across the three southern Californian sites, and between pairs of sites for direction and magnitude of tree response to dominant environmental gradients. For example, comparisons between Camp Paivika (CP) and Strawberry Peak (SP) provided insight on tree response with respect to a two fold change in $O_3$ exposure and N deposition under relatively mesic conditions (Table I). Comparisons between Lassen (LS), in northern California, and Camp Osceola (CO), on the eastern end of the San Bernardino Mountains, provided insight on tree response to the addition of moderate $O_3$ exposure in a droughty environment. Changes in tree attributes between mesic (1993, 1995) and xeric (1994) years elucidated the effects of drought (Grulke, 1999). Differences in tree response could also be attributed to genotypic differences between the four populations, but are unquantifiable at this time. The study reported here describes the relative importance of environmental stressors ($O_3$, drought) versus an enhancer (N deposition) on foliar retention, components of aboveground growth, and whole tree biomass allocation in a widespread, economically important tree species, ponderosa pine.

## 2. Methods

2.1. SITE LOCATIONS AND TREE SELECTION

Three sites were chosen along a pollution gradient within the Sierran mixed conifer zone (*sensu* Barbour 1988), in the San Bernardino Mountains east of Los Angeles, CA: Camp Paivika (CP) at the western-most end (34°14'05"N, 117°19'12"W, 1800 m), Strawberry Peak (SP) as an intermediate site (34°14'00"N, 117°08'12"W, 2240 m), and Camp Osceola (CO) at the eastern-most end (34°09'42"N, 116°51'00"W, 1820 m) (Table I). The fourth site was located in an atmospherically clean site near the southeastern corner of Lassen Volcanic National Park (LS) in the southern Cascade Mountains of California (40°20'11"N, 121°35'04"W, 1700 m), a northern extension of the forest type.

At each of the four sites, a regionally typical stand was chosen, a plot was established (1992) in a natural, multi-aged stand, and the 12 most uniformly average trees in a 21 to 60 yr tree age class were selected (Grulke et al., 1998; Grulke, 1998). Data from trees in two tree age classes (21 to 40 and 41 to 60 yr old) were combined when there was no significant difference in net photosynthetic rates or respiration (Grulke, 1998; 1999), leaf N, foliar retention, or needle and branch growth rates at the p ≤ 0.05 level (t-test). "Average" trees within the stand were selected on the basis of bole diameter, total height, percent chlorotic mottle of 1 yr old foliage (Miller et al., 1996), number of needle age classes, and distance to the nearest intra- and interspecific tree as a measure of inferred competition. Individual trees were widely spaced and generally had no canopy influence from another intensively chosen tree on site. Thus, individual whole trees were used as within-site replication.

TABLE I
Summary of environmental gradients across the four study sites.

|  | LS | CO | SP | CP |
|---|---|---|---|---|
| Ozone exposure, ppb per h |  |  |  |  |
| all hourly values | 38 [42][b] | 62 [64] | 69 [76] | 79 [80] |
| hourly values >40 ppb[a] | 2 [8] | 28 [33] | 38 | 56 [57] |
| N deposition, kg ha$^{-1}$[c] | n.d. | 6-9 | n.d. | 20-40 |
| Soil N, %[d] |  |  |  |  |
| $A_1$ | 0.01 | 0.08 | 0.11 | 0.20 |
| $B_1$ | 0.01 | 0.04 | 0.07 | 0.09 |
| Pre-dawn xylem potential, MPa[a] | -1.0 [-1.1] | -1.1 [-1.7] | -0.9 [-1.6] | -1.0 [-1.5] |
| Annual precipitation, cm[e] (1980-1992) | 161.1 | 89.7 | 95.8 | 98.0 |
| Cumulative degree days, °C[f] | 580 [680] | 1310 [1230] | 1980 [1820] | 2650 [2840] |
| Length of growing season, days[f] | 88 [105] | 133 [78] | 113 [105] | 181 [170] |

[a] Summarized from Grulke (1999) and Grulke et al. (1998).
[b] When two numbers are listed, the first number represents mesic years (1993, 1995) and the second represents the xeric year [1994].
[c] Values for N deposition are broad due to variations in canopy structure (summarized from Fenn and Kieffer 1999).
[d] Summarized from Grulke et al. (1998).
[e] Annual precipitation was averaged over the hydrologic years 1980 through 1997 (San Bernardino County Water Resources).
[f] Temperature data has been summarized from this study.

## 2.2. MICROENVIRONMENTAL MONITORING

Microenvironmental monitoring stations were installed at each of the sites and were operational from January 1, 1993 to October 15, 1995. Thermocouples at 1.5 m in the air were monitored in 5 min intervals, and data were averaged and stored at 1 h intervals. Cumulative degree days were calculated from daily mean temperatures from January 1. Growing season length is defined here from air temperature as the first day in the year followed by 5 consecutive days $\geq 15°C$ to the last day of the year preceeded by at least 5 consecutive days $\geq 15°C$. Methodology and values for other supporting microenvironmental data have been published elsewhere (Grulke et al., 1998; Grulke 1999).

## 2.3. $O_3$ INJURY, FOLIAR RETENTION, AND GROWTH MEASURES

Estimates of foliar chlorotic mottle (Miller *et al.*, 1996), number of live needle age classes, proportion of foliage retained within a needle age class (length of foliage/branchlet length within a whorl), and needle and primary lateral branch growth were measured as in Grulke and Lee (1997) and Grulke *et al.* (1998). Measures were taken monthly on five branches in the lower one third of the canopy on each of 12 trees in 1993 and 1994 and six trees in 1995, but are reported here at the time of maximum response to environmental stressors or enhancers generally occurring in mid September. Branchlet diameter was measured at the base of current year growth only in September, 1995 with digital calipers to the nearest 0.01 mm. For the above measurements, values from the five branches were averaged to yield a single value for each tree at each site. Tree height (measured with a meter tape or clinometer to the nearest 0.5 m) and bole diameter (measured at 1.4 m to the nearest 0.1 cm) were taken in 1992 at the time of plot establishment.

## 2.4. BIOMASS MEASURES

Whole tree biomass harvests were made on three trees (average for the total stand population of 21 to 60 yr old trees) at each site in early September, 1993. All coarse and fine roots were excavated to the best of our ability to bedrock, generally encountered at 1.5-2 m. Fine roots were at an annual minimum at the time of sampling (Grulke *et al*, 1998). In the field, foliage was stripped from branches by needle age class, and packaged in paper sacks by canopy thirds based on the length of the live crown. Foliage was placed in ovens at 105°C within 10 h of collection for 2 d, then weighed on a top loading balance to the nearest 0.1 g. Branches were also bundled and processed by upper, middle, and lower third of the canopy. In the laboratory, a subset of 40 branchlets per canopy third were measured for total length and diameter at the base of the current year growth. The trunk was cut into five parts for easier transport. Roots were separated into coarse ( $\geq 1$ cm in dia.), medium (> 2 mm, < 1 cm in dia.), and fine ( $\leq 2$ mm dia.) size classes, dried, and weighed. Branches, bole, and roots were dried in a large lumber drying oven at 105°C and weighed on a high capacity top loading scale (30 kg capacity) at 2 d intervals until no additional weight loss was observed.

# 3. Results

## 3.1. VISIBLE $O_3$ INJURY SYMPTOMS

Chlorotic mottle increased with increasing $O_3$ exposure (Table II), a result well known for this pollution gradient (Miller *et al.*, 1989). The data are presented here to place the 21- to 60-yr-tree age class within the context of data already published for mature trees. Chlorotic mottle was low and did not statistically differ at the two least polluted sites, Lassen (LS) and Camp Osceola (CO) in any of the three years of study. The two western-most and polluted sites in the San Bernardino Mountains, Strawberry Peak (SP) and Camp Paivika (CP), had significantly higher chlorotic mottle in 1993 (1992 and 1993 had higher than average precipitation; San Bernardino County Water Resources Division) relative to the less polluted sites. However, in the xeric year [1994] and in the following year (1995), the greater chlorotic mottle differed significantly only at CP.

## 3.2. FOLIAR RETENTION

The measure of foliar retention has three components: the number of live needle age classes (whorls) retained, the proportion of foliage retained within a whorl, and the relative biomass allocation to the multiple needle age classes. Across the pollution gradient, as chlorotic mottle increased, the number of whorls decreased (Table II). The least polluted site, LS, consistently had the greatest number of whorls, which was greatest in the mesic years. The least polluted site in southern California, CO, had a similar number of whorls retained in 1993 and 1994 as at LS, but in 1995 (the growing season following the drought) had one fewer whorl than the cleanest site. Trees at the western end of the San Bernardino Mountains consistently had fewer whorls, and CP had significantly fewer whorls than all other sites in both 1994 and 1995.

The proportion of foliage retained within a whorl increased slightly with increasing pollution: 0.60 (LS), 0.65 (CO), 0.70 (SP), and 0.75 (CP), but none of these differences were statistically significant in a mesic year (1993). The abrupt (interannual) loss of within-whorl needle retention, and the number of whorls retained is clearly illustrated (Figure 1). At the cleanest site (LS), drought in 1994 increased needle senescence two fold only in the oldest whorl (7 yr old). The needle age at which half the foliage was lost decreased with increasing pollution load (CO: half of 4-5 yr old needles; SP: half of 3 yr old needles; CP: half of 2 and 3 yr old needles).

With increasing pollution exposure, more of the canopy foliar mass (whole tree) was distributed over fewer needle age classes (Figure 2). At the least polluted site, LS, 92% of the needle mass was evenly distributed over 4 needle age classes. At CO, 80% of the needle mass at CO was evenly distributed over 2 needle age classes, with 3 yr old needles comprising most of the remaining needle mass. Approximately two thirds of the needle mass of trees at SP were current year needles, and one third of the mass was 1 yr old. At CP, the most polluted site, 95% of the needle mass was current year foliage, resulting in trees with only one overwintering needle age class. Although more needle age classes were retained at the latter two sites, their mass did not contribute significantly to that of the canopy. Within site error was high because only three trees were harvested per site.

## 3.3. ABOVEGROUND GROWTH

The effect of environmental influences on growth was assessed from elongation growth (needle and lateral branches) and measures of wood production (branch and bole diameter growth). Multiple year measurements of growth were conducted in the lower third of the canopy and extensive measures of growth were made within tree canopies in a mesic year (1993, second of two years of above-average precipitation; San Bernardino County Water Resources Division). Differences among sites were tested with an analysis of variance and differences between paired sites were tested with a t-test. Statistical significance was reported at $p \leq 0.05$.

Needle length in the lower third of the canopy was greatest at the two western-most sites in the San Bernardino Mountains, SP and CP, in all three years (Table II). Despite differences in heat sum and degree days (Table I), needle length did not differ significantly between the two sites. In 1993, needle length did not differ significantly across the southern Californian sites, but the northern site had shorter needles probably due to both heat sum and a short growing season (Table I). In 1994 and 1995, needle length at LS and CO did not differ significantly. In those years, SP and CP differed significantly from LS and CO.

Needle length in the lower third of the canopy was greatest at the two western-most sites in the San Bernardino Mountains, SP and CP, in all three years (Table II). Despite differences in heat sum and degree days (Table I), needle length did not differ significantly between the two sites. In 1993, needle length did not differ significantly across the southern Californian sites, but the northern site had shorter needles probably due to both heat sum and a short growing season (Table I). In 1994 and 1995, needle length at LS and CO did not differ significantly. In those years, SP and CP differed significantly from LS and CO.

Lateral primary branch length of current year growth measured in the lower third of the canopy showed similar trends to that of needle length. Branchlet length was significantly lower at LS and CO relative to SP and CP in all three years of study. During the drought year, branchlet length was low and did not significantly differ across the southern Californian sites. Needle and branchlet length was lowest (but not significantly) for the most xeric site (CO) in the year following the drought suggesting a possible residual effect. Lower branchlet length at CP in 1995 may reflect the combined effect of $O_3$ exposure and the residual effect of drought stress in the previous year. Branchlet length measured from whole tree harvests was the lowest in the lower third of the canopy at all sites. Tree height was greatest at CP relative to all other sites (Table II). Trees at CO, the most xeric southern Californian site, were significantly shorter relative to all other sites. Trees at LS and SP had similar height.

Table II.

Summary of foliar injury[a], number of needle age classes, and growth across the four sites.

| Characteristic: | LS | CO | SP | CP | F | p |
|---|---|---|---|---|---|---|
| *Chlorotic mottle* | | | | | | |
| 1993 | 1±0a | 8±2a | 19±5b | 24±5b | 7.53 | <0.001 |
| 1994 | 2±1a | 1±1a | 12±2a | 35±7b | 18.33 | <0.001 |
| 1995 | 2±0a | 2±0a | 8±1a | 28±5b | 24.97 | <0.001 |
| *Number of needle age classes* | | | | | | |
| 1993 | 5.6±0.2a | 5.6±0.4a | 3.2±0.2b | 2.6±0.2b | 33.81 | <0.001 |
| 1994 | 5.2±0.2a | 4.7±0.2a | 3.3±0.2b | 2.2±0.2c | 39.20 | <0.001 |
| 1995 | 5.9±0.3a | 4.6±0.1b | 4.0±0.2b | 2.8±0.2c | 31.56 | <0.001 |
| *Elongation growth:* | | | | | | |
| *Needle length*, mm (lower third of the canopy) | | | | | | |
| 1993 | 104±4a | 130±7bc | 151±5bd | 142±8b | 11.83 | <0.001 |
| 1994 | 116±7a | 113±7ac | 148±7bd | 152±4bd | 11.19 | <0.001 |
| 1995 | 105±10a | 95±10a | 130±5b | 150±6b | 9.52 | <0.001 |
| *Lateral primary branchlet length*, mm (lower third of the canopy) | | | | | | |
| 1993 | 35±3a | 48±5a | 57±5b | 64±12b | 3.26 | 0.030 |
| 1994 | 22±3a | 35±6ac | 41±6bc | 41±5bc | 3.05 | 0.038 |
| 1995 | 24±4a | 20±3ac | 61±8bd | 40±10ac | 7.47 | 0.002 |
| *Branchlet length*, mm (whole canopy measurements in 1993) | | | | | | |
| Upper 1/3 | 131±19a | 60±19a | 132±41a | 103±41a | 1.14 | 0.389 |
| Middle 1/3 | 69±15a | 32±5a | 60±23a | 46±16a | 1.08 | 0.410 |
| Lower 1/3 | 22±3a | 20±3a | 31±6a | 31±11a | 0.70 | 0.577 |
| *Tree height,* | | | | | | |
| m (1992) | 5.1±0.3a | 2.7±0.2b | 5.0±0.3a | 7.4±0.5c | 27.34 | <0.001 |
| *Enlargement growth:* | | | | | | |
| *Branchlet diameter*, mm (lower third of the canopy) | | | | | | |
| (1995) | nd | 5.9±0.3a | 7.3±0.3b | 7.2±0.3b | 7.81 | 0.002 |
| *Branchlet diameter*, mm (whole canopy measurements in 1993) | | | | | | |
| Upper 1/3 | 7.1±1.9a | 7.1±1.2a | 7.7±1.3a | 9.1±0.3a | 0.45 | 0.724 |
| Middle 1/3 | 4.7±0.7a | 5.7±0.7a | 5.1±0.5a | 5.6±0.2a | 0.63 | 0.614 |
| Lower 1/3 | 3.6±0.3a | 5.0±0.4b | 4.1±0.0b | 4.7±0.2b | 5.50 | 0.029 |
| *Bole diameter*, cm | | | | | | |
| (1992) | 3.4±0.2a | 2.2±0.2b | 3.9±0.2a | 5.5±0.3c | 16.77 | <0.001 |

[a] Differing letters indicate significant differences at the $p \leq 0.05$ level.

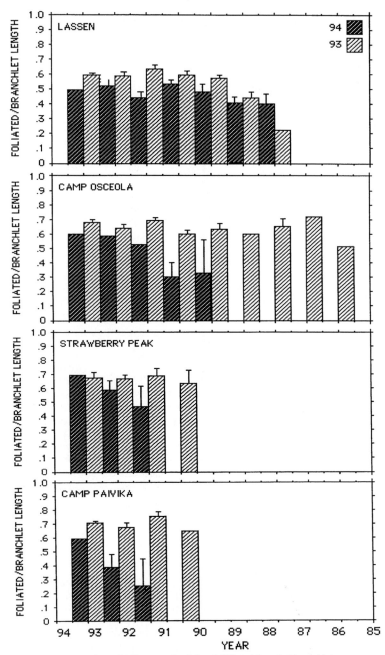

Fig. 1. The proportion of foliage retained (total foliated length / branchlet length, within a needle age class) is presented for 1993 (light hatching) and 1994 (dark hatching). The bars represent the average value for 12 trees, ± 1 S.E.

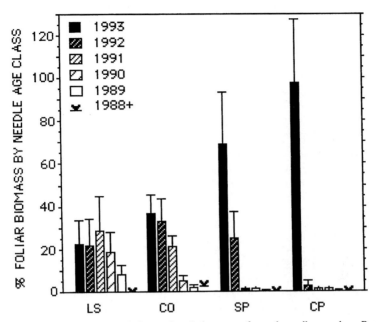

Fig. 2. Proportion of needle biomass relative to the whole canopy for each needle age class. Bars represent the average of three trees per site, ± 1 S.E.

Branchlet diameter measured in the lower third of the canopy in 1995 was significantly lower at CO relative to the other southern Californian sites (Table II). Similar to needle elongation growth, branchlet diameter measured from whole tree harvests in 1993 was lowest in the lower third of the canopy. Based on whole canopy measurements, there were significant differences in branchlet diameter between LS and the southern Californian sites. Mid-canopy branchlet diameter did not differ significantly across all sites. In the upper third of the canopy, branchlet diameter increased from LS and CO, to SP, to CP, but differences were not statistically significant due to high within-site variability. Bole diameter for 21 to 60 yr old trees was significantly greater at CP than SP and LS, and was significantly lower at CO relative to all other sites.

3.4. WHOLE TREE ALLOCATION OF BIOMASS

Total tree biomass varied seven fold across the four sites (Table III). Although the three trees harvested were average relative to the stand population of 21 to 60 yr old trees, at least one tree of the three trees at each site was ± 1 S.D. of the mean of the sample. Whole tree biomass was greatest at CP, similar at both LS and SP, and lowest at CO. Total canopy foliar mass varied ten fold across the four sites. The trees at the most polluted site had significantly greater foliar mass than at SP, and both CP and SP had significantly greater foliar mass than the two least polluted sites, LS and CO. Foliar mass in the upper and middle third of the canopy was significantly greater at CP relative to the other sites. Trends in branchlet mass within the canopy were similar to that of foliar mass, but differences between sites were not statistically significant.

Table III
Components of canopy biomass[a] for whole trees harvested in 1993.

| Canopy characteristic: | LS | CO | SP | CP | F | p |
|---|---|---|---|---|---|---|
| *Branchlet mass per branchlet*, g | | | | | | |
| Upper one third | 2.3±0.4a | 2.2±0.8a | 2.4±1.0a | 3.4±1.6 a | 0.31 | 0.815 |
| Middle one third | 0.8±0.3a | 0.6±0.2a | 0.9±0.2a | 0.8±0.3a | 0.27 | 0.844 |
| Lower one third | 0.2±0.0a | 0.3±0.1a | 0.3±0.1a | 0.4±0.1a | 1.07 | 0.414 |
| *Foliar mass*, g | | | | | | |
| Upper one third | 199±90a | 88±27a | 287±45a | 1475±500b | 1138 | 0.004 |
| Middle one third | 432±233a | 127±46a | 501±222a | 1628±227b | 10.99 | 0.003 |
| Lower one third | 221±112a | 116±22b | 608±261a | 1356±683a | 2.31 | 0.153 |
| *Total foliage*, kg | 0.9±0.4a | 0.3±0.1a | 1.4±0.5b | 4.0±1.2c | 5.44 | 0.025 |
| *Total tree mass*, kg | 31.2±19.2a | 6.6±1.6a | 35.3±4.8a | 47.1±17.1a | 1.685 | 0.247 |

[a] Differing letters indicate significant differences at the $p \leq 0.05$ level.

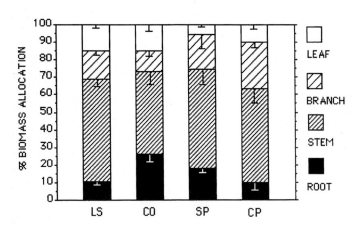

Fig. 3. Percent dry weight of root, leaf, branch, and bole biomass averaged for three whole trees harvested per site. Error bars are as described for Fig. 2.

Biomass allocation to leaf, branch, bole, and roots was also variable between individuals within a site (Figure 3). The proportion of biomass allocated to fine root biomass was a minor component (0.3 and 2%) at this mid-September harvest, and did not appear to be a major source of error. The proportion of biomass allocated to the bole was greatest at CP, followed by SP and LS. The xeric site in southern California, CO, had the lowest proportion of branch biomass of all the sites. Although CP had the greatest foliar biomass of all the sites (Table III), the proportion of foliage was among the lowest of the four sites. Root biomass was greatest at CO, and least at CP and LS. Foliar mass was greatest at CO and LS, and least at SP and CP (Figure 3).

## 4. Discussion

### 4.1. Foliar Injury and Retention

Chlorotic mottle of pine foliage is a definitive characteristic of exposure to oxidants (Miller et al., 1963; 1996), and is supported by long term studies of canopy health on mature trees (Miller et al., 1989). In this study, high oxidant exposure resulted in significant foliar injury along the western end of the pollution gradient, which was not mitigated by high N deposition. In a drought year, chlorotic mottle declined at three of the four sites, but was greatest at the most polluted site. After mid July in this year, gas exchange was low and no further decline in photosynthetic capacity was observed with increasing $O_3$ exposure (Grulke, 1998). Stomatal functioning was severely impaired (incomplete closure) at the most polluted site (Grulke, 1999), and trees at this site may have experienced much greater $O_3$ uptake over the growing season relative to the other sites.

The number of live needle age classes (whorls) is a well-established attribute inversely associated with high $O_3$ exposure in pine (Miller et al., 1996). High $O_3$ exposure reduced foliar retention in chamber studies with ponderosa pine and when combined with drought stress, reduced foliar mass (Beyers et al., 1992). However, N fertilization has also been found to promote senescence of the oldest needle age classes in pine (Oren and Schulze, 1989; Gower et al., 1993; Reich et al., 1995). If N is readily available, it may be more C cost-effective to drop older foliage that has lower photosynthetic capacity and accompanying respiratory losses. Older needles appear to function primarily as a repository for nutrient-containing compounds (Chapin et al., 1990), but if absent can limit new branchlet growth through a lowered branch C balance (Jonasson, 1989).

$O_3$ exposure, N deposition, and drought stress all lower some aspect of needle retention in pine. In this study, three components of needle retention were analyzed: the number of whorls retained, the proportion of needles retained within a whorl, and the proportion of foliar biomass in each needle age class. Across the pollution gradient, as chlorotic mottle increased, the number of whorls decreased. Whorl number also decreased in a xeric year relative to mesic years. Needle retention within a whorl did not vary across the pollution gradient in a mesic year, but was significantly reduced by drought stress. In the xeric year, needle loss within a whorl was exacerbated with increasing $O_3$ exposure and N deposition. Also with increasing $O_3$ exposure and N deposition, the proportion of the foliar mass was concentrated in increasingly fewer needle age classes. Despite the presence of multiple needle age classes at the most polluted site, their contribution to total canopy foliar mass was minor. At the most polluted site, 95% of the needle mass was current year foliage, resulting in trees with only one overwintering needle age class.

## 4.2. Growth and Biomass Allocation

Needle, branch, and bole growth was greatest at the most polluted site (CP), and declined with decreasing pollution exposure across the southern Californian gradient. Needle, branch, and stem mass were greatest at the most polluted sites (CP, SP). Of the environmental factors differing between the sites, this response might be expected in response to differences in ecosystem energy inputs (cumulative degree days, season length) or N deposition. Season length was similar between three of the four sites: the northern Californian site, and the eastern and intermediate sites in southern California (averaging LS: 97 d, CO: 105 d, and SP: 109 d, versus CP: 170 d). Despite the differences in cumulative degree days, some attributes of tree growth at the cleanest site in northern California were most comparable to the xeric site in southern California (needle length in a xeric year and the following year; branch length in the lower third of the canopy in any year), and others (needle length after 2 yr of above-average precipitation; tree height; bole diameter) were more similar to the intermediate pollution exposure site (SP). If cumulative degree days were dominating the growth response, a three fold difference between LS and SP would be expected to have a significant effect on these attributes. Despite high $O_3$ exposure, N deposition dominated the growth response at the most polluted sites (CP, SP).

As a single stressor, oxidant exposure in chamber studies generally reduces growth and biomass in pine (as reviewed in Reich, 1987; Darrall, 1989; Bytnerowicz and Grulke, 1992; Matyssek et al., 1994). Nitrogen fertilization increases foliage, branch, and bole mass in pines (*Pinus ponderosa*: Powers, 1983; Gholz et al., 1991; Raison et al., 1992; Gower et al., 1993; Tingey et al., 1996; *Pinus radiata*: Fife and Nambiar 1997). In studies of other conifers, needle length and specific leaf weight increased in response to fertilization (Chandler and Dale, 1990; Raison et al., 1992). Although foliar retention is generally reduced with N fertilization, some of the greater canopy biomass observed in long term studies could be accounted for by the retention of branches in the lower canopy (Powers, pers. comm.). The trees at the most polluted site had significantly greater foliar mass than at SP, and both CP and SP had significantly greater foliar mass than the two least polluted sites, LS and CO. Although CP had the greatest foliar biomass of all the sites, as a proportion, foliage was among the lowest of the four sites at CP. The least polluted sites had the highest proportion of whole tree biomass in foliage. The most polluted sites had the highest proportion of biomass allocated to the bole. Drought stress increased the proportion of whole tree biomass allocated to root biomass (CO compared to SP and CP). The low proportion of root biomass at the most polluted site supported measures of standing live root biomass in another study across the same gradient (Grulke et al., 1998).

Trees at the xeric site in southern California (CO), had the lowest whole tree biomass, height, and bole diameter, a function of drought and moderate $O_3$ exposure.

This site also had the lowest proportion of whole tree biomass allocated to branches. Needle and branchlet length was lowest for the most xeric site (CO) in the year following the drought suggesting a possible residual effect. The significant drought stress experienced in 1994 did not have as large an effect on needle and branch elongation growth as was expected. These data were taken in the lower third of the canopy, which may be the least responsive tissue to environmental influences external to the canopy due to shading and within-tree competition for resources. Branchlet length and diameter, and branchlet and foliar mass were greater in the upper two thirds of the canopy, but differences were significantly greater only for foliar mass at the most polluted site. Thus future studies of tree growth response to oxidant injury, N deposition, and drought stress should be conducted in the mid or upper canopy.

## 5. Conclusions

High $O_3$ exposure and its effects on chlorotic mottle of foliage at sites closest to the Los Angeles, California air basin are significant. This is a definitive response to oxidant pollution, and high N deposition did not eliminate the foliar $O_3$ injury. $O_3$ exposure, N deposition, and drought stress all lower the number of live needle age classes, and at sites where they may all be present, this tree attribute cannot be used to distinguish their singular effects. Retention of needles within a whorl was lowered by drought stress, and was further exacerbated when drought stress was combined with $O_3$ exposure. The significant increase in needle and branchlet elongation growth, and the increase in branchlet and bole enlargement growth are responses that result from long-term N deposition, and provide strong evidence that N deposition overrides $O_3$ exposure effects at the most polluted site. The effects of N deposition on elongation growth are overridden by drought stress. Based on whole tree harvests, root biomass was lowest at sites with the highest pollution exposure, confirming previous chamber exposure and field studies. Whole tree measures indicate that at the least polluted site, foliar biomass was evenly distributed across three of the five needle-age classes retained. At the most polluted site, 95% of the foliar biomass was found in the current year's growth. High N deposition and high $O_3$ exposure combine to shift biomass allocation in pine to that of a deciduous tree.

## Acknowledgments

Although the research described in this article has been funded in part by U.S. Environmental Protection Agency agreement, DW 12934530, it has not been subjected to the Agency's review, does not necessarily reflect the views of the Agency, and no official endorsement should be inferred. The field research effort was supported by Janet Kirsch, Patti Haggerty, Adam Cosner, Michella Watson, and Mark Tacheny.

## References

Barbour, M. G.: 1988, Pgs. 131-164, In: *North American Terrestrial Vegetation*, eds, Barbour, M. G., Billings, W. D. Cambridge University Press, NY.

Beyers, J. L., Riechers, G. H. and Temple, P. J.: 1992, *New Phyt.* **122**, 81-90.

Bytnerowicz, A. and Fenn, M. E.: 1996, *Env. Poll.* 92, 127-146.

Bytnerowicz, A. and Grulke, N. E.: 1992, Pages 183-233, In *The Response of Western Forests to Air Pollution*, eds. Olson, R., Binkley, D., Bohm, M. Springer-Verlag, NY, USA.

Chandler, J. W. and Dale, J.E.: 1990, *Tree Phys.* **6**, 41-56.

Chapin F. S., III, Schulze, E.-D. and Mooney, H. A.: 1990, *Ann. Rev. Ecol. Syst.* **21**, 423-47.

Darrall, N. M.: 1989, *Plant, Cell, Env.* **12**, 1-30.

Fenn, M. E. and Kiefer, J.W.: 1999, *Env. Poll.*, (*in press*).

Fenn, M. E. and Poth, M. A.: 1999, *Ecol. Stud.* **134**, 288-316.

Fenn, M. E., Poth, M. A. and Johnson, D. W.: 1996, *For. Ecol. Manage.* **82**, 211-230.

Fife, D. N. and Nambiar, E. K. S.: 1997, *For. Ecol. Manage.* **93**, 137-152.

Gholz, H. L. and Cropper, W. P.: 1991, *Can. J. For. Res.* **21**, 1742-1747.

Gholz, H. L., Vogel, S. A., Cropper, W. P., Jr., McKelvey, K., Owel, K. C., Teskey, R. O. and Curran, P. J.: 1991, *Ecol. Monogr.* **61**, 33-51.

Gower, S. T., Haynes B. E., Fassnacht, K. S., Running, S. W. and Hunt, E. R., Jr.: 1993, *Can. J. For. Res.* **23**, 1704-1711.

Grulke, N. E.: 1998, *Air Waste Manag. Assoc., 91st Annual Meeting Proc.*, TA-37.05.

Grulke, N. E.: 1999, *Ecol. Stud.* **134**, 126-163.

Grulke, N. E., Andersen, C.P., Fenn, M.E. and Miller, P.R.: 1998, *Envir. Poll.* **103**, 63-73.

Grulke, N. E. and Lee, E.H.: 1997, *Can. J. For. Res.* **17**, 1658-1668.

Haynes, B. E. and Gower, S.T.: 1995, *Tree Phys.*: **15**, 317-325.

Hermann, R. K. and Petersen, R.G.: 1969, *For. Sci.* **15**, 226-237.

Jonasson, S.: 1989, *Oikos* **56**, 121-131.

Kiefer, J. W. and Fenn, M.E.: 1997, *For. Ecol. Manage.* **94**, 47-59.

Matyssek, R., Reich, P., Oren, R. and Winner, W.: 1994: In *Ecophysiology of Coniferous Forests*, eds. Smith, W. K., Hinckley, T. M. Academic Press, San Diego, CA, USA.

McBride, J. R. and Miller, P. R.: 1987, Pages 217-228, In *Effects of Atmospheric Pollutants on Forests, Wetlands, and Agricultural Ecosystems*, eds. Hutchinson, T.C., K.M. Meema, K.M. Springer-Verlag, NY, USA.

McBride, J. R., Semion, V.P. and Miller, P.R.: 1975, *Calif. Agric.* **8**, 9.

Miller, P. R., Longbotham, G.J. and Longbotham, C.R.: 1983, *Plant Disease* **67**, 1113-1115.

Miller, P. R., McBride, J.R., Schilling, S.L. and Gomez, A.P.: 1989, Pages 309-323, In: *Effects of Air Pollution on Western Forests*, eds Olson, R.K., Lefohn, A.S., Transactions, Air Waste Manage. Assoc. Symposium, Anaheim, CA.

Miller, P. R., Parmeter, J. R., Taylor, O. C. and Cardiff, E. A.: 1963, *Phytopath.* **53**, 1073-1076.

Miller, P. R., Stolte, K. W., Duriscoe, E. and Pronos, J. (tech coor): 1996, *Monitoring ozone air pollution effects on western pine forests*. USDA Forest Service, Pacific Southwest Region, GTR-155, Albany, CA, USA.

Oren, R. and Schulze, E.-D.: 1989, *Ecol. Stud.* **77**, 425-443.

Peterson, D. L., Arbaugh, M. J. and Robinson, L.J.: 1991, *The Holocene* **1**, 50-61.

Powers, R. F.: 1983, Pages 388-397, In: *IUFRO Symposium on Forest Site and Continuous Productivity*, eds Ballard, R., Gessel, S.P. USDA FS Gen. Tech. Rep. PNW-163.

Raison, R. J., Myers, B. J. and Benson, M. L.: 1992, *Forest. Ecol. Manage.* **52**, 139-158.

Reich, P. B.: 1987, *Tree Phys.* **3**, 63-91.

Reich, P. B., Koike, T., Gower, S. T. and Schoettle, A. W.: 1995, Pgs. 225-254, In: *Ecophysiology of Coniferous Forests*, Smith, W.K., Hinckley, T.M., Academic Press, San Diego, CA.

San Bernardino County Water Resources Division, http://www.co.san-bernardino.ca.us/pwg/waterresources.

Tingey, D. T., Johnson, M. G., Phillips, D. L., Johnson, D. W. and Ball, J. T.: 1996, *Tree Phys.* **16**, 905-914.

# CHANGES IN FREE-RADICAL SCAVENGERS DESCRIBE THE SUSCEPTIBILITY OF *PINUS PONDEROSA* TO OZONE IN SOUTHERN CALIFORNIAN FORESTS

M. TAUSZ[1], A. BYTNEROWICZ[2], W. WEIDNER[1], M. J. ARBAUGH[2], P. PADGETT[2] and D. GRILL[1]

[1] *Institut für Pflanzenphysiologie, Universität Graz, Schubertstraße 51, A-8010 Graz, Austria;* [2] *U.S. Department of Agriculture Forest Service, Pacific Southwest Research Station, 4955 Canyon Crest Drive, Riverside CA 92507-6090, USA*

(Received 25 September 1998; accepted 26 February 1999)

**Abstract.** Damage of *Pinus ponderosa* Dougl. Ex P. and C. Laws. in the San Bernardino Mountains in Southern California is a well-known example for ozone-phytotoxic effects in forests. In October 1997, needles were collected at the severely impacted Dogwood plot, where pines showing severe crown thinning ('symptomatic') and asymptomatic ones grow together. Visible chlorosis was not or only mildly developed on the sampled current year's ($c$) and previous year's ($c+1$) needles at the time of collection. In sun exposed needles of symptomatic trees, but not in needles which were dark adapted overnight, significantly lower glutathione contents were found compared to asymptomatic trees. In asymptomatic trees sun-exposed and dark adapted needles contained 85 to 90% of the glutathione pool in the reduced state, whereas in sun exposed needles of symptomatic trees a significantly higher proportion of oxidised glutathione was observed (>20%). The redox state of the ascorbate pool remained unaffected, but needles of asymptomatic trees showed a significant decrease of total ascorbate upon darkening. Tocopherol contents were not significantly different between symptomatic and asymptomatic individuals. The results indicate oxidative alterations in needles of symptomatic trees. Since these changes are detectable before severe visual symptoms appear on the needles, they represent initial physiological damage and might be useful for an early risk assessment of ozone phytotoxic potential on pine trees in the field.

**Keywords:** *Pinus ponderosa*, ozone, ascorbate, tocopherol, glutathione, antioxidants, oxidative stress

## 1. Introduction

The adverse effects of air pollution on the ecologically prominent forest species *Pinus ponderosa* Dougl. ex P. and C. Laws. are well documented in Southern California. Visible symptoms like chlorotic mottle of the needles and premature needle loss are used to monitor ozone injury to trees at field plots (Miller *et al.*, 1996).

At some plots, symptomatic (canopy thinning and chlorotic mottle on older needles, Miller *et al.*, 1996) and asymptomatic trees grow in close vicinity. Symptomatic trees, too, have needle age classes without or with only mild visible symptoms which might be used for the search for early indicators of future injury.

Damage at the organ level is preceded by changes at the cellular and biochemical level. Stress conditions promote the formation of highly reactive oxygen species in particular in the photosynthetic apparatus of plants. These compounds initiate chlorophyll destruction, eventually leading to visible chlorosis, if not scavenged by the antioxidative system (Elstner and Osswald, 1994). Ascorbate, glutathione and tocopherol are low molecular weight antioxidants able to detoxify active oxygen species

(Foyer, 1997). Ascorbate is not only an efficient chemical quencher of dangerous free radicals (Smirnoff, 1996), it is also required for the formation of zeaxanthin, the most important quencher of excess light energy (Demmig-Adams and Adams, 1994) and for the regeneration of α-tocopherol (Fryer, 1992). During these processes, ascorbate is oxidised itself, and can be regenerated by glutathione, which is in turn regenerated by the reducing power of the photosynthesis (Foyer, 1997).

The capacity of the cellular antioxidant defence is determined by the pool sizes of the antioxidants and also by the ability of the metabolism to keep them in an active reduced state. Changes in the antioxidative system indicate early responses of plant cells to stress (Polle and Rennenberg, 1994).

The present study compares the antioxidative defence systems of symptomatic and asymptomatic *P. ponderosa* trees growing at the same forest site with the objective to find early stages of cellular injury and to verify their presumably oxidative nature.

## 2. Materials and Methods

### 2.1. SAMPLING SITE

The study was conducted at the "Dogwood A" plot of the San Bernardino Gradient Study (Arbaugh *et al.*, 1998) located in the western end of the San Bernardino Mountains (Southern California, USA) at 1725 m. The forest is a mixed conifer forest established by natural growth. Average daily ozone concentrations between May and October were 0.10 ppm which makes this plot moderately to highly impacted in this region. With respect to nitrogen deposition this plot is regarded as moderately impacted (Fenn *et al.*, 1996).

### 2.2. COLLECTION AND PREPARATION OF MATERIAL

Five symptomatic and 5 asymptomatic dominant *P. ponderosa* individuals were determined according to their crown thinning symptoms. This rough evaluation in the field was supported by results of the long-term assessment of ozone injury at these individuals according to Miller *et al.* (1996). Sun-exposed branches were cut in the field at a height of about 6-10 m from 11 a. m. to 2 p. m. on cloudless days (October 1997). Some fascicles were removed immediately, needles cut separately according to the age class (previous year's needles = $c+1$, current year's = $c$) and immersed in liquid nitrogen (within seconds, = light exposed needles). Since the production of free radicals in plant cells is mainly light dependent (Elstner and Osswald, 1994), a second sub-sample of needles was taken in a dark adapted stage: the remaining branches were kept well watered, at room temperature (20 °C) and in darkness overnight. Fascicles were taken from the branches the next day in the laboratory and frozen in liquid nitrogen (= dark exposed needles). The needle material was lyophilized and sealed in plastic bags. Lyophilized needles were ground in a dismembrator, the needle powder was stored frozen in humidity proof plastic vials before it was subjected to HPLC analysis.

## 2.3. BIOCHEMICAL ANALYSES

Ascorbate (AA) and dehydroascorbate (DHA) were determined according to Tausz et al. (1996), glutathione in its reduced (GSH) and oxidised (GSSG) form according to Kranner and Grill (1996), and tocopherols according to Wildi and Lütz (1996).

## 2.4. STATISTICS

Statistical evaluations were completed using Statistica (StatSoft, USA, 1994) software package. Figures show medians and median deviations which are most suitable for small sample sizes (Sachs, 1992, p. 336-337). Comparisons between light and dark exposed samples were calculated using the Wilcoxon matched pairs test, differences between symptomatic and asymptomatic trees by Mann-Whitney U-test (Sachs, 1992). $P<0.05$ was regarded significant.

## 3. Results and Discussion

Visible injury was only mild or not observed on the needle age classes sampled in this study. The changes in the antioxidative system of these non-symptomatic needles may thus be regarded as early stages of their forthcoming visible injury.

The concentrations of α-tocopherol (other tocopherols were not found) were higher in the $c+1$ needles than in current ones which is consistent with the published information on conifers (Polle and Rennenberg, 1994). In spite of reports of decreased tocopherol levels as an early indication of chlorophyll degradation, concentrations in needles of symptomatic trees did not differ significantly ($P>0.05$) from those of asymptomatic ones (Table 1).

Table I

α-Tocopherol concentrations of *P. ponderosa* needles in $\mu g\ g^{-1}$ DW. Medians ± median deviations of 5 individual trees. Asympt. = asymptomatic trees, Sympt. = symptomatic trees, sun light adapted (L) and darkened (D) samples of previous year's ($c+1$) and current year's ($c$) needles. Significant differences between D and L or between symptomatic and asymtpomatic trees were absent ($P>0.05$).

|  | $c+1$ needles | | $c$ needles | |
|---|---|---|---|---|
|  | D | L | D | L |
| Asympt. | 260 ± 39 | 226 ± 16 | 127 ± 10 | 126 ± 11 |
| Sympt. | 210 ± 36 | 224 ± 24 | 117 ± 10 | 144 ± 6 |

As in previously studied pine species (e. g. Jiménez et al., 1997), total ascorbate was generally higher in $c+1$ needles than in $c$ ones. The results for DHA accounting for 10-20% of the total ascorbate seem reasonable compared to the published data on various conifers (reviewed in Polle and Rennenberg, 1994) or recent results on *Pinus canariensis* (Jiménez et al., 1997). However, published data vary widely, and the natural occurence of DHA in tissues has even been denied (Morrell et al., 1997), an opinion that was rejected by other authors (Noctor et al., 1998). In the present study, significant changes in the redox state of ascorbate did not occur, but upon darkening reduced and total ascorbate decreased significantly in $c+1$ needles of asymptomatic trees, but not in those of symptomatic individuals (Figure 1).

Glutathione concentrations were roughly in the range found previously for other conifers (Polle and Rennenberg, 1994, Jiménez et al., 1997), without any needle age dependent changes. In the asymptomatic ponderosa pine trees glutathione content did not vary between light and dark, but in the symptomatic trees total and reduced glutathione even increased after storage in the dark. In the light, $c+1$ needles of symptomatic trees contained significantly less total (P=0.047) and reduced GSH (P=0.016) than the corresponding samples of asymptomatic trees (Figure 2). These changes contradicted the previously reported light induced increase of glutathione by about 200 nmol in spruce (Schupp and Rennenberg, 1989). The symptomatic, but not the asymptomatic trees exhibited a significantly increased proportion of oxidised glutathione in light exposed, but not in darkened needles (Figure 2). The normal proportion of oxidised glutathione in tissues is up to 10% (Polle and Rennenberg, 1994) and may reach 15% of the total glutathione pool in pines (Jiménez et al., 1997). A further increase in the oxidised proportion appears only under oxidising conditions (Polle and Rennenberg, 1994) or in visibly damaged needles (Schmieden et al., 1993). It reflects initial stages of biochemical damage to the antioxidative system in the symptomatic trees which may eventually lead to an overcharge of the protective systems. The affinity of glutathione for oxygen radicals under physiological conditions is orders of magnitude lower than that of ascorbate (Polle and Rennenberg, 1993). Hence, it is more likely that glutathione is mainly depleted by ascorbate regeneration in the Foyer-Halliwell-cycle (Foyer, 1997), rather than by direct oxidation through active oxygen species.

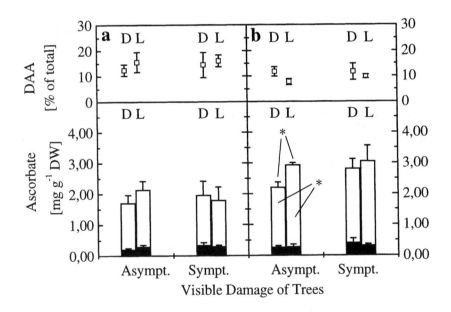

Fig. 1. The ascorbate system in needles of *Pinus ponderosa* trees with (Sympt.) and without (Asymptom.) severe crown thinning. a. Current year's ($c$), b. previous year's needles ($c+1$). Total ascorbate (columns) consists of reduced ascorbate (open part of the columns) and dehydroascorbate (closed part of the column). Columns show medians of 5 individual trees, error bars show median deviations. Dark adapted stage (D), light adapted stage (L). Asterisks indicate significant differences between light and dark adapted stage.

## 4. Conclusions

The results indicate a disturbance of the antioxidative system in symptomatic ponderosa pine trees at an early stage of damage, in particular when light energy is present. The oxidative nature of the cellular processes leading to the chlorotic mottling symptoms is most probable. Therefore, evaluation of the antioxidative systems of ponderosa pine is also a promising approach for development of an early risk assessment tool for the phytotoxic effects of ozone in field conditions.

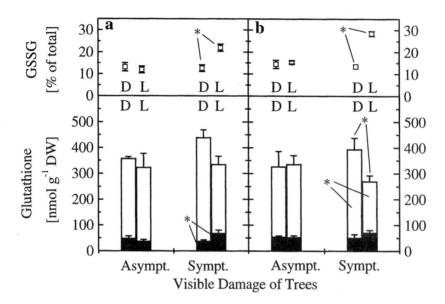

*Fig. 2.* The glutathione system in needles of *Pinus ponderosa* trees with (Sympt.) and without (Asymptom.) severe crown thinning. a. Current year's ($c$), b. previous year's needles ($c+1$). The total pool of glutathione (columns) consists of GSH (open part of the column) and GSSG (closed part of the column). Columns show medians of 5 individual trees, error bars show median deviations (omitted when within the symbol). Dark adapted stage (D), light adapted stage (L). Asterisks indicate significant differences between light and dark adapted stage.

## References

Arbaugh, M. J., Peterson, D. L. and Miller, P. R.: 1998, in *Oxidant Air Pollution Impacts in the Montane Forests of Southern California: A Case Study of the San Bernardino Moutnains*, Miller, P.R. and McBride, J.R. (eds.), pp. 179-207, Ecological Studies 134, Springer-Verlag, Berlin.
Demmig-Adams, B. and Adams, W. W.: III.: 1994, in *Causes of Photooxidative Stress and Amelioration of Defense Systems in Plants*, CRC Press Boca Raton, pp. 105-126.
Elstner E. F. and Osswald, W.: 1994, *Proc. Roy. Soc. Edinburgh*, **102B**, 131-154.
Fenn, M. E., Poth, M. A. and Johnson, D. W.: 1996, *For. Ecol. Manag.* **82**, 211-230.
Foyer C.: 1997, in *Oxidative Stress and the Molecular Biology of Antioxidant Defense*. Cold Spring Harbor Laboratory Press, Cold Spring Harbor, pp. 587-621.
Fryer, M. J.: 1992, *Plant Cell Environ.* **15**, 381-392.

Jiménez, M. S., Tausz, M., Zellnig, G., Peters, J., Grill, D. and Morales, D.: 1997, *Phyton (Austria)*, **37**, 109-114.

Kranner, I. and Grill, D.: 1996, *Phytochem. Anal.* **7**, 24-28.

Miller P. R., Stolte K. W., Duriscoe D. M. and Pronos J.: 1996, *Evaluating Ozone Air Pollution Effects on Pines in the Western United States, Gen. Tech. Rep. PSW-GTR-155,* U. S. Dept. of Agriculture, Albany, CA.

Morell S., Follmann, H., De Tullio, M. and Häberlein, L.: 1997, *FEBS Letters* **414**, 567-570.

Noctor, G., Arisi, A. M., Jouanin, L., Kunert, K. J., Rennenberg, H. and Foyer, C. H.: 1998, *J. Exp. Bot.* **49**, 623-647.

Polle, A. and Rennenberg, H.: 1994, in *Causes of Photooxidative Stress and Amelioration of Defense Systems in Plants,* Foyer, C. H. and Mullineaux, P. M. (eds.), pp.199-218, CRC Press, Boca Raton.

Sachs, L.: 1992, *Angewandte Statistik*, Springer Verlag, Berlin.

Schmieden, U., Schneider, S. and Wild, A.: 1993, *Environ. Pollut.* **82**, 239-244.

Schupp, R. and Rennenberg, H.: 1989, *Ann. Sci. For.* **46**, 837-841.

Smirnoff, N.: 1996, *Ann. Bot.* **78**, 661-669.

Tausz, M., Kranner, I. and Grill, D. 1996, *Phytochem. Anal.* **7**, 69-72

Wildi, B. and Lütz, C.: 1996, *Plant Cell Environ.* **19**, 138-146.

# VISIBLE OZONE INJURY ON FOREST TREES IN GREAT SMOKY MOUNTAINS NATIONAL PARK, USA

A. CHAPPELKA[1], G. SOMERS[1] and J. RENFRO[2]

[1]*School of Forestry, Auburn University, AL 36849, USA,* [2]*National Park Service, Division of Resource Management and Science, Great Smoky Mountains National Park, Gatlinburg, TN 37738, USA*

(Received 25 September 1998; accepted 26 February 1999)

**Abstract.** During the summer of 1991 ozone injury trend plots in Great Smoky Mountains National Park, USA, consisting of mature black cherry, sassafras and yellow-poplar were established near three ozone monitors, ranging in elevation from 597-1265 m. Beginning in mid-August 1991-1993, three exposed branches each from the upper- and mid- to lower-crown of each tree were collected and evaluated for ozone injury. Of the trees examined, 63%, 52% and 36% exhibited some amount of foliar injury in 1991, 1992 and 1993, respectively. Ozone injury across species was the greatest at Cove Mountain in all three years of the study. Overall, across sites and years, 11, 12 and 11% of all leaves examined exhibited visible injury for black cherry, sassafras and yellow-poplar, respectively. The percentage of injured leaves per branch was greater in the mid- to lower-canopy for black cherry, across all sites. Trees for each species that exhibited the greatest or least amounts of visible injury did so in all three years of the study, indicating a differential sensitivity within each species population. No significant ozone exposure-tree response relationships were observed with any variable tested. These data indicate that ozone concentrations are high enough to cause visible symptoms to selected trees within Great Smoky Mountains National Park, USA.

Keywords: ozone, ecosystem, visible injury, foliar symptoms, elevation.

## 1. Introduction

Ozone is considered an important phytotoxic air pollutant in the eastern United States, and can be transported long distances from urban sources to remote, forested areas (US EPA, 1996). Concentrations of ozone in the southern Appalachian Mountain region are as high or higher than in any other area in the country, excluding the South Coast Air Basin of California (Lefohn and Lucier, 1988). Major effects of ozone on terrestrial vegetation include visible injury (Manning, 1993), reductions in growth and productivity (US EPA, 1996), shifts in community diversity (Barbo *et al.*, 1998) and increased susceptibility to other abiotic or biotic stresses (Chappelka and Chevone, 1992).

Visible symptoms of ozone injury have been observed on sensitive plant species in many areas of the eastern USA at ambient concentrations (Chappelka and Samuelson, 1998). However, the relationship between foliar ozone injury and tree growth is unclear.

Great Smoky Mountains National Park (GRSM) is the most visited National Park in the USA (Shaver *et al.*, 1994). The Park encompasses over 200,000 hectares in Tennessee and North Carolina, and contains a wide diversity of tree species, herbaceous plants and wildlife. Therefore, any detrimental effects on vegetation within GRSM is of concern. The objectives of this study were to determine the incidence and severity of foliar ozone symptoms on mature black cherry (*Prunus serotina* Ehrh.), sassafras (*Sassafras albidum* Nutt.) and yellow-poplar (*Liriodendron tulipifera* L.). In addition, ozone exposure/plant response relationships were examined. A companion study using similar protocols was conducted during the same time-frame in the Shenandoah National Park in Virginia (Hildebrand et al., 1996).

## 2. Materials and methods

To observe the relationships between ambient ozone concentrations and tree responses, field plots were established in 1991 in the vicinity of three air quality monitoring stations: Cove Mountain (elev. 1265 m), Look Rock (elev. 823 m) and Twin Creeks (elev. 597 m). All monitors were located on the northern periphery of the Park (Tennessee). The presence of an ozone monitor at each site permitted correlations of ozone exposure statistics with observed biological effects. A yellow-poplar trend plot was not established at Cove Mountain until 1992. Sassafras was not sampled at Twin Creeks in 1993. Selection criteria have been detailed in Chappelka et al. (1994). Ozone exposure statistics are shown in Table I; Seasonal (May- September) 24 hr monthly mean (ppb), SUM0 = total cumulative ozone concentration (ppm-h), SUM06 = cumulative ozone concentration $\geq$ 0.06 ppm, and W126 = cumulative ozone concentrations weighted by a sigmoidal function (Lefohn et al., 1988).

Three exposed branches from the upper-crown and three branches from the mid- to lower-crown of each tree were collected and evaluated for visible ozone injury expressed as an adaxial leaf surface stipple and reddening; foliage was examined for incidence and severity of injury. If injury was observed, the percentage of injured leaves per branch was estimated. The Horsfall-Barratt rating scale was used to quantify the relative severity of symptoms on the injured leaves (Horsfall and Barratt, 1945). Before going to the field, crews were trained in ozone symptom recognition using a computer-based "expert system" (Nash et al., 1992).

The percent trees exhibiting injury were analyzed using analysis of variance (ANOVA) techniques (Steel and Torrie, 1960). Since there is only a single observation for each site and year, the test must therefore, assume no interaction between site and year, so the results have to be viewed with caution. Repeated measures analysis was used to determine the percent of leaves injured per branch and percent leaf area affected (Neter et al., 1990). Site and year were considered fixed factors at 3 levels. Trees were a random sub-sample of the conditions at each site. Canopy position (upper or lower canopy) was a fixed factor crossed with site and year. Using statistical contrasts the following hypotheses were tested: no difference between year; no differences between sites; no difference between canopy positions; and no interactions between site, year and/or canopy position. When the hypothesis between sites was rejected then Cove Mountain was tested against the average of Look Rock and Twin Creeks. To determine whether or not there was a possible genetic trend over time, i.e. whether the same trees were injured each year, trees were ranked across sites and years for incidence and severity of injury, and the non-parametric Kruskal-Wallis test was used to test for departures in ranking from a random assignment over time (Steel and Torrie, 1960). Regression analysis (Montgomery and Peck, 1982) was used to test if any ozone exposure (SUM0, SUM06, etc.) by tree response (incidence or severity of visible foliar injury) relationships existed.

## 3. Results and Discussion

3.1 INCIDENCE AND SEVERITY OF OZONE INJURY

Visible ozone injury was observed on trees of all species sampled at all three locations. Of the trees examined, 63%, 52% and 36% exhibited some amount of foliar injury in 1991, 1992 and 1993, respectively. Significant site effects were observed for both black cherry and sassafras ($p = 0.0089$ and $p = 0.026$, respectively). Using statistical contrasts it was determined that there was a significantly greater percentage of ozone-injured black cherry and sassafras at Cove Mountain than either of the other sites ($p = 0.0036$ and $p = 0.012$, respectively). Ozone concentrations also were the greatest at this site (Table I). These data support the hypothesis that ozone injury increases with elevation (Winner *et al.*, 1989; Chappelka and Samuelson, 1998; Chappelka *et al.*, 1999). Percent injured yellow-poplar, however, exhibited no differences among sites ($p = 0.80$) and years ($p = 0.31$) regarding visible injury. Yellow-poplar is a very site-demanding species and other factors such as soil moisture and microclimate may have influenced the results (Hildebrand *et al.*, 1996).

TABLE I

Ozone exposure statistics for monitors located in Great Smoky Mountains National Park from 1991-1993

| Location | Year | Exposure Statistics[a] | | | |
|---|---|---|---|---|---|
| | | 24 hr means (ppb) | SUM0 (ppm-h) | SUM06 (ppm-h) | W126 (ppm-h) |
| Cove Mountain | 1991 | 55 | 176.5 | 79.4 | 57.2 |
| | 1992 | 50 | 168.4 | 50.0 | 40.4 |
| | 1993 | 58 | 189.9 | 99.6 | 66.6 |
| Look Rock | 1991 | 47 | 146.7 | 38.3 | 29.0 |
| | 1992 | 45 | 148.2 | 35.0 | 30.6 |
| | 1993 | 56 | 144.4 | 66.0 | 47.0 |
| Twin Creeks | 1991 | 21 | 69.5 | 1.0 | 2.6 |
| | 1992 | 25 | 75.2 | 2.9 | 3.9 |
| | 1993 | 28 | 99.1 | 5.3 | 6.4 |

[a] 24 hr means = seasonal (May - September) 24 hr daily means (ppb)/month; SUM0 = total cumulative ozone concentration (ppm-h); SUM06 = cumulative ozone concentration $\geq 0.06$ ppm; W126 = cumulative ozone concentrations weighted by a sigmoidal function (Lefohn *et al.*, 1988); Data provided by the Monitoring and Data Analysis Branch, National Park Service.

In general, the % injured leaves and % leaf area affected exhibited greater injury for black cherry and sassafras at Cove Mountain compared with the other two sites ($p < 0.0001$). There was no difference detected for these variables between Look Rock and Twin Creeks for either species, except regarding black cherry % leaf area affected ($p = 0.00015$);

TABLE II

Mean incidence (% trees injured, TREEINJ) and severity [% leaves injured (PINJ) and % leaf area affected (PLAJ)] on three different tree species in Great Smoky Mountains National Park from 1991-1993

| Species[a] | Site | 1991 | | | 1992 | | | 1993 | | |
|---|---|---|---|---|---|---|---|---|---|---|
| | | TREE INJ | PINJ | PLAJ | TREE INJ | PINJ | PLAJ | TREE INJ | PINJ | PLAJ |
| BC | CM | 97 | 39 | 2 | 67 | 24 | 3 | 57 | 8 | 1 |
| | LR | 50 | 9 | 1 | 20 | 6 | 1 | 33 | 6 | 2 |
| | TC | 33 | 7 | <1 | 17 | 2 | <1 | 30 | 4 | <1 |
| SAS | CM | 93 | 47 | 2 | 93 | 19 | 2 | 43 | 2 | <1 |
| | LR | 70 | 13 | 1 | 37 | 4 | <1 | 15 | <1 | <1 |
| | TC | 50 | 8 | 1 | 27 | 6 | 1 | --- | --- | --- |
| YP | CM | --- | --- | --- | 63 | 19 | 3 | 33 | 1 | <1 |
| | LR | 80 | 23 | 2 | 73 | 24 | 4 | 27 | 1 | <1 |
| | TC | 33 | 2 | <1 | 67 | 12 | 1 | 50 | 6 | 1 |

[a] BC = black cherry; SAS = sassafras; YP = yellow-poplar; CM = Cove Mountain; LR = Look Rock; TC = Twin Creeks; TREEINJ = % injured trees; PINJ = % injured leaves (including symptomatic and asymptomatic branches); PLAJ = % leaf area affected (including symptomatic and asymptomatic branches).

there was a greater percentage of leaf area affected at Look Rock than at Twin Creeks (Table II). Regarding yellow-poplar, when Cove Mountain was included in the analysis (1992 & 1993) the overall site effect was only marginally significant for both % injured leaves and % leaf area affected (p = 0.062 and p = 0.079, respectively). When Cove Mountain was removed from the analysis, Look Rock significantly differed from Twin Creeks during all three years of the study for both variables (p < 0.001). As can be observed in Table II, these variables were confounded by year (p < 0.0001). These results were similar to those reported for the % injured trees. Overall, across sites and years, 11, 12 and 11% of all leaves examined exhibited visible injury for black cherry, sassafras and yellow-poplar, respectively.

Black cherry had significantly more % injured leaves and % leaf area affected in the mid- to lower-crown vs. upper-crown (p < 0.0001 and p = 0.0018, respectively). The % injured leaves per canopy position for black cherry are shown in Table III. Combining data across sites and years, 8% injured leaves were in the upper-canopy and 15% in the mid- to lower-canopy. Our results were similar to those reported by Hildebrand et al. (1996) and Fredricksen et al. (1995) for black cherry in Virginia and Pennsylvania, respectively. They both found that the % injured leaves and % leaf area affected was greater in the mid- to lower-canopy compared with leaves sampled in the upper-canopy. Fredricksen et al. (1995, 1996) postulated that these differences were a result of greater ozone uptake per net photosynthesis observed in shaded environments. When all sites were included in the analysis neither sassafras nor yellow-poplar exhibited significant effects regarding ozone injury at either crown position.

TABLE III

Mean % injured leaves by canopy position for black cherry in Great Smoky Mountains National Park from 1991-1993

| Site | Year | Upper canopy | Mid to lower canopy |
|---|---|---|---|
| Cove Mountain | 1991 | 34 | 45 |
|  | 1992 | 21 | 27 |
|  | 1993 | 3 | 13 |
| Look Rock | 1991 | 3 | 14 |
|  | 1992 | 5 | 9 |
|  | 1993 | 3 | 9 |
| Twin Creeks | 1991 | 4 | 10 |
|  | 1992 | 1 | 3 |
|  | 1993 | 1 | 7 |

3.2 DIFFERENTIAL SENSITIVITY

Using the Kruskal-Wallis test it was determined that the incidence and severity of visible foliar injury was not a random occurrence for all three tree species sampled ($p < 0.0001$). Trees that exhibited the greatest or least amounts of injury (% injured branches and % injured leaves) did so in all three years of the study. These data indicate a differential sensitivity to ozone for the three species observed. Hildebrand *et al.* (1996) in a companion study in the Shenandoah National Park found that 87% of the trees that had exhibited symptoms in 1993 did so in 1991 and 1992.

Somers *et al.* (1998) used the same trees as in our study to compare 5 and 10 yr radial growth increments with the amount of visible injury for black cherry and yellow-poplar. Significantly more radial growth was observed for the non-sensitive (little to no visible injury) than sensitive (severely injured) yellow-poplar. No significant relationship was observed with black cherry. These results (Somers *et al.*, 1998) indicated the possibility of ozone-induced growth losses in sensitive genotypes of some species. Further results are needed with many different genotypes and species before the relationship between visible injury and growth reductions due to ozone can be ascertained.

3.3 OZONE EXPOSURE/TREE RESPONSE

No significant ozone exposure/tree response relationships were detected with any species or variable used (data not shown) in our study. Hildebrand *et al.* (1996) found ozone injury on black cherry significantly correlated ($r^2 = 0.87$ and 0.81, respectively) with SUM06 and W126 in the Shenandoah National Park. They found no exposure-response relationship with yellow-poplar, and attributed this to differences in microsite factors (soil moisture). Chappelka *et al.* (1999) analyzed a combined data set from both Great Smoky Mountains and Shenandoah National Parks and found that the relationships with the exposure indices and response variables were strengthened. More data points were included in the analysis (doubled from 9 to 18). Both the SUM06 and W126 exposure statistics preformed well and were positively related ($p \leq 0.01$) to the % of injured trees. Data from these and other studies (Winner *et al.*, 1989; Showman, 1991) indicate that although ozone-exposures may be at phytotoxic levels, other factors such as low sample numbers, alterations in host sensitivity with elevation, site location and micro-site differences (water availability, light, etc.) cannot be discounted (Chappelka and Samuelson, 1998).

## Acknowledgments

The authors would like to thank Barton's Tree Service, E. Robbins, J. Coker and E. Mudd for help in data collection and analysis. The authors appreciate the reviews of an earlier version of this manuscript by two anonymous reviewers. This research was supported by funds provided by the United States Department of Interior National Park Service Cooperative Agreement 5460-2-9009.

## References

Barbo, D.N., Chappelka, A.H., Somers, G.L., Miller-Goodman, M.S. and Stolte, K.: 1998, *New Phytol.* **138**, 653.
Chappelka, A.H. and Chevone, B.I.: 1992, in *Surface Level Ozone Exposures and Their Effects on Vegetation*, Lefohn, A.S. (ed.), pp. 271-324, Lewis Publ. Inc, Chelsea, MI.
Chappelka, A.H., Renfro, J.R., and Somers, G.L.: 1994, 'Visible Ozone Injury on Native Plant Species Within Great Smoky Mountains National Park', *Proc. 87$^{th}$ Ann. Met. Of Air and Waste Mange. Assoc.*, **94-TA36.04**, Cincinnati, OH.
Chappelka, A.H., and Samuelson, L.J.: 1998, *New Phytol.* **139**, 91.
Chappelka, A., Skelly, J. Somers, G., Renfro, J. and Hildebrand, E.: 1999, *Water, Air, Soil Pollut.* (this volume).
Fredricksen, T.S., Joyce, B.J., Skelly, J.M., Steiner, K.C., Kolb, T.E., Kouterick, K.B., Savage, J.E. and Snyder, K.R.: 1995, *Environ. Pollut.* **89**, 273.
Fredricksen, T.S., Kolb, T.E., Skelly, J.M., Steiner, K.C., Joyce, B.J. and Savage, J.E.: 1996, *Tree Physiol.* **16**, 485.
Hildebrand, E.S., Skelly, J.M., and Fredricksen, T.: 1996, *Can. J. For. Res.* **26**, 658.
Horsfall, J.G. and Barratt, R.W.: 1945, *Phytopathology* **35**, 655.
Lefohn, A.S., Laurence, J.A., and Kohut, R.J.: 1988, *Atmos. Environ.* **21**, 1229.
Lefohn, A.S. and Lucier, A.A.: 1988, *J. Air Pollut. Contr. Assoc.* **38**, 1504.
Manning, W.J.: 1993, *Proc. 86$^{th}$ Ann. Meet. of Air and Waste Mange. Assoc.*, **93-WA80.01**, Denver, CO.
Montgomery, D.C. and Peck, E.A.: 1982, *Introduction to Linear Regression Analysis*, John Wiley & Sons, New York, NY
Nash, B.L., Saunders, M.C., Miller, B.J., Bloom, C.A., Davis, D.D. and Skelly, J.M.: 1992, *Can. J. For. Res.* **22**, 1770.
Neter, J., Wasserman, W. and Kutner, M.H.: 1990. *Applied Linear Statistical Models, 3$^{rd}$ edn.*, R.D. Irwin Inc., Homewood, IL.
Shaver, C.L., Tonnessen, K.A. and Maniero, T.G.: 1994, *Ecol. Appl.* **4**, 690.
Showman, R.E.: 1991, *J. Air Waste Mang. Assoc.* **41**, 63.
Somers, G.L., Chappelka, A.H., Rosseau, P. and Renfro. J.R.: 1998, *For. Ecol. Mgt..* **104**, 129.
Steel, R.G.D. and Torrie, J.H.: 1960, *Principles and Procedures of Statistics*, McGraw Hill Inc., New York, NY.
United States Environmental Protection Agency (US EPA): 1996, *Air Quality Criteria for Ozone and Other Photochemical Oxidants, Vol. II*, EPA/600/P-93/004bF.
Winner, W.E., Lefohn, A.S., Cotter, I.S., Greitner, C.S. Nellesen, J., McEvoy, L.R., Olson, R.L., Atkinson, C.J. and Moore, L.D.: 1989, *Proc. Natl. Acad Sci.* **86**, 8828.

# MATURE BLACK CHERRY USED AS A BIOINDICATOR OF OZONE INJURY

A. CHAPPELKA[1], J. SKELLY[2], G. SOMERS[1], J. RENFRO[3] and E. HILDEBRAND[2]

[1]*School of Forestry, Auburn University, AL 36849, USA,* [2]*Department of Plant Pathology, The Pennsylvania State University, University Park, PA 16802, USA,* [3]*National Park Service, Division of Resource Management and Science, Great Smoky Mountains National Park, Gatlinburg, TN 37738, USA*

(Received 25 September 1998; accepted 26 February 1999)

**Abstract.** Incidence and severity of foliar symptoms due to ambient ozone exposures were documented on mature black cherry (*Prunus serotina*) in two National Parks [Great Smoky Mountains National Park (GRSM) and Shenandoah National Park (SHEN)] in the Appalachian Mountains of the eastern USA during the summer of 1991-1993. Three plots in each park containing 30 trees each (Big Meadows in SHEN had 60 trees) with 90 and 120 trees total trees were evaluated in GRSM and SHEN, respectively. Plots were established at different elevations adjacent to ozone monitoring stations. Samples of foliage were collected and three exposed branches from the upper- crown and three branches from the mid-to-lower crown were examined for symptoms of foliar ozone injury. Incidence was greatest in 1991 at both locations; 60% and 45% for GRSM and SHEN, respectively. In 1992 and 1993, incidence was very similar in both parks, with approximately 33% of the trees affected. Black cherry at the highest elevations exhibited the greatest amount of symptoms in both parks all three years of the study. These sites also exhibited the highest ozone concentrations. In addition, the percent of trees injured by ozone was positively correlated with SUM06 and W126. These results along with forest surveys and open-top chamber studies indicate that black cherry may be a reliable bioindicator of foliar injury due to ambient ozone.

**Keywords:** black cherry, ozone exposure-plant response, national parks, bioindicator.

## 1. Introduction

A bioindicator is a vascular or non-vascular plant that exhibits a typical and verifiable response when exposed to an air pollutant (Manning, 1993). These plants can be used to detect the presence of a specific air pollutant in a specific location or region. Whether introduced (sentinels) or native (detectors), these plants can provide unique information regarding the ambient air quality in a particular area (Manning, 1993). Bioindicator plants are especially useful in areas where electrical power is not available to operate ozone monitors.

Ozone is considered as the most important phytotoxic air pollutant in the eastern United States and can be transported long distances from urban sources to rural, forested areas (US EPA, 1996). The major effects of ozone on terrestrial vegetation include visible injury, reductions in growth and productivity, changes in crop quality and increased susceptibility to other abiotic or biotic stresses (US EPA, 1996; Chappelka and Samuelson, 1998).

Visible symptoms of ozone injury have been observed on sensitive plant species in many areas of the eastern USA (Duchelle and Skelly, 1981; Neufeld *et al.*, 1992; Simini *et al.*, 1992; Chappelka *et al.*, 1997). Black cherry (*Prunus serotina* Ehrh.), has been identified as a sensitive native bioindicator of ambient ozone in the USA (Davis and Skelly, 1992; Simini *et al.*, 1992; Chappelka *et al.*, 1997), Mexico (Skelly *et al.*, 1997) and Europe (Innes *et al.*, 1996).

The purpose of this paper is to report three-year results (1991-1993) relating ambient ozone concentrations to visible foliar injury on mature black cherry in Great Smoky Mountains National Park (GRSM) and Shenandoah National Park (SHEN) in the eastern USA. The findings reported here were made during two independent investigations using the same protocols in GRSM (Chappelka et al., 1994) and SHEN (Hildebrand et al., 1996), and illustrate the usefulness of this species as a native bioindicator of ambient ozone over the mid-Atlantic region of the USA.

## 2. Materials and methods

During August-September 1991-1993 three mature canopy black cherry plots were established and subsequently measured in GRSM and in SHEN; each plot in each park containing 30 trees each (Big Meadows in SHEN had 60 trees). The presence of an ozone monitor at each site (3 per park) permitted correlations of ozone exposure statistics with observed biological effects. Selection criteria have been detailed in Hildebrand et al. (1996) and Chappelka et al. (1994).

In GRSM the monitors were located at Cove Mountain, Look Rock and Twin Creeks. All monitors were located on the north-west periphery of the park (Tennessee). The air quality monitoring stations in SHEN were located at Dickey Ridge, Big Meadows and Sawmill Run. These stations were located at the northern (mile post 5), central (mile post 50) and southern (mile post 95) regions of the park, respectively. Ozone exposure statistics are shown in Table I; SUM0 = total cumulative ozone concentration (ppm-h), SUM06 = cumulative ozone concentration $\geq 0.06$ ppm, and W126 = cumulative ozone concentrations weighted by a sigmoidal function (Lefohn et al., 1988).

Three exposed branches from the upper-crown and three branches from the mid- to lower-crown of each tree were collected and evaluated for visible ozone injury expressed as an adaxial leaf surface stipple; foliage was examined for incidence and severity of injury. If injury was observed, the percentage of injured leaves per branch was estimated. The Horsfall-Barratt rating scale was used to quantify the relative severity of symptoms on the injured leaves (Horsfall and Barratt, 1945). Before going to the field, crews were trained in ozone symptom recognition using a computer-based "expert system" (Nash et al., 1992). Although crews from both parks were not "cross-checked" during the study, before evaluations were started crew members from both parks met and discussed the evaluation system and symptom recognition, and co-evaluated several plants in the field for ozone injury. During the study, several meetings were held among the crews from both parks, where methodologies were discussed and revised if necessary.

The results for the individual parks have been presented elsewhere (Hildebrand et al., 1996; Chappelka et al., 1994, 1999). To compare results among the two parks Spearman's Rank Correlation Analysis was used (SAS, 1985). This is a non-parametric procedure. The variables used in this comparison were % injured trees, the total leaf area injured (% leaves injured x % leaf area affected), elevation, SUM0, SUM06 and W126.

TABLE I

Cumulative seasonal (May-September) ozone exposure statistics (ppm-h) for monitors located in Great Smoky Mountains National Park (GRSM) and Shenandoah National Park (SHEN) from 1991-1993

| Location | Elevation (m) | Year | Exposure Statistics (ppm-h) | | |
|---|---|---|---|---|---|
| | | | SUM0[a] | SUM06[a] | W126[a] |
| Cove Mountain (GRSM) | 1265 | 1991 | 176.5 | 79.4 | 57.2 |
| | | 1992 | 168.4 | 50.0 | 40.4 |
| | | 1993 | 189.9 | 99.6 | 66.6 |
| Big Meadows (SHEN) | 1067 | 1991 | 156.9 | 75.6 | 53.2 |
| | | 1992 | 137.1 | 49.3 | 36.6 |
| | | 1993 | 133.1 | 66.0 | 47.0 |
| Look Rock (GRSM) | 823 | 1991 | 146.7 | 38.3 | 29.0 |
| | | 1992 | 148.2 | 35.0 | 30.6 |
| | | 1993 | 144.4 | 66.0 | 47.0 |
| Dickey Ridge (SHEN) | 617 | 1991 | 133.1 | 51.3 | 38.8 |
| | | 1992 | 133.0 | 25.0 | 22.2 |
| | | 1993 | 159.6 | 37.8 | 30.9 |
| Twin Creeks (GRSM) | 597 | 1991 | 69.5 | 1.0 | 2.6 |
| | | 1992 | 75.2 | 2.9 | 3.9 |
| | | 1993 | 99.1 | 5.3 | 6.4 |
| Sawmill Run (SHEN) | 457 | 1991 | 137.1 | 29.6 | 24.4 |
| | | 1992 | 87.2 | 16.9 | 14.1 |
| | | 1993 | 92.5 | 27.0 | 21.5 |

[a] SUM0 = total cumulative ozone concentration (ppm-h); SUM06 = cumulative ozone concentration $\geq 0.06$ ppm; W126 = cumulative ozone concentrations weighted by a sigmoidal function (Lefohn et al., 1988); Data provided by Monitoring and Data Analysis Branch, National Park Service.

## 3. Results and Discussion

Black cherry proved to be an excellent species choice for comparison of ozone induced foliar injury on a spatial and temporal basis in two National Parks located in the eastern USA. It is important to note that foliar injury was easily observed on black cherry in both the SHEN and GRSM situated ca. 300 km distant from one another (Table II). Incidence was the greatest in 1991 at both locations; 60% and 45% for GRSM and SHEN, respectively. In 1992 and 1993, incidence was very similar in both parks, with approximately 33% of the trees affected. The % leaves injured and % leaf area affected varied with location, but overall were very similar all three years of the study (Table II).

The presence of ozone induced symptoms on black cherry during all three years of the investigation in both parks points to the sensitivity of this species to ambient ozone exposures commonly experienced in this region of the United States (Lefohn et al., 1997).

TABLE II

Incidence (% trees injured, TREEINJ), and % leaves injured (PI) and % leaf area affected (PA) on ozone symptomatic black cherry in Great Smoky Mountains National Park (GRSM) and Shenandoah National Park (SHEN) from 1991-1993

| Location[a] | 1991 | | | 1992 | | | 1993 | | |
|---|---|---|---|---|---|---|---|---|---|
| | TREEINJ | PI | PA | TREEINJ | PI | PA | TREEINJ | PI | PA |
| CM (GRSM) | 97 | 41 | 2 | 67 | 36 | 5 | 57 | 11 | 1 |
| BM (SHEN) | 87 | 29 | 8 | 75 | 26 | 3 | 82 | 27 | 5 |
| LR (GRSM) | 50 | 18 | 2 | 20 | 32 | 6 | 33 | 16 | 2 |
| DR (SHEN) | 40 | 13 | 2 | 13 | 20 | 1 | 13 | 35 | 6 |
| TC (GRSM) | 33 | 20 | 2 | 17 | 13 | 1 | 30 | 12 | 1 |
| SR (SHEN) | 7 | 7 | 2 | 10 | 6 | 1 | 0 | 0 | 0 |

[a] CM= Cove Mountain; BM = Big Meadows; LR = Look Rock; DR = Dickey Ridge; TC = Twin Creeks; SR = Sawmill Run.

Elevational gradients of ozone exposures and foliar injuries on black cherry were clearly evident, with increasing injury on higher elevation plots and significantly less injury on black cherry found at lower elevations (Figure 1). Highly significant ($p \leq 0.001$), positive correlations were found between elevation and % of injured trees, SUM0, SUM06 and W126 (Table III). These results compare favorably with many previous reports (e.g. Skelly et al., 1983; Winner et al., 1989) and further confirm the importance of understanding ozone exposure regimes at forested sites where the presence or absence of foliar symptoms is being evaluated.

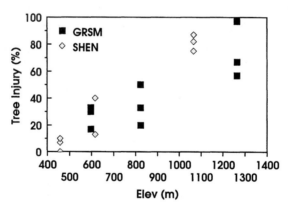

Fig. 1. Percent ozone-injured trees with increasing elevation in GRSM and SHEN from 1991-1993.

Hildebrand et al. (1996) observed significant ozone-exposure/plant response relationships for black cherry in SHEN. Chappelka et al. (1999) however, found no significant relationships with the ozone-exposure/plant response data for any species tested in GRSM. By combining data from both parks the relationships with the exposure indices and response variables were strengthened as shown in Figure 2. More data points were

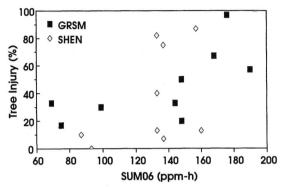

Fig. 2. Percent ozone-injured trees with increasing SUM06s in GRSM and SHEN from 1991-1993.

included in the analysis (doubled from 9 to 18). Both the SUM06 and W126 exposure statistics performed well and were positively related ($p \leq 0.01$) to the % of injured trees (Table III).

These results were expected since both variables were highly correlated with each other ($r = 0.99$). This suggests that higher ozone exposures are more important than lower exposures in eliciting symptoms on sensitive trees, similar to findings by other investigators (Hildebrand et al., 1996; Lefohn et al., 1988). However, PLA was significantly correlated with SUM0, indicating that symptom development can occur below a threshold of 0.06 ppm.

TABLE III

Spearman correlation coefficients and probabilities for comparing selected variables from Great Smoky Mountains National Park (GRSM) and Shenandoah National Park (SHEN) from 1991-1993

| Variable | TREEINJ | TLA | ELEV | SUM0 | SUM06 |
|---|---|---|---|---|---|
| TREEINJ[a] | --- | --- | --- | --- | --- |
| TLA | 0.55* | --- | --- | --- | --- |
| ELEV | 0.87*** | 0.58* | --- | --- | --- |
| SUM0 | 0.52* | 0.52* | 0.78*** | --- | --- |
| SUM06 | 0.72** | 0.40 | 0.82*** | 0.80*** | --- |
| W126 | 0.72** | 0.46 | 0.85*** | 0.82*** | 0.99*** |

[a] TREEINJ = % injured trees; TLA = total leaf area injured, calculated by multiplying % injured leaves by % leaf area; SUM0 = total cumulative ozone concentration (ppm-h); SUM06 = cumulative ozone concentration $\geq 0.06$ ppm; W126 = cumulative ozone concentrations weighted by a sigmoidal function (Lefohn et al., 1988); *** = $p \leq 0.001$; ** = $p \leq 0.01$; * = $p \leq 0.05$; n =18.

Heck et al. (1998) reported the results of a US EPA-sponsored workshop on identification of the state of knowledge and future research efforts, regarding ozone effect research in the USA. One of the major data gaps was the validation of ozone-exposure/plant response functions under ambient conditions. The data reported here will help alleviate this information gap.

More recent studies of black cherry sensitivity to ambient ozone exposures continue to demonstrate the valued use of this species as a bioindicator of ozone pollution (Fredericksen et al., 1995, 1996; Skelly et al., 1997; Innes et al., 1996; Lee et al.,1999). Clonal propagation of sensitive and tolerant genotypes may become important as introduced sentinels in the context of bioindicator system development (Manning, 1993). Results from our combined study indicate that although ozone-exposures may be at phytotoxic levels, other factors such as sample numbers, potential alterations in host sensitivity with elevation, and micro-site differences (water availability, light, etc.) cannot be discounted (Hildebrand et al., 1996; Chappelka and Samuelson, 1998; Chappelka et al., 1999).

## Acknowledgments

The authors would like to thank Barton's Tree Service, E. Robbins, J. Coker, J. Ferdindand, B. Moore (Trees for Life), E. Mudd, and J. Savage for help in data collection and analysis. This research was supported by funds provided by the United States Department of Interior National Park Service Cooperative Agreements 4000-9-8004, Supplementary Agreement 16, and 5460-2-9009.

## References

Chappelka, A.H., Renfro, J.R. and Somers, G.L.: 1994, *Proc. 87$^{th}$ Ann. Meet. Of Air and Waste Mange. Assoc.,* **94-TA36.04**, *Cincinnati, OH.*
Chappelka, A., Somers, G. and Renfro, J.: 1999, *Water, Air, Soil Pollut.* (this volume).
Chappelka, A., Renfro, J. and Somers, G.: 1997, *Environ. Pollut.* **95**, 13.
Chappelka, A.H. and Samuelson, L.J.: 1998, *New Phytol.* **139**, 91.
Davis, D.D. and Skelly, J.M.: 1992, *Water, Air, Soil Pollut.* **62**, 269.
Duchelle, S.F. and Skelly, J.M.: 1981, *Plant Dis.* **65**, 661.
Fredericksen, T.S., Joyce, B.J., Skelly, J.M., Steiner, K.C., Kolb, T.E., Kouterick, K.B., Savage, J.E. and Snyder, K.R.: 1995, *Environ. Pollut.* **89**, 273.
Fredericksen, T.S., Skelly, J.M., Steiner, K.C., Kolb, T.E. and Kouterick,K.B.: 1996, *Environ. Pollut.* **91**, 53.
Heck, W.W., Furiness, C.S., Cowling, E.B. and Sims, C.K.: 1998, *Environ. Mgt.* (**10**), 11.
Hildebrand, E.S., Skelly, J.M. and Fredricksen, T.: 1996, *Can. J. For. Res.* **26**, 658.
Horsfall, J.G. and Barratt, R.W.: 1945, *Phytopathology* **35**, 655.
Innes, J.L., Skelly, J.M., Landolt, W., Hug, C., Snyder, K.R. and Savage, J.E.: 1996, *Conf. Paper N. 15.* Vienna, Bundedministerium fur Umwelt, Jugend und Familie.
Lee, J.C., Skelly, J.M., Steiner, K.C., Zhang, J. and Savage, J.E.: 1999, *Environ. Pollut.* (In press).
Lefohn, A.S., Jackson, W., Shadwick, D.S. and Knudsen, H.P.: 1997, *Atmos. Environ.* **31**, 1695.
Lefohn, A.S., Laurence, J.A. and Kohut, R.J.: 1988, *Atmos. Environ.* **21**, 1229.
Manning, W.J.: 1993, *Proc. 86$^{th}$ Ann. Meet. of Air and Waste Mange. Assoc.,***93-WA80.01***, Denver, CO.*
Nash, B.L., Saunders, M.C., Miller, B.J., Bloom, C.A., Davis, D.D. and Skelly, J.M.: 1992, *Can. J. For. Res.* **22**, 1770.
Neufeld, H.S., Renfro, J.R., Hacker, W.D. and Silsbee, D.: 1992, in *Tropospheric Ozone and the Environment II*, Berglund, R.D. (ed.), pp. 594-617, Air and Waste Mange. Assoc., Pittsburgh, PA.
SAS Institute Inc.: 1985, *SAS/STATTM user's guide. Version 6.07*, Cary, NC.
Simini, M., Skelly, J.M., Davis, D.D. and Savage, J.E.: 1992, *Can. J. For. Res.* **22**, 1789.
Skelly, J.M., Yang,Y.S., Chevone, B.I., Long, S.J., Nellessen, J.E. and Winner, W.E.: 1983, in *Proc. Air Pollution and the Productivity of Forests*, Davis, D.D., Millen, A.A., Dochinger, L. (eds.), pp. 143-159, Izaak Walton League.
Skelly, J.M., Savage, J.E., deBauer, M. and Alvarado, D.: 1997, *Environ. Pollut.* **95**, 155.
United States Environmental Protection Agency (US EPA): 1996, *Air Quality Criteria for Ozone and Other Photochemical Oxidants, Vol. II,* EPA/600/P-93/004bF.
Winner, W.E., Lefohn, A.S., Cotter, I.S., Greitner, C.S., Nellessen, J., Mc Evoy, L.R., Olson, R.L., Atkinson, C.J. and Moore, L.D.: 1989, *Proc. Nat'l Acad. Sci.* **86**, 8828.

# FOLIAR SYMPTOMS AND GROWTH REDUCTION OF *Ailanthus altissima* Desf. IN AN AREA WITH HIGH OZONE AND ACIDIC DEPOSITION IN ITALY

E. GRAVANO[1], M. FERRETTI[2], F. BUSSOTTI[1] and P. GROSSONI[1]

[1]*Dept. of Plant Biology, Piazzale delle Cascine 28 50144 Firenze, Italy;* [2] *LINNÆA ambiente srl, Via Sirtori 37 50137 Firenze, Italy;*

(Received 25 September 1998; accepted 26 February 1999)

**Abstract.** Since 1985, severe defoliation and foliar symptoms have been observed on *Ailanthus altissima* Desf. ramets located at a site where high ozone concentration (up to 128 $nl\,l^{-1}$), strongly acidic drizzle and dew (to pH 1.4) were recorded. The behaviour of plants growing in an open field (directly exposed to the pollutants), and below the canopy of an old stone pine stand was compared. Compared to the below-canopy trees, the open field trees show a variety of foliar symptoms (diffuse yellowing, apical, marginal and spot-like necrosis and interveinal bronzing and stippling), defoliation and much less vegetative vigour. Experimental treatments with acidic mist caused the same necrosis observed in the field, but failed to reproduce bronzing and stippling. Since then, stipple and bronzing have been observed again at a variety of locations in the same area as well as throughout Italy, while necrosis has only seldom been recorded. Recently, surveys on symptoms have been supported by anatomical analysis by mean of light microscope and by experiments in fumigation chambers. On the whole, the results led us to suspect ozone sensitivity. These findings suggest that the symptoms observed since 1985 in the surveyed area were probably caused by the combined effects of both strongly acidic episodes and exposure to high ozone levels.

**Keywords**: acidic deposition, *Ailanthus altissima* Desf., bioindicators, foliar symptoms, ozone.

## 1. Introduction

In the early 80s, unusual and apparently unexplained yellowing, necrosis and subsequent leaf shedding were observed on *Ailanthus altissima* Desf. plants located in an area where strongly acidic precipitation was known to occur (e.g. Gellini *et al.*, 1985). In 1985 and 1986 studies were undertaken to clarify the nature of the acidic deposition and its relationship to the severe decline observed. Experimental investigations confirmed that strongly acidic episodes actually caused some of the symptoms observed in the field. Additional symptoms (bronzing, stippling) observed under field conditions were not reproduced by the experimental tests with acidic episodes (Ferretti, unpublished). Such symptoms were also repeatedly observed at several locations. Bronzing and stippling closely resemble the "ozone-like" symptoms recently observed (and subsequently experimentally reproduced) on a variety of broad-leaved trees in Southern Europe (e .g. Skelly *et al.*, 1998). Ozone is thought to be the most hazardous pollutant for vegetation in the Mediterranean region, where its formation and accumulation in the atmosphere is enhanced by the environmental and climatic conditions (Butkovic *et al.*, 1990; Fortezza *et al.*, 1993; Millán *et al.*, 1996). Researchers have known for some time that ozone damages photosynthesis processes, producing visible symptoms in the needles of several species of *Pinus* growing in the southern United States (Fox and

Mickler, 1995) and on the Mediterranean *Pinus halepensis* Mill. (Kärenlampi, 1987; Velissariou *et al.*, 1992; Davison *et al.*, 1995). For broad-leaved trees, symptoms for several forest species in the eastern United States have been described (Skelly *et al.*, 1987). The results of a survey carried out in southern Switzerland both in the open field and with open-top chambers (OTC) have also been reported (Skelly *et al.*, 1998). In broad-leaved trees symptoms usually consist of small red or brown stipples appearing on the interveinal portions of the leaf, which can extend to cover a large section of the lamina. This most recent evidence led us to reconsider the early findings in a different perspective, e.g. as a possible combination of symptoms, partly caused by strongly acidic episodes and partly by high ozone levels. With this in mind, the original unpublished findings, consisting of descriptions of symptoms, experiments, photographs and increment measurements have been re-interpreted in this paper. Deposition measurements and ozone levels for the study area were published by Barbolani *et al.* (1986) and Lorenzini *et al.* (1994). We refer to these papers for a better characterisation of the pollution climate in this area.

## 2. Materials and Methods

### 2.1 STUDY SITE AND SAMPLE PLANTS

The study area is located within the San Rossore Estate (Pisa, Italy). The characteristics of the area have been described in several papers (e.g. Gellini *et al.*, 1983; Barbolani *et al.*, 1986). It is a flat, sandy area near the Tyrrhenian coast, with the ground water table close to the surface. The mean annual precipitation is 908 mm with a Mediterranean regime; the mean annual temperature is 15.1°C. The sample plants were young root suckers 3-6 years old (ramets) within a stand growing at about 5 km from the coast. During the course of 2 growing seasons (1985 and 1986) two groups of 5 ramets each were measured monthly and assessed weekly: one of the two groups was obviously damaged and was located in a clearing where it was exposed to open field condition (group E), whereas the other group was apparently healthy and was sheltered by the canopy of a mature *Pinus pinea* L. stand (group S). The observations recorded included: longitudinal growth of main stem, the number of leaves and a description of the visible foliar symptoms.

### 2.2 ATMOSPHERIC PRECIPITATION AND OZONE

Table I shows the pH of drizzle and dew (volume <1 mm) in the open field and in throughfall. Strongly acidic events associated with wash-out processes were obvious, with HCl being the major contributor to the acidity (Barbolani *et al.*, 1986). It is evident that only the exposed trees (open field) were subjected to extreme episodes: the amount of deposition was very low in these events and, in the throughfall, these small rain volumes were completely retained in the pine canopy; moreover the pH was well buffered in the throughfall (Barbolani *et al.*, 1986). Because the pH is weighted by the amount of precipitation, at the end of the year the mean pH of the throughfall was not very different from the mean rainfall pH. Ozone levels were monitored by Lorenzini *et al.* (1994) over the period 1986-1991 (Table II). High ozone concentrations were recorded, and the occurrence of widespread foliar symptoms on the sensitive tobacco cultivar (*Nicotiana tabacum* Bel W3) confirmed that large areas of coastal Tuscany experienced ozone concentrations above 40 nl l$^{-1}$ (e.g. Lorenzini and Panattoni, 1986). However, ozone levels were generally fairly low below the forest canopy (Dell'Era *et al.*, 1998).

Table I
Individual pH values of the drizzle and dew collected at the two sites and the mean annual pH of the precipitation (all types included) in 1985.
S: Sheltered site; E: exposed site. -: no deposition in the collector (after Barbolani et al., 1986).

| Date | Open field (site E) | Throughfall (site S) |
|---|---|---|
| May 2 | 2.85 | - |
| June, 5 | 2.17 | 6.15 |
| June, 14 | 2.50 | - |
| September, 13 | 1.40 | - |
| September, 27 | 2.05 | - |
| October, 6 | 2.02 | - |
| Annual weighted mean | 3.48 | 3.79 |

Table II
Values of some ozone descriptors at a location 5 km south of the study site in the period 1986-1991.
M1: maximum hourly peak; M7: maximum mean concentration between 9 am and 4 pm
(after Lorenzini et al., 1994).

| Ozone descriptor | M1 max, nl l$^{-1}$ | M1 mean, nl l$^{-1}$ | M7 mean, nl l$^{-1}$ | No. of days above 100 nl l$^{-1}$ |
|---|---|---|---|---|
| | 128 | 60 | 53 | 17 |

## 2.3 MICROSCOPY

Light microscopy tests were performed on samples of leaves with and without symptoms using the following techniques (Jensen, 1962): Schiff reagent plus Aniline Black B for carbohydrates and proteins; Aniline Black B for callose; Fluoroglucinol for the lignified walls; primary blue light fluorescence (450-490 nm) for chlorophyll; primary UV fluorescence (350-390 nm) for lignin.

## 2.4 REPRODUCTION OF SYMPTOMS

The experimental reproduction of symptoms considered: (i) the possible role of extremely acidic events (such as those that have been observed at the study site): these tests were performed in 1986 by spraying 4 groups of young (3-6 year-old) root-ramets with asymptomatic leaves with deionized water (control) and 3 solutions with HCl at pH=2.5; 2; 1.5 (treatments were made once, until the leaves were fully wetted); (ii) the possible role of ozone: current year ramets of *Ailanthus altissima* were grown in a greenhouse under controlled conditions, and then exposed to ozone fumigation in 1998. The tests in fumigation chambers used 50 and 100 nl l$^{-1}$ of ozone for 5 hours a day for 15 days. These are still underway in collaboration with the DCDSL in Pisa, and only preliminary findings will be reported here.

## 3. Results

### 3.1 1985-1986 STUDY

The symptoms observed on the leaves developed in the following way: (1) marginal and apical chlorosis, (2) curling of the leaflets, (3) yellowing, (4) development of brown patches and (5) necroses. Whitening of the margins and of the apical portions of the leaflets was also observed. Chlorosis was already visible in late spring (May) in the leaves of the trees exposed directly to pollutants. The evolution of the symptoms was much slower and less severe in the sheltered trees. In some cases the development of the symptoms accelerated immediately after highly acidic drizzle. Starting in July, small brownish stipples developed in the interveinal parts of the abaxial side of the leaflets. The individual leaflets of the composite leaves were affected and were prematurely shed, starting from the older leaves. In general, the taller the ramet, the more pronounced were the symptoms (Table III). Damage was widespread in the area and its quantification was obvious: all the exposed trees were affected, whereas only a few leaves of the sheltered trees showed the same symptoms.

Table III
Type of symptom, its target and its first record in 1985. Lv=leaves; lf=leaflets

| Symptom | Target on the foliage | Target on the plant | Target on the population | Day of first observation |
|---|---|---|---|---|
| Chlorosis + whitening | marginal /apical parts of lf | Older lv/lf | tallest ramets | May, 11 |
| Idem + curling | marginal /apical parts of lf | Older lv/lf | tallest ramets | May, 16 |
| Yellowing + defoliation | entire lf and lv | Older lv/lf | tallest ramets | May, 29 |
| Necrosis | parts (spot-like) of lf | Older lv/lf | tallest ramets | June, 5 |
| Bronzing, stippling | interveinal parts of lf | Older lv/lf | tallest ramets | July, 22 |

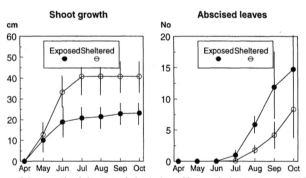

Fig.1 - Mean values of the growth of shoots and abscission of leaves during two years (1985-86) in individual trees, exposed and sheltered

Figure 1 shows the trend of shoot growth and leaf shedding. In exposed plants the longitudinal growth stopped prematurely and the length increment was reduced in comparison with the sheltered ones. Differences between exposed and sheltered trees became significant (P=0.003) in the late spring (June), and the significance of difference was maximum in July (P=0.0003). The number of new leaves produced in the trees of the two groups was basically identical (16-17), but in exposed trees leaves were shed earlier and more abundantly than in the sheltered ones. Loss of leaves began during the summer; differences between exposed and sheltered trees became significant in July(P=0.068); the significance was maximum in August (P=0.0016) and decreased during the fall (P=0.016 in October).

The effects of simulated acid mists are reported in Table IV. The pH 2 mist (i.e. at a level within the range observed in the open field) produced spot-like necroses after 48 hours in the interveinal areas, whereas pH 1.5 caused broad marginal necroses. However, stippling and bronzing were not recorded.

Table IV
Effects of simulated acidic mists on 3-6 years old Ailanthus altissima. 1996-1998 study

| Treatment | Symptoms after 48 hours | Target on the foliage |
|---|---|---|
| Control | no symptoms | - |
| pH=2.5 | curling | marginal parts of the leaflets |
| pH=2.0 | curling+spot necrosis | interveinal parts of the leaflets |
| pH=1.5 | curling+extensive necrosis | Interveinal, marginal and apical parts of the leaflets |

3.2 1996-1998 STUDY

A new survey was undertaken at several locations in and nearby the 1985-86 study site. Several visits were carried out in late summer 1996, 1997 and 1998. These visits concentrated on the brown stippling of the leaflets, which was recognised to be identical to that observed in 1985-86. Based on the 1996-97 observations, histology tests were performed in 1997 on the brownish leaflets, which was the symptom observed in the open field and not reproduced by the artificial mists. In symptomatic leaves, primary fluorescence tests revealed that it was principally the palisade mesophyll cells that were affected: in these cells the chlorophyll had been almost entirely degraded, the cytoplasm content had deteriorated and the organelles were no longer recognisable. A manifestation of damage was the presence of callose, deposited on the walls of the mesophyll cells, that coincided with necrotic areas. In the later stages the callose was no longer present and the walls were partially lignified.

The preliminary results of the ozone fumigation experiments suggest that exposure to 50-100 $nl \cdot l^{-1}$ for 5 hours a day for 15 days can produce ivory white stipples which later evolve into small brown stipples on the leaflets of current year ramets.

## 4. Discussion and Conclusions

Leaf symptoms observed in the open field were confirmed by experimental exposure of *Ailanthus* to simulated acidic mists and ozone fumigation. Yellow patches and marginal necroses appearing in late spring 1985 were attributable to the acidity of the fine drizzle and dew. We also believe that the necrotic spots which accompanied the first manifestations of the damage were also due to acidity. The possible contribution of gaseous HCl cannot be excluded. Bronzing and stippling were observed only later in the summer, at a time when ozone exposure was probably more important. Ozone exposure causes bronzing and stippling and can eventually speed up the premature shedding of the leaf (Treshow and Anderson, 1989).

We have shown that extremely acidic drizzle and dew is a very localised phenomenon (Barbolani *et al.*, 1986), whereas ozone is a regional-scale pollutant. The response of *Ailanthus altissima* to these pollutants, singly and in combination, led to a complex symptomatology. The preliminary findings on treatments performed in controlled environments confirm the fact that this species is highly sensitive to ozone exposure. Foliar symptoms occur at ozone doses that are quite common in the Mediterranean region. The structural features of *Ailanthus altissima* leaves (highly hydromorphic, with a simple palisade and abundant intercellular spaces) are considered typical of those species or individuals which are highly sensitive to ozone (Bennett *et al.*, 1992; Pääkkönen *et al.*, 1995), since they are typical of species with high gas exchange rate. In addition, its remarkable hardiness and the speed at which it grows suggest that it must have a continuous and intense metabolic activity even during the summer, thus being subjected to the absorption of high doses of ozone.

However, it appears likely that there is a genetic variation within the species that has genotypes with different levels of sensitivity. The identification of sensitive clones (which can easily be vegetatively propagated) would provide a valid bioindicator for ozone in Mediterranean-type areas.

## Acknowledgments

We thank the Regione Toscana and the San Rossore Estate for collaboration and logistic support. The authors are also grateful to Giacomo Lorenzini and Cristina Nali (DCDSL, University of Pisa) for the experiments in fumigation chambers.

## References

Barbolani, E., Del Panta, S., Ferretti, M. and Pantani, F.: 1986, *Ann.Chim.*, **76**, 365-374.
Bennett, J.P., Rassat, P., Berrang, P. and Karnosky, D.F.: 1992, *Environ. Exptl Bot.*, **32**, 33-41
Butkovic, V., Cvitas, T. and Klasing, L.: 1990, *Sci. Tot. Environ.* **99**, 145-51.
Davison, A.W., Velissariou, D., Barnes, J.D., Gimeno, B. and Inclan, R.: 1995, in *Bioindicators of Environmental Health,* Munawar, M. *et al.* (eds.), pp. 63-72, SPB Academic Publishing, Amsterdam.
Dell'Era, R., Brambilla, E. and Ballarin-Denti, A.: 1998, *Chemosphere*, **36**, 1083-1088.
Fortezza, F., Strocchi, V., Giovannelli, G., Bonasoni, P. and Georgiadis, T.: 1993, *Atmos. Environ.* **27A**, 2393-2402.
Fox, S. and Mickler, R.A. (eds.): 1995, *Impact of air pollutants on southern pine forests.* Ecological Studies 118. Springer Verlag, Berlin, Germany.
Gellini, R., Pantani, F., Grossoni, P., Bussotti, F., Barbolani, E. and Rinallo C.: 1983, *Eur. J. For. Pathol.* **13**, 296-304.
Gellini, R., Pantani, F., Grossoni, P., Bussotti, F., Barbolani, E. and Rinallo C.: 1985, *Eur. J. For. Pathol.* **15**, 147-157.
Jensen, W.A.: 1962, *Botanical Histochemistry. Principles and Practice.* W.A. Freeman and Company. San Francisco, USA.
Kärenlampi, L.: 1987, *Savonia*, **9**, 1-12.
Lorenzini, G., Nali, C. and Panicucci, A.: 1994, *Atmos. Environ.* 38, 51-59.
Lorenzini, G. and Panattoni A.: 1986, in *Environmental Quality and Ecosystem Stability*, Bar-Ilan University Press, Ramat-Gan (Israel). Vol. IIIB, 511-517.
Millán, M., Salvador, R., Mantilla, E. and Artiñano, B.: 1996, *Atmos. Environ.* **12**, 1909-24.
Pääkkönen, E., Metsärinne, S., Holopainen, T. and Kärenlampi, L.: 1995, *New Phytol.* **132**, 145-154
Skelly, J.M., Davis, D.D., Merrill, W., Cameron, E.A., Brown, H.D., Drummond, D.B. and Dochinger, L.S. (eds.): 1987, *Diagnosing injury to eastern forest trees.* University Park, Penn State University, College of Agriculture. Pennsylvania, USA.
Skelly, J.M., Innes, J.L., Snyder, K.R., Savage, J.E., Hug, C., Landolt, W. and Bleuler, P.: 1998, *Chemosphere*, **36**, 995-1000.
Treshow, M. and Anderson, F.K.: 1989, *Plant stress from air pollution.* John Wiley & Sons, Chichester, UK.
Velissariou, D., Davison, A.W., Barnes, J.D., Pfirmann, T., MacLean, D.C. and Holevas, C.D.: 1992, *Atmos. Environ.* **26**, 373-80.

# THE RELATIVE SENSITIVITY OF DIFFERENT MEDITERRANEAN PLANT SPECIES TO OZONE EXPOSURE

R. INCLÁN[1], A. RIBAS[2], J. PEÑUELAS[2] AND B. S. GIMENO[1]

[1] *Depto. Impacto Ambiental de la Energía, CIEMAT. Avda Complutense 22, 28040 Madrid, Spain.*
[2] *CREAF, Universitat Autonoma de Barcelona, 08193 Bellaterra, Spain.*

(Received 25 September 1998; accepted 26 February 1999)

**Abstract.** An experimental study was performed in open-top chambers to assess the relative $O_3$-sensitivity of the following Mediterranean woody plants: *Quercus ilex* ssp. *ilex*, *Quercus ilex* ssp. *ballota*, *Olea europaea* ssp. *sylvestris*, *Ceratonia siliqua* and *Arbutus unedo*. Assessment of $O_3$ 'sensitivity' was based on the development of $O_3$-induced visible injury and the extent of the reduction in relative growth rate induced by $O_3$. Two-year-old seedlings were exposed in open-top chambers to charcoal filtered air (CFA), Non-filtered air (NFA) or Non-filtered air supplied with 40 ppb $O_3$ from 09:00 to 18:00 hours (NFA+40). No clear relationships were found between the development of visible injury and effects on plant growth or accumulated biomass. *Olea europea* ssp. *sylvestris* and *Quercus ilex* ssp. *ballota* appeared to be the most $O_3$-sensitive tree species screened; a ten month exposure to ambient $O_3$ levels (AOT40 = 12.4 ppm.h) caused reductions in height and/or stem growth. Further research is needed to characterise the physiological, biochemical and anatomical characteristics that may underlie the 'sensitivity' of certain Mediterranean species to $O_3$

**Keywords**: Mediterranean, woody species, ozone, critical levels

## 1. Introduction

In the last ten years, several attempts have been made to refine ozone ($O_3$) critical levels for plant receptors. In 1996, at a UN/ECE workshop in Kuopio (Finland), the $O_3$ critical level for forest trees was set at an AOT40 value (accumulated exposure over a threshold of 40 ppb) of 10 ppm.h for a six-month period (Kärenlampi and Skärby, 1996), corresponding to a predicted 10% loss in plant biomass or growth. However, this critical level was based upon a very limited database for *Fagus sylvatica*, *Betula pendula*, *Picea abies*, *Pinus sylvestris* and *Quercus robur*, and it was acknowledged that there is a paucity of data relevant to assessments of the risk posed by $O_3$ to woody vegetation in southern Europe (Elvira *et al.*, 1995; Elvira and Gimeno, 1996; Elvira *et al.*, 1998; Inclán *et al.*, 1998). There is a clear need to fill this gap in knowledge, since ambient levels of $O_3$ commonly exceed UN-ECE critical level guidelines for the protection of vegetation in the Mediterranean region (Velissariou *et al.*, 1992; Gimeno *et al.*, 1996; Millán *et al.*, 1996). This paper therefore presents the results of a screening study aimed at assessing the sensitivity of several Mediterranean species common to the Iberian Peninsula.

## 2. Materials and Methods

One-year-old plants of *Quercus ilex* ssp. *ilex*, *Quercus ilex* ssp. *ballota*, *Olea europaea* ssp. *sylvestris*, *Ceratonia siliqua* and *Arbutus unedo*, were raised from seed originating from eastern Spain (*Quercus ilex* ssp. *ilex*; *Arbutus unedo*) and Southern Spain (*Quercus ilex* ssp. *ballota, Olea europaea* ssp. *sylvestris*). These species were selected due to their wide coverage in the Iberian peninsula. Seedlings were selected to ensure phenotypic homogeneity within a given taxon, and transferred in September 1996 to 225 $cm^3$ pots and enclosed in open-top chambers (OTCs). One month later seedlings were transplanted in to 3 $dm^3$ containers filled with 50% peat, 30% sand and 20% natural soil from the local area. Soil pH was adjusted to 7.0 by adding $CaCO_3$, and a slow-release fertilizer incorporated (NPK 15:8:11; Osmocote plus) plus additional magnesium, iron, copper, manganese, molybdenum, boron and zinc. Plants were irrigated twice weekly using a droplet irrigation system; soil volumetric water content was measured using Time Domain Reflectometry (TRIME, IMKO, Micromodultechnick, Germany).

Seedlings were exposed to Charcoal Filtered Air (CFA), Non-Filtered Air (NFA) or Non-Filtered Air plus 40 ppb (nl $l^{-1}$) $O_3$ from 09:00 to 18:00 hours (NFA+40). The latter level was targeted to achieve maximum $O_3$ concentrations close to 100 ppb, which are in the range of the levels recorded in the Spanish Mediterranean area (Elvira *et al.*, 1998). Supplemental $O_3$ was produced by passing oxygen through a high-voltage electrical discharge generator (Fisher mod. 503, Germany). Air quality inside the OTCs was continuously monitored using an automated time-sharing system which diverted the airflow from the chambers at regular intervals to a bank of regularly-calibrated monitors; $O_3$ (Dasibi model 1008 RS, USA), sulphur dioxide (Dasibi model 2108, USA) and nitrogen oxides (Dasibi model 4108, USA). Further information concerning ozone dispensing and monitoring systems has been described elsewhere (Elvira *et al.*, 1998).

Three replicate chambers per treatment, with 36 plants in total of each species, were employed. However, for *Arbutus unedo* only 18 plants per treatment were studied, distributed between OTCs supplied with either CFA or NFA+40 treatments.

The extent of $O_3$-induced visible injury was recorded at intervals throughout the experiment. Plant height and stem diameter were measured both on the introduction of the plants into the chambers and ten-months later (July 1997). Stem diameter was measured at two locations at the base of the shoot of each plant using a digital calliper (Digit Cal SM, model Tesa, USA). No bias in these two parameters was evident for any species prior to the onset of $O_3$ treatments.

The diameter of trees from each chamber was aggregated prior to performing statistical analyses. Normal distribution and homogeneous variance of the data were tested by Kolmogorov-Smirnov and Cochran's test, respectively, and subsequently data were subjected to ANOVA. Significant differences between means were determined using the least significance difference test (LSD) calculated at the 5 % level.

## 3. Results and discussion

Ambient $O_3$ concentrations surpassed the critical level for forest tree species in just 5 months (March to July) of exposure. The maximum ambient monthly AOT40 values were found in April and May, in the range of 3-4 ppm.h; the accumulated AOT40 value in the NFA treatment throughout the experimental period (September 1996 to July 1997) was 12.4 ppm.h (see Table I). The maximum hourly average $O_3$ concentration during this period was registered in July 1997 (91 ppb), but high $O_3$ concentrations in the range of 70-90 ppb were also experienced during the autumn and winter. Charcoal filtration efficiently reduced $O_3$ levels in the CFA chambers, with the result that the AOT40 and AOT60 values for this treatment were zero. Ozone AOT40 and AOT60 values in the NFA+40 treatment during the experimental period were 79.8 and 37.4 ppm.h, respectively, while the maximum hourly average $O_3$ concentration was 130 ppb. Seasonal average (January-July) $SO_2$ and NO and $NO_2$ concentrations were 1, 1.3 and 3 ppb, respectively.

TABLE I

Ozone exposure in the different ozone treatments from June 1996 to July 1997

| TREATMENT | AOT30 (ppm.h) | AOT40 (ppm.h) | AOT60 (ppm.h) | Maximum hourly mean (ppb) | 10h (07 – 17 h GMT MEAN (ppb) |
|---|---|---|---|---|---|
| CFA | 0.27 | 0 | 0 | 40 | 14 |
| NFA | 37.2 | 12.4 | 0.73 | 91 | 38 |
| NFA+40 | 119 | 79.8 | 37.4 | 130 | 61 |

Ozone-induced visible injury was observed (as dark pigmented stipples) on the adaxial surface of leaves of *Quercus ilex* ssp. *ballota* and *Arbutus unedo* seedlings in the NFA+40 treatment. However, reductions in plant biomass were poorly correlated with the development of visible $O_3$ injury. For example, $O_3$ exposure resulted in the development of visible injury on the leaves of *A. unedo* but resulted in no significant changes in growth. In contrast, no visible symptoms of $O_3$ injury developed on the leaves of *Olea europaea*, but its growth was found to be significantly reduced following $O_3$ exposure. This finding is in agreement with the poor correlation between visible $O_3$ injury and effects on growth reported by other researchers (see Chappelka and Samuelson, 1998; Davison and Barnes, 1998).

*Quercus ilex* ssp. *ballota* and *Olea europaea* ssp. *sylvestris* proved to be rather sensitive to $O_3$; exposure to ambient levels of $O_3$ (12.4 ppm.h) causing reductions in stem growth and plant height (see Table II). In fact, these species appeared as 'sensitive' as *Fagus sylvatica* (see Kärenlampi and Skärby, 1996), and considerably more sensitive than *Pinus halepensis* (Elvira et al., 1998). After ten months of exposure, the height of *Olea europaea* seedlings grown in the NFA (AOT40 = 12.4 ppm.h) and NFA+40 (AOT40 = 79.8 ppm.h) was reduced by 12% and 20 % ($p<0.05$), respectively, when compared to their counterparts raised in CFA. Similarly, seedlings grown in OTCs supplied with NFA+40 showed a 22% reduction in stem diameter when compared to those grown in CFA. After ten months of exposure, stem diameter was reduced by a 22-31% in *Quercus ilex* ssp. *ballota* plants grown in the NFA and NFA+40 chambers when compared to those in CFA chambers. Ozone exposure did not cause any significant changes in the growth of the remaining species (see Table II). These results indicate that $O_3$ levels in the range of those reported in Spain by several authors (Gimeno et al., 1996; Milán et al., 1996; Sanz and Millán, 1998) may result in adverse effects on the growth of sensitive plant species, confirming reports of the potential phytotoxicity of $O_3$ in Spain (Gimeno et al., 1995; Ribas et al., 1998)

The observed results are in agreement with the critical levels established in Kuopio for tree species i.e. an AOT40 value of 10 ppm.h (Kärenlampi and Skärby, 1996). The design of the experiment was intended to simulate field conditions, as near as possible, and thus soil water volumetric content ranged between 40 and 75% for Q. ilex ssp. ballota and 20-40% for Arbutus unedo, based on the very different seasonal and diurnal patterns of leaf water potential observed in the field for these species - attributed to differences in their rooting depths (Castell et al., 1994). The experimental conditions would not, however, account for the extreme drought conditions that Mediterranean plants must endure every 15-20 years.

TABLE II

Effects of a ten month ozone exposure on several Mediterranean woody species.

| Species | Treatment | Height increment (mm) | Increment in stem diameter (mm) |
|---|---|---|---|
| Olea. Europaea | CFA | 513 ± 23 a | 4.84 ± 0.20 a |
| | NFA | 452 ± 18 b | 4.48 ± 0.20 a |
| | NFA+40 | 411 ± 20 c | 3.77 ± 0.20 b |
| Ceratonia siliqua | CFA | 86 ± 9 a | 1.14 ± 0.14 a |
| | NFA | 72 ± 11 a | 1.11 ± 0.18 a |
| | NFA+40 | 76 ± 10 a | 0.86 ± 0.16 a |
| Quercus ilex ssp. ballota | CFA | 121 ± 9 a | 2.51 ± 0.12 a |
| | NFA | 115 ± 11 a | 1.97 ± 0.15 b |
| | NFA+40 | 112 ± 11 a | 1.73 ± 0.14 b |
| Quercus ilex ssp. ilex | CFA | 94 ± 10 a | 1.52 ± 0.12 a |
| | NFA | 102 ± 8 a | 1.38 ± 0.17 a |
| | NFA+40 | 96 ± 9 a | 1.58 ± 0.14 a |
| Arbutus unedo | CFA | 91 ± 16 a | 2.26 ± 0.39 a |
| | NFA+40 | 66 ± 14 a | 1.96 ± 0.33 a |

Values indicate means ± standard errors. Within a given taxon, any two means having the same letter in common are not significantly different at the 5% level (LSD), n=12 for all species except A. unedo (n=9)

It is interesting to note the difference in the impacts of $O_3$ on the two studied subspecies of Q. ilex; ssp. ballota is more adapted to continental conditions than ssp. ilex, (see CSIC, 1990). Additional studies are required to characterise the physiological, biochemical and anatomical characteristics that may underlie such intra-specific variations in $O_3$ sensitivity, as well as interactions with atmospheric/soil drought under Mediterranean conditions (see Barnes et al. 1999).

## Acknowledgements

This research was funded by the EU contract LIFE97/ENV/FIN/000336 and Spanish CICYT projects CLI97-0735-C0302 and CLI97-0344. Angela Ribas was supported by Fundacio Territori i Paisatge de la Caixa Catalunya.

## References

Barnes, J.D., Gimeno B.S., Davison, A.W., Busotti F., Velissariou D. and Gerant D.: 1999, in *Ecology, Biogeography and Management of Pinus halepensis and Pinus brutia Forest Ecosystems in the Mediterranean Basin*, G. Ne'eman and I. Izhaki (eds), Backhuys, Leiden. (In Press)
Castell, C., Terradas, J. and Tenhunen, J.D.: 1994, *Oecologia* **98**, 201-211.
Chappelka, A.H. and Samuelson, L.J.: 1998, *New Phytol.* **139**, 91-108.
CSIC, Real Jardín Botánico.: 1990, *Flora Ibérica – Plantas vasculares de la Península Ibérica y Baleares*. Vol II. p. 20.
Davison, A.W. and Barnes, J.D.: 1998, *New Phytol.* **139**, 135-151.
Elvira, S., Alonso, R., Inclan, R., Bermejo, V., Castillo, F.J. and Gimeno, B.S.: 1995, *Water Air Soil Pollut.* **85**, 1387-1392.
Elvira, S. and Gimeno, B.S.: 1996, in *Critical Levels for ozone in Europe: Testing and finalising the concepts*. Kärenlampi, L. and Skärby, L. (eds.), pp. 169-182, University of Kuopio, Dept. of Ecol. and Environ. Sci.
Elvira, S., Alonso, R., Castillo, F. and Gimeno, B.S.: 1998, *New Phytol.* **138**, 419-432.
Gimeno, B.S., Salleras, J.M., Porcuna, J.L., Reinert, R.A, Velissariou, D. and Davison, A.W.: 1995, in *Bioindicators of Environmental Health*. Munawar, M.H, Hänninen, O., Roy, S., Munawar, S. and Kärenlampi, L., Brown, D. (eds.), pp. 52-62, SPB Academic Publishing, Amsterdam, The Netherlands.
Gimeno, B.S., Cabal, H., García, C., Artiñano, B., Vilaclara, E. and Guardans, R.: 1996, in *Critical Levels for ozone in Europe: Testing and finalising the concepts*. Kärenlampi, L. and Skärby, L. (eds.), pp. 228-233, University of Kuopio, Dept. of Ecol. and Environ. Sci.
Inclán, R., Alonso, R., Pujadas, M., Terés, J. and Gimeno, B.S., 1998, *Chemosphere*, **36**, 685-690.
Kärenlampi, L. and Skärby, L. (eds.): 1996, *Critical Levels for ozone in Europe: Testing and finalising the concepts*. University of Kuopio, Dept. of Ecol. and Environ. Sci.
Millán, M.M., Salvador, R., Mantilla, E. and Artíñano, B.: 1996, *Atmos. Environ.* **30**, 1909-1924.
Ribas, A, Filella, I., Gimeno, B.S. and Peñuelas, J.: 1998. *Water Air Soil Pollut.* **107**, 347-365.
Sanz, M.J. and Millán M.M.: 1998. *Chemosphere,* **36**, 1089-1094
Velissariou, D., Davison, A.W, Barnes, J.D., Pfirrmann, T. and Holevas, C.D.: 1992, *Atmos. Environ.* **26**, 373-380.

# BIOCHEMICAL PROCESSES IN AN ECOSYSTEM: HOW SHOULD THEY BE MEASURED?

ROBERT L. HEATH

*Department of Botany* and *Plant Sciences; University of California; Riverside, CA 92521-0124 U.S.A.*

(Received: 25 September 1998; accepted 26 February 1999)

**Abstract.** Urban air basins produce many pollutants, such as oxides of nitrogen (NOx) and ozone, which move into rural areas. NOx itself has been shown not to be especially phytotoxic except at high levels. More problematical is the conversion of NOx into inorganic nitrogen that the plant can utilize. In ecosystems which are nitrogen- limited, excess nitrogen can radically increase plant growth by stimulating the nitrogen-handling metabolic pathways which then alters "normal" carbon (C) metabolism. On the other hand, ozone alters plant productivity by inducing a decline in growth, speeding developmental events to senescence, lowering reproductive rates, and impairing resistance to other stresses. Three major metabolic events induced by ozone exposure have been well studied: 1) increased turnover of antioxidant systems, 2) production of symptoms similar to tissue wounding, especially ethylene production; and 3) decline in photosynthesis. In order to understand the effect upon productivity of any pollutant, one must understand the mechanisms by which that effect is generated. Those mechanisms must not be merely descriptive statements, but must be firmly modeled with predictions. Any model will be highly complex; however here, the focus will be upon C metabolism. Both a closure of the stomata and a decline in the ability to fix $CO_2$ within the chloroplast can alter the photosynthetic rate of leaves. Yet the movement of C throughout the plant (translocation) seems also to be altered by pollutants. The loss of carbohydrate to roots and growing shoot tips has a profound effect upon the plant's ability to respond normally to the integrated effects of the total environment; loss may cause the most pronounced and long lasting dilemma for plants.

**Keywords**: Ozone, Plant Biochemistry, Stomatal Conductance, Oxides of Nitrogen, Antioxidants, Calcium and Signal Transduction.

## 1. Introduction

In order to evaluate any risk to a system, it is critical to define what is meant by the risk and to list all the processes which go into the formation of, reaction to, and counter-action to that risk. For biological organisms those processes can be multiple and complex. In addition, biological organisms have both plasticity and interactions with each other, so making the task of listing processes nearly impossible.

Air pollutants, under some concentrations, do pose a risk to plants and generate a stress response in them (Tausz *et al.*, 1998). What does that "stress" means to a plant? All living organisms live within an environmental range and so some production of "stress-related" responses must occur at the endmost portion of the range. A large deviation from that environmental range can overwhelm any restoration process and the organism. While these states of extreme deviation may be useful to give an understanding of the magnitude of the range, those states would be more pathological rather than physiological. Initially with any stress investigation, deviations leading to death are of use to define the extremes. However, current air pollution research should be concerned with the "physiology", which is that state which deviates from the "norm" but does not permanently or irreversibly injure the living system. The application of a physiological stress and the plant's response to it is most useful to define a protection program to lower forest risks. This discussion will focus upon the individual plant and attempt to define the range and risks of oxidant pollutants to it.

Stress induced by air pollutants begins with the entry of the pollutant into the plant through the leaf boundary layer and stomata into the substomatal cavity. The movement of ozone ($O_3$) into the plant has been well discussed (see Heath and Taylor, 1997) but the description of that movement is somewhat artificial in that it is modeled by a linear flux. The entry or influx of gases (j) into the internal space of a leaf is dependent upon physical and chemical processes in the gas phase and on surfaces, and is proportional to the gradient of concentration of gas from the outside ($C_O$) inwards ($C_i$), and is, in most cases, very non-linear (Parkhurst, 1984). The linear flux model is expressed as $j = g (C_O - C_i)$ where the proportionality coefficient is total conductance (g, but the conductance must be adjusted for the diffusion of each molecular species, by a measure of molecular weight, (Ball, 1987), through both the boundary layer and stomata. Here $C_i$ is an average internal concentration. For both water and $CO_2$, this formulation has been used for years (Farquhar and Sharkey, 1982). However, the internal concentration of $O_3$ is believed to be virtually zero (Laiski et al., 1989), most probably because $O_3$ is so reactive with cellular chemicals. Thus, the linear model should be well served. It is not clear however, whether a zero $C_i$ is true for $NO_x$ within the leaf. In the absence of any good internal measurements for $NO_x$, the effective delivery rate of both $O_3$ and $NO_x$ can be set to be [g . $C_O$] with total conductance being the major regulatory control (Taylor and Hanson, 1992; Amiro et al., 1984).

## 2. Biochemical Measurements

In order to place this discussion into an appropriate context, the following sequence of thought regarding biochemical measurements should be understood. A biochemical assay of a plant stress should not be done merely because it is possible, but rather the assay should be used to test a hypothesis. The superoxide dismutase assay is relatively easy but there are several isozymes which are located in different regions of the cell and so a general assay will measure the sum of all isozymes (Scandalios, 1994). For example, if one isozyme is increased while another is decreased by fumigation, the assay may measure no net change (e.g., Lee and Bennett, 1982; Polle and Rennenberg, 1991; Casano et al., 1994; Benes et al., 1995). Superoxide dismutase may be involved in the response to air pollutant exposure, but a general measurement will yield a false negative result.

### 2.1 TREATMENTS: WHAT EXPOSURE CONDITIONS ARE VALID?

Two types of fumigation protocol can be described: a high level for short term (normally less than a day) and a low, near-ambient level for long terms (typically weeks). The short term fumigation with experimental sampling within a day or two does not allow the plant to change developmentally; e.g., early senescence, while possibly initiated, cannot progress over a day. The long term fumigation is involved with developmental changes of the control plant and the possibility of accelerated aging of the fumigated plant. Thus, chronologically-similar plants, under different treatments, are no longer developmentally similar and the compared metabolisms will be naturally different.

Furthermore in short term fumigation, should plants be tested for biochemical events during the fumigation, immediately after the end of fumigation, or a day or so later? In each case a different metabolic response is found (see Langebartels et al., 1991, for a good demonstration of the time scale of sampling).

### 2.2 SAMPLING: HOW MANY PLANTS FOR EACH TREATMENT SHOULD BE USED?

These are statistical questions but have important ramifications for biochemistry. The extraction and assay may be so difficult that high rates of sampling cannot be effectively carried out on any reasonable time scale. Further, sampling in the field may not be easily done if the samples must be brought back into the laboratory. For example, plant materials are typically frozen in liquid nitrogen for transportation. While the $^{14}C$ assay for Rubisco works quite well on crop plants (Ward and Keys, 1989) even if the tissue is frozen, it fails with loblolly pine tissue if the tissue is frozen (Tissue et al., 1993; confirmed in Ponderosa pine by Heath and Miller, unpublished).

### 2.3 EXTRACTION: WHICH REGULATORY CONTROL POINTS OF METABOLISM MAY NOT BE OBSERVED?

More will be said later about the complexity of enzymes in metabolic pathways, but it is well accepted that most critical steps in a pathway are heavily regulated by small molecular weight compounds which are lost under typical grinding and assay conditions. The activity of the enzyme in vitro may not represent its actual activity in vivo and so assay results may be misleading. Again Rubisco provides an example. The enzyme must be activated by $CO_2$ and Magnesium ($Mg^{2+}$) for full activity, but there are several other activator systems working in vivo, such as enzymes, substrate/product binding, and pH (Stitt, 1996), which control the actual speed of carboxylation within the chloroplast. Another example of a key enzyme to be measured is nitrate reductase. With the possibility of more inorganic nitrogen (N) within the leaf or soil, nitrate reductase should increase and does so in crop plants (Pate, 1983). Yet nitrate reductase is highly regulated (Huber et al., 1994) and observing its activity as a ground powder does not necessarily represent its activity within the cell - again the regulators can be lost or diluted by grinding.

### 2.4 MEASUREMENT OF ASSAY: WHAT IS THE RATE OF ASSAY OR THROUGH-PUT AND THE PRECISION OF THE MEASUREMENTS?

Many enzyme assays (especially those using coupled or indirect reactions) can suffer from interference by other compounds (see Bergmeyer and Grassl, 1983; especially Chapter 2). Some of those compounds may be missing in certain species (allowing an easy and precise assay), but are present in other species (thus, giving a false reading of activity). This is especially critical in a fumigated system which has generated other compounds in reaction to the stress itself. The control sample would be assayed quite easily but the fumigated sample could give an invalid depressed reading because a fumigation-generated compound itself interfered with the assay.

### 2.5 FINAL INTERPRETATION AND UNDERSTANDING.

Too often a total metabolic process is not measured, even under laboratory settings. More usually the size of a metabolic pool is measured, such as the amount of ATP or sucrose. Not only is that a misleading measurement, for it blends the rates of influx to and efflux from the pool, but also there are several quasi-independent pools of the same compound within different organelles of an individual cell. The measurement then is a blend of all the pools of the cells within the extracted tissue. To be sure, some simple measurements are necessary due to a lack of knowledge of all the fluxes which affect a particular pool, or to a requirement for normalization (such as total protein or chlorophyll content of extract), but one should be careful of any generalization made from such a measurement.

A protocol can be outlined of what should be measured in order to make a pronouncement on what processes are being affected by a stress. Once the tissue for sampling is chosen, there are two initial steps of analysis:

(1) The extract must be normalized or standardized against some gross measure of the plant material extracted. For example, if N or protein flux is being studied, one might use total N or total soluble protein as a normalizing factor. If a photosynthetic processes is being examined, one might use chlorophyll. The normalization factor may not be the same for all assays.

(2) A marker of an unchanging process or enzyme should be used as an internal standard in order to ascertain that material in the extract was not altered during the extraction. For example, 18S-ribonucleic acid is often used as an unchanging standard for m-RNA. If one suspected that only Calvin Cycle enzymes were being changed, one might choose a mitochondrial marker enzyme such as succinate dehydrogenase as an unchanged parameter.

## 3. Rate of "detoxification" of an air pollutant— $O_3$

$O_3$ is very unstable near surfaces (Heath, 1987) and only interacts with the internal constituents of the leaf (although some argue that it can modify the surface waxes of the cuticle). It therefore does not accumulate in or on leaves. The accumulation of $NO_x$ is a different matter. For while the chemical form of N may change, nitrogenous compounds do accumulate on the leaf surface and in soil where they can be washed into the root zone from which they may enter the plant and be transported to the shoot. With N pollution the question is by what pathway(s) does N enter the plant. Others in this volume have discussed pathways and N accumulation, see Rennenberg and Gebler (1999).

With $O_3$ the external concentration cannot be used to give an indication of "dose" (Heath, 1980; Heath, 1994b). Rather a second level of knowledge must be used, that of stomatal conductance, to measure how much $O_3$ may gain entry into the tissues. But this information is insufficient because we do not know how deep within the air space of the leaf $O_3$ can penetrate; in essence, we do not know the concentration gradient of $O_3$ after it enters the boundary layer area. Our best measurement of dose, at the present time, is the amount of $O_3$ expected to penetrate to the tissue through the stomata (see above).

The expression of this dose is as a rate of delivery to a surface area (mol m$^{-2}$ s$^{-1}$) or total accumulation (mol m$^{-2}$, rate integrated over exposure time). It is unknown which measurement is the most critical for the development of any injury. Two final problems exist, whether the dose or accumulation [1] has a threshold (a level below which no observable effect, however defined, is noted) and [2] is non-linear with a given measured effect. The best current guess is that acute injury may relate to speed of $O_3$ entry into the tissue while early senescence may be related to accumulated dose.

Let us use a level of $O_3$ which has been measured in a forest system for over three decades—the San Bernardino National Forest (see Anon, 1991). These levels are extreme values under which ecosystems can live but yet are being modified (for a very complete review, see Miller and McBride, 1998). These values are useful to generate a maximum "stress" on a plant.

An annual average daily external concentration of 0.10 ppm for $O_3$ (over 24 hours) can be used to calculate an average rate of entry during the day. For stomatal conductance of about 50 mmol m$^{-2}$ s$^{-1}$ ($O_3$ conductance* of 30 mmol m$^{-2}$ s$^{-1}$), that rate is 30 mmol m$^{-2}$ s$^{-1}$ x 0.10 x$10^{-6}$ = 3 x $10^{-9}$ mol $O_3$ m$^{-2}$ s$^{-1}$. The maximum short term rate is for the highest annual maximum (0.30 ppm) and is equal to 9 x$10^{-9}$ $O_3$ mol m$^{-2}$ s$^{-1}$. If all the $O_3$ is directly reacted by an antioxidant, the rate of antioxidant regeneration should be about the maximum rate of $O_3$ delivery or 9 x$10^{-9}$ $O_3$ mol m$^{-2}$ s$^{-1}$. So the regeneration rate of an antioxidant must be at least 3 x$10^{-7}$ mol g-FW$^{-1}$ s$^{-1}$ or roughly 300 μM s$^{-1}$. This assumes that one molecule of antioxidant is used to detoxify one $O_3$ molecule. These calculations are probably not exactly correct but are only in error by a factor of 2 to 3 (see Dietz, 1997).

For comparison the maximum rate of photosynthesis can be 5 μmol m$^{-2}$ s$^{-1}$ or nearly 500 times that of antioxidant regeneration. Thus, the need for antioxidant regeneration is much less than photosynthetic rate and so regeneration, in itself, is not energetically demanding. To be sure, the regeneration rate for antioxidants to rid the cell of $O_3$ may be too slow and other toxic processes may come into play. However, the energetic requirements of antioxidant regeneration are too small to be measured relative to photosynthetic activity.

## 4. Oxides of Nitrogen influences on plant productivity?

High levels of oxides of N induce visible injury of plant foliage and alter plant productivity (see U.S. E.P.A., 1992; Wellburn, 1990). For a discussion of nitrogen metabolism see Miflin and Lea, 1990; Lam et al., 1996; Solomonson and Barber, 1990.

There is, however, an important point in the movement of N compounds between cells and organelles. Two of the compounds produced from the transformation of nitrate ($NO_3^-$) into amino acids can produce uncharged species which can move easily across membranes (Heath, 1980). Nitrate reductase produces the weak acid, nitrite ($H^+$

* See Farquhar and Sharkey, 1982; for units. Area is leaf area.

+ $NO_2^-$ ⇔ $HNO_2$) and nitrite reductase produces the weak base, ammonia, ($H^+$ + $NH_3$ ⇔ $NH_4^+$). As shown in Figure 1, the equilibrium of the uncharged species across a membrane can give rise to a gradient of concentrations of the charged species which, in turn, leads to osmotic imbalance (see Raven, 1988, and Heath and Leech, 1978 for more details). In fact, uncoupling of photo-phosphorylation by ammonia ($NH_3$) works in precisely this manner (Crofts, 1976). This should be recognized in any discussion of N compounds within the plant.

## 5. The Modification of Physiological Processes Induced by $O_3$

In the past, most discussions of $O_3$ induced injury to plants have been confused by discussions of at least two different questions which are asked: [1] How does this injury affect the economic importance of that plant, or stated another way, how can the political processes regulate $O_3$ levels such that economic loss does not occur and therefore make regulation politically feasible? [2] By what biological mechanism does $O_3$ alter the normal physiological and ecological processes within the plant? Discussions of both questions simultaneously do not necessarily lead to coherent answers to either. To be sure, one would hope that answers to question [2] would aid in the formulation of answers to question [1]; however, an answer to question [1] does not inevitably lead to an answer [2]. Political and economic solutions often do not require an understanding of the mechanisms of biology.

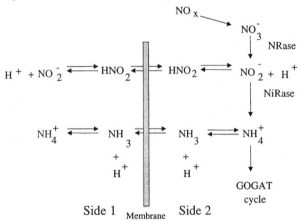

Fig. 1. The Movements of Weak Acids and Bases Across Membranes.
$NO_x$ generates nitrate which is converted into nitrite by nitrate reductase (NRase). The nitrite must move to the chloroplast in order to use light-derived energy to be converted ammonium (NiRase). Both nitrite and ammonium are in equilibrium with their uncharged counterparts, which pass through the membrane with ease. Depending upon the pH the charged forms can be in greatly differing concentrations across the membrane, leading to an osmotic potential difference (see arguments in Crofts, 1976 and Heath and Leech, 1978). GOGAT cycle is glutamine synthetase/glutamine oxoglutarate aminotransferase cycle, which takes in the ammonia unit using the organic acids, oxoglutarate and glutamate.

In this discussion I would like to focus only upon mechanisms. Under these conditions the species used to understand mechanisms are unimportant (thus making *Arapidopsis* an important plant for $O_3$ injury discussions, Sharma and Davis, 1995) but are related only to how they can give biological understanding. I am not trying to minimize the importance of large trees and understory plants; they are critical to understanding plant interactions at the ecosystem level. It is just that economically important trees may not be the best system to use to understand fundamental biological mechanisms.

There is not one single response to $O_3$ exposure (Guzy and Heath, 1993). At least three sequential processes combine to trigger each $O_3$ stress response from the movement of gases from the atmosphere to the sites of action within the leaf (Heath, 1980). The current hypotheses of $O_3$ induced alterations are presented in Table I. Gaseous diffusion is the first process. One must know how much gas has entered the tissue region to understand the mechanism of action (see above).

5.1 INTERCEPTION OF TOXIC PRODUCTS
There is an extensive literature relating to antioxidants, both within the apoplastic space and the cell itself, as protectants against $O_3$ injury (Schraudner *et al.*, 1992; Polle *et al.*, 1995; Dietz, 1997). The antioxidants either intercept and chemically alter the $O_3$ or its toxic products. Some of the possible products of $O_3$ breakdown are listed in Table II and include hydroxyl radical, superoxide and hydrogen peroxide as well as more unusual $O_3$ derivatives (see Heath, 1987; Heath and Taylor, 1997; Runeckeles and Vaartnou, 1997). Unfortunately the data are not clear as to possible products of $O_3$ or to the mechanism of antioxidant mechanisms within the apoplastic space. Ascorbate within the apoplastic space clearly can intercept the $O_3$ (Chameides, 1989; Luwe *et al.*, 1993) and so can prevent $O_3$ reaching the membrane. However, the ability of ascorbate to intercept $O_3$ depends upon: [a] the amount of antioxidant present within the apoplastic space and [b] the cell's ability to regenerate that pool of antioxidant after its reaction with $O_3$. Both processes depend upon the total amount of antioxidant needed.

Unfortunately it is very hard to measure regeneration since the oxidation of ascorbate is within the apoplastic space and the regeneration seems to require a movement of the oxidized ascorbate back into the cytoplasm or some other similar mechanism. Further, the net change in total cellular ascorbate may be very small under most conditions, as it would be maintained by other reducing agents within the cell (Dietz, 1997).

5.2 MOVEMENT OF PRODUCT(S) AND TRANSFORMATIONS
It is believed that the initial site of $O_3$ damage is the plasma membrane (Heath, 1987). $O_3$ is soluble in water but once having entered the aqueous phase, it rapidly breaks down. The mean free path before reaction is very short and, thus, only some oxidative products (see Table II) may be able to diffuse through the cell's membrane and react with internal metabolites and structural biochemicals.

TABLE I

Current theories of oxidant induced alterations of plants

1. Antioxidant protection against Superoxide, Hydrogen Peroxide, Hydroxyl Radical
    Superoxide Dismutase
    Peroxidases and Catalase
    Ascorbate and Glutathione
    Tocopherol
2. Wounding Response
    Wounding Proteins (e.g., chitinase, β-glucan synthase)
    Activated Oxygen
    Ethylene Production
3. Loss of Photosynthetic Capacity
    Stomatal Response
    Photosynthetic Process
    Photosystems
    Carboxylation
    Translocation
4. Membrane Dysfunction
    Loss of Ion Channels ($K^+$, $Ca^{2+}$)
    Loss of Permeability
   Alteration of Normal Ionic Channels
    Membrane Structure
    Loss of Signal Transduction Receptors
5. Premature Senescence
    Early Loss of Productivity
    Inability to Remobilize Nutrients

TABLE II

Mechanisms of toxic product production by $O_3$ within the cellular space

Possible $O_3$ Produced Compounds (see Heath, 1987)

$O_3 + HO^- \rightarrow HO_2^{\cdot} + O_2^{-}$ (superoxide)
$O_3 + O_2^{-} \rightarrow O_2 + O_3^{-}$
$O_3 + H^+ \rightarrow HO_3^{\cdot}$
$\quad\quad HO_3^{\cdot} \rightarrow O_2 + HO^{\cdot}$ (hydroxyl radical)
$O_2^{-} + H^+ = O_2^{\cdot}$ (peroxyl radical)
$O_2^{-} + HO_2^{\cdot} + H_2O \rightarrow H_2O_2 + O_2 + OH^-$

Detoxification of Radicals

| | |
|---|---|
| $2 O_2^{-} + 2 H^+ \rightarrow H_2O_2 + O_2$ | Superoxide Dismutase |
| $2 H_2O_2 \rightarrow O_2 + 2 H_2O$ | Catalase |
| $2 H_2O_2 +$ Ascorbate $\rightarrow$ Dehydroascorbate $+ 2 H_2O$ | Ascorbate Peroxidase |
| $2 H_2O_2 + $ GSH (Glutathione) $\rightarrow$ GSSG $+ 2 H_2O$ | Glutathione Peroxidase |
| $2 H_2O_2 + $ NADH $\rightarrow$ NAD$^{\cdot} + O_2^{-} + H^+$ | Peroxidase |
| $\{$NAD$^{\cdot} + O_2$  NAD$^+ + O_2^{-}\}$ | |
| GSH S-transferase: Regeneration of Antioxidants | |
| Dehydroascorbate $+ 2$GSH $\rightarrow$ GSSG + ascorbate | Ascorbate Reductase |

Certainly, membrane functions, such as membrane fluidity (Pauls and Thompson, 1981), permeability (Elkiey and Ormrod, 1979), potassium ($K^+$) exchange via ATPase reactions

(Dominy and Heath, 1985) and calcium ($Ca^{2+}$) exclusion (Castillo and Heath, 1990), are rapidly altered by $O_3$ exposure. The similarity of $O_3$-induced membrane disruption to wounding responses (Langebartels *et al.*, 1991; Kangasjärvi *et al.*, 1994) would suggest a membrane trigger, perhaps a change in $Ca^{2+}$ movements (Heath and Taylor, 1997), leading to the induction of normal wound-regulated genes (Mehlhorn *et al.*, 1991; Ecker and Davis, 1987; Kärenlampi *et al.*, 1994).

Many enzymatic processes decrease in response to $O_3$ exposure, while others increase (see Heath, 1996a, b; Heath and Taylor, 1997). For example, photosynthesis is inhibited perhaps by: [1] a toxic product of $O_3$ migrating through the cytoplast to react with photosynthetic processes directly within the chloroplast, [2] a spurious signal generated at the membrane affecting some control process or signal transduction pathway, such as an activation of a protein kinase, or [3] an ionic shift due to the loss of membrane selectivity changing chloroplast responses. Many of the metabolic events are controlled by changes in gene expression, which can be modulated by changes in energy and fixed C availability. It seems reasonable to suggest that alleviating the effects of air pollutant exposure must occur through modified gene expression to counter alterations in the level and/or activation of specific proteins (Narváez-Vásques *et al.*, 1994; Harris and Bailey-Serres, 1994).

5.3 IONIC FLOWS MAINTAINED BY THE PLASMA MEMBRANE

One hypothesis is that $O_3$ induces a loss of membrane permeability and an inhibition of transport of some ions. In particular a change in membrane function leading to a rise of intracellular $Ca^{2+}$ would lead to alteration of all sorts of intracellular metabolism, similar to that observed in the wounding of plant tissue (see Figure 2). Many of these same metabolic steps are observed during $O_3$ induced injury. More importantly, Castillo and Heath (1990) demonstrated that the plasma membrane $Ca^{2+}$-transport system (efflux pump and influx permeability assayed after isolation) were altered by $O_3$ exposure *in vivo* (efflux pump was inhibited and influx increased). Heath (1987) also argued that the breakdown of $O_3$ in water can produce $H^+$ and so alter the wall space pH. Such a scheme can be fitted into the sequence of reactions induced by an elicitor, as shown in Figure 2. The loss of $K^+$ from and gain of $Ca^{2+}$ into the cell's interior coupled with an increased amount of $K^+$ outside and shift in apoplastic pH could increase the loss of a wide range of metabolites, induce enzyme activation and alter normal gene transcription. Thus, $O_3$ interactions at the plasma membrane with little or no production of toxic intermediates within the cell could change the ionic balances within the cell and a rise in internal $Ca^{2+}$ could trigger a series of very diverse responses in the cell's biochemistry.

The hypothesis is that $O_3$ induces $Ca^{2+}$ increase within the cell. One might then ask if $Ca^{2+}$ was increased in the tissue. Certainly field samples of leaf/needle tissue are easily taken for future assay. Yet the levels of $Ca^{2+}$ in different regions of the tissue are such that no conclusions can be made from this simple experiment. A typical cell has a diameter of 20 μm yielding a volume of about $4 \times 10^{-15}$ $m^3$. Most of the cell is vacuole, about 10% is cytoplasm. The wall region is about 1% of this total cell volume. The triggering response of the $Ca^{2+}$ within the cytoplasm is a jump from $10^{-7}$ M to $2 \times 10^{-6}$ M, a twenty fold increase. How much total $Ca^{2+}$ does this represent? Not much, only $2 \times 10^{-6}$ M $\times$ 10% $\times$ $4 \times 10^{-12}$ L = $8 \times 10^{-19}$ moles (up from $4 \times 10^{-20}$ moles) or an increase of $7.6 \times 10^{-19}$ moles.

This $Ca^{2+}$ must come from the wall regions if the plasma membrane is made leaky. The wall space has a $Ca^{2+}$ concentration of ca. 5 mM, or $5 \times 10^{-3} \times 0.01 \times 4 \times 10^{-12}$ L = $2 \times 10^{-16}$ moles.

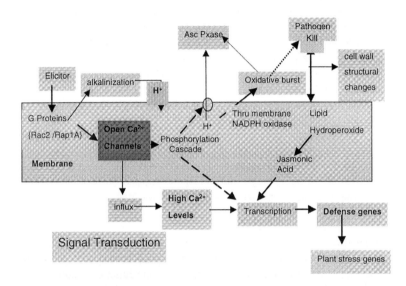

Fig. 2. A Schematic of the Generation of the Wounding Response within a Leaf by an Elicitor.
(Adapted from Ward et al., 1991, Zentgraf and Hemleben, 1996, Lamb and Dixon, 1997, and Simon-Plas et al., 1997). The initial event of the Elicitor response is linked to an activation of G Proteins. It is this protein which seems to activate ionic responses which may be similar to $O_3$ induced injury—an opening of $Ca^{2+}$ channels and an alkalization of the apoplastic wall region. Most of the above events after these two events have been studied for elicitor events, many of these, although not all, have been observed in $O_3$ induced injury. Abbreviations used: Asc Pxase, ascorbate peroxidase.

By removing $7.6 \times 10^{-19}$ moles $Ca^{2+}$ means that the wall concentration drops by 0.4%. First of all, the total amount of $Ca^{2+}$ within the tissue does not change; there is only movement between spaces. But even if the amount within the wall could be measured, the change would be too small to determine experimentally. Therefore, the $Ca^{2+}$ level must be measured within the cytoplasm, which can be accomplished within the laboratory but not with stored or frozen field samples.

Under many circumstances quite diverse metabolic processes can be altered by a cascade effect beginning at a fundamental level. There are several newly found protein kinase/phosphatase* activation systems in plants (Spanu et al., 1994), which may be

* A protein kinase is simply an enzyme which phosphorylates another enzyme with a nucleotide triphosphate thereby activating or deactivating it. The related phosphatase is the kinase's counter part, the phosphate group on the enzyme is removed by the phosphatase thereby deactivating or activating it, respectively (see Stone and Walker, 1995).

responsible for some of the metabolic imbalances seemingly caused by $O_3$ exposure (compare Langebartels et al., 1991, with Ward et al., 1991, and Lamb and Dixon, 1997).

Thus, one cannot escape the conclusion that $O_3$ is acting as the initial trigger to alter the many diverse pathways of metabolism. The cascade into so many pathways requires that the trigger would be a fundamental and initial process of normal signal transduction and so we should be looking for that initial transducing event. For now $Ca^{2+}$ balance would seem to be the most promising lead to the transducing event, but there may be another $O_3$-modified biochemical which is the real trigger. Mehlhorn and Wellburn (1987) showed that an inhibitor of ethylene formation, AVG (aminoethoxyvinyl glycine, an inhibitor of ACC synthase) would block ethylene formation and inhibit visible injury production. However, the ethylene pathway is highly regulated and also involves a series of protein kinases/phosphatases. In fact, this concept may be the reason why high levels of oxidant induce ethylene release and visible injury, while lower levels for longer periods induce early senescence of needles and leaves (Heath, 1994a; Heath and Taylor, 1997).

5.4 DEVELOPMENTAL AGE OF TISSUE

For biochemical analysis there is a continuing question as to which developmental age of the tissue to sample. Many of the classical biochemical analyses (from about 1960 to 1980) were done on young crop plants and often used the cotyledons or first true leaves. Yet it was recognized that developmental age does play a role, as shown by the studies of Tingey et al. (1976) and Dugger and Ting (1970), and summarized by Heath (1996b). Young, recently initiated leaves are different biochemically from mature, actively exporting leaves and old, senescent leaves (Casano et al., 1994), yet biochemists often conveniently forget this when trying to understand fundamental mechanisms of pathway control or signal transduction. However, when studying the potential effect of a stress (such as air pollution) on the biochemistry of a process in which the regulation is at least roughly understood (such as photosynthesis), the developmental age of an organ must be taken into account.

For example, the loss of Rubisco activity is thought to be an important and critical event in $O_3$ induced injury (Reddy et al., 1993; Pell et al., 1994). Pell's group showed that the most easily measured change was when the leaf was beginning to senesce (Eckardt and Pell, 1994; Wiese and Pell, 1997). In older leaves, Rubisco declined in unfumigated plants but $O_3$ induced a more rapid decline. It was difficult to measure a net loss of Rubisco in young leaves (control compared with fumigated) but easy to measure that loss in older leaves.

For a young crop plant before it begins to branch, the problem of how to express the developmental age of an organ or plant has been studied and has resulted in the formulation of a plastochron index (PcI, Erickson, 1976). The PcI is related to the organ's size when compared to a standard size (generally a small size that can be measured accurately). For example, the physiological age of a leaf is set to zero when its size is at 10% of its maximum size. The PcI of the plant would be the leaf number of the smallest leaf plus a scaled ratio of that leaf's size to the standard size. To the author's knowledge no one has tried to develop a coherent PcI for a multiply branched plant, such as an older tree.

The problem can be stated more clearly for a tree such as Ponderosa pine. Figure 3 shows a somewhat typical major branch with secondary branches. The question is: which whorls correspond to each other biochemically? Do we assume that all current year whorls are identical, when they are illuminated under the same environment? If they are not, to what do they correspond? Under most circumstances researchers sample tissues perceived to be identical.

In Figure 3 the primary branch has at least 3 whorls with needles attached, as do all the secondary branches, and so all are presumably at least 3 years old. Is the secondary branch (4$^{th}$ from top) with four whorls similar to the others? Under perfect conditions whorl 0 (current year's whorl) on a secondary branch would correspond to whorl 0 on the primary, in terms of age from the beginning of the primary branch. Yet secondary branches are not all initiated at the same equivalent physiological time. In the figure the year that each whorl or needle-less whorl has been initiated has been estimated from the age of the branch (by ring count) and from the number of nodes on the branch. The initiation time varies between 1 and 4 years. For biochemical sampling, these questions require good physiological answers.

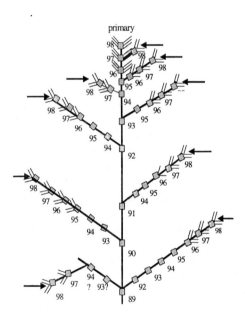

Fig. 3 Branching in Ponderosa pine
A branch was collected from a Ponderosa pine in the San Bernardino Mountains from a tree about 40 years old. It was determined by counting the rings of the branch that it had been first formed in 1989. The primary branch is at the top (see primary). There were, in general, about 3 whorls retained by each growing branch (1998=current and indicated by arrow, 1997=one year old, and 1996=two years old). In several cases the 1996 whorls were chlorotic and had lost many of their needles. Branches from the main branch were formed at varied times after the initial formation of the whorl. The years of formation of the whorl (even without needles) could be observed by the closely spaced needle scars.

## 6. $O_3$ Injury and Carbohydrate Status

Studies on carbohydrates were begun early in the investigations of $O_3$ effects on plants (summarized by Dugger and Ting, 1970, see Heath and Taylor, 1997). Sucrose applied to leaves resulted in protection, but the effect on susceptibility was evident only when the level of sugars was within a defined range. Further, the levels of starch and of sugars were roughly correlated with leaf damage. Thus, there is a relationship between sugar content and visible injury, but cause and effect has not yet been established.

Stomatal conductance and assimilation (ability to fix $CO_2$) are affected by $O_3$ exposure (Saxe, 1991, Heath, 1996b). Generally the ability to assimilate efficiently (whether due to stomatal closure or inhibition of $CO_2$ fixation) is reduced (Manning and Keane, 1988). It follows then, that the levels of sugars (both soluble and starch) should be reduced. The questions arise of [1] whether ability to make carbohydrate by direct photosynthetic assimilation is altered to a greater extent than [2] the ability to move the carbohydrates within the cell (e.g., inability to move triose phosphate out of the chloroplasts so that more starch is made in the chloroplast, but less sucrose can be made in the cytoplasm) and/or [3] whether translocation of the sugars out of the leaf is inhibited.

6.1 PHOTOSYNTHESIS, CHLOROPHYLL AND STOMATAL CONDUCTANCE

Saxe (1991) has done an excellent job of summarizing what is known about changes in photosynthesis upon varied air pollutant exposures. For $O_3$ the effect on photosynthesis of long term exposure at low levels is not very clear. One can find increases, no effects, and decreases for varied conditions. The majority of the data suggest that photosynthesis is lowered without a concurrent fall in stomatal conductance. Most data suggest that there is a lessened ability to fix $CO_2$ and the simplest explanation is a lowered amount of Rubisco, although Keller and Häsler (1987) suggest that the stomata do not respond normally and become "sluggish". Poorly functioning stomata could alter both the water use and assimilation, probably not to the benefit of the plant. Heath (1994a) has suggested that the guard cells, being closest to the stoma, are subject to the full concentration of $O_3$ and so should be most affected. Thus, a simple measurement may not yield meaningful results; the experimenter must find values for assimilation, stomatal conductance, and internal $CO_2$ level. Interestingly, changes in all three parameters can be found in the literature. For a more complete discussion see Heath (1996b).

Often the chlorophyll concentration in $O_3$ exposed tissue falls after long term fumigation (see Robinson and Wellburn, 1991). Most probably this is due to an early senescence of the tissue, but one must be careful of measurements and instrument calibration. For example, chlorophyll A to chlorophyll B ratios are used to describe changes in the photosystems of the chloroplast, but as Figure 4 shows, if the spectrophotometer is not calibrated correctly, this ratio can vary. If the absorption of the extract is not measured at the wavelengths of the peaks, but is in error by only 2nm, (off by only 2 nm), the ratio can vary by nearly 50%. Furthermore, alterations of the chlorophylls, due to acidity or alkalinity of the extract, can shift the peaks of the pigments (Heath, 1989).

The conductance of almond trees in the field can vary a great deal (see Figure 5A) and so one must not assume that a few measurements will yield the correct results. Again when to take the measurement is a large problem - many researchers take the measurements of conductance in the morning when the measurement is more stable. In the afternoon conductance often declines due to a problem with water balance within the leaf/needle (Weyers and Meidner, 1990).

It is of interest to take a normal distribution of stomatal conductance (e.g., from Figure 5A) and calculate what happens if $O_3$ accumulation is proportional to the stomatal conductance. The amount of $O_3$ delivered to the tissue at the high end of the range of $g$ (in Figure 5A at 0.8 cm sec$^{-1}$) will be twice as great as that delivered at the low end of the range (0.4 cm sec$^{-1}$). If we postulate that the low levels of $O_3$ can be either detoxified by antioxidants or accommodated by some sort of repair, then there will be a threshold of $O_3$ flux below which no injury will occur. This threshold will be

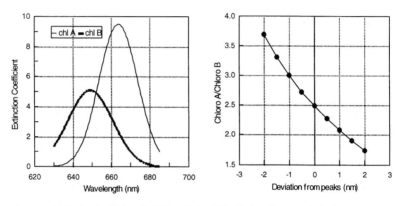

Fig. 4. A Calculated Shift of Chlorophyll A/B Ratio due to the Mis-calibration of a Spectrophotometer.
The left hand graph shows idealized chlorophyll spectra for chlorophyll A and B (the peaks and shapes with extinction coefficient units in m$^3$ g$^{-1}$ m$^{-1}$ are for a 80% acetone solution, from Vernon, 1960; Wellburn and Lichtenhaler, 1985). These spectra are used to calculate what A/B ratios can be obtained if the calibration of the wavelengths read by the spectrophotometer used to measure the absorption at two wavelengths (645 and 663 nm, peaks in 80% acetone) is in error by 0 to $\pm$ 2 nm.

proportional to a limiting stomatal conductance (for a given time and external $O_3$ level). If we also postulate that for a dose above this level, $g$ will suffer a decline as a percentage of the total amount of $O_3$ (or % decline is proportional to $g$ above the threshold). A model can be built and evaluated upon these concepts, shown in Figure 5B. Notice that for normal variation of $g$, the decline in $g$ for the higher values generates a new "apparent" normal curve which can be fitted to a normal distribution. The difficulties in interpretation of data such as this are many fold even when examining all the experimental data. For example, there is no effect for very low fluxes (below the threshold) but the depression in $g$ is 40% at high values (above the threshold). Yet the mean of the distributions only declines by 15%. Both the non-linearity in the $O_3$ effect and the scatter conspire to confuse the final conclusion.

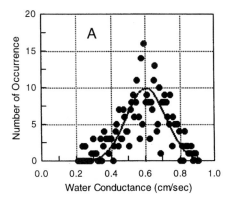 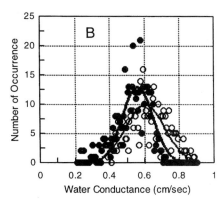

Fig. 5. Conductance Variations Measured with The Dual Isotope Label Porometer.
A. Distribution of Data on Stomata Conductance. Almond trees (10-20 years old growing in a commercial field near Bakersfield, CA) were measured by I.P. Ting and co-workers, using the Dual Isotope Porometer (see Heath, 1994b).

B. The Change in Distribution due to an $O_3$ Effect upon the Conductance. The data are taken from Figure 5A, but a 5-point running average was used on the distribution to smooth the actual curve (indicated by open circles) in the bottom curve. A best fit curve of the distribution was given as the dotted line. The $O_3$ response has been set to be for a threshold of 0.35 cm sec$^{-1}$ with a depression of conductance of 40% above the threshold (proportional to the difference in the measured and threshold conductances) as in the top curve (see text). Each point of the initial distribution was re-calculated based upon the $O_3$ response, given above, and re-plotted as closed circles. The best fit curve for this distribution was given as the solid line.

Another model can be constructed to show that the time when measurements are made might alter the perception of an $O_3$ effect, similar to that argued by Musselman and Massman (1999). This relies upon the diurnal fluctuation of the stomatal conductance. In Figure 6, the steady state photosynthesis in the morning is 7.8 μmol m$^{-2}$ sec$^{-1}$ and so if there is no decline, photosynthesis measured in the late afternoon would be the same (0% loss). But if in the afternoon conductance begins to decline, photosynthesis can suffer a 60% loss for a moderate decline, and 92% loss for a severe decline. What does this do to total productivity during the day? If the no-decline rate is set to 100%, a moderate decline lowers the productivity by 17% while a severe decline lowers it by 32%. Yet no change, in any case, would be measured, if the data were taken only in the morning. Also the total dose of $O_3$ would be quite different for the three cases, especially if the high levels of $O_3$ in the atmosphere were only in the afternoon.

## 6.2 ASSIMILATION AND RUBISCO

The production of fixed C is critical to productivity and resistance to other stresses. The question remains of how to rank the varied events that may lower that photosynthetic response. Visible injury results in the loss of photosynthetically active tissue, but much of that loss can be compensated for by increased light absorption by other leaves and by preventing senescence of the functioning leaves.

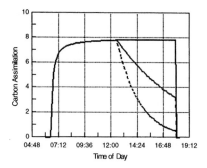

Fig. 6. Model of a Midday Decline in Conductance and Total Daily Carbon Production. The model has a rapid rise when sun comes up (within one hour). After 12:30 PM an exponential decline starts. The half time of decline is varied to produce different assimilation rates at the end of the day (indicated by solid line, no decline; dashed and dotted lines). The total area under the assimilation curve for the day is integrated to give the total amount of assimilation.

The loss of Rubisco is difficult to measure as Rubisco is present in very high concentration (see above). Another method to follow Rubisco changes is by measuring the concentration of its mRNA leading to the production of the protein (rbcS, the message for its small subunit is nuclear transcribed). Pell et al. (1994; see their Figure 1) showed a rapid, but not complete, loss of the message within a hour or so of exposure and a recovery a day after the exposure ceased. Yet there is an influence of the level of carbohydrate in the leaf upon rbcS (Heath and Taylor, 1997). It may be that this decline in Rubisco is due to the production of control metabolites, such as Fructose 2,6-bisphosphate, which leads to a shut down of the important sugar production pathways (Stitt, 1996).

To make matters worse for experiments, each functional enzyme may be controlled by several genes (although the coding regions are identical), as for Rubisco (Peters and Silverthorne, 1995; Cheng et al., 1998) which has multiple isoforms. Each isoform produces the same functional enzyme but the control of transcription is by different regulatory sites. Superoxide dismutase is even more complex—there are at least three forms of SOD usually denoted by their individual metal (Cu-Zn, Fe, and Mn). All seem to have multiple isoforms. Over- or under-expression of one isoform may or may not have an effect upon $O_3$ induced injury (Slooten et al., 1995; Van Camp et al., 1996). The concept that the amount of an enzyme indicates a single regulated process is certainly not correct, and now the species of mRNA may not necessarily indicate a specific regulated process, but rather be the totality of several gene products.

Interestingly Williams et al. (1994) have developed a correlation between the levels of ABA after water stress in *Arapidopsis thaliana* leaves and the loss of rbcS. This work suggests that a water stress may alter the relationships much more than merely closing the stomata. If ABA is lowering rbcS by whatever mechanism, rbcS may not be worth much as a marker of $O_3$ fumigation except under very controlled conditions.

Yet even with these data, many unanswered questions exist due to the high abundance of Rubisco normally present in the leaf (regulated or not). How many molecules of Rubisco can be lost from how many leaves before productivity is affected?

How flexible is the organism with regard to the total amount of Rubisco? How balanced or compensated are regulated pathways of photosynthesis in normally functioning individuals? In essence, a quantitative sequencing of loss of message to the loss of enzyme to the loss of productivity is hard to come by.

## 7. Summary

As to biochemistry induced by $O_3$ exposure, one could suggest the following hypotheses about injury:
- The varied antioxidants (metabolites or enzymes), if present in sufficient amounts, can eliminate the oxidant or its products. Too rapid oxidant entry can overwhelm the antioxidant response.
- The membrane is altered upon exposure to $O_3$, principally via protein changes, such that permeability, transport, and triggering mechanisms are no longer correct for the state of the cell. E.g., the rate of ion movements or sensitivities to messages could be too "slow", "fast" or "reactive" to maintain homeostasis. The cellular pools of $Ca^{2+}$ /$K^+$/$H^+$ are the primary suspects for such distortions.
- Visible injury is caused by the interaction of $O_3$ with stress-induced ethylene either by direct chemical transformation of ethylene to a toxic product or by alteration of the ethylene/binding site relations. Clearly this reaction will not occur if antioxidants eliminate $O_3$ or $Ca^{2+}$ changes are too small to trigger ethylene release.
- One product of $O_3$-induced membrane changes induces a decline in the m-RNA for Rubisco (*rbc*S and *rbc*L) such that the level of Rubisco declines within the chloroplast, leading to a lowered rate of $CO_2$ fixation and lowered productivity, especially in older tissue.
- $O_3$ alters the normal relations of guard cells and subsidiary cells such that the stomata close and so limit the $CO_2$ fixation patterns. Long term exposure to $O_3$ may alter a balanced regulation of the stomata upon another subsequent stress, such as drought.
- One of the most sensitive sites of $O_3$ exposure is the translocation of sugars such that even a mild exposure will alter the ability to transform carbohydrate within the cell and change sugar transport out of the source cell to sinks.

There are a series of "carbon costs" (some large and some small) to the plant by pollutant exposure. One of the smaller costs for $O_3$ seems to be the increased antioxidant needed and ethylene production. A larger cost would seem to be decreased production and pool sizes of sugar in leaves/needles. A potentially larger cost in productivity may be due to "false triggers" internal to the cell leading to: [1] increased retention of C in leaves (starch); [2] decreased allocation to roots and decreased pools of sugars in roots; and [3] decreased root/shoot ratio thereby signaling an imbalance in nutrient to carbon status.

There are two methods by which progress can be made in the study of forests and general ecosystems. One is to compare two equivalent ecosystems to find what is different between them. If the climate, water availability and soil types are the same but they differ in the distribution of species, then perhaps atmospheric pollutants may be the culprits. The question asked under these conditions would be: is the current pollutant load causing a loss of some of the species?

But that method is not enough. The second method of study must also come into play. The questions asked must be expanded to include: by what mechanism do pollutants alter the species distribution? This question leads into the possibility by which future physiology can be predicted for each pollutant load. In the final analysis field observations, studies of biochemical mechanisms, and model building must all be used together to give an answer that will be useful for political and economic risk evaluations and for the ultimate, full understanding of all biological processes.

## References

Amiro, B.D., Gillespie, T.J., and Thurtell, G.W.: 1984, *Atmos. Environ.* **18**, 1207-1215.
Anon: 1991; *Summary of Air Quality in California's South Coast* and *Southeast Desert Air Basin*, South Coast Air Quality Management District Governing Board, for the Crestline Station #5181.
Ball, J.T.: 1987, in: Zeiger, E., Farquhar, G.D., and Cowan, I.R. (eds) *Stomatal function*. Stanford Press, Stanford CA, pp 445-476.
Benes, S.E. Murphy, T.M. Anderson, P.D. and Houpis, J.L.J.: 1995, *Physiol. Plant.* **94**, 123-134.
Bergmeyer, J. and Grassl, M. (eds.): 1983, in: *Methods in Enzymatic Analysis, Vol. II: Samples, Reagents,* and *Assessment of Results.* 3$^{rd}$ Edition, Verlag Chemie, Weinheim Press, Deerfield Beach, FA, U.S.A., 549 pp.
Casano, L. M., Martín, M. and Sabater, B.: 1994, *Plant Physiol.* **106**, 1033-1039.
Castillo, F.J. and Heath, R.L.: 1990, *Plant Physiol.* **94**, 788-795.
Chameides, W.L.: 1989, *Environ. Sci. Tech.* **23**, 595-600.
Cheng, S.-H., d. Moore, B. and Seemann, J.R.: 1998, *Plant Physiol.* **116**, 715-723.
Conklin, P.L. and Last, R.L:. 1995, *Plant Physiol.* **109**, 203-212.
Crofts, A.R.: 1976, *J. Biol. Chem.* **242**, 3352-3359.
Dietz, K.-J.: 1997, *Progr. Bot.* **58**, 221-261.
Dominy, P.J. and Heath, R.L.: 1985, *Plant Physiol.* **77**, 43-45.
Dugger, W.M. and Ting, I.P.: 1970, *Annu. Rev. Plant Physiol.* **21**, 215-234.
Eckardt, N.A. and Pell, E.J.: 1994, *New Phytol.* **127**, 741-748.
Ecker, J.R. and Davis, R.W.: 1987, *Proc Natl Acad Sci USA* **84**, 5202-5206.
Elkiey, T. and Ormrod, D.P.: 1979, *Atmos. Environ.* **13**, 1165-1168.
Erickson, R. O.: 1976, *Annu. Rev. Plant Physiol.* **27**, 407-434.
Farquhar, G.D. and Sharkey, T.D.: 1982, *Annu. Rev. Plant Physiol.* **33**, 317-345.
Grulke, N.: 1998, in, *Oxidant Air Pollution Impact on the Montane Forests of Southern California: A Case Study of San Bernardino Mountains*, Miller, P., and McBride, J.R. (ed.), Springer-Verlag, New York, pp 126-163.
Gupta, A.S., Webb, R.P., Holaday, A.S. and Allen, R.D.: 1993, *Plant Physiol.* **103**, 1067-1073.
Guzy, M.R. and Heath, R.L.: 1993, *New Phytol.* **124**, 617-625.
Harris, M.J. and Bailey-Serres, J.N.: 1994, in: *Stress Induced Gene Expression in Plants*. Harwood, Basra, A.S. (ed.), Academic Press, Switzerland, p 185-207
Heath, R.L.: 1980, Annu. *Rev. Plant Physiol.* **31**, 395-431.
Heath, R.L.: 1984, in, *Gaseous air pollutants* and *plant metabolism*. Koziol, M.J., and Whatley, F.R. (eds), Butterworths, London, pp 275-290.
Heath, R.L. 1987. The biochemistry of ozone attack on the plasma membrane of plant cells. *Adv. Phytochem.* **21**: 29-54.
Heath, R.L.: 1989, in, *Biological Markers of Air-Pollution Stress* and *Damage in Forests*. U.S. National Academy Press, Washington DC. pp 347-356.
Heath, R.L.: 1994a, in, *Plant responses to the gaseous environment*. Chapman and Hall, London, pp 121-146.
Heath, R.L.: 1994b, *Photosyn. Res.* **39**, 439-451.

Heath, R.L.: 1996a, in, *Plant Response to Air Pollution*. John Wiley and Sons Ltd, Chichester, UK. pp 353-374.
Heath, R.L.: 1996b, in *Photosynthesis and the Environment*. Baker, N.R. (ed), Adv. in Photosynthesis, Kluwer Academic Publishers, Amsterdam. pp 409-433.
Heath, R.L., and Leech, R.M.: 1978, *Arch. Biochem. Biophys.* **190**, 221-226.
Heath, R.L., and Taylor, Jr., G.E.: 1997, in, *Forest Decline* and *Ozone: A Comparison of Controlled Chamber* and *Field Experiments*. Springer-Verlag, Berlin, pp. 317-368.
Huber, J.L., Redinbaugh, M.G., Huber, S.C., and Campbell, W.H.: 1994, *Plant Physiol.* **106**, 1667-1674.
Imsande, J., and Touraine, B.: 1994, *Plant Physiol.* **105**, 3-7.
Kangasjärvi, J., Talvinen, J., Utriainen, M., and Karjalainen, R.: 1994, *Plant Cell. Environ.* **17**, 783-794.
Kärenlampi, S.O., Airaksinen, K., Miettinen, A.T.E., Kokko, N.I., Holopainen, J.K., Kärenlampi, L.V., and Karjalainen, R.O.: 1994, *New Phytol.* **126**, 81-89.
Keller, T., and Häsler, R.: 1987, *Trees* **1**, 129-133.
Laiski, A., Kull, O., and Moldau, H.: 1989, *Plant Physiol.* **90**, 1163-1167.
Lam, H-M., Coschigano, K.T., Oliveira, L.E., Mela-Oliveira and Coruzzi, G.M.: 1996, *Annu. Rev. Plant Physiol. Molec. Biol.* **47**, 569-593.
Lamb, C., and Dixon, R.A.: 1997, *Annu. Rev. Plant Physiol. Mole. Biol.* **48**, 251-275.
Langebartels, C., Kerner, K., Leonard, S., Schraudner, M., Trost, M., Heller, W., and Sandemann, Jr., H.: 1991, *Plant Physiol.* **95**, 882-889.
Lee, E.H., and Bennett, J.H.: 1982, *Plant Physiol* **69**, 1444-1449.
Luwe, M.W.F.; Takahama, U.; and Heber, U.: 1993, *Plant Physiol.* **101**, 969-976.
Manning, W.J., and Keane, K.D.: 1988, in: *Assessment of crop loss from air pollutants,* Elsevier Applied Science, NY, p 365-386
Mehlhorn, H., O'Shea, J.M., and Wellburn, A.R.: 1991, *J. Exp. Bot.* **42**, 17-24.
Mehlhorn, H.; and Wellburn, A.R.: 1987, *Nature* **327**, 417-418.
Miflin, B.J., Lea, P.J. (eds.): 1990. *Intermediary Nitrogen Metabolism*. in: *The Biochemistry of Plants: A Comprehensive Treatise*. Vol. 16, Academic Press, NY, NY, 401 pp.
Miller, P., and McBride, J.R. (ed.): 1998, *Oxidant Air Pollution Impact on the Montane Forests of Southern California: A Case Study of San Bernardino Mountains*,Springer-Verlag, New York, 235pp.
Musselman, R.C., and Massman, W.J.: 1999, *Atmos. Environ.* **33**, 65-73.
Narváez-Vásquez, M., Orozco-Cárdenas, L., and Ryan, C.A.: 1994, *Plant Physiol.* **105**, 725-730.
Parkhurst, D.F.: 1984, *Can J. Bot.* **62**, 163-165.
Pate, J.S.: 1983, *Plant Physiology: A Treatise*, **8**, 335-395.
Pauls, K.P., and Thompson, J.E.: 1981. *Physiol. Plant.* **53**, 255-262.
Pell, E.J., Eckardt, N., and Glick, R.E.: 1994, *Photosyn. Res.* **39**, 453-462.
Peters, J.L., and Silverthorne, J.: 1995, *The Plant Cell* **7**, 131-140.
Polle, A., and Rennenberg, H.: 1991, *New Phytol* **117**, 335-343.
Polle, A.; Wieser, G.; and Havranek, W.M.: 1995, *Plant Cell Environ* **18**, 681-688.
Raven, J.A.: 1988, *New Phytol.* **109**, 1-20.
Reddy, G.N., Arteca, R.N., Dai, Y.R., Flores, H.E., Negm, F.B., and Pell, E.J.: 1993, *Plant Cell Environ* **16**, 819-826.
Robinson, D.C., and Wellburn, A.R.: 1991, *New Phytol.* **119**, 251-259.
Runeckeles, V.C., and Vaartnou, M.: 1997, *Plant Cell Environ* **20**, 306-314.
Saxe, H.: 1991, *Adv. in Botanical Res* .**18**, 1-128
Scandalios, J.G.: 1994, in, *Plant Response to the Gaseous Environment: Molecular, Metabolic* and *Physiological Aspects*. Chapman and Hall publishers, London, pp 147-164.
Schraudner, M., Ernst, D., Langebartels, C., and Sandermann, Jr., H.: 1992, *Plant Physiol.* **99**, 1321-1328.
Sharma, Y.K., and Davis, K.R.: 1995, *Plant Mol Biol* **29**, 91-98.
Simon-Plas, F., Rustérucci, C., Milat, M.-L., Humbert, C., Montillet, J.-L., and Blein, J.-P.: 1997, *Plant Cell Environ* **20**, 1573-1579.
Slooten, L., Capiau, K., Van Camp, W., Van Montagu, M., Sybesma, C., and Inzé, D.: 1995, *Plant Physiol.* **107**, 737-750.
Solomonson, L.P., Barber, M.J.: 1990. *Annu. Rev. Plant Physiol. Mol. Biol.* **41**, 225-253.
Spanu, P., Grosskopf, D.G., Felix, G., and Boller, T.: 1994, *Plant Physiol* **106**, 529-535.
Stitt, M.: 1996 *Photosynthesis* and *the Environment*. Kluwer, Amsterdam. pp 151-190.
Stone, J.M., and Walker, J.C.: 1995, *Plant Physiol* **108**, 451-457.
Tausz, M., Stabentheiner, E., Wonisch, A., and Grill,D.: 1998, *Environ. Sci. and Pollut. Res.* **1**, 96-100.

Taylor, Jr., G.E., and Hanson, P.J.: 1992, *Agric. Ecosystem Environ.* **42**, 255-273.
Tingey, D.T.; Standley, C.; and Field, R.W.: 1976, *Atmos. Environ.* **10**, 969-974.
Tissue, D.T., Thomas, R.B., and Strain, B.R.: 1993, *Plant Cell Environ* **16**, 859-865.
U.S.A. E.P.A.: 1992, *Air Quality Criteria for Oxides of Nitrogen,*. US Environmental Protection Agency, Environmental Criteria and Assessment Office, Research Triangle Park, NC, EPA-600/8-91/049aA, Vol. II, Chap. 9.
Van Camp, W., Capiau, K., Van Montagu, M., Inzé, D., and L. Slooten, L.: 1996, *Plant Physiol.* **112**, 1703-1714.
Vernon, L.P.: 1960, *Anal. Chem.* **32**, 1144-1150.
Ward, D.A., and Keys, A.J.: 1989, *Photosynth. Res.* **22**, 167-171.
Ward, E.R., Uknes, S.J., Williams, S.C., Dincher, S.S., Wiederhold, D.L., Alexander, D.C., Ahl-Goy, P., Métraux,J.-P., and Ryals, J.A.: 1991, *The Plant Cell* **3**, 1085-1094.
Wedin, D.A., and Tilman, D.: 1996, *Science* **274**, 1720-1723.
Wellburn, A.R., and Lichtenhaler, H.: 1985, *Adv. In Photosynth Res.* W. Junk, The Hague, **II**.1.9-12.
Wellburn, A.R.: 1990, *New Phytol.* **115**, 395-429.
Weyers, J., and Meidner, H.: 1990) in, *Methods in Stomatal Research*, Longman, Essex, pp 32-35.
Wiese, C.B., and Pell, E.J.: 1997, *Plant Cell Environ* **20**, 1283-1291.
Williams, J., Bulman, M.P., and Neill, S.J.: 1994, *Physiol Plant* **91**, 177-182.
Zentgraf, U., and Hemleben, V.: 1996, *Prog Bot* **57**, 218-232.

# ENVIRONMENTAL LIMITATIONS TO $O_3$ UPTAKE - SOME KEY RESULTS FROM YOUNG TREES GROWING AT ELEVATED $CO_2$ CONCENTRATIONS

M.S.J. BROADMEADOW[1], J. HEATH[2] and T.J. RANDLE[1]

[1]*Forest Research, Alice Holt Lodge, Wrècclesham, Farnham, Surrey, GU10 4LH, UK.* [2]*Division of Biological Sciences, Institute of Environmental and Biological Sciences, Lancaster University, Lancaster, LA1 4YQ, UK.*

(Received 13 November 1998; accepted 26 February 1999)

**Abstract.** Elevated carbon dioxide concentrations and limited water supply have been shown to reduce the impact of ozone pollution on the growth and physiology of *Quercus petraea* in a long-term factorial experiment. These responses can be explained by observed reductions in stomatal conductance, and thus potential ozone exposure of 28% and 40% for $CO_2$ and drought treatments respectively. However, parameterisation of a stomatal conductance model for *Quercus robur* and *Fagus sylvatica* grown under ambient and elevated $CO_2$ concentrations in a separate experiment has demonstrated that elevated $CO_2$ also reduces the responsiveness of stomata to both saturation deficit (LAVPD) and soil moisture deficit ($\psi$) in beech, and to a lesser extent, in oak. Season-long model simulations of ozone fluxes suggest that LAVPD and $\psi$ conductance parameters derived at ambient $CO_2$ concentrations will lead to these fluxes being underestimated by 24% and 2% for beech and oak respectively at 615 ppm $CO_2$.

**Keywords:** ozone, carbon dioxide, oak, beech, modelling, soil moisture.

## 1. Introduction

Across Europe as a whole, it is suggested that ozone pollution today probably results in yield losses for forest trees of approximately 10% (Semenov and Koukhta, 1996), although the accuracy of these estimates are uncertain (Skärby and Karlsson, 1996). It is difficult to predict ozone induced reductions in forest productivity because of uncertainty in the mode of action, and episodic nature of ozone pollution, together with the dependency of ozone mediated damage on climatic variables. Predictions for the future are further complicated by the physiological responses of trees to rising atmospheric carbon dioxide concentrations and changing climate.

The growth response of woody species to elevated atmospheric $CO_2$ is highly variable and dependent on environmental conditions, with increases in biomass of 10-50% commonly observed (Ceulemans and Mousseau, 1994). These increases are usually accompanied by reductions in water loss on a leaf area basis as a result of stomatal closure in response to high internal $CO_2$ concentrations (Morison, 1985; Curtis and Wang, 1998). Since the detrimental effects of ozone on plant growth are largely mediated through the stomata, this stomatal closure in response to elevated $CO_2$ would be expected to reduce the flux of ozone to the cellular or chloroplast membranes. An ameliorating effect of elevated $CO_2$ concentrations on ozone induced damage has been reported in terms of both growth (*eg* Mortensen, 1995; *Betula pubescens*) and biochemical integrity (*eg* Kellomaki and Wang, 1997; *Pinus sylvestris*), although the reverse has also been reported for an ozone sensitive *Aspen* clone (Kull *et al.*, 1996).

Clearly, further increases in ambient atmospheric $CO_2$ concentrations may benefit forest health if stomatal responses significantly limit ozone uptake. However, there are still major uncertainties regarding the stomatal response to $CO_2$, both in terms of the physiology/mechanisms and of vegetation-atmosphere exchanges (Morison, 1998). Trees in particular have shown very variable responses to experimental elevation of $CO_2$. In a recent survey and meta-analysis of experiments conducted on woody species, Curtis and Wang (1998) found that stomatal responses to elevated $CO_2$ ranged from strongly negative to strongly positive irrespective of cultural and stress factors. They concluded that there was a small (10%) non-significant reduction in stomatal conductance, much less than the significant 20% reduction reported across a wide range of woody and herbaceous species by Drake et al. (1997). The reported variability in stomatal response of trees to elevated $CO_2$ may result from the fact that stomata respond to a wide variety of external and endogenous factors. Any one or more of these factors (eg light, temperature, humidity, soil and plant water status) could interact to alter the apparent sensitivity of the stomata to $CO_2$ concentration. Whilst a large number of studies have dealt with interactions between $CO_2$ and temperature or nutrition on the growth response of trees, very little is known about the interactive effects of specific factors with $CO_2$ concentration on stomatal conductance. Certainly, we are in no position to make generalised predictions of how stomatal responses to elevated $CO_2$ will be modified by other environmental factors in a future climate. We therefore need to identify the basis of the wide variations between species and between separate experiments to make generalised predictions about the potential of $CO_2$ enrichment to protect against ozone-related damage.

The aims of this paper are firstly, to demonstrate how ozone exposure and its impact on tree growth and physiology is dependent on environmental variables mediated through stomatal conductance; secondly, to identify the impact of elevated $CO_2$ on the relationships between these environmental variables (specifically the saturation deficit, LAVPD and soil moisture content) for two contrasting UK forest tree species; finally, the implications for long-term ozone exposure modelling of these modifications to stomatal response functions will be assessed.

## 2. Materials and methods

2.1 EXPERIMENTAL FACILITIES

*Solardomes*
Nine 2-year-old seedlings of European beech (*Fagus sylvatica*), sweet chestnut (*Castanea sativa*) and pedunculate oak (*Quercus robur*) were transplanted in March 1997 to 4 l pots in a 4:1 mix of *Sphagnum* peat:granite grit with 1 g $l^{-1}$ slow release fertiliser in four chambers with duplicated $CO_2$ concentrations of ambient (365 ppm) or ambient + 250 ppm (615 ppm) (see Heath, 1998 for details). $CO_2$ concentrations were regulated by adding $CO_2$ to the in-flowing air-stream (2.5 - 3 air changes per minute). A more detailed description of the system is given elsewhere (Townend, 1993). In an additional experiment using the same system, ten trees per duplicate chamber were grown from seed in their respective $CO_2$ concentrations, with 2-year-old beech in 8 l

pots subjected to an imposed 4-week period of drought in July 1995 (see Heath and Kerstiens, 1997 for details). Soil water potential was measured with ceramic cup potentiometers.

*Open top chambers*

Sixteen open top chambers in the south-east of England have been used to identify the physiological interactions between ozone, elevated $CO_2$ and water supply on young trees in a factorial experiment, and the consequences of those interactions for biomass production (Crookshanks *et al.,* 1998). Twelve plants of sessile oak (*Quercus petraea*) were planted in March 1994 in the ground (in native soil, a highly cultivated humo-ferric podzol) as 1+1 seedlings in duplicate chambers of each treatment. Factorial treatments of ambient or elevated (700 ppm) $CO_2$, ambient or elevated (20 ppb overnight rising to 100 ppb for four hours in the middle of the day from 10 April to 20 September) ozone, and droughted (irrigation initiated at a soil moisture deficit, $\psi$, of 150 kPa) or well watered (irrigation initiated at 10 kPa $\psi$) were imposed. All combinations of the three individual treatments were represented. Elevated $CO_2$ concentrations were maintained by adding $CO_2$ to the in-flowing air-stream (0.5 $m^3$ $s^{-1}$; 2 air changes per minute) under positive feedback control (see Barton *et al.,* 1993). Ozone concentrations were controlled using the same software, with ozone produced from air (with water scrubber to remove $N_2O_5$) by variable voltage electric discharge (model B063 Odotrol: Wallace and Tiernan, Kent, UK). The elevated ozone treatment represented an AOT40 exposure of 48 ppm.h compared to 2.4 ppm.h for the ambient treatment in 1995. Irrigation was controlled using a data-logger (DT500: DataElectronics, Victoria, Australia) to operate solenoid valves when $\psi$ fell below 10 and 150 kPa (25% and 10% v/v soil moisture) for the irrigated and droughted treatments, respectively. Activation of the irrigation system resulted in the supply of 8 l of water applied through 36 drip heads per chamber. Droughted and well watered outside plots were also included for comparison (data not shown).

2.2 EXPERIMENTAL PROCEDURES

*Conductance measurements*

A porometer (PMR-1; PP-Systems Ltd., Hitchin, Herts., UK) was used in the experiment investigating the effects of drought on beech (Heath and Kerstiens, 1997). All other measurements were made with LCA3 plant gas exchange analysis systems (ADC Ltd., Hoddesdon, Herts., UK) using a PLC3-B leaf cuvette with an area of 6.2 $cm^2$. Boundary layer conductance was assumed as 30 mmol $m^{-2}$ $s^{-1}$. Measurements were made within one minute of placing the leaf within the cuvette, and conditions assumed to be those measured within the solardome/open top chamber, rather than within the leaf cuvette itself.

*Photosynthetic capacity*

Measurements were made for *Q.petraea* between 0800 and 1200 hours GMT on four dates in early September 1995 using an LCA3 plant gas exchange analysis system as described above. Air was supplied at 1700 ppm $CO_2$ by mixing ambient compressed air at 15 l per minute with $CO_2$ injected through a needle valve at a rate of approximately 30 ml per minute. Fine control was achieved using the integral soda-lime column of the gas analyser. Measurements were made at 800 µmol m$^{-2}$ s$^{-1}$ PAR supplied by a 12V halogen lamp (PP-systems, Stotfold, UK), and at cuvette temperatures between 25°C and 30°C.

*Foliar chlorophyll content*
Fully exposed, expanded leaves from the previous flush were collected on 8 September 1995 and immediately stored on dry ice. Four discs (approximately 100 mg) were removed with an 8 mm corer, and frozen with liquid nitrogen in a mortar. The tissue was ground to a paste with a pestle and suspended in 10 ml of 80 % acetone. The cellular debris was removed by centrifuging at 4000 rpm for one minute. The absorbance of the extract was measured at 664 and 647 nm (model 254 colorimeter; Corning Science Products, New York), and the chlorophyll concentration calculated according to Vernon (1960).

2.3 EXPERIMENTAL DESIGN AND STATISTICAL ANALYSIS

The Headley experiment was designed as a randomised split plot with duplicate factorial combinations of $O_3$, $CO_2$ and irrigation treatment. In addition, four outside plots were used, with water supply the only imposed treatment. Analysis of variance was carried out using Genstat 5.3 to test for main treatment effects and interactions (9 degrees of freedom for treatments, 10 for residuals), with all data log$_e$ transformed to stabilise the variance.

Data from the duplicate solardomes at Lancaster were analysed by analysis of variance. Individual trees were treated as the experimental unit.

## 3. Results

3.1. IMPACT STUDIES AT HEADLEY

Stomatal conductance of *Quercus petraea* was measured in the open top chambers on seven days during the 1995 growing season (Figure 1).

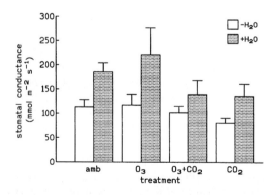

Fig. 1. Stomatal conductance of *Q. petraea*. Values represent the mean of seven days' measurements during the 1995 growing season, with ten individual measurements made per treatment per day (+/- 1 sd).

Mean stomatal conductance was reduced in both the droughted and elevated $CO_2$ treatments, whilst ozone alone led to a small increase in conductance, particularly in the well-watered plants.

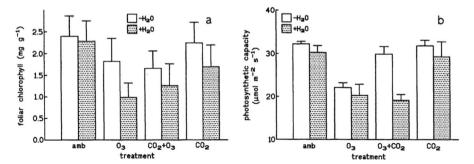

Fig. 2. (a) Foliar chlorophyll content of *Q. petraea* (fresh weight basis). Measurements were made between 1 and 10 September 1995 and represent the mean of four determinations per duplicate chamber (+/- 1 sd). (b) Light and $CO_2$ saturated photosynthetic rate ($A_{max}$) of *Q. petraea*. Values represent the mean of four days measurements towards the end of the 1995 growing season, with ten individual measurements made per treatment per day (+/- 1 sd).

Foliar chlorophyll content was measured after five months of ozone fumigation (September 1995; Figure 2a) and showed a significant interaction between ozone induced damage and $CO_2$ treatment. The largest $O_3$-induced reduction was observed in the irrigated ambient $CO_2$ treatment, with both reduced water supply and elevated $CO_2$ providing protection. Similar trends were observed in photosynthetic capacity expressed as $CO_2$ saturated ($C_a$=1700 ppm) photosynthetic rate ($A_{max}$; Figure 2b, Broadmeadow and Freer-Smith, 1999); here, a negligible $O_3$-induced reduction in $A_{max}$ was observed in the droughted elevated $CO_2$ treatment (compared to the ambient droughted control), which contrasted with the 40% reduction in the irrigated elevated $CO_2$ treatment.

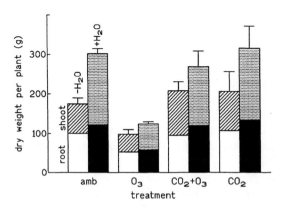

Fig. 3. Total biomass production for oak after three years growth under the experimental conditions (December 1996). Values represent the mean of twelve trees per duplicate chamber (+/- half the duplicate chamber mean difference).

These impacts at the physiological level were evident in the response of total biomass production of *Q. petraea* for the various treatments (Figure 3) with highly significant effects of both $CO_2$ and $O_3$. There were also significant interactions between $CO_2$ and $O_3$, and $O_3$ and irrigation (p<0.01 and p<0.05 respectively), with $CO_2$ reducing and irrigation increasing the response to ozone fumigation.

## 3.2. STOMATAL CONDUCTANCE PARAMETERISATION

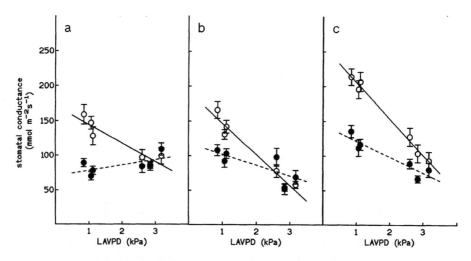

Fig. 4. Stomatal conductance of (a) beech, (b) chestnut and (c) oak on six days of contrasting leaf-to-air vapour pressure deficit (LAVPD) in July 1997. Open circles: ambient air; solid circles: elevated $CO_2$; error bars represent 1 se. Data from Heath (1998).

Research carried out using young trees growing in closed chambers (Solardomes) at Lancaster has analysed stomatal responses to elevated $CO_2$ in more detail, identifying how these responses are modified by other climatic variables. This work concentrated on three members of the Fagaceae: European beech (*Fagus sylvatica*), sweet chestnut (*Castanea sativa*) and pedunculate oak (*Quercus robur*).

When trees were well-watered and LAVPD was moderate or low, the stomatal conductance of beech, chestnut and oak was always significantly lower (by up to 50%) in elevated than in ambient $CO_2$ (Figure 4). However, when LAVPD was high, the degree of stomatal closure observed in elevated $CO_2$ was much reduced or completely disappeared, particularly in beech and chestnut (Figure 4). The result was that, while oak always maintained some degree of stomatal closure in elevated $CO_2$, beech and chestnut did not reduce stomatal conductance at all - or even increased it in some cases - on days when LAVPD was high.

Soil drought was an equally strong modifier of the stomatal response to $CO_2$, with large species differences evident in this interaction. During a period of progressive soil moisture deficit in 1997 (data not shown), stomatal conductance of beech was similar in ambient and elevated $CO_2$ treatments (Heath, 1998). In contrast, the stomatal conductance of oak was reduced by an average of 50% in elevated $CO_2$ during this period (Heath, 1998). Over a 4-week period of imposed drought in July 1995 (Figure

5a), beech maintained a consistently higher (approximately double) stomatal conductance in elevated $CO_2$, in spite of soil water potential becoming much lower than for trees grown in ambient air. Thus, beech is likely to have been more susceptible to ozone damage during this period of soil moisture deficit as a result of an increased physiologically effective dose in the elevated $CO_2$ treatment. Pearson and Mansfield (1993) also found that exposure to ozone itself reduced the degree of stomatal closure occurring in beech during drought, whilst elevated $CO_2$ has been shown to exacerbate the effects of $O_3$ pollution on photosynthetic integrity (Kull et al., 1996). Thus, interactions with other environmental variables cannot be ignored when attempting to make predictions of effective exposure to ozone.

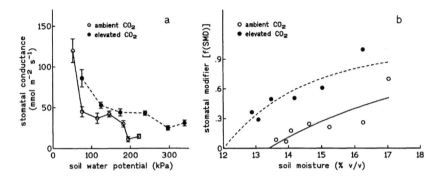

Fig. 5 (a) Stomatal conductance of beech over a 4-week period of drought in July 1995, plotted against soil water potential for ambient (open symbols) and elevated (closed symbols) $CO_2$ concentrations: error bars represent 1 se. Data from Heath and Kerstiens (1997). (b) Stomatal response function of beech derived from (a).

Stomatal responses of oak and beech to soil water potential and LAVPD have been parameterised using data from the solardomes at Lancaster for the Jarvis model of stomatal conductance (Jarvis, 1976; equations 1 to 5) (Table I). No temperature function is included here, since the response to temperature alone is difficult to distinguish from that to LAVPD.

$$g_s = \min \{[g_{smax}\, f(I)\, f(SMD)\, f(LAVPD), f(CO_2)], [g_{s0}]\} \quad \text{(eqn. 1)}$$

where:

$$f(I) = (Q_z\, I)/(g_{smax} + (Q_z\, I)) \quad \text{(eqn. 2)}$$
$$f(LAVPD) = 1-((LAVPD-d_1)/(d_2-d_1)) \quad \text{(eqn. 3)}$$
$$f(SMD) = 1- \exp(k_{sm}((sm_{sat} - sm)-(sm_{sat} - sm_{wilt}))) \quad \text{(eqn. 4)}$$
$$f(CO_2) = 1-(c_{cor}(C_a-350)) \quad \text{(eqn. 5)}$$

where $g_s$ is stomatal conductance to water vapour, $f(I)$, $f(LAVPD)$, $f(CO2)$ and $f(SMD)$ are the stomatal functions to light, LAVPD, $CO_2$ concentration and soil moisture deficit, $g_{s0}$ is minimum stomatal conductance, $Q_z$ is the slope of the light response curve when I=0 (=0.002 mol μmol $PFD^{-1}$), $d_1$ and $d_2$ are the values of LAVPD at which $f(LAVPD)$ = 1 and 0, $sm_{sat}$ and $sm_{wilt}$ are the soil moisture content at field capacity and wilting point,

$k_{sm}$ is the exponential coefficient of the soil moisture deficit response, $C_a$ is $CO_2$ concentration and $c_{cor}$ is the slope of the linear response of $g_s$ to $C_a$.

$g_{smax}$ was derived as the y-intercept of the $g_s$ vs LAVPD response, $d_1$ was assumed to be zero and $d_2$ was LAVPD when $g_s = g_{s0}$. In the case of beech under elevated $CO_2$, $f$(LAVPD) was assigned a value of 1, and $g_{smax} = 0.086$ mol m$^{-2}$ s$^{-1}$ (based upon the mean value of $g_s$ observed at high and low LAVPD). $c_{cor}$ was calculated from $g_{smax}$ at ambient and elevated $CO_2$. Soil water potentials from the beech drought experiment were converted to volumetric soil moisture content using a moisture retention curve (data not shown). $sm_{sat}$ is 0.3 v/v (for the humo-ferric podzol in the Headley open top chambers), and $sm_{wilt}$ and $k_{sm}$ were derived by fitting the observed relationship between $sm$ and $f$(SMD) by least sum of squares, resulting in the relationships shown in Figure 5b. $f$(SMD) was assigned a value of 1 for oak, since corresponding experimental data are not available, but the stomatal response of oak to soil moisture does not appear to be affected by $CO_2$ treatment (Heath, 1998).

TABLE I.
Parameters for the Jarvis model of stomatal conductance used to estimate physiologically effective ozone dose.

| parameter | beech | | oak | |
|---|---|---|---|---|
| | 365 ppm $CO_2$ | 615 ppm $CO_2$ | 365 ppm $CO_2$ | 615 ppm $CO_2$ |
| $g_{smax}$ | 171 | | 259 | |
| $g_{s0}$ | 10 | 10 | 10 | 10 |
| $c_{cor}$ | 1.99 x 10$^{-3}$ | | 1.76 x 10$^{-3}$ | |
| $d_1$ | 0 | 0 | 0 | 0 |
| $d_2$ | 6.04 | 100 | 4.76 | 5.84 |
| $sm_{sat} - sm_{wilt}$ | 16.6 | 18.0 | 16.6 | 17.6 |
| $k_{sm}$ | 0.019 | 0.042 | 0.019 | 0.019 |

The Jarvis conductance model was then run within the framework of the Forestry Commission individual tree growth model GROMIT (Randle and Ludlow, 1998), for the 1997 growing season using weather and ozone concentration data obtained from the Headley open top chambers (one hour time interval). Trees from all treatments were assumed to have identical parameters apart from the conductance parameters listed in Table I. For clarity, ozone was assumed to have no direct effect on stomatal conductance. Ozone flux was calculated as the product of ozone concentration and stomatal conductance to water vapour, corrected for the difference in diffusive coefficient of water vapour and ozone ($j_{O3}/j_{H2O} = 0.61$; Taylor and Constable, 1994).

The model was run assuming ambient (365 ppm) or elevated (615 ppm) $CO_2$ at ambient or elevated (ambient + 2°C) temperatures, with results expressed as cumulative ozone flux (Table 2) for the period 1 May to 30 September for upper and lower canopy leaves (total leaf area index =2.0). For oak (*Quercus* spp.), this modelling exercise demonstrates the dramatic reduction (36%) in physiologically effective ozone flux that is predicted under conditions of elevated $CO_2$ alone, and that the modification of $f$(LAVPD) by elevated $CO_2$ has little effect on the model output (Table II, Figure 6a). In the case of beech, if ambient $CO_2$ stomatal response functions are assumed, elevated $CO_2$ would be expected to lead to a larger, 43% reduction in ozone dose over the course of the growing season. However, if the modified stomatal response functions derived here are assumed, the predicted physiologically effective ozone dose is only reduced by 29% in elevated $CO_2$ (Table II; Figure 6b).

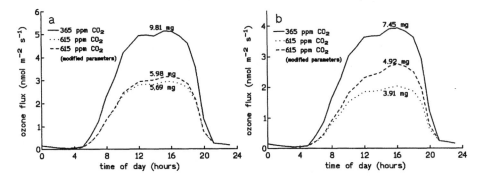

Fig. 6. Modelled ozone dose for (a) oak and (b) beech on 22 May 1997. The solid and dotted lines represent the flux at ambient and ambient +250 ppm $CO_2$ using identical (ambient) conductance parameters, whist the dashed line represents the flux in elevated $CO_2$ assuming the modified conductance parameters from Table II.

TABLE II.
Model output expressed as cumulative ozone dose and total transpirational water loss for the 1997 growing season (1 May - 30 September).

| scenario | cumulative ozone dose (mg m$^{-2}$) | | | | transpiration (mm) | |
|---|---|---|---|---|---|---|
| | beech | | oak | | beech | oak |
| | upper canopy | lower canopy | upper canopy | lower canopy | | |
| ambient parameters | | | | | | |
| ambient | 507 | 462 | 614 | 531 | 127 | 148 |
| +250 ppm $CO_2$ | 290 | 266 | 394 | 342 | 91 | 121 |
| +2°C | 492 | 449 | 592 | 511 | 133 | 155 |
| +250 ppm $CO_2$;+2°C | 285 | 260 | 383 | 331 | 98 | 127 |
| modified parameters | | | | | | |
| +250 ppm $CO_2$ | 359 | 329 | 403 | 349 | 109 | 123 |
| +250 ppm $CO_2$;+2°C | 354 | 324 | 392 | 339 | 117 | 129 |

Table II also demonstrates the predicted effect of incident light on ozone flux, indicating that upper canopy leaves are likely to receive a higher ozone flux than those from the lower canopy, although other factors such as branch hydraulic resistance (Chappelka and Samuelson, 1998) may affect this. The difference between upper and lower canopy fluxes are likely to be greater still in forest canopies where leaf area indices are generally higher than the 2.0 assumed in these simulations, although ozone concentration gradients present in forest canopies (Samuelson and Kelly, 1997) are not modelled here. A temperature rise of 2°C leads to a small reduction in modelled cumulative ozone dose in both species (2-3%), as a result of stomatal closure. This effect of increased temperature is predicted at both ambient and elevated $CO_2$ concentrations.

Total water use of oak is predicted to be markedly lower (18%; Table II) at elevated $CO_2$ concentrations, even assuming a 2°C increase in temperature (9% lower), and the effects of the modified stomatal parameters at elevated $CO_2$ concentrations are negligible. In the case of beech, elevated $CO_2$ alone is predicted to result in a larger,

28% reduction in water use, although the modified stomatal functions limit this reduction to 15%. Under conditions of elevated temperature and $CO_2$ concentrations, the effect of the modified stomatal parameters is more dramatic, with the reduction in water use falling from 23% to 8% when compared with current conditions. These modified stomatal responses are therefore highly relevant to the modelling of growth responses of trees to changing climate on a larger scale, and also to the feedback between tree function and climate change.

## 4. Discussion

In general terms, we subscribe to the widely held view that both elevated $CO_2$ and drought protect the biochemical integrity of a plant against ozone through reducing the physiologically effective ozone dose, as shown by the biomass and physiological response data reported here. The mechanism behind this conclusion is based upon stomatal closure commonly observed at elevated $CO_2$ concentrations (Eamus and Jarvis, 1989), although the magnitude of this effect varies widely (Curtis and Wang, 1998; Drake et al., 1997). In addition, other ontogenetic, structural, and biochemical factors also limit ozone mediated damage (Wieser and Havranek, 1993). The data presented here demonstrate how other environmental conditions may modify this stomatal response to elevated $CO_2$, and thereby alter predictions of how environmental and climate change may affect yield losses attributable to ozone. The fact that the stomata of certain species (eg beech) may fail to close in response to elevated $CO_2$ during periods of high evapo-transpirational demand and/or soil drought also has wide-reaching implications for climate modelling and the widely-held view that increased water use efficiency under elevated atmospheric $CO_2$ will result in improved drought tolerance.

Accurate predictions of stomatal responses to elevated $CO_2$ may be confounded by difficulties in obtaining precise parameterisations of stomatal response functions. The fundamental parameter for the conductance model used here (Jarvis, 1976) is $g_{smax}$, a quantity that is difficult to measure precisely. The values of $g_{smax}$ presented here are derived by extrapolating the LAVPD response to zero LAVPD, which may show diurnal variation or be affected by short term variation in the moisture content of the rooting zone. For beech, the $CO_2$ induced reduction in $g_{smax}$ is much larger than reported in other studies, both on beech (-30%; Overdieck and Forstreuter, 1994), and across 38 studies of woody plants (-10%; Curtis and Wang, 1998). If the value of the $CO_2$ response of $g_{smax}$ reported by Overdieck and Forstreuter (1994) is used in the model simulation (95 and 68 mmol m$^{-2}$ s$^{-1}$ at 350 and 700 ppm $CO_2$; $f(CO_2)=0.019$), a very different result with respect to cumulative ozone dose is predicted; at 615 ppm $CO_2$, the total ozone flux would be 2% above that at ambient $CO_2$ concentrations if the modified stomatal responses, $f(LAVPD)$ and $f(SMD)$ are assumed. In this case, the response to elevated $CO_2$ would therefore be opposite to that expected from gas exchange measurements reported to date. This therefore demonstrates both the importance of accurate parameterisations of the conductance models, and the relevance of the modifications to stomatal responses reported here to the modelling of tree and forest function.

Although the general response of stomata to environmental variables has been described in some detail, most experiments have measured stomatal conductance relative to only one variable, and usually at a single (usually relatively low) value of LAVPD. However, particularly in relation to ozone pollution episodes, we need to model the behaviour of stomata during periods of high evaporative demand and/or drought. Reduced stomatal sensitivity to elevated $CO_2$ at high, compared to low LAVPD has been reported in other woody species: in *Quercus ilex* (Tognetti *et al.*, 1998); in *Cercis canadensis* (Will and Teskey, 1997); and in *Pinus radiata* and *Pseudotsuga menziesii* (Hollinger, 1987). There is clearly a need for further investigation of the effects of LAVPD on the magnitude of the stomatal response to $CO_2$, since it appears possible that measurement of LAVPD might account for much of the variability of responses reported in the literature. These studies of stomatal conductance must be reported more systematically, providing parameters for the various models of stomatal conductance. Furthermore, stomatal responses of seedlings may differ from those of mature trees (Samuelson and Edwards, 1993), and the effect of ozone on tree growth also varies with both age and size (Kolb *et al.*, 1997). Only when we fully understand the basis of the variability in response (between species and between experiments) and the underlying mechanisms will we be able to accurately predict the likely effects of future environmental change on the susceptibility of trees and forests to ozone damage.

Returning to the question of modelling physiologically effective ozone doses, it should be remembered that episodes of ozone pollution generally coincide with periods of high LAVPD and soil moisture deficit when the stomatal response to elevated $CO_2$ is at a minimum. Hence, during these critical periods, it is quite possible that any protective effect of elevated $CO_2$ mediated through a reduction in physiologically effective ozone dose will be less than expected, or disappear altogether in some species. Therefore, in spite of widely-held assumptions to the contrary, elevated $CO_2$ may be unable to protect against the compounding effects of drought, excessive transpiration rates and ozone stress in some tree species.

## 5. Conclusion

Elevated $CO_2$ concentrations have clearly been shown to reduce the impact and magnitude of ozone uptake in *Q. petraea* through reductions in stomatal conductance, and thus ozone uptake. However, long-term model simulations may overestimate the ameliorating effect of increasing $CO_2$ concentrations on the impact of ozone pollution as a result of the limited responsiveness of stomata to both saturation and soil moisture deficits at elevated $CO_2$, particularly for beech. This should be taken into account when predicting future forest yield losses attributable to ozone.

## Acknowledgements

This work was funded by the UK Forestry Authority (Forest Practice Division), the European Union (ECOCRAFT - contract ENV4-CT95-0077) and the UK DETR (Carbon sequestration in vegetation and soils, EPG 1/1/39). The authors thank Sue Benham for providing ozone concentration data, Samantha Jackson for undertaking much of the gas-exchange analysis at Headley and Tracy Houston for statistical advice and analysis. Thanks are also due to Terry Mansfield for helpful discussions throughout the duration of the work.

## References

Barton, C.M.V., Lee, H.S.J. and Jarvis, P.G.: 1993, *Plant, Cell Environ.* **16**, 1139-1148.
Broadmeadow, M.S.J. and Freer-Smith, P.H.: 1999, *Irish Forestry,* In Press.
Ceulemans, R. and Mousseau, M.: 1994, *New Phytol.* **127**, 425-446.
Chappelka, A.H. and Samuelson L.J.U.: 1998, *New Phytologist* **139**, 91-108.
Crookshanks, M., Taylor, G. and Broadmeadow, M.: 1998, *New Phytol.* **138**, 241-250.
Curtis, P. S. and Wang, X.: 1998, *Oecologia* **113**, 299-313.
Drake, B.G., Gonzalez-Meler, M. and Lomg, S.P.: 1997, *Ann. Rev. Plant Physiol. Plant Molec. Biol.* **48**, 609-639.
Eamus, D. and Jarvis, P.G.: 1989, *Adv. Ecol. Res.* **19**, 1-55.
Heath, J.: 1998, *Plant, Cell Environ.* **21**, 1077-1088.
Heath, J. and Kerstiens, G.: 1997, *Plant, Cell Environ.* **20**, 57-67.
Hollinger, D. Y.: 1987, *Tree Physiol.* **3**, 193-202.
Jarvis, P.G.: 1976, *Phil. Trans. Roy. Soc. London, ser. B* **273**, 593-610.
Kellomaki, S. and Wang, K-Y.: 1997, *Environ. Pollut.* **97**, 17-27.
Kolb, T.E., Frederickson, T.S., Steiner, K.C. and Skelly, J.M.: 1997, *Environ. Pollut.* **98**, 195-208.
Kull, O., Sober, A., Coleman, A.D., Dickson, R.E., Isebrands, J.G., Gagnon, Z. and Karnosky, D.F: 1996, *Can. J. For. Res.* **26**, 639-648.
Morison, J.I.L.: 1985, *Plant Cell Environ.* **8**, 467-474.
Morison, J.I.L.: 1998, *J. Exptl Bot.* **49**, 443-452.
Mortensen, L.V.: 1995, *Environ. Pollut.* **87**, 337-343.
Overdieck, D. and Forstreuter, M.: 1994, *Tree Physiol.* **14**, 997-1003.
Pearson, M. and Mansfield, T. A.: 1993, *New Phytol.* **123**, 351-358.
Randle, T.J. and Ludlow, A.R.: 1998, in *The green spruce aphid in western Europe: Ecology, status, impacts and prospects for management.* Forestry Commission Technical Paper 24. Forestry Commission, Edinburgh. pp 32-40.
Samuelson, L.J. and Edwards, G.S.: 1993, *New Phytol.* **125**, 373-379.
Samuelson, L.J. and Kelly, J.M.: 1997, *New Phytol.* **136**, 255-264.
Semenov, S. and Koukhta, B.: 1996, in *Critical Levels for Ozone in Europe: Testing and Finalising the Concepts, UN-ECE Workshop Report,* Kärenlampi, L. and Skärby, L. (eds), pp. 96-107, University of Kuopio, Dept. of Ecol. and Environ. Sci.
Skärby, L. and Karlsson, P.E.: 1996, in *Critical Levels for Ozone in Europe: Testing and Finalising the Concepts, UN-ECE Workshop Report,* Kärenlampi, L. and Skärby, L. (eds), pp. 72-85, University of Kuopio, Dept. of Ecol. and Environ. Sci.
Taylor, G.E. and Constable, J.V.H.: 1994, in *Air Pollutants and the Leaf Cuticle.* Percy, K.E., Cape, J.N., Jagels, R. and Simpson, C.J. (eds), pp15-37, NATO ASI series G, vol 36. Springer-Verlag, Heidelberg.
Tognetti, R., Longobucco, A., Miglietta, F. and Raschi, A.: 1998, *Plant, Cell Environ.* **21**, 613-622.
Townend, J.: 1993. *Tree Physiol.* **13**, 389-399.
Vernon, L.P.: 1960, *Anal. Chem.* **32**, 1144-1150.
Wieser, G. and Havranek M.: 1993, *Trees* **7**, 227-232.
Will, R. E. and Teskey, R. O.: 1997, *J. Exptl Bot.* **48**, 2095-2102.

# EFFECTS OF TROPOSPHERIC $O_3$ ON TREMBLING ASPEN AND INTERACTION WITH $CO_2$: RESULTS FROM AN $O_3$-GRADIENT AND A FACE EXPERIMENT

D.F. KARNOSKY[1], B. MANKOVSKA[2], K. PERCY[3], R.E. DICKSON[4], G.K. PODILA[5], J. SOBER[1], A. NOORMETS[1], G. HENDREY[6], M.D. COLEMAN[7], M. KUBISKE[8], K.S. PREGITZER[1], and J.G. ISEBRANDS[4]

[1] *School of Forestry and Wood Products, Michigan Technological University, 101 U.J. Noblet Forestry Building, 1400 Townsend Drive, Houghton, Michigan 49931-1295.* [2] *Forest Research Institute, T.G. Masaryka 22, 960 92 Zvolen, Slovakia.* [3] *Natural Resources Canada, Canadian Forest Service, P.O. Box 4000, Fredericton, New Brunswick, Canada E3B 5P7.* [4] *U.S. Forest Service, North Central Forest Experiment Station, Forestry Sciences Laboratory, 5985 Highway K, Rhinelander, Wisconsin 54501.* [5] *Department of Biological Sciences, Michigan Technological University, 1400 Townsend Drive, Houghton, Michigan 49931-1295.* [6] *Brookhaven National Laboratory, Biosystems and Process Sciences Division, 1 South Technology Street, Upton, New York 11973.* [7] *USDA Forest Service, Savannah River Institute, P.O. Box 700, Building 760-15G, New Ellenton, South Carolina 29809.* [8] *Department of Forestry, Mississippi State University, Box 9681, Mississippi State, Mississippi 39762-9681.*

(Received 25 October 1998; accepted 25 February 1999)

**Abstract.** Over the years, a series of trembling aspen (*Populus tremuloides* Michx.) clones differing in $O_3$ sensitivity have been identified from OTC studies. Three clones (216 and 271[($O_3$ tolerant] and 259 [$O_3$ sensitive]) have been characterized for $O_3$ sensitivity by growth and biomass responses, foliar symptoms, gas exchange, chlorophyll content, epicuticular wax characteristics, and antioxidant production. In this study we compared the responses of these same clones exposed to $O_3$ under field conditions along a natural $O_3$ gradient and in a Free-Air $CO_2$ and $O_3$ Enrichment (FACE) facility. In addition, we examined how elevated $CO_2$ affected $O_3$ symptom development. Visible $O_3$ symptoms were consistently seen (5 out of 6 years) at two of the three sites along the $O_3$ gradient and where daily one-hour maximum concentrations were in the range of 96 to 125 ppb. Clonal differences in $O_3$ sensitivity were consistent with our OTC rankings. Elevated $CO_2$ (200 ppm over ambient and applied during daylight hours during the growing season) reduced visible foliar symptoms for all three clones from 31 to 96% as determined by symptom development in elevated $O_3$ versus elevated $O_3$ + $CO_2$ treatments. Degradation of the epicuticular wax surface of all three clones was found at the two elevated $O_3$ gradient sites. This degradation was quantified by a coefficient of occlusion which was a measure of stomatal occlusion by epicuticular waxes. Statistically significant increases in stomatal occlusion compared to controls were found for all three clones and for all treatments including elevated $CO_2$, elevated $O_3$, and elevated $CO_2$ + $O_3$. Our results provide additional evidence that current ambient $O_3$ levels in the Great Lakes region are causing adverse effects on trembling aspen. Whether or not elevated $CO_2$ in the future will alleviate some of these adverse effects, as occurred with visible symptoms but not with epicuticular wax degradation, is unknown.

**Key words:** ozone, carbon dioxide, FACE, aspen, greenhouse gases, climate change, gradients

## 1. Introduction

Global atmospheric concentrations of the greenhouse gases carbon dioxide ($CO_2$) and tropospheric ozone ($O_3$) are increasing at the rate of 1 to 2% per year (Keeling *et al.*, 1995; Mohnen *et al.*, 1993). While elevated $CO_2$ tends to enhance tree photosynthesis and growth and to increase water use efficiency (Ceulemans and Mosseau, 1994), $O_3$ is a highly phytotoxic gas that impacts forest trees (Barnard *et al.*, 1991; Chappelka and Samuelson, 1998; Hogsett *et al.*, 1997). Treshow (1970) and Treshow and Stewart

(1973) were the first to demonstrate that trembling aspen (*Populus tremuloides* Michx.) is sensitive to $O_3$. The strong genetic control involved in aspen's response to $O_3$ was first shown by Karnosky (1976, 1977). Subsequent aspen studies have: documented decreased growth under ambient $O_3$ conditions (Wang et al., 1986); developed dose response relationships for growth responses (Karnosky et al., 1992a, b; Karnosky et al., 1996; Karnosky et al., 1998); characterized the physiological mechanisms for $O_3$ responses of aspen (Gagnon et al., 1992; Coleman et al., 1995a, b; Coleman et al., 1996); documented population differences in $O_3$ tolerance (Berrang et al., 1986, 1989, 1991); and implicated superoxide dismutase in $O_3$ tolerance (Sheng et al., 1997; Karnosky et al., 1998).

In contrast to their response to $O_3$, *Populus* species generally respond positively to increases in $CO_2$ concentrations. Increases in individual leaf area, whole crown leaf area, leaf area duration, and leaf area index have been observed in *Populus* trees grown in elevated $CO_2$ (Ceulemans et al., 1994; Curtis et al., 1995). Aspen grown in twice ambient $CO_2$ had greater photosynthetic rates and a greater whole plant photosynthesis because of increased photosynthetic rates in the lower half of the crown (Kubiske et al., 1997).

Given that both tropospheric $O_3$ and $CO_2$ are likely to continue to increase into the foreseeable future, the need for a better understanding of the complex interaction of $CO_2$ and $O_3$ on forest ecosystems is obvious (Volin and Reich, 1996). However, relatively little research has been done to characterize the interacting effects of $O_3$ and $CO_2$ for trees (Barnes and Wellburn, 1998). Kull et al. (1996) reported that elevated $CO_2$ (150 ppm over ambient) increased $O_3$ susceptibility of two aspen clones exposed to $O_3$ as determined by photosynthetic responses. However, subsequent evaluation of growth responses of the same aspen clones suggested that elevated $CO_2$ (150 ppm over ambient) did not exacerbate the $O_3$ effects as predicted from Kull's work, but it also did not ameliorate the adverse effects of $O_3$ on plant growth over a three-year time period with plants growing in natural soil in OTCs (Karnosky et al., 1998). In contrast, Volin and Reich (1996) and Volin et al. (1998) reported that elevated $CO_2$ (300 ppm over ambient) ameliorated $O_3$ effects for young aspen plants grown for 58 or 101 days in pots in growth chambers.

Two novel methods for studying the effects of $O_3$ on aspen ecosystems are described in this paper. First, aspen clones differing in $O_3$ sensitivity were established in plantations at three locations differing in their ambient $O_3$ characteristics (41 to 70.4 ppmh in SUM0 [the sum of daylight hourly $O_3$ concentrations] from June 1 to August 30, 1996). Second, a Free-Air $CO_2$ and $O_3$ Enrichment (Aspen FACE) experiment near Rhinelander, Wisconsin, was established where aspen clones differing in $O_3$ and/or $CO_2$ sensitivity were planted in a FACE system capable of accurately and reliably dispensing $CO_2$ and/or $O_3$.

## 2. Methods

### 2.1. THE $O_3$ GRADIENT STUDIES

In 1993, we established three types of aspen plantations at each of three locations (Rhinelander, Wisconsin - low $O_3$; Kalamazoo, Michigan - intermediate $O_3$; and Kenosha, Wisconsin - high $O_3$) along a natural $O_3$ gradient in the Lake States (Fig. 1).

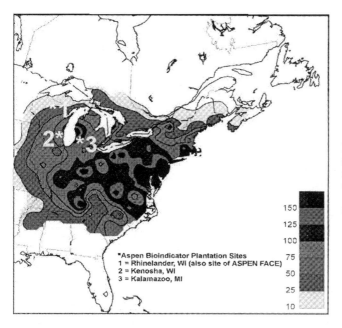

Fig. 1. Spatial distribution of average hours with ozone >82 ppb (1986-1993) (from: Dann and Summers, 1997) and locations of aspen plantations (1, 2, 3).

At each location, we established three experiments. The first was a "common garden" experiment of 7 aspen clones differing in $O_3$ sensitivity (clones 1, 253 and 259 = $O_3$ sensitive; clones 10, 216, 221, and 271 = $O_3$ tolerant [Karnosky, 1976; Karnosky et al., 1992a]). This experiment consisted of individual trees of each clone planted at 2 m x 2 m spacing and there were 10 replicates at each site. In the second experiment, we established a "growth and yield" trial with $O_3$ tolerant clone 216 and $O_3$ sensitive clone 259 planted in 16 tree (4 x 4) blocks at 2 m x 2 m spacing between trees and with six replicates. The final experiment at each site was a "competition trial" between clones 216 and 259 where trees were planted at 0.5 m x 0.5 m between trees and 100 tree (10 x 10) blocks were established in either pure clonal blocks or mixed clonal blocks (the $O_3$ sensitive and tolerant clones were alternately spaced in this part of the plantation). Again, six replicates were used at each site. Two border rows of clone 271 were established around each of the three plantations at each site.

Weeds were controlled around all trees by herbicide application in the first season and then mowing in subsequent years until weed competition was no longer a problem. All trees were measured annually and also observed at least once per season for visible foliar symptoms. All three sites were old field sites that had not been in agricultural use for several years previously. The sites all have sandy loam soils with relatively high fertility. Temperature and rainfall patterns from the three sites suggest that they are not widely different in summer climate (Table I). Each site was enclosed in a 3 m tall deer fence. $O_3$ has been monitored independently and continuously at the Kenosha site and Rhinelander site by the Wisconsin Department of Natural Resources. The Kalamazoo site was monitored for $O_3$ by the Pharmacia-Upjohn Company about 0.5 km from our aspen site. The $O_3$ data was quality assured and was part of the AIRS (Aeometric Information and Retrieval Systems of the U.S. Environmental Protection Agency) network. A summary of the $O_3$ values for 1995-1997 are shown in Table II.

TABLE I

Comparisons of monthly mean temperatures (°C) and seasonal total rainfall (cm) at our three aspen sites along a natural $O_3$ gradient (from: http://www.acdc.noaa.gov/online).

| | Average Monthly Temperatures (deg. C) | | | | | | | | |
|---|---|---|---|---|---|---|---|---|---|
| | 1995 | | | 1996 | | | 1997 | | |
| | Kala | Keno | Rhine | Kala | Keno | Rhine | Kala | Keno | Rhine |
| June | 23.38 | 21.57 | 21.98 | 22.38 | 17.95 | 18.00 | 21.97 | 17.98 | 19.60 |
| July | 25.37 | 23.37 | 21.83 | 22.95 | 20.75 | 17.88 | 23.05 | 20.61 | 19.55 |
| Aug | 25.89 | 23.92 | 21.78 | 24.35 | 22.33 | 20.08 | 21.68 | 19.26 | 16.81 |

| | June through August Rainfall (cm) | | |
|---|---|---|---|
| Date | Kalamazoo | Kenosha | Rhinelander |
| 1995 | 26.58 | 19.64 | 20.17 |
| 1996 | 30.48 | 39.10 | 35.43 |
| 1997 | 29.58 | 38.17 | 24.12 |

TABLE II

Summary of $O_3$ values at the three locations where we have aspen bioindicator plots for 1995 to 1997.

| | 1995 | | | 1996 | | | 1997 | | |
|---|---|---|---|---|---|---|---|---|---|
| Site | 1-hr Max (ppb) (June-Aug) | SUM 0 ppmh (June-Aug) 8 a.m.-8 p.m. | SUM 06 ppmh (June-Aug) 24 hr. | 1-hr Max (ppb) June-Aug) | SUM 0 ppmh (June-Aug) 8 a.m.- 8 p.m. | SUM 06 ppmh (June-Aug) 24 hr. | 1-hr Max (ppb) (June-Aug) | SUM 0 ppmh (June-Aug) 8 a.m.-8 p.m. | SUM 06 ppmh (June-Aug) 24 hr. |
| Rhinelander, WI | 79 | 37.1 | 4.1 | 85 | 41.0 | 2.5 | 80 | 40.6 | 4.1 |
| Kalamazoo, MI | 125 | 48.0 | 20.6 | 106 | 47.3 | 17.3 | 96 | 53.8 | 25.7 |
| Kenosha, WI | 120 | 59.5 | 29.8 | 122 | 70.4 | 15.1 | 112 | 58.7 | 10.7 |

We evaluated foliar injury on all leaves on a subset of the trees representing all the clones at the $O_3$ gradient sites. Foliage from a subset of the trees representing all clones at the $O_3$ gradients sites was collected for wax composition analyses in August of 1995, 1996, 1997, and 1998. Leaf segments (5 mm x 10 mm) were air dried, gold-coated on a cold stage, and examined under a JSM 6400 SEM. Two hundred stomates per clone, per site were evaluated for occlusion.

## 2.2 THE ASPEN FACE STUDY

The Forest Atmospheric Carbon Transfer and Storage (FACTS 2) Free-Air $CO_2$ and $O_3$ Enrichment (Aspen FACE) facility was located at the USDA Forest Service, Harshaw Experimental Farm near Rhinelander, Wisconsin. The study encompassed 32 ha of land with twelve 30 m diameter treatment rings spaced 100 m apart within a deer-fenced area. The 12 rings were composed of 3 control rings, 3 rings with elevated $O_3$, 3 rings with elevated $CO_2$, and 3 rings with elevated $O_3$ + elevated $CO_2$. Ozone was administered during the daylight hours according to the 1.5x profile for the lower Great Lakes region as described by Karnosky et al. (1996). This profile was a modified ambient profile developed from the 1987 $O_3$ data for Washtenaw County, Michigan. It was modified to more closely fit the 6-year averages documented by Pinkerton and Lefohn (1987). We only fumigated during appropriate weather conditions, i.e. we avoided fumigating on cool days (<15°C) or during rain, fog, or dew events. Ozone was monitored by Thermo Environmental Instruments Inc. Model 49C monitors adapted for rapid response times (4 sec) and calibrated weekly. $CO_2$ was administered during daylight hours at 560 ppm, which was about 200 ppm above the ambient level of $CO_2$. $CO_2$ was monitored with Licor Inc. LI-6252 $CO_2$ analyzers and calibrated weekly.

The Aspen FACE gas dispensing apparatus was modified from that developed by Lewin et al. (1994) and Nagy et al. (1994). In order to accommodate both $CO_2$ and $O_3$ from our vertical vent pipes, we modified our gas injection system in two ways compared to that system used at the FACTS I (Duke) experiment with loblolly pine (Hendrey et al., 1998). First, to accommodate a larger flow rate to dilute $O_3$ concentrations coming out of the vertical vent pipe (to minimize $O_3$ concentrations near the vertical vent pipes), we changed from a hole configuration to a slot. We oriented the vertical vent pipe so the slot was pointed away from the center of the ring and placed a set of baffles to redirect air coming from the vertical vent pipes into the FACE ring. Finally, we added a flashing on top of the baffles to direct air down and into the ring (Walklate et al., 1996). Together, these modifications improved the stability and distribution of both $CO_2$ and $O_3$ in our FACE rings and minimized $CO_2$ usage. $CO_2$ was distributed from the central storage tanks to each FACE ring through copper pipe. $O_3$ was generated from pure oxygen with a Praxair $O_3$ generator capable of producing 18-30 lbs. of $O_3$ per day. The $O_3$ was dispensed to the FACE rings through stainless steel tubing. Three controlling microcomputers were used to control the opening and closing of the valves at each ring and to control release of $CO_2$ and/or $O_3$ into the rings. $CO_2$ and $O_3$ were dispensed from the upwind side of each ring and monitored at plot centers.

Each Aspen FACE ring was divided into two halves by a central walkway. One half of each ring was planted at 1 m x 1 m with 1-year-old plants of five trembling aspen genotypes differing in $O_3$ sensitivity (8L, 216 and 271 = $O_3$ tolerant; and 42E and 259 = $O_3$ sensitive). Each clone was planted in two-tree plots randomly located in the one half ring. This resulted in 30 to 35 trees per clone being in the sweet spot where $CO_2$ and $O_3$ were close to target values (±10% for 1 minute averages). The other one half of the ring was divided into two sections: one planted with aspen clone 216 ($O_3$ tolerant) and sugar maple (*Acer saccharum* Marsh.) seedlings intermixed at 1 m x 1 m spacing, and the other with aspen clone 216 and paper birch (*Betula papyrifera* Marsh.) intermixed at 1 m x 1 m spacing. There are five border rows of aspen around each FACE ring. An irrigation system was used to supplement rain events during the first two growing seasons. It was used on four occasions in 1997 and six times in 1998 after periods of extended dry conditions to assure establishment of the trees. Soil tensiometers were used to evaluate when watering was needed. All trees were planted from 1-year-old stock in July, 1997 and treatments with $CO_2$ were run from August 15 to September 30, 1997. $CO_2$ treatments resumed on May 1, 1998 and $O_3$ treatments started on May 15, 1998. Both $CO_2$ and $O_3$ treatments were stopped on October 12, 1998. The treatments were timed to have $CO_2$ exposures from budbreak to budset for trembling aspen.

We thoroughly characterized all plants in our FACE study. They were all measured at the time of planting and at the end of the first growing season. A subset of intensively sampled trees were used for characterizing seasonal growth, crown architecture, and physiological characteristics. We sampled foliage from sample leaves of clones 216, 259 and 271 for SEM in August of 1997 and 1998 as previously described. Visible foliar symptoms were scored on June 1, July 1, August 1, and September 1, 1998.

## 3. Results and Discussion

Visible foliar symptoms for aspen have not been detected at the clean-air (Rhinelander) gradient plot or in the control and elevated $CO_2$ treatments in the FACE experiment. However, visible foliar symptoms were seen in five out of six years (no symptoms were seen in 1998) at the higher $O_3$ (Kalamazoo and Kenosha sites) and in the $O_3$ and $O_3 + CO_2$ FACE treatments in 1998. Symptomatic leaves were those showing black bifacial necrosis, chlorosis, or upper leaf surface black or red stipple (Karnosky *et al.*, 1996). Young leaves did not show symptoms and recently mature leaves (LPI 6-10, Larson and Isebrands, 1970) showed most pronounced symptoms. Older leaves (LPI 15-20 or more) often showed multiple symptoms and secondary pests so they were more difficult to diagnose.

Foliar symptom data for a typical year at the gradient plots is shown in Table III and at the FACE site is shown in Table IV. At both of the higher $O_3$ gradient sites and in the FACE experiment, clone 259 was the most sensitive clone. This is consistent with previous research from OTCs (Karnosky *et al.*, 1996). Relatively, symptoms were worse in the $O_3$ treatments in the FACE experiment. The seasonal 12-hr. daylight $O_3$ doses that induced the symptoms ranged from 47.3 to 70.4 ppmh at the gradient sites and from 57.6 to 60.7 ppmh (AOT 40 values at the FACE study were 25.8 to 30.7 ppmh) at the FACE experiment (Table V). These data are consistent with previous OTC results where visible foliar injury was seen on 50% of all leaves on $O_3$-sensitive clones at 50 to 60 ppmh (Karnosky *et al.*, 1996). Elevated $CO_2$ in the FACE experiment resulted in a decreased amount of visible foliar symptoms in all three clones. This is consistent with the results of Volin and Reich (1996) and Volin *et al.* (1998) who found that elevated $CO_2$ ameliorates the effects of $O_3$ on photosynthesis and growth in aspen. It is contrary to our previous results which suggested no amelioration (Kull *et al.*, 1996) in OTCs. The level of $CO_2$ was higher in our FACE experiment (ambient + 200 ppm) versus our OTC work (ambient + 150 ppm). In addition, the soil at the FACE site is considerably higher in N than that in our OTC site. Either of these two factors could have caused the difference between our OTC and FACE results with the same clones.

## TABLE III

Summary of the percentage of leaves per tree showing visible injury (% L.I.) for $O_3$ and presence of sooty mold fungus (*Alternaria* spp.) on three aspen clones scored in August, 1996 at the three $O_3$ gradient sites. Means are of 20 trees per clone. Clones with the same first letter were not different ($p < 0.05$) from one another within each site and clones with the same second letter were not different ($p < 0.05$) between sites as determined by the Student-Newman-Keuls multiple comparisons test.

| Location | Clone | % L.I.[1] ($O_3$) ± S.E. | % L.I.[1] (Sooty mold) ± S.E. |
|---|---|---|---|
| Rhinelander | 216 | 0 ± 0a, x | 0 ± 0 a, x |
| | 259 | 0 ± 0a, x | 0 ± 0 a, x |
| | 271 | 0 ± 0a, x | 0 ± 0 a, x |
| Kalamazoo | 216 | 8.2 ± 1.8 a, y | 0 ± 0 a, x |
| | 259 | 34.7 ± 2.8 b, y | 21.0 ± 3.3 b, y |
| | 271 | - | - |
| Kenosha | 216 | 5.0 ± 1.9 a, y | 10.6 ± 2.6 a, y |
| | 259 | 43.1 ± 4.2 c, y | 55.6 ± 4.8 b, z |
| | 271 | 16.5 ± 1.8 b, y | 13.0 ± 1.8 a, y |

[1] Percentage of leaves affected by $O_3$ or sooty mold fungus.

## TABLE IV

Percent leaves showing visible injury on August 1, 1998 for three trembling aspen clones varying in $O_3$ tolerance in the $O_3$ or $O_3$ + $CO_2$ treatments of the FACTS 2 (Aspen FACE) project in Harshaw, Wisconsin. Each replicate is the mean of at least 30 trees. Clones with the same first letter were not different ($p < 0.05$) from one another for that treatment and clones with same second letter were not different between treatments as determined by the Student-Newman-Keuls multiple comparisons test.

| Clone | $O_3$ Tolerance | Ozone | | | | Ozone + $CO_2$ | | | |
|---|---|---|---|---|---|---|---|---|---|
| | | Rep 1 | Rep 2 | Rep 3 | $\bar{x}$ | Rep 1 | Rep 2 | Rep 3 | $\bar{x}$ |
| 216 | tolerant | 17.8 | 12.7 | 24.7 | 18.4 b, y | 5.2 | 4.5 | 8.1 | 5.9 b, x |
| 259 | sensitive | 55.3 | 51.1 | 57.8 | 54.7 c, y | 33.1 | 37.7 | 41.9 | 37.6 c, x |
| 271 | tolerant | 11.5 | 4.5 | 30.0 | 15.3 a, y | 0 | 0 | 1.6 | 0.5 a, x |

We found evidence of a secondary pathogen on trees impacted by $O_3$ at two of our $O_3$ gradient plots. A sooty mold fungus (*Alternaria* spp.) occurred on a large number of leaves on the $O_3$ sensitive clone 259 at both the Kalamazoo and Kenosha sites. We believe this may have been associated with the degradation of the leaf epicuticular waxes visible under SEM (Figure 2). Epicuticular wax deposits on all three clones were unaltered at the Rhinelander site but modified at the Kalamazoo and Kenosha sites. Waxes were platelike rather than crystalline and stomates were occluded by amorphous deposits on leaves collected from these sites. Effects appeared to be most dramatic at the Kenosha site in all three years and enhanced on the most $O_3$ sensitive clone (259). This modification resulted from changes in wax synthesis induced by $O_3$ (Percy *et al.*, 1998).

TABLE V

Mean ozone concentration (ppb) for the 7 a.m. to 7 p.m. time interval and AOT40 (ppmh) for 24 hrs per day for ambient air and the 3 $O_3$ and 3 $O_3$ + $CO_2$ treatments of the FACTS 2 (Aspen FACE) project in Harshaw, Wisconsin, for the period from May 1 to August 31, 1998.

| Treatment | Rep | May | | June | | July | | August | | Season | |
|---|---|---|---|---|---|---|---|---|---|---|---|
| | | Mean (ppb) | AOT 40 (ppmh) | Mean (ppb) | AOT 40 (ppmh) | Mean (ppb) | AOT 40 (ppmh) | Mean (ppb) | AOT 40 (ppmh) | Mean (ppb) | AOT 40 (ppmh) |
| $O_3$ | 1 | 52 | 5.6 | 55 | 6.7 | 57 | 7.6 | 59 | 6.9 | 54.7 | 26.8 |
| $O_3$ | 2 | 54 | 6.5 | 57 | 7.8 | 58 | 8.3 | 58 | 8.1 | 56.3 | 30.7 |
| $O_3$ | 3 | 53 | 6.2 | 55 | 6.9 | 56 | 7.1 | 59 | 7.0 | 54.7 | 27.2 |
| $O_3$+$CO_2$ | 1 | 52 | 6.1 | 55 | 7.3 | 55 | 7.1 | 64 | 9.2 | 54.0 | 29.7 |
| $O_3$+$CO_2$ | 2 | 53 | 6.3 | 57 | 7.9 | 57 | 7.8 | 62 | 8.3 | 55.7 | 30.3 |
| $O_3$+$CO_2$ | 3 | 50 | 5.2 | 55 | 6.7 | 57 | 7.4 | 57 | 6.5 | 54.0 | 25.8 |
| Ambient | | 41 | 2.2 | 33 | 0 | 30 | 0 | 29 | 0.2 | 34.6 | 2.4 |

To quantify the degree of epicuticular wax degradation, we examined leaves from the three aspen clones (216, 259, 271) in the four Aspen FACE treatments and at the highest $O_3$ (Kenosha) gradient plot (Figure 3). We developed a coefficient of occlusion based on the degree of stomates covered by amorphous wax deposits with 0 = no occlusion and 4 = total occlusion. The Kenosha site had the highest coefficient of occlusion for all three clones. This was particularly interesting because no visible foliar symptoms were present in 1998 at the Kenosha site. This shows the value of epicuticular waxes as a very sensitive bioindicator of $O_3$ and/or $CO_2$ effects. At the FACE site, we saw significantly increased stomatal occlusion with all three treatments ($CO_2$ + $O_3$, $CO_2$, and $O_3$) as compared to controls. While $O_3$ has been commonly reported to induce wax degradation (Percy et al., 1990) and stomate occlusion (Huttunen, 1990), we believe this is the first report of increased wax occlusion of stomates in elevated $CO_2$.

We compared the relative photosynthetic rates under control (ambient $CO_2$) and elevated $CO_2$ treatments in the FACE treatments in 1997 (Table VI). As can be seen, elevated $CO_2$ significantly increased the maximum light saturated photosynthesis. We compared aspen from our FACE site to OTC data (Coleman et al., 1995), field-grown seedlings (Roden and Pearcy, 1993), and to potted plants grown in the biotron (Volin et al., 1998). Our FACE rates of photosynthesis were much higher than our previous OTC results. In addition, the biotron-grown plants (Volin et al., 1998) were the least productive photosynthetically. The differences could be due to different light intensities. Light levels are generally lower in the biotron (Olsyk et al., 1986) and light is also decreased significantly (about 12% lower on average) in OTCs (Heagle et al., 1996). The N levels in the various studies could also have affected the photosynthetic rates as photosynthesis is linearly related to N levels in aspen (Coleman et al., 1998).

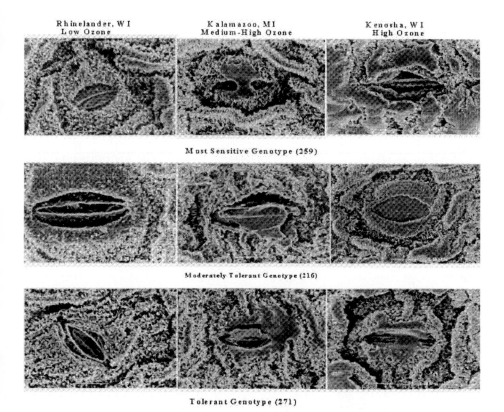

Fig. 2. Relationship between 1997 leaf surface wax structure and ozone levels in three aspen genotypes of differing $O_3$ sensitivities growing in 4-year-old plantations we established along a natural ozone gradient through Wisconsin and Michigan. *Note*: There was no visible foliar injury on the leaves in this composite photograph.

## 4. Summary and Conclusions

We present evidence suggesting that ambient $O_3$ concentrations in the lower Great Lakes region of the United States are inducing visible foliar symptoms and epicuticular wax degradation on trembling aspen. This evidence is from two novel experimental methods. First, we examined the response of three aspen clones varying in $O_3$ sensitivity and that had been established in replicated trials along a natural $O_3$ gradient. Second, we examined the response of the same three aspen clones in a Free-Air $CO_2$ and $O_3$ Enrichment (FACE) facility. The $O_3$-induced visible foliar symptoms were consistent among and between clones for the two experiments and consistent with our previous rankings. Elevated $CO_2$ in the FACE study decreased visible foliar injury suggesting an amelioration of $O_3$ effects by $CO_2$ for aspen. However, for epicuticular wax degradation, elevated $O_3$ at the two elevated $O_3$ sites and in the FACE experiment consistently degraded the aspen wax surface and this effect was not ameliorated by elevated $CO_2$. In fact, elevated $CO_2$ alone appeared to negatively impact aspen epicuticular waxes.

TABLE VI

Photosynthesis light response parameters (means ± se) of *Populus tremuloides* trees grown in the FACTS 2 experiment under ambient field conditions (ambient) and under free-air $CO_2$ enrichment (elevated). For comparison, other published results are shown for these same *P. tremuloides* clones grown under elevated $CO_2$ in OTCs (Coleman et al., 1995b), and for unspecified, biotron-grown seedlings (Volin et al., 1998) and unspecified field trees of *P. tremuloides* (Roden & Pearcy, 1993). Parameters are: light saturated photosynthesis rate ($A_{sat}$, µmol m$^{-2}$s$^{-1}$), dark respiration rate ($R_d$, µmol m$^{-2}$s$^{-1}$), light compensation point (LCP, µmol m$^{-2}$s$^{-1}$), apparent quantum yield (f, µmol mol$^{-1}$). Significant differences (p < 0.05) among treatments from the FACTS 2 experiment are indicated by (*) with n = 3 as determined by a simple t test in Systat.

|  | Volin et al., 1998 (Biotron) |  | Ambient |  | FACTS 2 (FACE) |  |
|---|---|---|---|---|---|---|
|  | Ambient $CO_2$ | Elevated $CO_2$ | Coleman et al., 1995 (OTCs) | Roden & Pearcy, 1993 (field) | Ambient $CO_2$ | Elevated $CO_2$ |
| $A_{sat}$ | 5.8 | 7.8 | 8.5 ± 0.6 | c. 15 | 21.6 ± 1.1 | 30.8 ± 1.4 * |
| $R_d$ |  |  | 1.8 ± 0.1 | 1.7 ± 0.2 | 3.8 ± 0.2 | 2.9 ± 0.2 * |
| LCP |  |  | 48 ± 6 | 71 ± 7 | 33 ± 3 | 21 ± 1 * |
| φ |  |  | 51 ± 5 | 27 ± 2 | 68 ± 4 | 74 ± 6 |

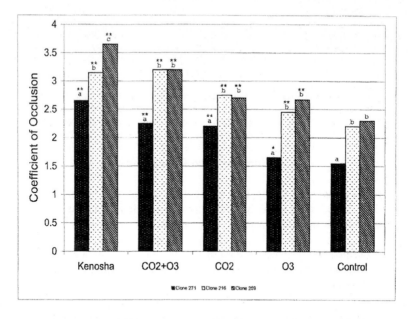

Fig. 3. Coefficient of occlusion as determined by scanning electron microscopy for the four treatments at the Aspen FACE ($CO_2 + O_3$, $CO_2$, $O_3$, and control) and at the Kenosha site for three trembling aspen clones sampled in August, 1998. 0 = no occlusion; 4 = almost totally occluded by leaf surface waxes. Each FACE treatment value was the mean of 100 stomates per clone for each of 3 replicates and the Kenosha data is from 100 stomates of each clone. Significant differences from control values are shown by *(p < 0.05) and **(p < 0.01) and significant differences (p < 0.05) between clones for a given treatment are indicated by letters as determined by the nonparametric Kolmogorov-Siminov test (Myslivec, 1957 and Smelko, 1991).

## Acknowledgments

This research was partially supported by the USDA Forest Service Global Change Program, the U.S. Department of Energy (DE-FG02-95ER62125), the National Science Foundation (DBI-9601942; IBN-9652675), the National Council of the Paper Industry for Air and Stream Improvement (NCASI), Michigan Technological University, the Slovakian Forest Research Institute, and the Canadian Forest Service. The authors appreciate the AIRS network $O_3$ values provided for the $O_3$ gradient plots by Teague Pritchard. The authors acknowledge the thorough and helpful comments of reviewers Lucy Sheppard, Andrew R. McLeod, and M. Gundhardt-Goerg.

## References

Barnard, J.E., Lucier, A., Johnson, A.H., Brooks, R.T., Karnosky, D.F. and Dunn, R.H.: 1991, NAPAP State of Science/Technology Report 16.
Barnes, J.D. and Wellburn, A.R.: 1998, In: L.L. De Kok and I. Stulen (Eds.), Responses of plant metabolism to air pollution. Backhuys Publishers. Leiden, The Netherlands. pp. 147-164.
Berrang, P.C., Karnosky, D.F. and Bennett, J.P.: 1989, *Can. J. For. Res.* **19**, 519-522.
Berrang, P.C., Karnosky, D.F.and Bennett, J.P.: 1991, *Can. J. For. Res.* **21**, 1091-1097.
Berrang, P.C., Karnosky, D.F., Mickler, R.A. and Bennett, J.P.: 1986, *Can. J. For. Res.* **16**, 1214-1216.
Ceulemans, R. and Mosseau, M.: 1994, *New Phytol.* **127**, 425-446.
Chappelka, A.H. and Samuelson, L. J.: 1998, *New Phytol.* **139**, 91-108.
Coleman, M.D., Dickson, R.E. and Isebrands, J.G.: 1999, *Physiol. Plant.* (In Press).
Coleman, M.D., Dickson, R.E., Isebrands, J.G. and Karnosky, D.F.: 1995a, *Tree Physiol.* **15**, 593-604.
Coleman, M.D., Dickson, R.E., Isebrands, J.G. and Karnosky, D.F.: 1995b, *Tree Physiol.* **15**, 585-592.
Coleman, M.D., Dickson, R.E., Isebrands, J.G. and Karnosky, D.F.: 1996, *Tree Physiol.* **16**, 145-152.
Curtis, P.S., Vogel, C.S., Pregitzer, K.S., Zak, D.R. and Teeri, J.A.: 1995, *New Phytol.* **129**, 253-263.
Dann, T. and Sumners, P.(Editors): 1997, Canadian 1996 $NO_x$/VOC Science assessment. Report of the Data Analysis Working Group. 259 pp.
Gagnon, Z.E., Karnosky, D.F., Dickson, R.E. and Isebrands, J.G.: 1992, *Amer. J. Bot.* **79**, 107.
Heagle, A.S., Reinert, R.A. and Miller, J.E.: 1996, *J. Environ. Qual.* **25**, 273-278.
Hendrey, G.R., Lewin K., Nagy, J. and Ellsworth, D.: 1998, Global Change Biol. **5**, 101-105.
Huttunen, S.: 1996, In: K.E. Percy, J.N. Cape, R. Jagels and C.J. Simpson (Eds.) *Air Pollutants and the Leaf Cuticle*, NATO ASI Series, Vol. G36, Heidelberg. pp. 81-96.
Hogsett, W.E., Weber, J.E., Tingey, D., Herstrom, A., Lee, E.H. and Laurence, J.A.: 1997, *Environ. Manage.* **21**, 105-120.
Karnosky, D. F.: 1976, *Can. J. For. Res.* **6**, 166-169.
Karnosky, D. F.: 1977, *Can. J. For. Res.* **7**, 435-436.
Karnosky, D.F., Gagnon, Z.E. and Reed, D.D.: 1992a, *Can. J. For. Res.* **22**, 1785-1788.
Karnosky, D.F., Gagnon, Z.E., Reed, D.D. and Witter, J.A.: 1992b, *Water, Air, and Soil Pollut.* **62**, 189-199.
Karnosky, D.F., Pregitzer, K.S., Kubiske, M., Reed, D.D., Isebrands, J.G., Lindroth, R.L., Zak, D.R., Host, G.E., Hendrey, G.: 1995, U.S. Department of Energy, Program for Ecosystem Research, Research Summaries 1995. pp. 28-29.
Karnosky, D.F., Gagnon, Z.E., Dickson, R.E., Coleman, M.D., Lee, E.H. and Isebrands, J.G.: 1996, *Can. J. For. Res.* **16**, 23-27.
Karnosky, D.F., Podila, G.K., Gagnon, Z.E., Pechter, P., Akkapeddi, A., Sheng, Y., Riemenschneider, D.E., Coleman, M.D., Dickson, R.E. and Isebrands, J.G.: 1998, *Chemos.* **36**, 807-812.
Keeling, C.D., Whort, T.P., Wahlen, M. and van der Plicht, J.: 1995, *Nature* 375, 666-670.
Kubiske, M.E., Pregitzer, K.S., Mikan, C.J., Zak, D.R., Maziasz, J.L. and Teeri, J.A.: 1997, *Oecologia* **110**, 329-336.
Kull, O., Sober, A., Coleman, M.D., Dickson, R.E., Isebrands, J.G., Gagnon, Z. and Karnosky, D.F.: 1996, *Can.J. For. Res.* **16**, 639-648.
Larson, P.R. and Isebrands, J.G.: 1971, *Can. J. For. Res.* **1**, 1-11.
Mohnen, V.A., Goldstein, W. and Wang, W.C.: 1993, *Air and Waste* **43**, 1332-1344.
Myslivec, V.: 1957, Czechoslovak Academy of Science. State Agricultural Publishing House. No. 654.

Prague, Czechoslavakia, pp. 486-495.
Olsyk, D.M., Bytnerowicz, A., Kats, G., Dawson, P.J., Wolf, J. and Thompson, C.R.: 1986, *J. Environ. Qual.* **15**, 417-422.
Percy, K.E., Krause, C.F. and Jensen, K.F.: 1990, *Can. J. For. Res.* **20**, 117-120.
Percy, K.E., Karnosky, D.F. and Mankovska, B.: 1998, Proc. 18th International Meeting for Specialists in Air Pollution Effects on Forest Ecosystems. Edinburgh. p. 98.
Sheng, Y., Podila, G.K. and Karnosky, D.F.: 1997, *For. Gen.* **4**, 25-33.
Smelko, S.: 1991, Technical University, Forestry Faculty, Technical University Publishing. Zvolen, Slovakia. pp.179-183.
Treshow, M.: 1970, *Environ. Pollut.* **1**,155-161.
Treshow, M. and Stewart, D.: 1973, *Biol. Conserv.* **5**, 209-214.
Volin, J.C. and Reich, P.B.: 1996, *Physiol. Plant.* **97**, 674-684.
Volin, J.C., Reich, P.B. and Givnish, T.J.: 1998, *New Phytol.* **138**, 315-325.
Walklate, P.J., Xu, Z.G. and McLeod, A.B.: 1996, *Global Change Biol.* **2**:75-78.
Wang, D., Karnosky, D.F., Bormann, F. H.: 1985, *Can. J. For. Res.* **16**, 47-55.

# RESPONSES OF YOUNG TREES (FIVE SPECIES IN A CHAMBER EXPOSURE) TO NEAR-AMBIENT OZONE CONCENTRATIONS

M. S. GÜNTHARDT-GOERG, S. MAURER, J. BOLLIGER, A. J. CLARK, W. LANDOLT and J. B. BUCHER

*Swiss Federal Institute of Forest, Snow and Landscape Research, Zürcherstrasse 111, CH-8903 Birmensdorf ZH, Switzerland*

(Received 30 October 1998; accepted 26 February 1999)

**Abstract.** This investigation is intended to add to the quantitative information about the sensitivity of European deciduous tree species to environmentally realistic ozone ($O_3$) exposure in respect to the established critical level of an AOT40 of 10 ppm.h (6-month growing season, daylight hours). Cuttings (without leaves) of *Fagus sylvatica* L., *Sorbus aucuparia*, L., *Carpinus betulus* L.., *Fraxinus excelsior* L., and cuttings and seedlings of *Prunus serotina* Ehrh. were exposed during one growing season either to filtered air, to which 50 % of ambient $O_3$ concentration was added, or to 50 % + 30 ppb $O_3$, resulting in a final AOT40 of 0.3 and 20.7 ppm.h, respectively. The foliage formed per tree varied between and within the species, but was not significantly modified by $O_3$, whereas the number of symptomatic leaves per tree significantly increased in the 50 % + 30 ppb $O_3$ regime. By mid July light-green spots appeared in the leaves (except *C. betulus*), which developed into stippling (*F. excelsior*), red (*P. serotina*) or necrotic spots by September. The $CO_2$-assimilation rate decreased more with increasing visual symptoms (earliest in 20- day-old leaves) than with age-dependent leaf discoloration in *F. sylvatica*, *F. excelsior*, and *P. serotina* (not in *C. betulus*). The dark-adapted photosystem II quantum yield (Fv/Fm) slightly declined with leaf age, but a considerable reduction became apparent in 107-day-old leaves with $O_3$-symptoms only. By late morning yield reductions in light-adapted leaves were exaggerated in such leaves from the 50 % + 30 ppb $O_3$ regime. The above mentioned responses were not reflected in significant changes of the net biomass production during the experiment. The critical level, therefore, is based on visual and functional symptoms rather than on production.

**Keywords:** AOT40, assimilation rate, autumn leaf discoloration, biomass, chlorophyll fluorescence, critical level, visible symptoms

## 1. Introduction

Many different effects of $O_3$ on trees (Sandermann *et al.*, 1997) and the trees' defence mechanisms (Polle *et al.*, 1999) have been reported based on at least double-ambient $O_3$ concentrations, but the derivation of a critical level for near-ambient $O_3$ concentrations is still a controversial issue. A provisional critical level for $O_3$ has been established (Fuhrer and Achermann, 1994; Lucas and Skärby, 1994) as 10 ppm.h, calculated as a cumulative exposure above the threshold of 40 ppb (AOT40). The AOT40 was calculated on a 24h basis because there is some evidence that stomata may not close entirely during night-time (Matyssek *et al.*, 1995) and that mechanisms responsible for detoxification of $O_3$ or its reaction products may be reduced in darkness. It was further accepted that a seasonal time of 6 months should cover the most sensitive period of the trees and that plants ought to be protected from a 10 % loss in net biomass production by this critical value, even though the latter has been determined on the data from only a small number of experiments. The provisional critical level of 10 ppm.h has been exceeded in 'non-smog' areas like southern Sweden (data from 1990 - 93, Kindbom et al., 1995) and also in the mountainous regions of Switzerland, where $O_3$-concentrations almost doubled between the fifties and the nineties (Staehelin et al., 1994). The calculation of the critical level has been modified since by confining the exceedance to the daylight hours only (calculated global clear-sky radiation above 50 $Wm^{-2}$) similar to that for agricultural plants (Skärby and Skeffington, 1996). New experimental data were provided in 1996 to support the AOT40 of 10 ppm.h from a significant regression of biomass production of 0-3 year-old *Fagus sylvatica* versus AOT40 (daylight hours), established by normalising data from three different projects (Braun and Flückiger, 1995, Steingröver et al., 1995, Küppers et al., 1994). Because this data base is still insufficient, the present investigation has been designed to add to the quantitative information about the sensitivity of different deciduous tree species to realistic $O_3$ concentrations in respect to the critical level established in 1996. The $O_3$ scenarios therefore were established in the present study to represent a background concentration (50 % ambient) vs. a near-ambient AOT40 of twice the critical level of 10 ppm.h (characteristic for higher altitudes of 1600-1800 m a.s.l. in Switzerland).

## 2. Materials and Methods

The experiment took place in two large (10 $m^2$ each, wind speed 0.5 $ms^{-1}$) walk-in climate chambers, where the climatic parameters (temperature, humidity and light hours) reflected the means of ambient climate measurements at the experimental site (Zürich, Switzerland) from May to November over a period of ten years (modified from Günthardt-Goerg et al., 1996). The chambers were supplied with charcoal and Purafil filtered air, to which 50 % of the current year ambient $O_3$ concentration (area within the buildings of the institute) was added ($O_3$ generated from pure oxygen), or 50 % + 30 ppb, resulting in a final AOT40 of 0.3 (**A-**), and 20.7 ppm.h (**A+**) respectively (Fig. 1). $O_3$ concentrations within the chambers were adjusted by mass flow controllers and monitored as in ambient air in the area of the buildings and at the forest edge using Monitor Labs Model 8810 instruments, which were calibrated against each other. *Sorbus aucuparia*, L. grafted stock, *Carpinus betulus* L., *Fraxinus excelsior* L., *Fagus sylvatica* L., and *Prunus serotina* Ehrh. rooted cuttings (without leaves, 16 per species, 8 each from 2 clones, 4 per clone and chamber) and 16 *P. serotina* seedlings taken from a native forest in the south of Switzerland, were weighed, potted (10 l / plant) and exposed from May 5th to November 15th. Nutrients were applied by a slow-release fertiliser with a life span of 6 months (Osmocote, Sierra Chemical Europe, The Netherlands, initial N:P:K = 315:98:221). Phenology of each leaf per tree cutting or seedling was estimated monthly for visible symptoms (spots or stippling) and leaf colour.

One leaf of each tree was labelled during flushing and measured at the age of 20 (4 August), 67 (20 September) and 107 (31 October) days. During daylight hours (9 a.m. to 5 p.m.) leaf gas exchange was measured at the light conditions of the chambers and at light-saturated conditions with a portable $CO_2/H_2O$ porometer (Model CQP-130, Walz, Effeltrich, Germany) at ambient (chamber) temperature, air humidity, and $CO_2$ concentration (leaf area enclosed into the cuvette: 7.5 $cm^2$). The dark adapted photosystem II quantum yield (Fv/Fm, calculated as 1 - Fo/Fm; measured from 3 a.m. to before dawn) and the light adapted photosystem II quantum yield ($\delta$F/Fm', calculated as 1 - Fo'/Fm'; measured from 10 a.m. to 12 a.m.) were determined for the same leaves using the 'pulse-amplitude modulation' method for chlorophyll fluorescence (Model PAM-2000, Walz, Effeltrich, Germany). For each leaf, four measurements were made by applying saturation pulses of 1000 µmol photons $m^{-2}$ $s^{-1}$ and 0.8 s in duration. The

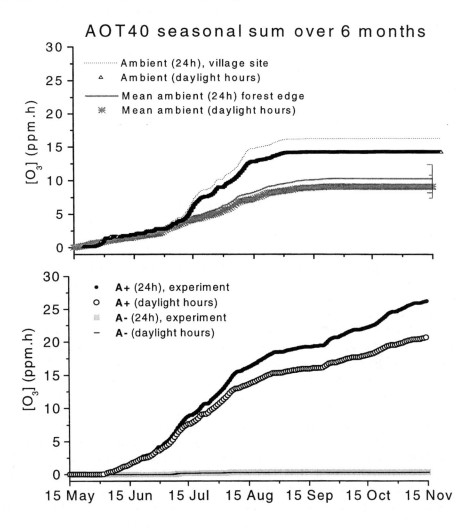

Fig. 1: AOT40 was calculated as the sum of hourly $O_3$ concentrations above 40 ppb during 24h or daylight hours (ambient > 50 W $m^{-2}$, experiment $\geq$ PPFD 400 µmol $m^{-2}$ $s^{-1}$) over the 6-month growing season of the experiment.
Ambient air: AOT40 for ambient air within the area of the buildings of our institute 1995; mean ambient ($\pm$ SE) $O_3$ concentrations from the forest edge, Birmensdorf, Switzerland (1991 - 1995).
Experiment: **A-** = AOT40 as 50 % ambient 1995, **A+** = AOT40 as 50 % ambient + 30 ppb.

photon flux density of the measuring light provided by the PAM-2000 unit was < 1 µmol m$^{-2}$ s$^{-1}$. In the final harvest, foliage area and net production of foliage, wood and root dry mass were determined for each plant.

Pooled data were analysed for each measured parameter with multiple factor ANOVA, but comparative tests were made between date, species (both clones, because no difference was found between the clones), light conditions for gas exchange and $O_3$-regimes with one-way ANOVA (GLM procedure, SAS Institute Inc. 1985). Where significant treatment effects were detected, means were separated using the HSD test. AOT40 was calculated over the 6-month growing season of the experiment by the sum of hourly $O_3$ concentrations exceeding 40 ppb during either 24h or daylight hours. The sum of daylight hours was calculated (according to Bruck et al., 1985) using light intensities > 50 W m$^{-2}$, which for the chambers was equivalent to a PPFD equal to or greater than 400 µmol m$^{-2}$ s$^{-1}$.

## 3. Results and Discussion

Usually at a forest edge in the neighbourhood of our institute the seasonal mean AOT40 for $O_3$ is below the 10 ppm.h limit (Fig. 1, mean ambient = 9.1 ppm.h) and visible leaf symptoms in trees are rarely found, whereas less than 500 m from the forest edge ambient AOT40 within the area of the buildings amounts to 14.1 ppm.h reflecting a rural village site in the Swiss midland (Fig. 1, ambient). The latter measurements were used for our experiment, namely the 50 % ambient $O_3$-regime (Fig. 1 **A-**), which did not exceed the limit of 40 ppb, and the 50 % ambient + 30 ppb $O_3$-regime, which often did. All exceedance summed up over the season reached an AOT40 of 20.7 ppm.h (Fig. 1, **A+** on a daylight hour basis).

The deciduous trees used in the experiment had different growth habits that were dependent on the species and propagation. *P. serotina* vegetatively propagated from cuttings produced 5 times more foliage area than *P. serotina* propagated from seeds, and 16 times more than *F. sylvatica* cuttings. The foliage of *C. betulus* cuttings entirely originated from primary buds which were present when the cuttings were exposed; other species formed new secondary buds during the season. This 'secondary foliage' tended to be increased by **A+** (vs. **A-**), but great variations in foliage area did not allow a significant change to be detected (Fig. 2). In *P. serotina,* from seed, autumnal leaf discoloration (yellowing, reddening or bronzing) had already started in July, significantly earlier than in the other trees including *P. serotina* grown from cuttings (Fig. 3, white bar parts). Leaf discoloration and shedding, although differing by species and propagation did not significantly change by the addition of 30 ppb $O_3$ (Fig. 3, white and black bar parts, **A+** vs. **A-**). Independent of leaf discoloration, a wide variety of leaf symptoms (spots) was formed particularly in the **A+**. By mid July light-green spots appeared in the leaves of *F. sylvatica* cuttings, *P. serotina* seedlings and cuttings, and *S. aucuparia* cuttings, whereas *C. betulus* and *F. excelsior* still showed green leaves. These symptoms developed, into necrotic spots in *F. sylvatica*, and *S. aucuparia*, and into red spots in *P. serotina* by September. In September *F. excelsior* initially exhibited light green

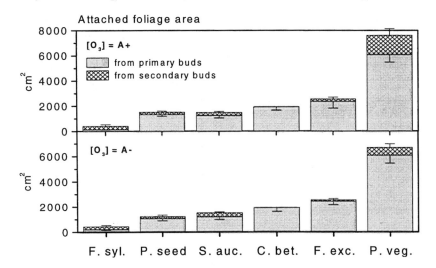

Fig. 2: Attached foliage area at the harvest date November 15th in the **A-** and **A+** $O_3$-regime of the experiment. Primary buds = buds formed when the cuttings were exposed in May, secondary buds = buds formed during the exposure time. Means are plotted from 8 trees each, standard error is shown positively for the cross-hatched bars representing foliage from secondary buds, but negatively from primary buds (shaded bar). Abbreviations: *F. syl.* = *Fagus sylvatica*, *P. seed* = *P. serotina* propagated from seeds, *S. auc.* = *Sorbus aucuparia*, *C. bet.* = *Carpinus betulus*, *F.exc.* = *Fraxinus excelsionr*, and *P. veg.* = *Prunus serotina* propagated from cuttings.

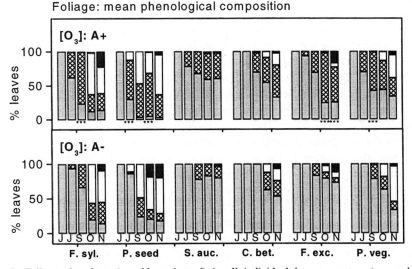

Fig. 3: Foliage phenology (monthly estimated) in all individual leaves per tree (tree order with increasing foliage production and species abbreviation according to Fig. 2). The stacked bars represent the mean percent foliage composition (means from 8 trees each) per species or propagation in each $O_3$-regime from June to November. Green leaves are plotted light grey, leaves with $O_3$-symptoms (or in the **A-** with '$O_3$-like' symptoms) crosshatched, leaves with autumnal discoloration white, and shed leaves black. *** in **A+** indicate a significant difference ($p < 0.05$) of the green and symptomatic foliage to the corresponding means in **A-**.

leaf tips and margins, which later developed into stippling, in *C. betulus* leaf bronzing only began at the margins in September. Leaf sensitivity to $O_3$ in terms of early appearance of leaf symptoms was at the highest in *F. sylvatica* and *P. serotina* from seeds (Fig. 3, cross hatched bar parts, beginning in July) which both had formed the smallest foliage area, in contrast to *C. betulus* and *F. excelsior*, although their foliage was older (originated from primary buds, Fig. 3 vs. Fig. 2). However, in vegetatively propagated *P. serotina*, the indeterminate growth habit had contributed to more green foliage relative to the symptomatic foliage at the end of the season than in *P. serotina* grown from seeds (Fig. 3, **A+**). Similar differences between seedlings, saplings and canopy trees have been observed in Pennsylvania (USA) by Fredericksen et al. (1996).

Nevertheless the number of green leaves (Fig. 3, light grey bar parts) per tree decreased in favour of more leaves with $O_3$-symptoms (cross-hatched) in the **A+** (significant effects, p = 0.001 for each of the factors $O_3$- regime, month and species). The difference of the means had to be large to overcome the great variability among the individuals. Depending on the species, the largest difference in symptomatic and green foliage occurred in July, September or October. Particularly in *P. serotina* and *F. sylvatica*, $O_3$-symptoms were combined with autumnal leaf discoloration at the end of the season.

Leaf gas exchange and chlorophyll *a* fluorescence emission values differed between the species, leaf age, and $O_3$-regimes, with *F. excelsior* showing the highest assimilation rate. Overall (pooled data) leaf age and species were significant factors at p = 0.0001; $O_3$-regime was significant at p = 0.0001 for transpiration rate and p = 0.011 for assimilation rate, and not significant for Fv/Fm or δF/Fm'. In *S. aucuparia* and *C. betulus*, which showed no significant difference in the number of symptomatic leaves per tree (Fig. 2), the leaf gas exchange and chlorophyll fluorescence properties also did not change with increasing leaf age or $O_3$-regime. In contrast, functional changes due to **A+** (vs. **A-**) and increasing leaf age were similar in leaves of *F. sylvatica*, *F. excelsior* and *P. serotina*, though the development of their visible symptoms differed.

The assimilation rate of green 20-day-old leaves of *F. sylvatica*, *F. excelsior* and *P. serotina* (**A-** and **A+**, e.g. *F. sylvatica*, Fig. 4) was higher than that of leaves with visible light-green spots (in **A+** only). The assimilation rate tended to decline further with leaf age, when the symptoms developed into necrotic spots in **A+** relative to **A-**. A small decline was also apparent in the oldest leaves with age-dependent discoloration. The transpiration rate tended to be lower in the older leaves with established visible symptoms in **A+** vs. **A-**. There was no significant difference between the gas exchange measured at the light conditions of the chambers and that measured at light-saturation.

Fv/Fm, the maximum potential quantum yield, represents photosystem II efficiency for primary photochemistry. The values measured predawn indicate a relatively recovered state of the photosystem II during night-time. Fv/Fm remained nearly unchanged with increasing leaf age in the **A-** in all species (for example *F. sylvatica*, Fig. 5), but clearly declined in 107-day-old leaves with $O_3$-symptoms (**A+**) of *F. sylvatica* and *F. excelsior* *(not P. serotina)*. In the measurements of light adapted leaves (with biochemical limitations on the electron transport), δF/Fm' tended to decrease in all species with leaf age and discoloration in the **A-**, but this trend was exaggerated in the 107-day-old leaves with $O_3$-symptoms (e.g. *F. sylvatica*, Fig. 5, **A+**).

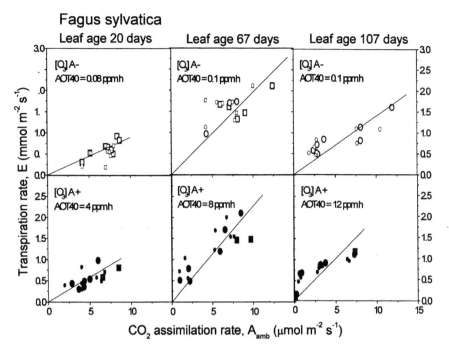

Fig. 4: *F. sylvatica* (for example): Leaf gas exchange (details see methods) from leaves which had been labelled after flushing and measured when 20-, 67- and 107-day-old at light-saturation and at the chamber light conditions (smaller symbols). AOT40 was calculated for those leaf ages (daylight hour basis). **A** - open squares = green leaves, open circles = leaves with autumnal discoloration. **A** + solid squares = green leaves, solid circles = leaves with $O_3$-symptoms (light-green spots when 20-day-old, later developing to necrotic spots).

The fact that the decline in the $CO_2$ assimilation rate had already appeared in 20-day-old leaves in the **A+**, whereas the fluorescence measurements of photosystem II only became apparent in 107-day-old leaves, demonstrated that the quantum yield of photosystem II can be more stable than $CO_2$ assimilation (similar to findings for birch, Maurer et al., 1997). However, both assimilation rate and quantum yield of photosystem II measured in leaves with conspicuous $O_3$-symptoms still showed photosynthetic activity, but at a reduced level compared with green leaves. Visible symptoms and functional decline occur due to changes at the cellular level (Günthardt-Goerg et al., 1997). Isolated or groups of injured cells are unevenly distributed within a leaf, as are other cells which appear to be highly efficient and balance the injured cells.

The % change of the mean values of different parameters for **A+**, relative to **A-** are conspicuous (Fig. 6). They show:
1) less green leaves (grey bars), but more leaves with $O_3$-symptoms (cross-hatched).
2) larger leaf area formed from secondary buds (horizontally hatched) in *F. sylvatica*, *F. excelsior* and vegetatively propagated *P. serotina*.
3) reduced assimilation rate (left hatched) and transpiration rate (right hatched), except in *C. betulus*.
4) a small, not significant reduction of the net whole tree biomass (black bars), except in vegetatively propagated *P. serotina* with their prominent second flush.

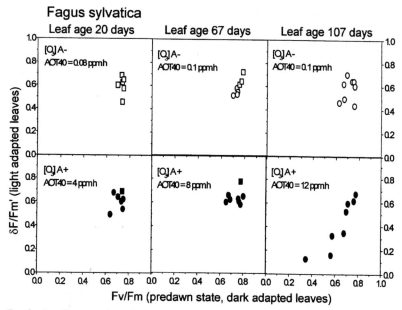

Fig. 5: *F. sylvatica* (for example): Maximum potential quantum yield, representing photosystem II efficiency (Fv/Fm in dark-adapted leaves, δF/Fm' in light adapted leaves), the same leaves as Fig. 4.

Neither foliage, stem, root biomass, nor their ratios were significantly changed by A+ with AOT40 = 20.7 ppm.h (data not included). A 10% biomass reduction for AOT40 = 10 ppm.h cannot be confirmed by this experiment or by a similar study (Landolt et al., 1996). Because earlier experiments have not been designed and calculated for the evaluation of critical levels, the present critical level of 10 ppm.h is still based on a few open-top chamber experiments with *F. sylvatica*: (1) an experiment with *F. sylvatica* in unfiltered vs. filtered ambient air with prevailing $O_3$ pollution (seedlings from the cotyledon stage over a 1-3 season period, Braun and Flückiger, 1995), (2) an experiment with simulated ambient pollution mixtures (2-year-old *F. sylvatica* seedlings over a 2 season period, Küppers et al., 1994), (3) an experiment with different $O_3$-concentrations (final AOT40 = 0, 20, 40 and 80 ppm.h, 3-year-old old *F. sylvatica* saplings, Steingröver et al., 1995). Other experiments with juvenile deciduous trees report significant biomass reductions for $O_3$-sensitive species or genotypes (*Populus tremuloides* Michx. cuttings and seedlings, AOT40 = 50 - 92 ppm.h, seasonal exposure, Karnosky et al., 1996; *P. serotina* seedlings, AOT40 = 28.3 or 40.4 ppm.h, seasonal exposure, Neufeld et al., 1995; *Betula pendula* Roth. clones, AOT40 = 18 ppm.h over a 2 season period, Pääkkönen et al., 1997). To support or reject the present critical level of AOT40 = 10 ppm.h in respect to significant biomass reduction in young deciduous trees, more experimental data with AOT40 near 10 ppm.h are needed.

At least in Europe it is difficult to find any growth reductions in mature trees due to $O_3$. Forest inventories show increasing growth trends of mature trees, which appear to be based on land use history and forest management (Spiecker, 1999). However, visible $O_3$-symptoms may correlate to reduced growth; recently significant intraspecific radial growth reductions over a 5- and 10-year period have been reported in mature

*Liriodendron tulipifera* L. with visible foliar injury, as compared to non symptomatic trees, and in contrast to no significance within *P. serotina* in the Great Smoky Mountains, USA (Somers *et al.*, 1998).

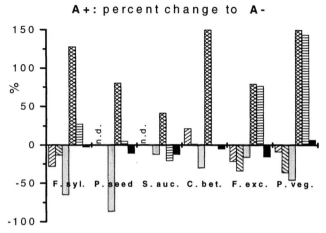

Fig. 6: Means of different parameters from the A+ plotted as percent change relative to those from the A- (Fig. 2 for species abbreviation). Left hatched bars = assimilation rate, right hatched bars = transpiration rate (both pooled over all leaf ages, not determined = n.d. in *P. serotina* from seeds and *S. aucuparia*). Grey bars = green foliage (number of leaves), crosshatched bars = foliage with leaf symptoms in September (more than 150 % increase for *C. betulus* and *P. serotina* cuttings. Horizontally hatched bars = leaf area originating from secondary buds, black bars = net whole tree biomass production by November 15th.

## 4. Conclusions

The seasonal AOT40 of ambient air can vary within a 500 m distance by 5 ppm.h.
Visible symptoms and reductions in leaf gas exchange of *F. sylvatica, F. excelsior* and *P. serotina* have been detected in the present study between an AOT40 of 4 and 12 ppm.h, and reductions in the quantum yield of photosystem II (two European species only) at AOT40 of 12 ppm.h. Certain deciduous tree species, therefore, appear to be more sensitive than shown in earlier experiments (40 ppm.h < *F. sylvatica* < 80 ppm.h, Lippert *et al.*, 1996). However, in the Swiss midland with about 30-50 'summer smog days' (Brönnimann and Neu, 1997) the AOT40 of 10 ppm.h is not regularly achieved, and $O_3$-symptoms are rare (except in certain regions in the south, Skelly *et al.*, 1999) or occur late in the summer, when they are mixed up with autumnal discoloration.

In the present study leaf discoloration was not related to the $O_3$ regime. Discoloration therefore should be separated from the estimation of visible species-specific $O_3$ symptoms (spots and stippling).

A realistic critical level should take into account that the leaf sensitivity to $O_3$ varies not only within a tree species, within genotypes, and individual trees, but is also dependent on tree propagation. The single leaf sensitivity to $O_3$ varies greatly between species, but can be independent of the foliage productivity or the growth habits of the trees.

Visible symptoms and concomitant functional reductions are not necessarily reflected in significant changes of net biomass production. Young and even older trees may have a surplus of leaves, and cope with a certain premature foliage loss without significant consequences for growth parameters. The present critical level for $O_3$, therefore, can be based on visual and functional symptoms rather than net biomass production.

## Acknowledgements

We gratefully acknowledge the technical assistance of D. Mathies, P. Bleuler and D. Tarjan.

## References

Braun, S. and Flückiger, W.: 1995, *New Phytol.* **129**, 33-44.
Brönnimann, S. and Neu, U.: 1997, *Atmos. Environ.* **31,** 1127-1135.
Bruck, M., Hammer, N., Neuwirth, F. and Schaffar, S.: 1985, *Meteorologische Daten und Berechnungsverfahren*, 3. Auflage, dbv-Verlag, Wien.
Fredericksen, T.S., Skelly, J.M., Steiner, K.C., Kolb, T.E. and Koutrick, K.B.: 1996, *Environ. Pollut.* **91**, 53-63.
Fuhrer, J. and Achermann, B. (eds.): 1994, *Critical levels for ozone: a UN-ECE workshop report.* Schriftenreihe der FAC Liebefeld, **16**, FAC Liebefeld, Switzerland.
Günthardt-Goerg, M. S., Schmutz, P., Matyssek, R. and Bucher, J. B.: 1996, *Can. J. For. Res.* **26**, 649-657.
Günthardt-Goerg, M.S., McQuattie, C.J., Scheidegger, C., Rhiner, C. and Matyssek, R.: 1997, *Can. J. For. Res.* **27**, 453-463.
Kärenlampi, L. and Skärby, L. (eds.): 1996, *Critical levels for ozone in Europe-Testing and finalizing the concepts UN-ECE workshop report.* University of Kuopio, Kuopio, Finland.
Karnosky, D.F., Gagnon, Z.E., Dickson, R.E., Coleman, M.D., Lee, E.-H. and Isebrands J.G.: 1996, *Can. J. For. Res.* **26**, 23-37.
Kindbom, K., Lövblad, G., Peterson, K. and Grennfelt, P.: 1995, *Ecol. Bull.* **44**, 35-42.
Küppers, K., Boomers, J., Hestermann, S., Hanstein, S. and Guderian, R.: 1994, in *Critical levels for ozone a UN-ECE workshop report.* Fuhrer, J. and Achermann, B. (eds.), pp. 98-110. *Schriftenreihe der FAC Liebefeld*, vol. **16**, FAC Liebefeld, Switzerland.
Landolt, W., Bucher, J. B., Pfenninger, I. and Bleuler, P.: 1996, in *Critical levels for ozone in Europe: Testing and finalizing the concepts.* Kärenlampi, L. and Skärby, L. (eds.), pp. 244-248, University of Kuopio, Finland.
Lippert, M., Steiner, K., Payer, H. D., Simons, S., Langebartels, C. and Sandermann, H.Jr.: 1996, *Trees* **10**, 268-275.
Lucas, P. and Skärby, L.: 1994, in *Critical levels for ozone a UN-ECE workshop report*, Fuhrer, J. and Achermann, B. (eds.), pp.14-16, *Schriftenreihe der FAC Liebefeld*, vol. **16**, FAC Liebefeld, Switzerland.
Matyssek, R., Günthardt-Goerg, M.S., Maurer, S. and Keller, T.: 1995, *Tree Physiol.* **15**, 159-165.
Maurer, S., Matyssek, R., Günthardt-Goerg, M. S., Landolt, W. and Einig, W.: 1997, *Trees* **12**, 1-10.
Neufeld, H.S., Lee, E.H., Renfro, J.R., Hacker, W.D. and Yu, B.H.: 1995, *New. Phytol.* **130**, 447-459.
Pääkkönen, E., Holopainen, T. and Kärenlampi, L.: 1997, *Environ. Pollut.* **95**, 37-44.
Polle, A., Matyssek, R., Günthardt-Goerg, M.S. and Maurer, S.: 1999, in *Environmental pollution and plant responses*, Agrawal,S.,Agrawal,M. and Krizek,D.T.(eds.),CRCPress/Lewis Publishers, New York, in press
Sandermann, H. Jr., Wellburn, A.R. and Heath, R.L.: 1997, *Ecological Studies* **127**, Springer, Berlin.
Skärby, L. and Skeffington, R.: 1996, in *Critical levels for ozone in Europe:Testing and finalizing the concepts UN-ECE workshop report.* Kärenlampi, L. and Skärby, L. (eds.), pp.18-23, University of Kuopio, Kuopio, Finland.
Skelly, J., Innes, J. L., Savage, J. E., Snyder, K. R., Vanderheyden, D., Zhang, J. and Sanz, M. J.: 1999, *Water Air Soil Pollut.* (this volume)
Spiecker, H. : 1999, *Water Air Soil Pollut.* (this volume)
Somers, G. L., Chappelka, A. H., Rosseau, P. and Renfro, J. R.: 1998, *For. Ecol. Manage.* **104**, 129-137.
Staehelin, J., Thudium, J., Buehler, R., Volz-Thomas, A. and Graber, W.: 1994, *Atmos. Environ.* **28**, 75-87.
Steingröver, E., Dueck, T. and van der Eerden, L.: 1995, in *Acid Rain Research: Do we have enough answers ?* G.J. Heij and W.J. Erisman (eds.), pp. 213-222, Elsevier Science BV, Amsterdam.

# PIGMENT CONCENTRATIONS AND RATIOS OF ALEPPO PINE SEEDLINGS EXPOSED TO OZONE

S. MANNINEN[1], D. LE THIEC[2], C. ROSE[2], G. NOURRISSON[2], F. RADNAI[2], J. P. GARREC[2] and S. HUTTUNEN[1]

[1]*Department of Biology, University of Oulu, P.O. Box 3000, 90401 Oulu, Finland.* [2]*INRA-Centre de Recherches Forestières, Unité d'Ecophysiologie Forestière-Equipe de Pollution Atmosphérique, 54280 Champenoux, France*

(Received 25 September 1998; accepted 26 February 1999)

**Abstract.** Two-year-old Aleppo pine (*Pinus halepensis* Mill.) seedlings were exposed to ambient air+50 ppb $O_3$ in open-top chambers (24 hours/day, 7 days/week) during May-October 1997 and to ambient air+70 ppb $O_3$ from May 1998 onwards. One growing season fumigation with ozone did not affect the pigment concentrations of the current-year (c) needles, nor were there any differences in photosynthesis or stomatal conductance. In May 1998, however, a marked carry-over effect was seen in the chlorophyll *a* and *b* and total carotenoid concentrations of the $O_3$-fumigated one-year-old (c+1) needles. The chlorophyll *a* and *b* and total carotenoid concentrations of newly flushed needles of the $O_3$-fumigated seedlings also seemed to be slightly decreased, as was their net photosynthesis when compared to the values of the filtered-air control needles. The chlorotic mottle and the changes in chloroplast pigments and photosynthesis of the c and/or c+1 needles of the NFA+$O_3$ seedlings in May 1998 indicate that frequent episodes of ozone concentrations of ≥100 ppb, especially when they also occur during the evening and night hours, as in some areas in southern Europe, may result in visible needle damage on Aleppo pine.

**Keywords**: *Pinus halepensis*, chlorophyll, carotenoid, net photosynthesis, visible damage

## 1. Introduction

Aleppo pine is the most common pine in the Mediterranean area and one of the few European pine species that currently shows visible damage (Gimeno *et al.*, 1992; Velissariou *et al.*, 1992). The chlorotic mottle of Aleppo pine needles observed in the field is identical to the ozone symptoms described in other conifers and the symptoms produced experimentally (Velissariou *et al.*, 1992). Although no effects on growth have yet been observed in the fumigation experiments with seedlings, physiological alterations, such as changes in chlorophyll content and assimilation rate, may ultimately lead to growth reduction (Dizengremel and Gérant, 1997). Tree biomass and growth are directly related to carbon acquisition and allocation, which, in turn, are related to these physiological parameters. However, the processes involved in pigment alterations triggered by ozone exposure are not completely established. According to Sasek and Richardson (1989), chlorophyll degradation may be induced through enzyme-mediated processes, which could be related to the acceleration of senescence (Lucas *et al.*, 1993) rather than through direct oxidation, as suggested by some authors (e.g. Elstner *et al.*, 1985). Chlorosis is not a primary result of ozone exposure, but rather a secondary effect due to impaired photosynthetic capacity, since it is unlikely that ozone or its reactive products reach the chloroplast to induce injury (Heath and Taylor, 1997).

In the present paper, the validity of pigment concentrations of Aleppo pine needles and their ratios, as indicators of the harmful impact of ozone, are discussed in relation to measurements of stomatal conductance and photosynthetic rate, and to the critical AOT40 of 10 ppm.h (accumulated exposure over the threshold concentration of 40 ppb) proposed to protect European forest trees (Kärenlampi and Skärby, 1996).

## 2. Material and methods

Two-year-old seedlings of Aleppo pine (*Pinus halepensis* Mill., provenance Languedoc) growing in pots (22 l) were exposed to ambient air+50 ppb $O_3$ (NFA+$O_3$) in two open-top chambers (24 hours/day, 7 days/week) during May-October 1997 at Col du Donon (48°29'N, 7°05'E), France. The seedlings were grown in a greenhouse during the winter and brought back to the experimental field at the end of April. The fumigations were continued in May 1998, when the NFA+$O_3$ seedlings were exposed to ambient+70 ppb $O_3$. The ozone concentration was changed, because the ambient ozone concentrations are higher in spring than in summer and autumn (see Dizengremel and Gérant, 1997), and we wanted to make the experimental conditions correspond better to the natural seasonal variation in ambient ozone concentrations. The control treatments included two charcoal-filtered air chambers (CFA) and one non-filtered air chamber (NFA), in addition to which there was one ambient air plot (AA).

Current- (c) and/or previous-year (c+1) needles were sampled from 10 seedlings/chamber and AA plot in mid-September 1997 and at the end of May 1998. The needles for pigment analyses were frozen with liquid nitrogen and put into a freezer (-70°C) prior to analysis. The pigments were extracted with dimethyl sulfoxide (DMSO) (Hiscox and Israelstam, 1979), and their absorbances were measured at 470, 646 and 663 nm with a Beckman DU®-64 spectrophotometer. The concentrations of chlorophyll *a* (Chl *a*), chlorophyll *b* (Chl *b*) and total carotenoids (Car) were calculated according to Wellburn (1994). For carbon (C) and nitrogen (N) analysis, dried, ground needle material was combusted at 1000°C in an EA-1110 CHNS-O elemental analyser (CE Instruments) equipped with a Porapack PQS column and a TC detector. Eager 200 for Windows™ (Ver. 1.00) software was used to calculate the results.

A Licor 6200 photosynthesis system was used to measure net assimilation rate (A) and stomatal conductance to water vapour ($g_w$) of the c needles in mid-September 1997 and at the end of May 1998 (Dixon *et al.*, 1998). The measurements were carried out on sunny days between 8.00 and 10.00 GMT. There were no differences in photosynthetically active radiation, vapour pressure deficit, relative humidity and temperature between the treatments during the measurements. The photon flux density of photosynthetically active radiation was 1063±103 µmol m$^{-2}$ s$^{-1}$ in September 1997 and 1093±113 µmol m$^{-2}$ s$^{-1}$ in May 1998. The vapour pressure deficit was 10.98±0.66 and 17.69±1.29 kPa, relative humidity 47.2±2.3 and 48.1±2.5%, and the temperature of the chamber 19.9±1.2 and 21.0±1.0°C in September 1997 and May 1998, respectively. In both September 1997 and May 1998, 8 seedlings/chamber and AA plot were each measured for a period of 20 s. All gas exchange measurements are expressed as per needle surface area. The needle surface area was calculated using an equation that relates area to dry mass. For the determination of the needle surface area, 200 needles were cut into millimetre sections and the area of each section was traced out with a scanning electron microscope. This method allowed calculation of a relationship between dry weight (DW, mg) and surface area (S, cm$^2$). The formula determined was as follows: $S = 0.086 \times DW + 0.478$ ($r^2=0.98$).

Analysis of variance (ANOVA) and Fisher's PLSD as a *post hoc* test were used to assess the differences between the treatments, using individual chambers as the unit of replication.

## 3. Results

The average ozone concentrations during daylight hours were around 40-50 ppb in the NFA chamber, while those in the NFA+$O_3$ chambers were around 90-100 ppb and 110-120 ppb in summer 1997 and May 1998, respectively. The highest 1-h ozone concentration of 172 ppb was measured in the NFA+$O_3$ chambers in May 1998. The ozone exposure resulted in an AOT40 (daylight hours) of 110.5 ppm.h in May-October 1997 and a value of 32.6 ppm.h in May 1998 before the sampling (Table I).

TABLE I

Cumulative ozone exposure (AOT40, ppm.h) in the different treatments. The ranges for 1-h mean ozone concentrations (ppb) are given in parenthesis.

|                  | CFA | NFA         | NFA+$O_3$       | AA          |
|------------------|-----|-------------|-----------------|-------------|
| May-October 1997 | 0   | 13.4 (4-108)| 110.5 (54-158)  | 13.4 (4-108)|
| May 1998         | 0   | 5.8 (25-102)| 32.6 (95-172)   | 5.8 (25-102)|

No statistically significant effects of ozone on pigment concentrations and ratios, net photosynthesis and stomatal conductance, and C and N concentrations between the treatments were observed in c needles in mid-September 1997 (Figures 1 and 2, Tables II and III). In May 1998, however, both c and c+1 needles of the NFA+$O_3$ seedlings showed chlorotic mottle, and a marked carry-over effect was observed in the chlorophyll *a* (p=0.014), chlorophyll *b* (p=0.043) and total carotenoid (p=0.039) concentrations of c+1 needles when compared to the CFA controls. No differences were seen in the chlorophyll *a*/*b* or chlorophyll *a*+*b*/total carotenoid ratios of c+1 needles between the NFA+$O_3$ and CFA or NFA controls, however. The chlorophyll and carotenoid concentrations of c needles of the NFA+$O_3$ seedlings also suggested a decreasing trend due to the ozone exposure, as did net assimilation, but the differences between the NFA+$O_3$ seedlings and the CFA controls were not statistically significant. In May 1998, the highest stomatal conductance was measured from c needles of the CFA seedlings. In c+1 needles, the lowest pigment concentrations and ratios of chlorophyll *a*+*b*/total carotenoid were found in AA seedlings, i.e. the pigment concentrations of the seedlings growing in OTCs were higher due to the chamber effect (higher temperature).

## 4. Discussion

According to Wellburn *et al.* (1996), Aleppo pine can be regarded as a fairly $O_3$-tolerant species. However, the visible chlorotic symptoms found on needles of Aleppo pine in Spain (Gimeno *et al.*, 1992) and Greece (Velissariou *et al.*, 1992) indicate that trees growing in the field may be damaged by the present ambient ozone concentrations. It is known that chlorophyll content may reveal the effects of ozone the season after the

### TABLE II
Net assimilation rate (A, $\mu mol\ m^{-2}s^{-1}$) and stomatal conductance ($g_w$, $mol\ m^{-2}s^{-1}$) of the c needles (mean±SD). Significant differences (Fisher's PLSD, $p<0.05$) between the treatments are indicated by different letters in each row.

|  |  | CFA | NFA | NFA+$O_3$ | AA |
|---|---|---|---|---|---|
| September 1997 | A | 7.14±0.13a | 7.99a | 7.62±0.23a | 8.34a |
|  | $g_w$ | 0.137±0.002a | 0.160a | 0.162±0.003a | 0.190a |
| May 1998 | A | 3.39±0.45a | 2.87b | 2.37±0.37b | 2.73b |
|  | $g_w$ | 0.063±0.001a | 0.038b | 0.042±0.008b | 0.043b |

### TABLE III
Carbon and nitrogen concentrations (% DW) of the c needles in September 1997 (mean±SD). Significant differences (Fisher's PLSD, $p<0.05$) between the treatments are indicated by different letters in each row.

|  | CFA | NFA | NFA+$O_3$ | AA |
|---|---|---|---|---|
| C | 48.3±0.13a | 48.9a | 48.3±0.15a | 48.9a |
| N | 3.24±0.08a | 3.07a | 3.21±0.20a | 2.74a |

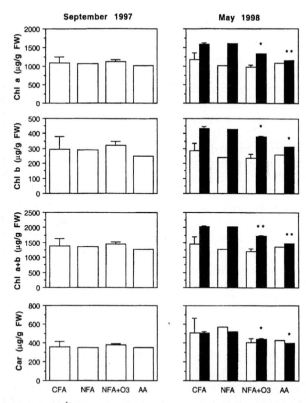

Fig. 1. Pigment concentrations ($\mu g\ g^{-1}$ FW) of c (white bar) and c+1 (black bar) needles in September 1997 and May 1998. The significance levels for differences (ANOVA) between the CF and other treatments are indicated as follows: ***$p<0.001$, **$p<0.01$, *$p<0.05$.

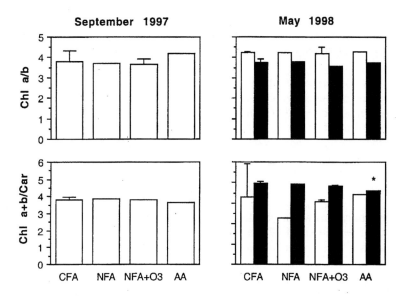

Fig. 2. Pigment ratios of c (white bar) and c+1 (black bar) needles in September 1997 and May 1998. The significance levels for differences (ANOVA) between the CF and other treatments are indicated as follows: ***p<0.001, **p<0.01, *p<0.05.
------*/////////*

exposure, and there may be cumulative effects of ozone exposure on chlorophyll content over several seasons (Skärby and Karlsson, 1996; Fuhrer et al., 1997). In accordance with the present results, Elvira et al. (1998) found a less complete winter recovery of chlorophyll levels in c+1 needles of $O_3$-fumigated (NFA+40 ppb) Aleppo pine seedlings. However, they did not observe a reduction in chlorophyll levels until the end of the second fumigation period, whereas the present results show marked differences in the chlorophyll and carotenoid concentrations of c+1 needles between the NFA+$O_3$ seedlings and the CFA controls as early as the end of May 1998. Moreover, both the chlorophyll and the carotenoid concentrations, as well as the net photosynthesis of the c needles seemed to be slightly reduced in May 1998 after only three weeks of fumigation. Takemoto et al. (1997), who also fumigated continuously, as we did, reported more marked needle injuries and a decrease in the net photosynthesis of the c needles of ponderosa pine at the end the first fumigation period in October under twice-ambient ozone concentrations, whereas no decrease in the chlorophyll and carotenoid concentrations of the needles was found until the second growing season. The 24-h mean ozone concentrations in the 2x$O_3$ treatment were 96 and 135 ppb during the two growing seasons (Takemoto et al., 1997) (i.e. equal to the concentrations in the NFA+$O_3$ treatment).

The chlorotic mottling and the reductions in the chlorophyll and carotenoid concentrations and/or net photosynthesis of the c and c+1 needles of the NFA+$O_3$ seedlings in May 1998 show that ozone levels of ≥100 ppb are high enough to damage (young) needles under conditions where frequent episodes of such peak ozone concentrations occur. This means that ozone episodes during which the 1-h ozone concentrations may reach 170 ppb (cf. Velissariou et al., 1992) can induce visible needle damage in Mediterranean Aleppo pines. The results may also point to the importance of night-time high ozone levels as a factor enhancing the damage (Matyssek et al., 1995). In the field in Spain and Italy, the most abundant needle damage has been found in the areas where the ozone concentrations are elevated even during the evening and night (see Gerosa et al., 1999; Sanz, personal communication). The deleterious effects of ozone on the chlorophyll concentrations and $CO_2$ assimilation rate of Aleppo pine needles reported by Elvira and Gimeno (1996) were related to AOT40 values ranging from 44 to 54 ppm.h. In the present study, the AOT40 value calculated for the c needles of the NFA+$O_3$ seedlings reached 33 ppm.h in May 1998.

## Acknowledgements

The present study was funded by INRA, the Academy of Finland and the Jenny and Antti Wihuri Foundation. Ms Pirita Haapavatia is thanked for her assistance with the pigment analyses. The English was revised by Mrs Sirkka-Liisa Leinonen.

## References

Dixon, M., Le Thiec, D. and Garrec, J.P.: 1998, *Environ. Exp. Bot.* **40**, 77-91.
Dizengremel, P. and Gérant, D.: 1997, *Interactions between ozone, climatic and nutritional factors on coniferous tree physiology, Final Scientific Report, Synthesis Report*, INRA, Nancy.
Elstner, E.F., Osswald, W. and Youngman, R.J.: 1985, *Experientia* **41**, 591-597.
Elvira, S. and Gimeno, B.S.: 1996, in *Critical Levels for Ozone in Europe: Testing and Finalizing the concepts. UN-ECE Workshop Report*, Kärenlampi, L. and Skärby, L. (eds), pp. 169-182, University of Kuopio, Dept. of Ecol. and Environ. Sci.
Elvira, S., Alonso, R., Castillo, F. and Gimeno, B.: 1998, *New Phytol.* **138**, 419-432.
Fuhrer, J., Skärby, L. and Ashmore, M.R.: 1997, *Environ. Pollut.* **97**, 91-106.
Gerosa, G., Spinazzi, F. and Ballarin Denti, A.: 1999, *Water Air Soil Pollut.* (this volume)
Gimeno, B.S., Velissariou, D., Barnes, J.D., Inclán, R., Peña, J.M. and Davison, A.W.: 1992, *Ecología* **6**, 131-134.
Heath, R.L. and Taylor Jr., G.R.: 1997, in *Forest Decline and Ozone*, Sandermann, H., Wellburn, A.R., Heath, R.L. (eds.), pp. 317-368, Ecological Studies 127, Springer, Berlin Heidelberg.
Hiscox, J.D. and Israelstam, G.F.: 1979, *Can. J. Bot.* **57**, 1332-1334.
Kärenlampi, L. and Skärby, L. (eds): 1996, *Critical Levels for Ozone in Europe: Testing and Finalizing the concepts. UN-ECE Workshop Report,* University of Kuopio, Dept. of Ecol. and Environ. Sci., 363 pp.
Lucas, P.W., Rantanen, L. and Mehlhorn, H.: 1993, *New Phytol.* **124**, 265-275.
Matyssek, R., Günthardt-Goerg, M.S., Maurer, S. and Keller, T.: 1995, *Tree Physiol.* **15**, 159-165.
Sasek, T.W. and Richardson, C.J.: 1989, *For. Sci.* **35**, 745-755.
Skärby, L. and Karlsson, P.E.: 1996, in *Critical Levels for Ozone in Europe: Testing and Finalizing the concepts. UN-ECE Workshop Report*, Kärenlampi, L. and Skärby, L. (eds.), pp. 72-85, University of Kuopio, Dept. of Ecol. and Environ. Sci.
Takemoto, B.K., Bytnerowicz, A., Dawson, P.J., Morrison, C.L. and Temple, P.J.: 1997, *Can. J. For. Res.* **27**, 23-30.
Velissariou, D., Davison, A.W., Barnes, J.D., Pfirrmann, T., MacLean, D.C. and Holevans, C.D.: 1992, *Atmos. Environ.* **26**, 373-380.
Wellburn, A.R.: 1994, *J. Plant Physiol.* **144**, 307-313.
Wellburn, F., Lau, K.K., Milling, P. and Wellburn, A.R.: 1996, *J. Exp. Bot.* **30**, 1361-1367.

# PASSIVE OZONE MONITORING FOR FOREST HEALTH ASSESSMENT

R. M. COX and J. W. MALCOLM

*Natural Resources Canada, Canadian Forest Service - Atlantic Centre, P. O. Box 4000, Fredericton, N.B. E3B 5P7.*

(Received 25 September 1998; accepted 26 February 1999)

**Abstract:** Critical levels of tropospheric ozone, established for the protection of crops and other plants, are now reported as being exceeded over large forested areas, giving rise to the need for an extensive monitoring program to confirm ambient levels within the forest and to detect related forest health effects. The requirement for an inexpensive monitor that can be used in remote locations prompted the development of the Can Oxy Plate™ passive ozone monitor and a monitoring protocol by the air pollution research group of the Canadian Forest Service, Forest Health Network. The monitors underwent initial trials in 1996 and operational trials during 1997 that involved two 2-3 week mid summer exposures in the canopy at selected forest health monitoring plots across Canada, and at adjacent forest openings. In both trials monitors were also co-located with the nearest instrumental ozone monitor. This allowed for the production of a field calibration for quality assurance assessment under field conditions. Results from 1996 indicate highly significant correlations with accumulated ambient ozone concentrations from the instrumental monitors at the co-located sites ($r=0.88$, $p=0.0002$). However, no such relationship was found between these sites and the forest plots which were up to 200 kilometres away. This may indicate spatial heterogeneity in ozone exposure between the continuous air quality monitoring sites and the forest plots. This information, together with our knowledge that strong gradients of ozone exposure are found within the canopy, underlines the importance of *in situ* monitoring of ozone exposure of forest health plots at risk to ozone effects.

**Keywords:** ozone, passive monitor, indigo, forest health

## 1. Introduction

Interest in passive monitors has recently been rekindled by the need to monitor ozone in remote areas. Several chemicals have been evaluated for use in a passive system (reviewed by Williams, 1994), but relatively few have been utilized to assess ozone exposure under field conditions. The earliest measurements of ground-level ozone concentrations were made by the Swiss chemist Schönbein in the mid-1800's, using a test paper impregnated with potassium iodide (London, 1985). Indigo dye was chosen as the reagent used in the monitor described here because of its known chemistry with, and high specificity for ozone (Grosjean *et al.*, 1993), simple methods of preparation and analysis, and low cost. Indigo and Indigo carmine have also been tested and used previously for passive ozone monitoring (Grosjean *et al.*, 1993; Grosjean and Hisham, 1992; Werner and von Schönborn, 1989; Williams, 1994 for Indigo, and Grosjean *et al.*, 1995; Bytnerowicz *et al.*, 1993; Grosjean *et al.*, 1993; Grosjean and Hisham, 1992; Grosjean and Williams, 1992, for Indigo carmine, respectively.)

Passive monitors integrate exposure to give cumulative exposure values, although hourly fluctuations in ozone concentrations are not evident using this technique. The low cost and flexibility of placement make them attractive alternatives for assessing ozone exposure in remote areas where instrumental monitoring may be limited, and for locations that are difficult to access, such as within the forest canopy.

*Water, Air, and Soil Pollution* **116:** 339-344, 1999.
© 1999 *Kluwer Academic Publishers. Printed in the Netherlands.*

The objectives of this study were: (1) to test and calibrate a lightweight, cost-effective passive monitor for ozone exposure, that has simple methods of analysis and preparation (mostly off-the-shelf components): and (2) to obtain linear exposure-response relationships over a range of cumulative exposures within the ozone polluted regions of Canada.

## 2. Materials and Methods

### 2.1. THE MONITOR (Can Oxy Plate™)

Isatin, a product of the ozonolysis of indigo, is easily extracted for spectrophotometric analysis (Grosjean *et al.*, 1988; Werner and Schönborn, 1989). Gel Blot papers (Schleicher & Schuell, grade GB003, 0.8 mm thickness) were used as supports for the indigo reagent. Papers cut into 34 mm squares were immersed for two minutes in a water suspension of 2 or 3 g l$^{-1}$ indigo (Sigma, 95% dye content) (1996 and 97 respectively) in an ultrasonic bath which maintained the temperature at 30°C. Average dye loading per paper was 0.008g for 1996 batches and 0.014g for 1997. Papers were dried in a 50-60°C oven under slightly negative pressure, while circulating internal air through an activated carbon filter to remove oxidants. Once dry, the papers were dyed and dried an additional two times. Papers with non-uniform dye distribution were discarded.

The monitor components (Figure 1) were housed in a 48 mm diameter polystyrene petri dish (Millipore). Components were placed between 1.5 mm thick tight-fitting retaining rings (acrylonitrile, butadiene, styrene copolymer). The reagent paper was supported by, and separated from the outer diffusion membrane (0.5$\mu$m pore size, Teflon membrane) by the retaining rings.

Shelters for monitors used in field exposures were constructed from polyvinyl chloride (PVC) piping and sheeting (Figure 1). Each shelter housed two monitors held in place with Velcro strips. Shelters were constructed to allow turbulent transfer of air to the surface of the diffusion membrane while providing protection from rain and tree branches.

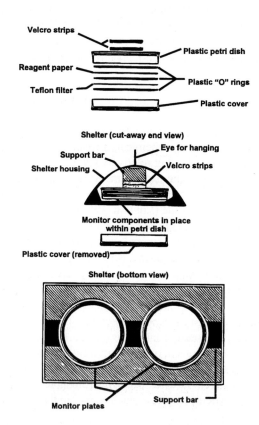

Fig. 1. Diagram of the Can Oxy Plate™ Passive ozone monitor

## 2.2. ANALYSIS OF EXPOSED MONITOR PAPERS

Reagent papers were removed from the monitors and isatin was extracted by shaking in warm ethanol (60°C) in sealed 500 ml Erlenmeyer flasks. The eluates were filtered through type A/E glass fibre filters (Gelman Scientific) to remove suspended material prior to making up to standard volume. Absorbance of the eluate was determined at 408 nm, the $\lambda_{max}$ for isatin (Mangini and Passerini, 1952). Absorbances were converted to molar quantities of isatin extracted using a standard curve (isatin source: Sigma). Values from unexposed control papers were subtracted from exposed values before response relationships were evaluated.

## 2.3. CALIBRATIONS

Laboratory calibrations were carried out with the use of Canadian Forest Service laminar flow fumigation chambers, to determine linearity of response, and as quality assurance against saturation of the indigo sink under field conditions. Exposures during the laboratory calibrations were from 4 hours to 12 days with a daily range of ozone concentrations from 50-270 ppb (TECO model 49 ozone analyser). Field calibrations were produced by co-locating passive monitors with instrumental ozone analysers operated by Provincial and Federal air quality monitoring networks. The field calibrations were used to determine the exposure values in associated forest plots. These calibration sites were also part of the quality assurance procedure of the operational field protocol developed for Forest Health Monitoring Plots across Canada. Co-located monitors remained exposed for 1 to 6 weeks.

## 2.4. EFFECTS OF STORAGE

Isatin was extracted from newly constructed unexposed monitors, and from monitors stored for 10 months in a zip-sealed plastic bag with activated charcoal in the dark at room temperature. These results were compared with known indigo loadings and measured monitor response at high exposure values, to determine the apparent loss of the available indigo sink due to storage.

## 2.5. FIELD DEPLOYMENT

The Can Oxy Plate monitors were placed out in the forest plots for 2 to 3 weeks, in 1996 and 1997 at 13 and 19 selected Forest Health Monitoring plots for the respective years across Canada. They were placed in locations that allowed free exposure to ambient air, by hanging from a branch, in the upper third of the green canopy, using wire or cord. Field blanks (monitor with two unexposed plates) were also located in the canopy at each site. Various methods for easily retrieving the shelter for removal and replacement of the Can Oxy Plates were used by the individual field personnel depending on stand characteristics. Monitors were also placed approximately 3 m above the ground, in the nearest forest clearing to each site, to monitor the ambient air outside the forest stand.

For field calibrations, Can Oxy Plate monitors were placed within 2 m of, and at the same height as the intake of the instrumental analyser. A field blank monitor containing unexposed plates was also set up at the co-located site. Hourly mean ozone concentrations from the instrumental air quality monitors were obtained from the operating authorities, and summed for the time of exposure. Seven co-located sites across Canada were included in the field calibration. All field plates were transported and stored in a zip-sealed plastic bag, also containing activated charcoal in a mesh pouch.

An investigation was also made into the mean ozone distribution (19 days) through a mixed forest stand at the University of New Brunswick wood-lot in Fredericton. Passive monitors were placed in the upper, middle, and lower canopy and at breast height of dominant trees, approximately 50 m apart. The selected trees were located in a transect through a forest stand tangential to the predominant wind direction along an overgrown trail (closed canopy overhead). The monitors were placed in their positions and tree heights were measured using a truck fitted with a telescopic arm and bucket.

## 3. Results and Conclusions

### 3.1. MONITOR CALIBRATIONS

Comparisons of the field and laboratory calibrations are shown in Figure 2. All regressions are linear and highly significant ($p<0.0001$). The slopes of the field calibrations are higher than those of the lab calibrations. This may be due to the higher wind speeds under field conditions, presumably reducing boundary layer resistance to diffusion of ozone to the monitoring plate surface. These results indicate that the monitors do not saturate under field conditions pertaining at the sites (1996-97) even up to 6 weeks exposure. Calibration slope differences between years are likely due to dye loadings (see Section 2.1).

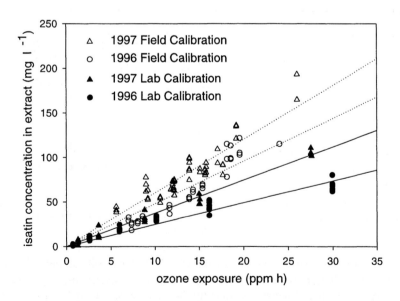

Fig. 2. Field and laboratory calibrations of Can Oxy Plate passive ozone monitors. Field calibrations carried out at ambient air quality monitor sites across Canada.

## 3.2. STORAGE EFFECTS.

Isatin extracted from unexposed plates increased by an average of 0.405 μmoles (n=4) following 10 months' storage, compared with new plates. Assuming that isatin is produced in a 1:1 molar ratio with indigo ozonolysis (Grosjean *et al.*, 19), the average loss of indigo was 1.4% of the total loaded on each paper (0.008 g or 29.0 μmoles). This loss represents 2.7% of the indigo sink required for a 27.0 ppm.h exposure (near the maximum of the linear calibration range). This indicates that storage for a year still allows for exposures near the limit of the linear response range of the monitor (i.e. 6 weeks at peak ozone concentrations in the Southern Maritime Region). However it must be emphasized that inclusion of unexposed plates as controls is necessary to account for storage.

## 3.3. FIELD RESULTS

The relationship of the mean concentration of ozone (ppb) indicated by the passive monitors with the mean hourly data from the co-located instrumental air quality monitor (field calibration) indicated a highly significant relationship (p=0.0002; Figure 3a). These instrumental monitoring data showed no such relationship with results from passive monitors placed in the canopy or in near-by forest clearings (figure 3b). This clearly shows the efficacy of monitoring ozone *in situ* at the forest health monitoring site, as there is obvious spatial variation in ozone exposure between the forest plots, and the instrumental monitors, which are generally located in urban environments to monitor human exposure. With the use of passive monitors, ozone exposure can be assessed at positions in the canopy where foliage samples can be taken to assess foliar injury and retention.

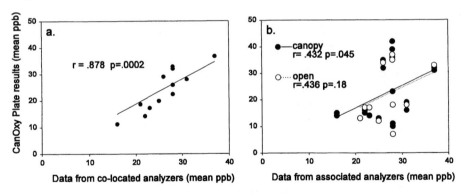

Fig. 3. Comparison of the relationship between instrumental ozone data and results from Can Oxy Plates : (a) co-located with the instrumental monitors, and (b) located in the canopy or in the open at associated forest plots.

Results from the tree-stand study are shown in Figure 4. Statistical analysis (protected Duncan's Multiple Range Test) indicated significant decreases in ozone exposure from top to the middle crown, and from the middle to lower crown, as well as from lower crown to breast height. An average ozone depletion of nearly 50% from the canopy top to breast height occurred within the stand, whereas the edge tree showed less of a gradient, presumably due to lateral penetration of the air-mass. These results indicate position in the

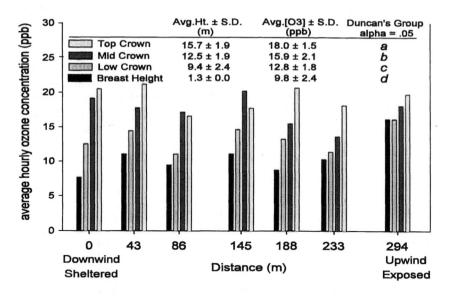

Fig. 4. Ozone profiles determined by Can Oxy Plate passive monitoring in seven trees through a mixed-wood stand over 19 days.

canopy and stand characteristics play an important role in foliage exposure to ozone and that passive monitors have the potential to play an important role in the empirical determination of mean foliar exposures and ozone depletions over the forested landscape.

## Acknowledgments

We would like to thank Bruce Pendrel for his assistance in co-ordinating the field operations together with the Forest Health Unit Leaders of the Canadian Forest Service's Forest Health Network and their staff across the country. We would also like to thank Environment Canada and the various provincial authorities that supplied the ozone monitoring data. Special thanks to Dennis Doucette whose technical assistance was invaluable.

## References

Bytnerowicz, A., Manning, W. J., Grosjean, D., Chmielewski, W., Dmuchowski, W., Grodzinska, K. and Godzik, B.: 1993, *Environ. Pollut.,* **80**, 301-305.
Grosjean, D., Grosjean, E. and Williams, E. L.: 1993, *Atmos. Environ.,* **27A**, 765-772.
Grosjean, D. and Hisham, M. W. M.: 1992, *J. Air Waste Manage. Assoc.,* **42**, 169-173.
Grosjean, D., Whitmore, P. M., and Cass, G. R.: 1988, *Environ. Sci. Technol.,* **22**, 292-298.
Grosjean, D. and Williams, E. L.: 1992, *Atmos. Environ.,* **26A**, 1407-1411.
Grosjean, D., Williams, E. L and Grosjean, E.: 1995, *Environ. Pollut.,* **88**, 267-273.
London, J. :1985, in *Ozone in the free atmosphere.* Whitten, R. C. and Prasad, S. S. (eds.), pp. 11-65, Van Nostrand Reinhold Company, New York.
Mangini, A. and Passerini, R.: 1952, *Chem. Abstr.* **46**, 350-351.
Werner, H. and von Schönborn, A.: 1989, *Proc. Statusseminar der PBWU zum Forschungsschwerpunkt "Waldschaden" GSF,* GSF, München-Neuherberg , 10 pp.
Williams T. P. W.: 1994, *Use of Passive Monitors to Assess Plant Bioindicators for Ground Level Ozone.* MSc Thesis Dept of Biology, University of New Brunswick, 133 pp.

# TROPOSPHERIC OZONE IN ALPINE FOREST SITES: AIR QUALITY MONITORING AND STATISTICAL DATA ANALYSIS

G. GEROSA[1,2], F. SPINAZZI[3], A. BALLARIN DENTI[1,2]

[1] *DIFCA, Dept. of Plant Physiology and Agricultural Chemistry, University of Milan, via Celoria 2, MILANO, Italy (ballarin@unimi.it),* [2] *FLA, Lombardy Foundation for the Environment, Foro Buonaparte 2, MILANO, Italy,* [3] *DISAT, Dept. of Environmental Sciences, University of Milan, via Emanueli 2, MILANO, Italy*

(Received 25 September 1998; accepted 26 February 1999)

**Abstract.** Plants represent one of the major sinks for tropospheric ozone that, at high concentrations, can affect plants' physiological activity with consequent serious damage. A research project has been promoted by the Lombardy Foundation for the Environment to investigate the effects of air pollution on forest ecosystems. The areas of study are located on the southern slopes of the Italian Alps in two valleys, only 10 km apart, selected because of their different plant injury: Val Gerola and Val Masino. Air quality ($O_3$, $NO_x$, $SO_x$, VOC) and meteorological parameters were monitored during four summer seasons (1994-97) using automatic sampling devices providing hourly mean values for each variable. Data analysis showed very different ambient ozone concentrations at the two sites, with average concentration values observed in the more damaged valley (Val Gerola) twice those measured at the other site. Multivariate data analyses have been used to interpret the observed differences in long-term $O_3$ exposure between the two sites and to identify possible underlying processes,.

**Keywords**: ozone, mountain forests, multivariate statistical techniques

## 1. Introduction

This work is part of a multidisciplinary research project aimed at the evaluation of forest ecosystems suffering from environmental stress. It was carried out in two forest areas (Figure 1) in Valtellina (Sondrio, Italy) with different degrees of forest decline (Table I): Val Gerola (1300 m a.s.l.) and Val Masino (1200 m a.s.l.). Previous investigations of both forest ecosystems have excluded most of the commonly responsible factors for forest decline, with the exception of ozone concentrations (Ballarin Denti *et al.*, 1998a).

Table I
Evaluation of visible injury in selected species of the two sample areas expressed as mean defoliation degree (%)
Source: *Regional Forestry Agency of Milan, 1996*

| Species | Val Gerola | | Val Masino | |
|---|---|---|---|---|
| Year | 1994 | 1995 | 1994 | 1995 |
| *Picea abies* | 35.7 | 32.1 | 24.2 | 24.7 |
| *Fagus sylvatica* | 36 | 33.5 | 16.4 | 13.6 |
| *Larix decidua* | 30 | 25 | 7.8 | 10.7 |
| *Abies alba* | - | - | 20 | 16.7 |
| *Castanea sativa* | - | - | 20 | 25 |

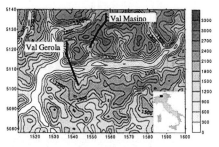

Fig. 1. Location of Valtellina in Italy and topography of the study area

*Water, Air, and Soil Pollution* **116**: 345-350, 1999.
© 1999 *Kluwer Academic Publishers. Printed in the Netherlands.*

Air quality ($O_3$, $NO_x$, $SO_x$, VOC) and meteorological parameters (temperature, pressure, relative humidity, solar radiation, wind speed and direction) were therefore monitored during the whole 1994-96 summer seasons, with the aid of both active (continuous analysers and mobile laboratories) and passive (diffusion tubes for ozone and VOC) (Hangartner et al., 1996) sampling devices providing hourly and daily mean values respectively. Multivariate statistical analysis of hourly mean values allowed the relationships between atmospheric pollutants and the main meteo-climatic factors to be studied, and contributed to a more detailed understanding of ozone micro and mesoscale distribution in a specific region, the alpine area, characterised by high orographic and circulatory complexity (Valli and Vecchi, 1996; Ballarin Denti et al.,1998b).

## 2. Methods

Data processing involved hourly mean frequency and maximum distributions for ozone concentration, Principal Component Analysis (PCA) on the correlation matrix (Joliffe, 1986) and Hierarchical Cluster Analysis (HCA) (Everitt and Graham, 1991).

PCA produces a set of linear combinations (principal components, PC) of observed variables along the directions of maximum variance, in decreasing order, of the data space considered. All the PC are mutually orthogonal and so they individually describe linearly independent (not linearly correlated) sources of variation with the associated amount of the total variance explained. PCA was chosen because it is a widely used exploratory multivariate technique, it involves no model, and it allows a representation of the relationships between all the variables of interest to be obtained at the same time. HCA helps in finding non-random structures (groups of observations) in a data set by analysing the distances between all observations. HCA methods differ in the type of distances (Euclidean, city-block) computed and in the algorithms used to analyse them. We chose the Euclidean distance on normalized variables, and Ward's algorithm (Ward, 1963).

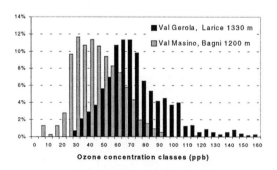
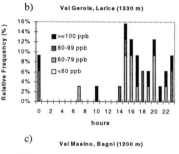
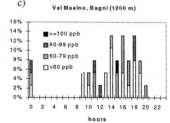

Fig. 2. a) Frequency distribution of hourly mean ozone concentration recorded at both Valtellina sites during summer 1996. b) and c) Time of daily ozone peaks. Relative frequency is the percentage of days the maximum concentartion has been measured in that hour. The shading represents the strength of the peak with respect to the indicated concentration class.

## 3. Results

The critical role of ozone for the stability of the two forest ecosystems is highlighted by the frequency distribution of hourly mean ozone concentrations (Figure 2a). In the more stressed forest (Val Gerola), 94% of hourly means are above 40 ppb, the phytotoxicity threshold for chronic exposures (AOT40=60 ppm·h), while this concentration is exceeded for only 51% of the time in Val Masino (AOT40=16 ppm·h). Moreover, the peaks in concentration are mostly present in the late evening (Figure 2b,c). These maxima cannot be explained simply in terms of local photochemical production because this would cease at sunset. This and other major differences in the meteo-chemical properties of both sites have been elucidated by means of PCA, as represented in Figure 3. The two systems differ in atmospheric circulation (WS), ozone and nitrogen oxide concentrations; the centroids of the hourly data are shifted relative to one another along the direction that includes these variables.

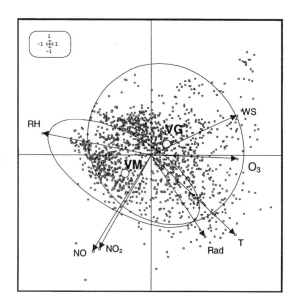

Fig. 3. First and second principal components from all 1996 data from both valleys. Loadings are plotted as arrows. Ellipsoids contain 95% of data scores for each valley.
Variables correlated with each other share the same direction of their loadings on the graph.

WS = Wind speed (ms$^{-1}$);
T = temperature (°C);
Rad = solar radiation (Wm$^2$);
RH = relative humidity (%);
NO, NO$_2$, O$_3$ = Nitrogen oxides and ozone concentrations (ppb)

A separate PCA on the hourly data for each site was used to formulate hypotheses about the origin of ozone, by studying the relationship between this pollutant and other meteo-chemical factors (Figure 4). In addition to a source of ozone variability due to local photochemical production (1st component: ozone correlated with solar radiation), the analysis shows the existence of a second source (3rd component), present only in Val Gerola, that is not explained by this process, but is negatively correlated with solar radiation and uncorrelated with temperature. The fact that this source is active both in the evening and at night suggests the existence of long-range transport mechanisms. How important is this transport mechanism in relation to the other ozone sources? Does it occour during the day? To answer these questions we used HCA, trying to separate and visualize the different atmospheric processes in terms of typical diurnal cycles for each site.

For this analysis we used all the observations over the 24 hours as the sample unit, extending the approach of Lavecchia *et al.* (1997). Three clusters of diurnal profiles

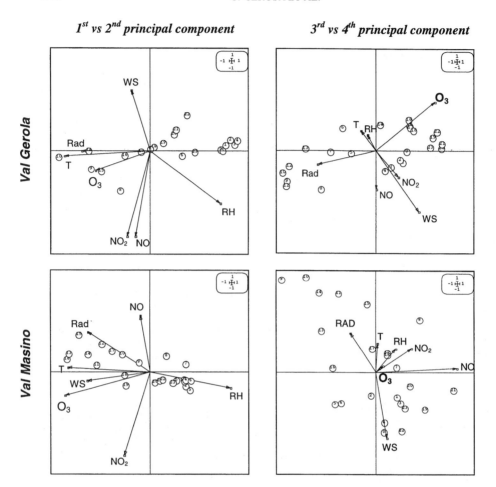

Fig. 4. PCA made separately for each valley. Variable loadings and hourly centroids of scores are plotted. Variance explained by each component: Val Gerola: 30%, 29%, 14%, 12%; Val Masino: 42%, 18%, 13%, 11%.

were identified in Val Gerola and two in Val Masino. Figure 5 shows the diurnal profile of the centroid for each group. A meteorological situation could be associated with each of them: a) and b) represent 'good' weather days; c) and d) 'bad' weather days; e) 'transition' days. It can be seen that while in Val Masino the most frequent daily ozone profile shows the typical bell-shaped pattern (revealing the existence of local photochemical production), in Val Gerola this pattern exists only during transition days. In Val Gerola the most frequent situation is characterised by an ozone concentration increasing during the day. The evening peak is matched by an increase in mountain breezes (SW quadrant), by an increase in the $NO_2/NO$ ratio and by a weakening of the increasing trend in humidity. The latter observations led us to suppose the influence of drier chemically-aged air masses from outside the valley.

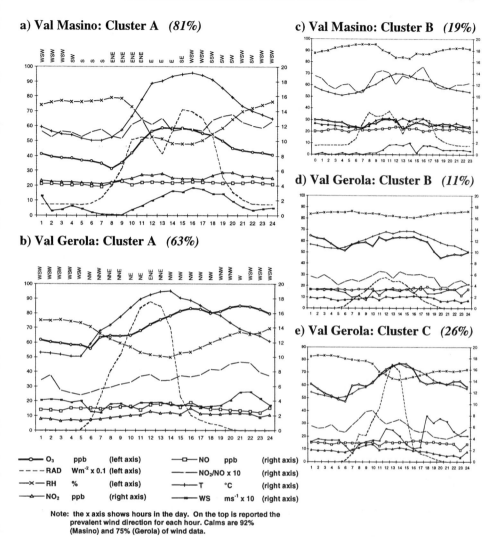

Fig. 5. Daily profiles resulting from Cluster Analysis. The number between parentheses indicates the percentage of days belonging to each cluster.

## 4. Discussion and conclusions

High ozone levels at higher altitude sites have been reported by many authors (Lefohn, 1994; Millán et al., 1996; Puxbaum et al.,1991; Aneja et al., 1994; Van Ooy et al., 1995). These are related to an increase in the $O_3$ background level due to the accumulation of photochemically produced ozone at a wider (regional and synoptic) scale and also to redistribution and exchange processes between the boundary layer and the free troposphere (Bacci et al., 1990). Thick ozone vertical stratifications, including persistent high concentration ozone air masses aloft, have already been observed above the southern alpine slope (Schlager et al., 1992). The daily profile analysis suggests that Val Gerola is immersed in an atmospheric layer characterised by an ozone background of 60 ppb compared to 30 ppb observed in Val Masino. This latter value is close to the estimated natural ozone background (Altshuller, 1987).

The differences in ozone levels observed in both valleys must be ascribed mainly to the different orographic profile, to the different atmospheric circulation and to the presence of ozone transferred from outside in different ways and, usually, in different amounts. In Val Gerola, the latter process involves the highest portion of the valley basin and we are probably dealing with ozone which originated in the urbanised areas of both the plain and the pre-alpine belt and which is blown during the day by breezes ascending along the southern and western slopes (Prévôt et al., 1997). This ozone, which accumulated at high altitudes because of minimal chemical scavenging, after subsidence and vertical mixing due to catabatic winds, is transported down and adds to locally produced ozone, thus generating the peak episodes which occur during the evening and night. In Val Masino, these processes occur more rarely, since they are obstructed by orography, and the observed ozone appears to have mainly a local origin, or to derive from the Valtellina valley bottom (Table II).

Table II
Summary of ozone processes found in the two valleys.

| Processes involving $O_3$ | Val Masino | Val Gerola |
|---|---|---|
| Local photochemical production | + | + |
| Scavenging | + | - |
| Valley breezes transport | + | + |
| Mountain breezes Transport | - | + |
| = Net accumulation | - | + |

## References

Altshuller, A. P.: 1987, *J. Air Pollut. Control Ass.* **37**, 1409-1417.
Aneja, V. P., Claiborn, C. S., Zheng Li and Murthy, A.: 1994, *Atmos. Environ.* **28**, 1781-1790.
Bacci, P., Sandroni, S. and Ventura, A.: 1990, *Sci. Total Environ.* **96**, 297-312.
Ballarin Denti, A., Cocucci, S., M. and Di Girolamo, F.: 1998a, *Chemosphere* **36**, 1049-1054.
Ballarin Denti, A., Brambilla, E. and Dell'Era, R.: 1998b, *Chemosphere* **36**, 1083-1088
Everitt, S. B. and Graham, D.:1991, *Applied Multivariate Data Analysis*, Arnold, New York.
Hangartner, M., Kirchner, M. and Werner, H.: 1996, *Analyst* **121**, 1269-1272
Joliffe, I. T.: 1986, *Principal Component Analysis*, Springer-Verlag, New York.
Lavecchia, C., Angelino, E., Bosio, C., Musitelli, A. and Bernasconi, A.: 1997, in *Proc. Air Pollution Modelling, Monitoring and Management*. Bologna, September 1997, pp. 597-606
Lefohn, A (ed):1994, *Atmos. Environ.* **28**, 1-173
Millán, M., Salvador, R. and Mantilla, E.: 1996, *Atmos. Environ.* **30**, 1909-1924.
Prévôt, A. S. H., Staehelin, J., Kok, G. L., Schillawski, R. D., Neininger, B., Staffelbach, T., Neftel, A., Wernli, H. and Dommen, J.: 1997, *J. Geophys. Res.* **102**, 23375-23388.
Puxbaum, H., Gabler, K., Smidt, S.and Glattes, F.: 1991, *Atmos. Environ.* **25A**, 1759-1666.
Schlager, H., Graf, J., Krautstrunk, M. and Brünner, M.: 1992, *Final MEMOSA PROJECT Report*, D.L.R., Institut für Physik der Atmosphäre, Oberpfaffenhofen (D)
Valli, G. and Vecchi, R.: 1996, in *Proc. Ozono e foto ossidanti in aree pedemontane, prealpine e alpine.* Saronno, 27 June 1995, pp. 70-76.
Van Ooy, D. J. and Carroll, J. J.: 1995, *Atmos. Environ.* **29**, 1319-1330.
Ward, J., H.: 1963, *J. Amer. Statist. Ass.* **58**, 236-244

# SHORT-TERM CHANGES OF RESPONSE INDICATORS OF ECOSYSTEM STATUS IN BROADLEAVED FORESTS IN TUSCANY (CENTRAL ITALY)[1]

M. FERRETTI[1], I. BONINI[3], F. BUSSOTTI[4], C. CELESTI[3], E. CENNI[1], A. CHIARUCCI[3], A. COZZI[1], V. DE DOMINICIS[3], P. GROSSONI[4], C. LEONZIO[3]

*[1]LINNÆA ambiente Srl, Firenze, Italy*
*[3]Università di Siena, Dipartimento di Biologia Ambientale, Siena, Italy*
*[4]Università di Firenze, Dipartimento di Biologia Vegetale, Firenze, Italy*

(Received 25 September 1998; accepted 25 February 1999)

**Abstract.** The status of different response indicators of forest condition were measured and assessed between 1995 and 1997 at 6 Permanent Monitoring Plots (PMPs) in Tuscany (central Italy), where beech, holm oak and Turkey oak are the most frequent tree species. Foliage transparency, leaf damage and crown dieback have changed significantly over the monitoring period. Leaf area, length of the current year shoots, and total leaf area changed as well, but only for Turkey oak. Changes were consistent between and within the plots. Different indices of plant diversity showed marked changes, apparently linked to natural dynamics within individual ecosystems. These rapid and contrasting fluctuations in the various indicators of forest ecosystem make it difficult to derive a synthesis about the general condition of ecosystems and – especially – about the effects of air pollution.

**Keywords**: biodiversity, crown condition, crown productivity, forest health, long-term monitoring, permanent plots, response indicators

## 1. Introduction

In recent years, considerable concern has been given to the long-term monitoring of forest health, with special emphasis placed on the effects of air pollution (EC-UN/ECE, 1998). A long-term monitoring program presents a variety of design-related and intepretation-related problems. Although design-related problems have been successfully addressed interpretation-related problems still remain (Innes, 1998). This is because there is no unique measurement of the ecosystem status, suggesting that a number of different response indicators should be measured to better understand ecosystem condition and changes (Ferretti, 1997). The EC-UN/ECE Intensive Monitoring (Level II) of Forest Ecosystem, adopted tree condition, increment and ground vegetation as response indicators of the impact of atmospheric pollution (EC-UN/ECE, 1998). The background to this approach is that changes in pollutant loads should be reflected by unambiguous changes in the indicators . However, it is known that dynamics within forest ecosystems take place over different temporal and spatial scales (Innes, 1998), depending on a variety of factors, such as external disturbances. Thus, faint, long-term trends in the condition of a forest ecosystem - as those expected to occur in response to chronic atmospheric pollution - can be strongly influenced by rapid changes occurring in the short-term following an acute stressor. The question raised here is that structural response indicators of the status of forest ecosystems - as those chosen by the EC-UN/ECE program - are subjected to a variety of forcing factors which make it difficult to distinguish «between healthy ecosystems, recovering from a natural disturbance, and those ecosystems that have lost their resilience due to

---

[1] MON.I.TO paper no. 10. This paper has been prepared within the project 92.60.IT.009.0, *MONitoraggio Intensivo foreste TOscane (MON.I.TO)*, cofinaced by the European Commission under the EC Reg. 3528/86 and 2157/92.

anthropogenic stress» (Rapport et al., 1998). In particular, we show how some of the indicators used by the international programs are likely to interfere with the interpretation stage, when their response will be tentatively linked to the the effects of air pollution.

## 2. Materials and Methods

The data provided by different response indicators of ecosystem status at 6 Permanent Monitoring Plots (PMPs) over a three year monitoring period in Tuscany, central Italy, are considered. Beech (*Fagus sylvatica* L.), holm oak (*Quercus ilex* L.) and Turkey oak (*Q. cerris* L.) are the most frequent species, and plots were selected in order to have 2 individual case studies for the same main species, with each of the two located under different conditions (Table I) (see Bartolozzi et al., 1996).

TABLE I

*Plots of the MON.I.TO program. N deposition is referred to the open field.*

| PMP NO. | 01 | 02 | 03 | 04 | 05 | 06 |
|---|---|---|---|---|---|---|
| Altitude (m asl) | 1350 | 1170 | 250 | 440 | 650 | 5 |
| Aspect | W-NW | NE | NE | W-NW | N-NW | W-NW |
| Slope (%) | 17 | 10 | variable | 9 | variable | 5 |
| Bedrock | Sandstone | Sandstone | Clay schists | Clay | Clay | Sand |
| Mineral soil, pH range ($H_2O$) | 3.46-3.99 | 3.50-4.00 | 5.60-6.00 | 5.90-6.78 | 3.88-4.40 | 4.80-5.80 |
| Mean precipitation (mm) | 2090 | 1324 | 978 | 873 | 728 | 637 |
| N deposition 1996 (kg/ha) | - | 16.16 | 14.66 | - | - | 17.56 |
| Mean temperature (°C) | 9.3 | 10.1 | 15.1 | 12.9 | - | 14.7 |
| Main tree species | beech | beech | holm oak | Turkey oak | Turkey oak | holm oak |
| Stems per hectar | 348* | 392* | 1454* | 1536** | 2061** | 2080* |
| Basal area, $m^2$/ha | 35.38* | 50.58* | 33.92* | 17.99** | 29.00** | 31.22* |

*: DBH threshold: 7.5 cm; **: DBH threshold: 5.0 cm

For the purpose of this paper, tree condition, crown productivity and vascular plant diversity were considered as response indicators. Tree condition was assessed on 30 trees per plot by scoring foliage transparency, extent of discoloration, occurrence of dieback on the large branches and frequency of leaf damage; estimates of crown productivity were made on 5 trees per plot, by measuring the dry weight of 100 leaves per tree, current year (c.y.) shoot length and total leaf area of the c.y. shoot. The diversity in vascular plant species was assessed at the scale of sub-plots of 100 $m^2$ by estimating the percent coverage of three vegetation layers (tree, shrub and herb) and the number of species. Full details on sampling design, scoring system, laboratory procedures and QA/QC are available in Bartolozzi et al. (1996). The methods applied (plot selection criteria, scoring systems, sample sizes) reflect those given in the UN-ECE Manuals.

## 3. Results

TREE CONDITION

The temporal trend in the tree condition indices is given in Figure 1 (upper). In the beech plots, at PMP 01 there were significant changes of crown transparency (decrease) and frequency of trees with leaf damage (holes and frayed edges caused by various insects) (increase). At PMP 02, crown transparency remained at the same level, while insect leaf damage changed significantly between years. No dieback was recorded. In the holm oak plots, at PMP 03 there was only a significant increase in transparency, while the opposite occurred at PMP 06, where there was also a significant increase in leaf damage caused by insects. Only a few cases of crown dieback were recorded. In the Turkey oak plots, no significant changes in transparency were recorded. At both plots, however, there were significant increases in leaf damage. Dieback increased significantly at PMP 04. Snow breakages (not reported here) took place at PMP 03 and 04 as a consequence of a wet snow episode in the late 1995 creating uneven openings in the canopy.

CROWN PRODUCTIVITY

Beech and holm oak showed site specific trends (Figure 1, middle). In PMP 01 the c.y. leaf area and the leaf dry weight increased significantly. In PMP 02 the maximum shoot-length was observed in 1996, followed by a sudden decrease in 1997. Foliage dry weight was lowest in 1997. The two areas with holm oak (PMPs 03 and 06) showed opposite trends in all the measured parameters, but the only significant difference was in the dry weight which was greater in 1996 at PMP 03. The two areas with Turkey oak (PMPs 04 and 05) showed the same trend for shoot length and leaf area: these two indices decreased in both PMPs in 1996 and recovered partially in 1997. Leaf dry weight showed a negative trend in PMP 04.

VASCULAR PLANT DIVERSITY

With the exception of the beech plots, the percent cover of the three main vegetation layers changed at all the PMPs over the three years of monitoring (Figure 1, bottom). A significant reduction in the tree canopy was observed at PMPs 03 and 04 caused by crown breakage following an exceptional snowfall during winter 1995-96. The ground cover of shrub and herbaceous layers showed an increase at the Turkey oak PMP 04, as a direct response to the higher photon flux density. At the holm oak PMP 03 a significant reduction in the shrub layer was observed, because, in contrast to the PMP 04, this layer was almost exclusively formed by sciophilous species. Species richness was similar in relation to forest type and rather constant during the three years, with the Turkey oak plot woods being the most species rich. The only significant change was observed at PMP 04, in relation to the above mentioned snowfall event (Chiarucci *et al.*, 1998).

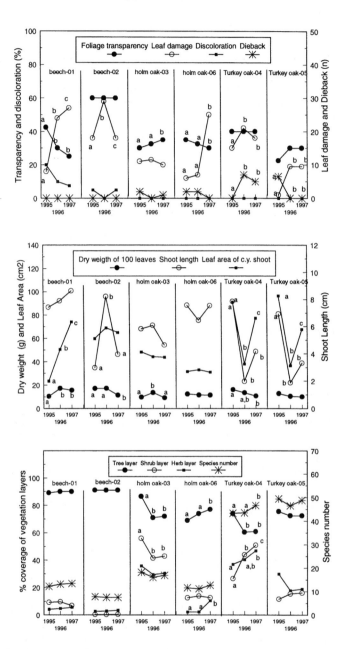

Fig. 1 - Trend of different indices of ecosystem status. Upper: Tree condition. Foliage transparency and discoloration (median values) were tested by the Wilcoxon matched pairs test. Leaf damage and dieback (frequency of symptomatic trees) were tested by the McNemar test for nominal/ordinal scales. Different letters indicate significant differences between subsequent years. Middle, crown productivity. Mean plot values. Bottom, vascular plant diversity. Percent coverage of each vegetation layer (herb, shrub and tree) and n. of the species. In the middle and bottom different letters indicate significant differences ($P<0.01$) by LSD test after ANOVA.

## 4. Discussion and Conclusions

A summary of significant directional changes for each indicator is given in Table II. In general, each of the plots behaved independently, which is not unexpected since they were individual case studies, and it is difficult to identify a common pattern, in the sense of a consistent association between indicator responses. In some cases tree condition and crown productivity indicators did behave consistently, as in PMP 01, where a significant decrease of transparency was associated with a significant increase of leaf area.

TABLE II

*Significant directional changes for individual indicator/index at each plot between 1995-1996 (referred to as 96 in the headline) and 1996-1997 (97 in the headline). +: significant increase; -: significant decrease.*

| Indicator/index | PMP 01 | | PMP 02 | | PMP 03 | | PMP 06 | | PMP 04 | | PMP 05 | |
|---|---|---|---|---|---|---|---|---|---|---|---|---|
| | 96 | 97 | 96 | 97 | 96 | 97 | 96 | 97 | 96 | 97 | 96 | 97 |
| Tree condition | | | | | | | | | | | | |
| - transparency | - | | | | + | | - | - | | | | |
| - discoloration | | | | | | | | | | | | |
| - leaf damage | + | + | + | - | | | | + | + | | + | |
| - dieback | | | | | | | | | + | | | - |
| Crown productivity | | | | | | | | | | | | |
| - leaf weight | + | | | - | + | - | | | | - | | |
| - current year leaf area | + | + | | | | | | | - | + | - | + |
| - lenght c. y. shoot | | | + | - | | | | | - | | - | |
| Diversity | | | | | | | | | | | | |
| - % coverage/herb layer | | | | | | | | + | + | + | | |
| - % coverage/shrub layer | | | | | - | | | | + | + | | |
| - % coverage/tree layer | | | | | - | | | + | - | | | |
| - n. species | | | | | | | | | | + | | |

However, even in this case, decreased transparency was accompanied by a significant increase in leaf damage by insects. In other cases, there are contrasting evidences. At PMP 04, there was an increase in crown dieback and leaf damage (a worsening in the tree condition) recorded also in the crown productivity indices. At the same time there was a significant increase in the herbaceous and shrub layer cover and in the number of species (an increase of the vascular plant diversity) as a consequence of the openings made in the canopy by the wet snow. However, the same episode caused a reversed effect at PMP 03, where the diversity indices decreased as a consequence of the different ecological attribute of understory species. Obviously, indicators were exposed to large underlying variation due in part to factors related to the original design of the study (e.g. the ability of the indices used to cope with the huge variety of environments and the ability of the sampling design to fully describe the variability at plot level - Breckendridge *et al.*, 1998) although this is consistent with the European Level II program directives. In any case, it seems that - in the forests considered here - episodes connected to the natural dynamics within the ecosystem can easily obscure any other (e.g. chronic pollution-induced) adverse effect. For example, changes in vegetation due to anthropogenic input were reported for both acidic and calcareous soils (Thimonier *et al.*, 1994): however, it is unlikely that deposition will cause such dramatic year to year changes as those reported between subsequent years in PMP 04. Conversely, it is very likely that the response of the system to deposition loads will be affected by the new dynamic that has started after such changes. Different indicators described different components of the forest ecosystem. Therefore, when multiple indicators/indices are used, an adequate conceptual and operational model is needed to correctly interpret the data (Muir and McCune, 1987). The concept «health» itself seems questionable as a conceptual and operational descriptor for an ecosystem (Suter II, 1993; Wicklum and Davies, 1995). However, when used under an operational perspective, the term health should be explicitly linked to a clear approach, like the Ecosystem Distress Syndrome (EDS, Rapport, 1998; Whitford et al., 1998). This approach is based on *functional* (besides structural) indicators and on the identification of common patterns among stressed ecosystems. Unfortunately, no attempt was made in this direction by the evaluation strategy proposed for the international programmes. Rather, major emphasis was placed on correlative studies between (many) chemical parameters and (few) response indicators where a series of individual case-study are managed as a probability sample (EC-UN/ECE, 1998). The data collected in Tuscany (i.e. on a geographical scale with much more homogeneity than Europe), with rapid and contrasting (between- and within-plots) directional changes question the suitability of the indicators adopted by the international programs designed to monitor the effects of air pollution on forests and - more generally - the current, common (ab)use of the term «forest health».

## Acknowledgements

The authors are grateful to two anonymous referees for helpful and stimulating comments. Simona Maccherini helped with the statistical analysis of the crown condition data.

## References

Bartolozzi L., Bussotti F., De Dominicis V. and Ferretti M. (Eds.) 1996. Regione Toscana-Giunta Regionale Publisher, Firenze: 92 ps.
Breckenridge, R. P., Kepner W. G. and Mouat D. A.: 1995. *Environ. Monit. Assess.,* 36: 45-60.
Chiarucci A., Bonini I., and De Dominicis V., 1998. *Stud in Plant Ecol.,* 20: 98.
EC-UN/ECE, 1998. Brussels, Geneva, 104 pp.
Ferretti M. 1997. *Environ. Monit. Assess.,* 48: 45-72.
Innes J. L.: 1998. In: Peterson D. L., Parker V. T. (Eds.) Ecological Scale: 429-457.
Muir, P. S. and McCune, B.: 1987. *Plant Disease,* **71,** 558-565.
Rapport D. J., Whitford W. G. and Hilden M.: 1998. *Environ. Monit. Assess.,* 51: 171-178.
Suter II, G. W.,: 1993. *Environ. Toxicol. Chemi.* 12: 1533-1539.
Thimonier A., Dupouey J. L., Bost F., and Becker M.: 1994. *New Phytol.,* 126: 533-539.
Wicklum D and Davies R. W.:1995. *Can. J. Bot.,* 73:997-1000.
Whitford W. G., De Soyza A.G., Van Zee J. W., Herrick J. E. and Havstad K. M., 1998. *Environ. Monit. Assess.,* 51: 179-200.

# GROWTH OF MATURE BEECH IN RELATION TO OZONE AND NITROGEN DEPOSITION: AN EPIDEMIOLOGICAL APPROACH

SABINE BRAUN[1], BEAT RIHM[2], CHRISTIAN SCHINDLER[3] AND WALTER FLÜCKIGER[1]

[1] *Institute for Applied Plant Biology, Sandgrubenstrasse 25, 4124 Schönenbuch, Switzerland*
[2] *Meteotest, Fabrikstrasse 14, 3012 Bern, Switzerland*
[3] *Institute of Social and Preventive Medicine of the University of Basel, Switzerland*

(Received 20 October 1998; accepted 25 February 1999)

**Abstract.** Stem increment of mature *Fagus sylvatica* L. from 57 plots in Switzerland was analysed with respect to environmental factors using multilinear regression. Nitrogen (N) deposition was positively correlated with stem growth, with an increase by 41.8% (confidence interval 38.9-44.2) per 10 kg N $ha^{-1}$ $a^{-1}$. There was a negative relationship between ozone ($O_3$) dose and diameter increment. The fit was better for maximum rather than average exposure values. Coefficients indicated growth changes of -22.5% (14.3-28.6 confidence interval) and -35.1% (17.8-47.3) respectively per 10 ppm hours. These changes exceed the -6.1% found in experiments with beech seedlings. This difference is mainly attributed to the fact that the seedling data are calculated for a single season's exposure whereas the stem increment data extend over four years, representing the product of four annual reductions. Increased sensitivity of mature trees compared to seedlings, however, is also suggested.

**Keywords.** Fagus sylvatica, stem increment, ozone, nitrogen deposition

## 1. Introduction

There is much concern over forest health in Europe (UN/ECE, 1997). Although many data have been collected during the past 15 years, few studies have shown relationships between forest health and anthropogenic stress factors by means of multivariate statistical methods. In Norway, Nellemann and Esser (1998) observed enhanced discoloration of Norway spruce foliage on sites where pH of the humus layer was <3.75. In Canada, Desrochers *et al.* (1999) demonstrated an interaction between stem growth of sugar maple and an exceedance of the critical load of acidity. Using a regression approach, McLaughlin and Downing (1995) were able to correlate short-term rates of stem expansion in mature trees of *Pinus taeda* with ozone ($O_3$) exposure as well as interactions with low soil moisture and high air temperatures. Epidemiological studies may help to fill the gap between field conditions and laboratory studies as experiments with large trees require expensive facilities and can, therefore, only be conducted with few individuals. The aim of this investigation was to identify factors which could explain stem growth in beech observation plots in Switzerland, especially anthropogenic influences such as $O_3$ pollution and the deposition of nitrogen (N) and acidity.

## 2. Materials and Methods

In Switzerland, observations of *Fagus sylvatica* L. have been made in 57 permanent forest plots since 1984/1989 (Figure 1). The plots extend over an altitude range of 260-1120 m asl and an age range of 65-175 years (Table I). Stem increment was determined non-destructively by measuring diameter at breast height (DBH) at marked stem points in autumn 1991 and 1995. Social position in the stand was scored into three classes (1=dominant, 2=codominant and 3=suppressed) and used as a continuous variable. Position within the stand was grouped into 0 (stand) and 1 (edge or clearing). Soil samples were taken in winter 1995/96 and 1996/97 by horizon (one combined sample per layer and plot). Base saturation was measured in $NH_4Cl$-extracts of air-dried soil and averaged over the top 40 cm. For data analysis, base saturation was classified into two groups: 0 at ≤ 40% and 1 at > 40%. Available soil water (0-40 cm depth) was calculated according to Benzler et al. (1982). Evapotranspiration after Penman (1956) and soil water calculations after Thornthwaite and Mather (1957) were combined to produce a site characteristic figure defined as the number of days when soil water content was below 70% field capacity. These calculations were based on climate data from 17 automatic weather stations for the years 1992-1995 supplied by the Swiss Meteorological Institute.

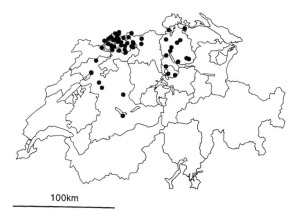

Fig. 1: Map of Switzerland showing the permanent observation plots used in this study

Deposition was modelled after the methods outlined in BUWAL (1994). $O_3$ dose (product of time and concentration with a threshold value of 40 ppb, in ppm hours) was calculated for daylight hours according to the critical level definition given by Fuhrer et al. (1997). The methods used for $O_3$ mapping are described in Achermann and Rihm (1997). $O_3$ dose values were available for all the years within the growth measurement period, 1992-1995. The average and the highest annual dose of the four years were used in the analysis.

A multivariate data analysis was performed using a mixed model (function lme in S-PLUS (S-PLUS4 1997)) to take into account the clustered pattern of the dataset. Nonsignificant parameters were eliminated in a stepwise procedure to minimize the AIC. The following site predictors were tested in the model: age, altitude, soil water regime, topography (aspect and slope), site fertility, humus form, base saturation, $O_3$

TABLE I

Plot characteristics – soil type, altitude and stand characteristics – age, height and stocking density.

| Soil type | Nos/category | Altitude (m.asl) | Nos/category |
|---|---|---|---|
| Regosol | 1 | ≤ 500 | 20 |
| Rendzine | 23 | >500-1000 | 34 |
| Canibisol | 11 | >1000 | 3 |
| Luvisol | 18 | | |
| Gleysol | 2 | | |
| Other | 2 | | |

| Stand | Mean | Minimum | Maximum |
|---|---|---|---|
| Tree age (yrs) | 115 | 65 | 175 |
| Height of dominant tree (m) | 33 | 26 | 40 |
| Number of trees/hectare | 290 | 120 | 590 |

dose (average and maximum dose over the four years period), N deposition, N-free acid deposition and foliar nutrient concentrations. Single tree predictors included crown transparency, initial diameter, crown projection area, social position and position within the stand. Crown transparency was arcsin and crown projection area root transformed. Both diameter and crown transparency showed a nonlinear behaviour, therefore the squares were included as well. Stem increment was transformed to the power of 0.7 to ensure normal distribution of the residuals. Base saturation and position within the stand were calculated as categorical variables. Residuals were analyzed for outliers and plotted against independent variables to check the linearity of the relationship. The figures were prepared by plotting the residuals of the regression (growth) without the independent variable in question ($O_3$ or base saturation) against the residuals of a regression of the independent variable against the other independent variables (partial residual plot). The mean of the corresponding variable was added to the residuals to give a more easily interpretable value. As the mixed model does not return $r^2$ values, the variance proportions given in the results section were calculated with normal regression using the same set of variables.

## 3. Results

The full model explained 12% of the total variance. If, however, only plot means were examined with site factors, 32% of the remaining variance was explained by four significant predictors which include soil factors (base saturation), deposition (both N and acid deposition) as well as $O_3$. The detailed statistics is given in Table II.

## TABLE II

ANOVA table of the multivariate analysis with the highest annual $O_3$ dose of the period 1992-1995. Dependent variable: Stem increment, transformed to the power of 0.7. *p<0.05, **p<0.01, ***p<0.001. 1583 observations in 57 plots. The t-value of a regression estimate is defined as the ratio between this estimate and its standard error. For large sample sizes, this ratio approximately follows a standard normal distribution.

|  | Coefficient | Std.Error | t-value | Significance |
|---|---|---|---|---|
| (Intercept) | 0.0790 | 0.681 | 0.116 |  |
| $O_3$ dose (maximum), (ppm h) | -0.0636 | 0.0145 | -4.381 | *** |
| N deposition (kg N ha$^{-1}$ yr$^{-1}$) | 0.0433 | 0.0102 | 4.241 | *** |
| Base saturation (0,1) | 0.262 | 0.0798 | 3.281 | ** |
| Acid deposition (keq ha$^{-1}$ yr$^{-1}$) | -0.965 | 0.377 | -2.560 | * |
| Diameter 1991 (mm) | 0.008062 | 0.00118 | 6.846 | *** |
| (Diameter 1991)$^2$ | -0.00000688 | 0.00000118 | -5.839 | *** |
| Social position (1-3) | -0.175 | 0.0605 | -2.887 | ** |
| Position within the stand (0,1) | 0.103 | 0.0208 | 4.931 | *** |
| Crown projection area (m$^2$)$^{0.5}$ | 0.00602 | 0.00132 | 4.551 | *** |
| Crown transparency (arcsin (%)) | 3.62 | 1.57 | 2.305 | * |
| (Crown transparency)$^2$ | -4.57 | 1.54 | -2.980 | ** |

### 3.1. TREE FACTORS

Tree related parameters explaining stem growth were factors describing the space available for growth (crown projection area, crown transparency, position within the stand) and age (Table II). Both DBH and crown transparency showed a non linear behaviour: at large DBH's the effect of DBH became negative. The effect of crown transparency was largest at high transparency values (>40% mean transparency over the assessment period) (Figure 2).

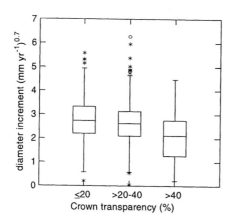

Fig. 2: Diameter increment of beech trees with different crown transparencies (average over the period 1992-1995). SYSTAT®-Boxplot, with the median as horizontal line, the box extending from the 25% to the 75%-quantile and whiskers indicating non outlier minima and maxima. Outliers are indicated by asterisks and circles. For statistics see Table II (partial correlation plot).

### 3.2. SOIL

Base saturation was a significant predictor for stem increment: growth was decreased at sites with base saturation values ≤40% (Figure 3). This is an interesting observation, since low base saturation values are often found on highly productive sites (till soils) in Switzerland. Confounding factors in the case of base saturation were mainly concentrations of $SO_2$ and of $NO_2$. However, their concentrations are well below the critical levels. Annual averages ranged between 6-14 µg/m$^3$ for $SO_2$ and 10-40 µg/m$^3$ for $NO_2$. They were therefore omitted from the model. Small intercorrelations of base saturation were also observed with N deposition (negative correlation) and with altitude (positive). If these interactions were important, they would both lead to increased growth estimates at low base saturation. Thus, they cannot explain the observed effect.

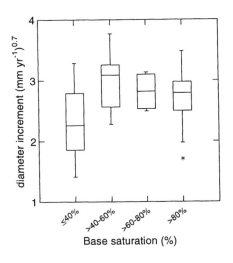

Fig. 3: Diameter increment on sites with different base saturation (mean of 0-40 cm soil depth). Because there were no significant differences above 40% base saturation, only two classes (≤40%, >40%) were used in the multivariate model (see Table II). Diameter increment data are residuals (partial correlation plot with plot means).

3.3. DEPOSITION

As N deposition acts both as nutrient and as acidifying compound, acid deposition was separated into N and acidity, the latter calculated from sulfur (S) deposition minus the deposition of base cations. Whereas N-free acid deposition was negatively correlated with growth, there was a positive correlation with N deposition (Figure 4). Due to intercorrelations among the deposition parameters, the effect of acid deposition is difficult to assess; the values of N-free acid deposition ranged between 0 and 0.89 keq ha$^{-1}$ a$^{-1}$. In the case of N deposition, a growth increase of 41.8% (confidence interval 38.9-44.2) per 10 kg N ha$^{-1}$ a$^{-1}$ is calculated from the regression coefficients for the range between 30 and 50 kg N ha$^{-1}$ a$^{-1}$. There are indications of an interaction between N deposition and P nutrition: at low P nutrition, growth is less correlated with N deposition (see also Flückiger and Braun 1999). The main confounding factor for N deposition is altitude; there is a highly significant correlation between N deposition and altitude ($r^2$=0.31). However, altitude does not become significant even when N deposition is taken out of the model. Thus, the relationship between growth and N deposition cannot be explained as an altitude effect.

3.4. OZONE

$O_3$ was negatively correlated with stem increment (Figure 5). The maximum $O_3$ dose for the 4 year period was a better predictor for growth than the average dose (t=-4.38 and -3.38, respectively), although both were significant at the p<0.001 level. Confounding factors may be altitude or the concentration of other pollutants such as $NO_2$ or $SO_2$. Whereas the average daylight dose shows some altitude dependence ($r^2$= 0.14), there was no relationship between altitude and the highest annual dose used in the

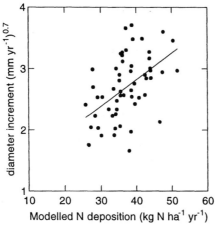

Fig. 4: Correlation between diameter increment of beech between 1992 and 1995 and modelled N deposition. Partial correlation plot with plot means (see Table II).

calculations. Both $NO_2$ and $SO_2$ correlated with cumulative $O_3$ dose. These two factors were omitted from the model as already discussed in the soil section. Age was also intercorrelated with $O_3$. This could be a potential problem since stem increment is also expected to decrease with age.

Calculations showed, however, that for this dataset the age spread corresponds to 6.5 ppm hours $O_3$ (maximum). Age itself was not a significant predictor in the model. The slope of the $O_3$ correlation with the highest annual dose was $-0.0636 \pm 0.0145$ (standard error) (Table II). When reverting to the untransformed stem increment, a growth reduction of 22.5% (confidence interval 14.3-28.6) for 10 ppm hours $O_3$ dose is found, calculated as the difference between growth at 10 and 40 ppm hours. In the case of average dose, the slope is $-0.107 \pm 0.0316$, resulting in a predicted growth reduction of 35.1% (confidence interval 17.8-47.3) per 10 ppm hours $O_3$ dose between 10 and 25 ppm hours.

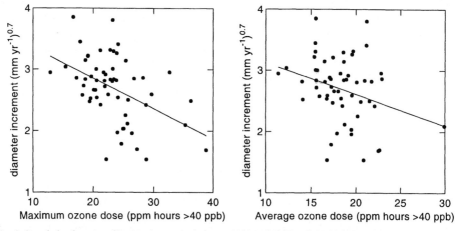

Fig. 5: Correlation between diameter increment between 1992 and 1995 and the highest annual $O_3$ dose During this period (left) and the average $O_3$ dose (right). Partial correlation plot with plot means. Although both correlations are significant at the $p<0.001$ level, the left one is better (t=-4.38 and -3.38, respectively) (see Table II).

## 4. Discussion

The positive correlation between N deposition and stem increment suggests that N may be involved in the increased forest growth observed in Europe during the last decades (Spiecker et al., 1996). The reduced stem increment at sites with low base saturation indicates that soil acidification should be monitored carefully to ensure sustainable forest growth. The correlation between $O_3$ and stem growth had a significanty higher slope than the relationship determined using beech seedlings (Fuhrer et al., 1997). A biomass change of 6.10% (95% confidence interval 2.23–9.97) per 10 ppm hours annual dose was calculated for these seedlings. When interpreting the apparent difference to the present data, one has to take into account that this value was calculated for a single exposure season. Experiments with seedlings have shown that growth reductions by $O_3$ may accumulate during at least three consecutive years (Braun and Flückiger, 1995). An annual growth decrease of 6.1% may thus result in a growth loss of 22.3% within four years which is in good agreement with the coefficient for the highest annual dose in the epidemiological study. Despite a poorer fit, the slope for the average $O_3$ dose exceeds the four-year estimate for seedlings, suggesting a somewhat higher sensitivity for adult beech. An increased sensitivity of mature trees compared to seedlings was found for physiological parameters in *Quercus rubra* (Samuelson and Edwards, 1993; Wullschleger et al., 1996) and for visible injury in *Prunus serotina* (Fredericksen et al., 1995). However, decreased $O_3$ sensitivity of mature trees has been described in *Pinus taeda* (Sasek and Flagler, 1995) and *Sequoiadendron giganteum* (Grulke and Miller, 1994).

## 5. Conclusion

The present data suggest a substantial influence of anthropogenic factors on stem growth of *Fagus sylvatica*. They also show that it is possible to demonstrate anthropogenic effects on forests using epidemiological methods when there is a uniform dataset with enough scatter among the parameters of interest. In the forest, growth reductions by $O_3$ is masked by the simultaneous growth stimulation by N deposition. It has, however, to be borne in mind that in spite of their opposite effect on the growth of aboveground biomass, both N (Wallenda et al., 1996) and $O_3$ (Spence et al., 1990) affect carbohydrate transport to the roots and hence root growth. Thus, while aboveground growth may seem quite normal, belowground growth may be severely affected. The consequences of these anthropogenic influences on root growth and vitality require further investigation.

## Acknowledgements

This project was supported by the Federal Office of Environment, Forest and Landscape (FOEFL) and by the Cantons of AG, BL, BS, BE, SO, ZG and ZH. We would like to thank B. Achermann, H. Balsiger, Dr. R. Eichrodt, J. Froelicher, W. Giss, Dr. H. Kasper, H. Schmid and Dr. R. Volz for their support and interest in our work. We are also indebted to B. Möcklin and I. da Costa for plant and soil analysis, Dr. H. Flückiger and A. Mebert for technical assistance and Dr. E. Hiltbrunner for helpful discussions. The Swiss Meteorological Institute kindly supplied us with climate data.

## References

Achermann, B. and Rihm, B.: 1997, *Calculation and Mapping of Critical Thresholds in Europe - Status Report 1997*, eds. Posch, M., Hettelingh, J.-P., de Smet, P. A. M. and Downing, R. J., UN/ECE, 1-163.
Benzler, J. H., Finnern, H., Müller, W., Roeschmann, G., Will, K. H. and Wittmann, O.: 1982, *Bodenkundliche Kartieranleitung*. E. Schweizerbart'sche Verlagsbuchhandlung, Stuttgart
Braun, S. and Flückiger, W.: 1995, *New Phytol.* **129**, 33-44.
BUWAL: 1994, Environmental Series **234**. Bern,1-68
Desrochers, P., Lambert, M. C. and Bernier-Cardou, M.: 1999. *Water Air Soil Pollution* (in press)
Flückiger, W. and Braun, S.: 1999. *Water Air Soil Pollution* (in press)
Fredericksen, T. S., Joyce, B. J., Skelly, J. M., Steiner, K. C., Kolb, T. E., Kouterick, K. B., Savage, J. E. and Snyder, K. R.: 1995, *Environ.Pollut.* **89**, 273-283.
Fuhrer, J., Skärby, L. and Ashmore, M. R.: 1997, *Environ.Pollut.* **97**, 91-106.
Grulke, N. E. and Miller, P. R.: 1994, *Tree Physiol.* **14**, 659-668.
McLaughlin, S. B. and Downing, D. J.: 1995, *Nature* **374**, 252-254.
Nellemann, Ch. and Esser, J. M.: 1998, *Ambio* **27**, 143-147.
Penman, H. L.: 1956, *Trans.Amer.Geophys.Union.* **37**, 1.
Samuelson, L. J. and Edwards, G. S.: 1993, *New Phytol.* **125**, 373-379.
Sasek, T. W. and Flagler, R. B.: 1995, *Impacts of Air Pollutants on Southern Pine Forests*, eds. Fox, S. and Micker, R. A., Springer Verlag, New York, USA, 424-466.
Spence, R. D., Rykiel, E. J. and Sharpe, P. J.: 1990, *Environ.Pollut.* **64**, 93-106.
Spiecker, H., Mielikäinen, R., Köhl, M. and Skorgsgaard, J. P.: 1996, *Growth Trends in European Forests*. Springer-Verlag, New York, p. 372
Thornthwaite, C. W. and Mather, J. R.: 1957, *Publications in Climatology* **10**, 185-243
UN/ECE: 1997, *Ten Years of Monitoring Forest Condition in Europe*. p. 386
Wallenda, T., Schaeffer, C., Einig, W., Wingler, A., Hampp, R., Seith, B., George, E., Marschner, H.: 1996, *Plant and Soil* **186**, 361-369.
Wullschleger, S. D., Hanson, P. J. and Edwards, G. S.: 1996, *Plant Cell and Environment* **19**, 577-584.

# DIFFERENCE IN NEEDLE LENGTH - A NEW AND OBJECTIVE INDICATOR OF POLLUTION IMPACT ON SCOTS PINE (*Pinus sylvestris*)

M. V. KOZLOV[1] and P. NIEMELÄ[2]

[1]*Section of Ecology, University of Turku, FIN-20014 Turku, Finland.* [2]*Faculty of Forestry, University of Joensuu, Box 111, FIN-80101 Joensuu, Finland.*

(Received 25 September 1998; accepted 26 February 1999)

**Abstract.** Needle fluctuating asymmetry (FA) in Scots pine (*Pinus sylvestris*) was studied south of the Severonikel nickel-copper smelter in the Kola Peninsula, Russia. We measured the difference in length of two needles of the same pair, and calculated FA as the ratio between this difference and the average length of the two needles of this pair. Needle FA did not depend on tree age, distance from the nearest neighbour or branch position within the crown, but increased with an increase in branching order. Needles originating in different years (1993-1997) demonstrated among-year variation in FA, which, however, was not related to annual fluctuations in $SO_2$ emission by the smelter. FA increased with decreasing distance from the smelter, and in the sites proximate to the smelter FA was double that found at the most distant (background) sites. The increase in FA was primarily due to an increased difference in the length of the two needles, and this difference may serve as a practicable indicator of pollution-induced stress in Scots pine.

**Keywords:** fluctuating asymmetry, pollution, Kola peninsula, tree age, branching order, competition.

## 1. Introduction

Fluctuating asymmetry (FA), i.e. small non-directional departures from the anticipated bilateral symmetry (Palmer and Strobeck, 1986; Polak and Trivers, 1994), reflects the developmental stability of the organism and increases with environmental stress (Zakharov, 1990; Parsons, 1990; Clarke, 1992; Graham et al., 1993). Historically, the study of FA has focused on animals, but recently it has been demonstrated that leaf FA in many plants (*Acer, Aegopodium, Betula, Convolvulus, Epilobium, Matricaria, Robinia, Sorbus, Fucus*) increases due to environmental contamination (Freeman et al., 1993; Kozlov et al., 1996; Kryazheva et al., 1996; Møller, 1998), herbivory (Zvereva et al., 1997a, b) and competition (Rettig et al., 1997).

The health status of coniferous forests has for some time been evaluated on the basis of needle coloration, crown depletion and other subjective indices (e.g. Hanisch and Kilz, 1988; Alexeyev, 1989; Anonymous, 1994). The maintenance of large-scale monitoring programs requires the periodical training of recording personnel to obtain comparable results (Ferretti et al., 1995). Thus a practicable and objective method of health assessment in conifers is needed (Cape, 1990; Ferretti, 1998).

The purpose of this paper is (i) to elaborate a practical method for the sampling and measurement of needle asymmetry in Scots pine (*Pinus sylvestris* L.), (ii) to evaluate the needle FA response to environmental pollution.

## 2. Material and Methods

The study was conducted in the early spring of 1998 in the Kola Peninsula, Northwestern Russia. The area surrounding the city of Monchegorsk (67° 52' N, 32° 48' E) is one of the most polluted in the boreal forest zone. Vast quantities of sulphur and heavy metals emitted by the Severonikel smelter since 1939 have caused widespread destruction of soils and vegetation (Kozlov and Haukioja, 1995). Coniferous forests have completely vanished up to 10 km south and 15 km north of the smelter, and have been replaced by willow- and birch-dominated communities or industrial barrens (Koroleva, 1993).

Ten sites were chosen, located 5 to 65 km south of the city of Monchegorsk along the road from Murmansk to St. Petersburg, and situated at least 50 m from the roadside. The proximate sites represent heavily polluted industrial barrens, where Scots pine is quite infrequent; the survivors demonstrate a creeping growth form, short (1-2 years) needle longevity and prominent needle discoloration. The most distant sites have nearly background levels of pollution; the pines in these sites look healthy, with a needle longevity of 4-6 years. For a map of the study area, see Kozlov et al. (1995); additional characteristics of the study sites are given by Koroleva (1993), Kozlov et al. (1995, 1996), Kozlov and Haukioja (1995), Zvereva et al. (1997b).

Since distinguishing between the three kinds of bilateral asymmetry requires an analysis of the left-minus-right values of the symmetrical character (Palmer and Strobeck, 1986), in five samples (with 20 needle pairs each) we marked the 'left' needle (as seen from the tip of the branch) in each pair prior to sampling, and conducted the statistical tests with signed values. In all other samples the needles were not marked, so that we were operating with the absolute values of differences in needle length.

To optimize the sampling strategy, we investigated variation in FA with tree age, competition with neighbours, branch position within the crown, branching order and the year of needle formation. In each of two 'intensive' study sites (15 and 47 km south of Monchegorsk) we selected five trees in each of four age classes (8-12, 17-25, 35-45, >80 yrs, as estimated from the number of annual whorls). The selected trees were 2-4 m apart from their nearest neighbour. Additionally, in the 'intensive' sites we selected five 17-25 year old trees which were suppressed by their larger neighbours (situated <0.5 m apart).

The effects of branch position in the crown, branching order, and the year of needle formation were investigated in five non-suppressed trees aged 17-25 years in each of the two 'intensive' sites, one branch per part of the crown (upper, medium, and lower). From the mid-crown branch we collected (i) five samples (20 needle pairs in each) of needles formed in different years (1993-1997) on the 2nd order axis (i.e. a branch arising from tree trunk), and (ii) two samples of needles (formed in 1997) from branches of the 3rd and 4th branching orders. From the other (upper and lower) branches, only one sample of needles formed in 1997 on the 2nd order branch was collected. From all the other trees, only one sample was collected, of needles formed in 1997 on the 2nd order mid-crown branch.

The response to pollution was studied by using five trees aged 17-25 years in each of ten study sites, taking samples of 20 needle pairs formed in 1997 on the 2nd order mid-crown branch.

The needle pairs were mounted on strong paper in such a way that the two needles comprising the pair were pressed together. The difference in the length of the two needles in the pair was measured under a dissecting microscope with an ocular scale to the nearest

0.05 mm (twice by different persons), and the length of the needle pair was measured with a ruler to the nearest 1 mm. FA was calculated as the ratio between the difference in needle length and the average needle length; the sample-specific values were averaged from the measurements of 20 needle pairs.

The presence of asymmetry in our data set (3400 needle pairs) was confirmed by the Kruskal-Wallis test. The type of asymmetry was revealed by testing the distribution of left-minus-right values in the five samples (100 needle pairs) where we distinguished between 'left' and 'right' needles in the pair. The effects of tree age, competition, branch position in the crown, branching order and year of needle formation were investigated by the Kruskal-Wallis test; the effects of pollution were detected by Pearson correlation coefficients with the log-transformed distance from the emission source (SAS Institute, 1990).

## 3. Results

Two independent measurements of the absolute difference in the length of two needles produced similar results ($r = 0.98$, $n = 3400$, $p < 0.0001$), indicating high repeatability. The differences in length between longer and shorter needles in the pair were significant ($\chi^2 = 8.76$, $df = 1$, $p = 0.0031$), suggesting the occurrence of asymmetry.

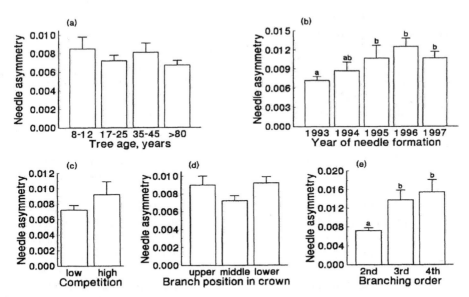

Fig. 1. Sources of among-tree and within-tree variation in needle fluctuating asymmetry (samples from two 'intensive' study sites; for more details, see text). a, tree age ($\chi^2 = 1.72$, $p = 0.63$); b, year of needle formation ($\chi^2 = 11.8$, $p = 0.02$); c, competition with the neighbours ($\chi^2 = 0.46$, $p = 0.50$); d, branch position within the crown ($\chi^2 = 4.07$, $p = 0.13$); e, branching order ($\chi^2 = 13.9$, $p = 0.001$). Bars indicate standard errors; sample size $N = 10$. Bars marked with different letters differ from each other at the probability level $p = 0.05$.

An absence of anti-symmetry (where asymmetry is the norm but the side containing the larger character varies) was revealed by the normality of distributions of left-minus-right values (p = 0.56), while the zero mean value (p = 0.42) confirmed the absence of directional asymmetry (where a character on one side usually has a larger value than its counterpart). Thus the observed asymmetry was classified as FA

We found no difference in FA between the two 'intensive' study sites, nor did the site interact with the other variables investigated (p > 0.05). The data from these two sites were therefore pooled to study within-crown and between-tree variation.

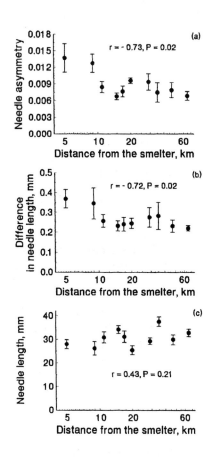

Fig. 2. Changes in needle characteristics along the pollution gradient south of the Severonikel smelter. a, fluctuating asymmetry; b, absolute difference in the length of two needles of the same pair; c, needle length. Bars indicate standard errors; sample size N = 5.

FA depended neither on tree age (Figure 1, a) nor on branch position within the crown (Figure 1, d) but increased with an increase in branching order (Figure 1, e). Among-year variation in FA (Figure 1, b) was not related to annual fluctuations in $SO_2$ emissions from the Severonikel smelter (r = -0.69, n = 5 years, p = 0.20).

FA increased with a decrease in distance from the smelter (Figure 2, a). In the sites nearest to the smelter FA was double compared to the most distant (background) sites. The increase in FA was due to an increase in the absolute difference in the length of the two needles (Figure 2, b), rather than to a decline in needle length (Figure 2, c).

## 4. Discussion

We demonstrated the presence of asymmetry in our data set as suggested by Pomory (1997), and then classified this asymmetry as FA using the protocol by Palmer and Strobeck (1986). The absence of both antisymmetry and directional asymmetry in our samples allowed us to simplify the sampling procedure considerably by omitting the attribution of needles to the 'left' or 'right' position in the pair.

An important point in our method is the direct measurement of the difference in needle length, instead of measuring the lengths of two needles and calculating the absolute difference, as in broad-leaved plants such as birch (Kozlov *et al.*, 1996; Kryazheva *et al.*, 1996). The use of the absolute size (length) of the needle excludes one possible source of spurious correlation (Sullivan *et al.*, 1993).

We demonstrated that branching order and the year of needle formation have to be taken into account when sampling needles for FA measurements, while tree position in the stand, tree age, and branch position within the crown did not contribute to the variation in needle FA. Since annual fluctuations in FA did not correlate with fluctuations in emission, we suggest that FA (which is a non-specific stress indicator) may appear sensitive to climatic variables. However, longer time series or manipulative studies are needed to explicitly test this hypothesis.

Environmental contamination in our gradient decreased nearly hyperbolically with an increase in distance from the smelter, approaching the regional background level in the most distant of our sites (Kozlov *et al.*, 1995). The Severonikel nickel-copper smelting complex emits dozens of substances with different toxicity and migratory abilities (Kozlov and Haukioja, 1995); habitat deterioration due to the long-lasting pollution impact imposed additional environmental stresses, due for instance to changes in the microclimate (Kryuchkov, 1993; Kozlov and Haukioja, 1997). Since it was impossible to determine the causal links between a certain pollutant and FA from our data set, we have chosen distance from the smelter as the measure of cumulative effect of pollution.

An increase in needle FA with an increase in pollution was predictable and agrees with earlier observations on deciduous plants (Freeman *et al.*, 1993; Graham *et al.*, 1993; Kozlov *et al.*, 1996; Kryazheva *et al.*, 1996; Møller, 1998). The novelty of our study is in the practical possibility of measuring needle FA in conifers. Since symmetrical individuals generally have faster growth, higher fecundity and better survival than more asymmetrical individuals (Polak and Trivers, 1994; Möller, 1997), an increase in FA can be expected to be linked with a decline in fitness.

The increase in FA in our samples was mainly related to an increase in an absolute asymmetry (= difference in needle length), whereas the mean length of the needle pair remained constant along the pollution gradient. It is thus possible to further simplify the measurements, and to use differences in needle length as a handy indicator of stress in Scots pine and possibly in other pines with a similar pattern of needle growth.

## Acknowledgements

We thank A. Blashkevich, E. Melnikov and V. Zverev for their assistance in needle sampling and measurement. E. Zvereva and two anonymous referees provided constructive criticism of an earlier draft of the manuscript. E. Valle kindly checked the language. The work was supported by the European Commission (research grant 97.60.SF.007.0) and the Academy of Finland.

## References

Alexeyev, V. A.: 1989, *Lesovedenie* **0** (4), 51-57 (in Russian).
Anonymous: 1994, *Manual on Methods and Criteria for Harmonized Sampling, Assessment, Monitoring and Analysis of the Effects of Air Pollution on Forests*, 3rd ed., UN-ECE, Progr. Coord. Centres, Hamburg and Prague.
Cape, J. N.: 1990, in *Proc. XIX World Congress IUFRO, Montréal, Canada, 5-11.8.1990*, Burley, J. (ed.), pp. 38-47, Oxford Univ., Oxford.
Clarke, G. M.: 1992, *Acta Zool. Fenn.* **191**, 31-35.
Ferretti, M.: 1998, *Chemosphere* **36**, 1031-1036.
Ferretti, M., Cenni, E. and Cozzi, A.: 1995, *Assessment of Tree Crown Transparency and Crown Discoloration as Performed by Surveyors from Six European Countries during the 7th International EC/ECE Mediterranean Intercalibration Course, Cagliari, Italy, 1994* (cited by Ferretti, 1998).
Freeman, D. C., Graham, J. H. and Emlen, J. M.: 1993, *Genetica* **89**, 97-119.
Graham, J. H., Emlen, J. M. and Freeman, C.: 1993, *Ecotoxicology* **2**, 185-195.
Hanisch, B. and Kilz, E.: 1988, in *Scientific Basis of Forest Decline Symptomatology (Air Pollution Report 15)*, J. N. Cape and P. Mathy (eds.), pp. 9-30, CEC, Brussels.
Koroleva, N.: 1993, in *Aerial Pollution in Kola Peninsula: Proceedings of the International Workshop, 14--16 April 1992, St. Petersburg*, Kozlov, M. V., Haukioja, E. and Yarmishko, V. T. (eds), pp. 339-345, Kola Science Centre, Apatity, Russia.
Kozlov, M. V. and Haukioja, E.: 1995, in *The Contaminants in the Nordic Ecosystem: the Dynamics, Processes and Fate*, Munawar, M. and Luotola, M. (eds.), pp. 59-69, S. P. B. Academic Publishing, The Netherlands.
Kozlov, M. V. and Haukioja, E.: 1997, in *Ecosystems and Sustainable development* (Advances in Ecological Sciences, vol. 1), Uso, J. L., Brebbia, C. A. and Power, H. (eds.), pp. 603-614, Computation Mechanics Publ., Southampton.
Kozlov, M. V., Haukioja, E., Bakhtiarov, A. V. and Stroganov, D. N.: 1995, *Environ. Pollut.* **90**, 291-299.
Kozlov, M. V., Wilsey, B. J., Koricheva, J. and Haukioja, E.: 1996, *J. Appl. Ecol.* **33**, 1489-1495.
Kryazheva, N. G., Chistyakova, E. K. and Zakharov, V. M.: 1996, *Ekologia*, **0** (6), 441-444 (in Russian)
Kryuchkov, V. V.: 1993, *Ecol. Appl.* **3**, 622-630.
Møller, A. P.: 1997, *Am. Nat.* **149**, 916-932.
Møller, A. P.: 1998, *Oikos* **81**, 444-448.
Palmer, A. R. and Strobeck, C.: 1986, *Ann. Rev. Ecol. Syst.* **17**, 391-421.
Parsons, P. A.: 1990, *Trends Ecol. Evol.* **5**, 315-317.
Polak, M. and Trivers, R.: 1994, *Trends Ecol. Evol.* **9**, 122-124.
Pomory, C. M.: 1997, *Anim. Behav.* **53**, 225-227.
Rettig, J. E., Fuller, R. C., Corbett, A. L. and Getty, T.: 1997, *Oikos* **80**, 123-127.
SAS Institute: 1990, *SAS/STAT User's Guide, Version 6.0*, SAS Institute, Cary, NC.
Sullivan, M. S., Robertson, P. A. and Aebischer, N. A.: 1993, *Nature* **361**, 409-410.
Zakharov, V. M.: 1990, in *Bioindications of Chemical and Radioactive Pollution*, Krivolutsky, D. (ed.), pp. 187-198, Mir Publishers, Moscow / CRC Press, Boca Raton.
Zvereva, E. L., Kozlov, M. V., Niemelä, P. and Haukioja, E.: 1997a, *Oecologia* (Berlin) **109**, 368-373.
Zvereva, E. L., Kozlov, M. V. and Haukioja, E.: 1997b, *J. Appl. Ecol.* **34**, 1387-1396.

# IMPLEMENTATION OF QUALITY ASSURANCE PROCEDURES IN THE ITALIAN PROGRAMS OF FOREST CONDITION MONITORING

M. FERRETTI[1], F. BUSSOTTI[2], E. CENNI[1], A. COZZI[1]

[1] Linnaea-ambiente srl, Via Sirtori 37 50137 Firenze, Italy;
[2] Dept. of Plant Biology, Piazzale delle Cascine 28 50144 Firenze, Italy

(Received 25 September 1998; accepted 25 February 1999)

**Abstract.** In 1996 the assessment of tree condition in Italy was subjected to a Quality Assurance (QA) program. The QA program consisted in (i) the adoption of Standard Operating Procedures (SOPs), the set of Measurement Quality Objectives (MQOs) expressed by Data Quality Limits (DQLs), (iii) a national training and intercalibration course, and (iv) field checks, performed on a number of areas spread over the country. In general, DQLs were achieved for a number of tree condition indices. However, problems were identified with crown transparency, the most used index in the international reports. Additional problems arose when considering the data quality at plot level as a result of the combined achievement of DQLs for the different tree condition indices.

**Keywords:** crown condition, field surveys, forest monitoring, permanent plots, Quality Assurance.

## 1. Introduction

Quality Assurance (QA) is central in long term environmental monitoring programs (Cline and Burkman, 1989; Millers et al., 1994). In addition to proper design, QA programs ensure a documentation of the quality of the field data and allow a data quality evaluation (Quality Control - QC).

Since 1987, tree crown transparency, discoloration and the occurrence of a series of damaging agents were recorded in Europe by the joint UE-UN/ECE systematic sample survey (Level I) on forest condition carried out according to a nominal 16x16 km grid. More recently, a series of intensive forest monitoring plots (Level II) were established under the auspices of the same program in order to collect data on various indicators and processes within the forest ecosystems.

In Italy, the Level I network has incorporated about 200 sample plots since 1988. In 1996, an intensive (level II) forest monitoring program was launched by the Italian Ministry for Agricultural Policy. The program, *CONtrollo ECOsistemi FORestali* (CON.ECO.FOR) considers 20 permanent monitoring plots. Tree condition assessment is recorded on both Level I and Level II plots, where trees are still the most important response indicators (Ferretti, 1997).

The procedures, the results of tree condition surveys and their interpretation have been often criticized (e.g. Innes, 1993; Skelly, 1992; Ferretti, 1997). A major problem in data collection is the subjectivity of the visual tree crown assessment, an estimate which throws into question the spatial and temporal comparability of the data (Innes, 1988; Ferretti *et al.*, 1994; 1995). In order to keep a proper documentation of the internal consistency and reproducibility of the data, a QA program was implemented in Italy in 1996 for tree condition assessment on Level I and II plots.

## 2. QA Procedures

The QA program consisted in (i) the adoption of Standard Operating Procedures (SOPs), (ii) the establishment of Measurement Quality Objectives (MQOs) expressed by Data Quality Limits (DQLs, Table 1), (iii) a national training and intercalibration course, and (iv) field checks, performed on a number of plots spread over the country and previously assessed by several field crews.

The SOPs consist of field manuals (Cenni et al., 1995; Müller and Stierlin, 1990; Ferretti et al., 1994). The MQOs and DQLs were set for each tree condition index according to the system in use by the US/EPA-EMAP-FHM (Talent-Halsell, 1994).

TABLE I

*Quality objectives (MQOs) expressed as quality limits of data (DQLs). @ = agreement.*

| Abbrev. | Parameter | Unit of measure | DQLs |
|---|---|---|---|
|  | Species | UE cod. | 99 % @ |
| a | Social status | 6 classes | 90 % @ ± 1 class |
| b | Crown freedom | 5 classes | 90 % @ ± 1 class |
| c | Unexplicable transparency, value*** | 21 classes | 90 % @ ± 10 % |
| d | Explicable transparency, value*** | 21 classes | 90 % @ ± 10 % |
| e | Branching, type | 6 classes | 90 % @ |
| f | Leaf colour, type | 5 classes | 90 % @ |
| g | Leaf colour, diffusion* | 4 classes | 90 % @ ± 1 class |
| h | Leaf damage, diffusion | 4 classes | 90 % @ ± 1 class |
| i | Curling | 4 classes | 90 % @± 1 class |
| j | Size of leaves | 4 classes | 90 % @ |
| k | Flowering effect | 4 classes | 90 % @ |
| l | Fruiting | 4 classes | 90 % @ ± 1 class |
| m | Crown regeneration, diffusion | 4 classes | 90 % @± 1 class |
| n | Stem regeneration, diffusion | 4 classes | 90 % @± 1 class |
| o | Insects | 5 classes | 90 % @ |
| p | Fungi | 5 classes | 90 % @ |
| q | Epiphytes | 5 classes | 90 % @ |
| r | Game | 5 classes | 90 % @ |
| s | Fire | 5 classes | 90 % @ |
| t | Anthropic damage | 5 classes | 90 % @ |
| u | Meteorological damage | 5 classes | 90 % @ |
| v | Acute pollution | 5 classes | 90 % @ |

*level I and II; ** in level I: total transparency.

## 3. Intercalibration of the field crews

The purpose of this phase was to provide surveyors with standard operational guidelines and to obtain the highest possible evaluation homogeneity. It has been common practice to attempt to achieve these results by providing an intercalibration course where the surveyors - after their initial training - all simultaneously evaluate a certain number of trees; supervising team then examines differences between the individual surveyors' assessments. The 1996 intercalibration course involved 32 teams and evaluated 160 trees belonging to the following species: *Fagus sylvatica* L. (levels I and II), *Quercus cerris* L. (levels I and II), *Quercus pubescens* Willd. (level I), and *Picea abies* Karst. (levels I and II). Fig. 1 shows how frequently observations (trees) fell within the DQLs.

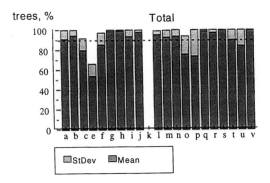

*Fig.1 - Frequency of observations (trees) within the DQLs for the different survey teams. The dotted line show the achievement limit of MQOs.*

Crown transparency, branching type, leaf colour type, interactions with defoliating insects, leaf invading fungi, and meteorological events are all additional confounding factors which make MQOs difficult to achieve.

## 4. Field checks

4.1. LEVEL I

Between mid-August and early September 1996 an independent surveyor examined 20 Level I plots; the frequency of individual observations in relation to deviations from controls is presented in Figure 2 (surveyor - control differences). In the case of crown transparency, values are in agreement in 22.4% of cases; the difference is ± 5% in 55.5% of cases, while in 77% of cases the difference is ± 10%. These results do not meet DQL requirements (Table I). Overall distribution of values does not reveal systematic deviations from controls, these values are also the result of "internal compensation" mechanisms between surveyors who tend to underestimate and others who tend to overestimate. Discoloration assessment may be more susceptible to subjectivity than crown transparency, however, the control findings suggest that DQL requirements have been satisfactorily met, with more than 95% of values being considered acceptable. This is probably due to the low frequency of discoloration in the sample trees. Finally, only in few plots the assessment values of both transparency and discoloration reach the DQLs (Figure 3).

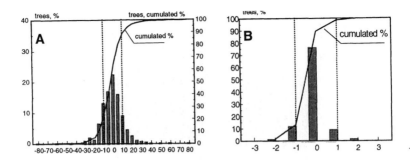

*Fig.2 - Frequency and cumulated frequency of observations (trees) according to the deviation to the control (x axis). Data refer to 20 level I plots and 522 trees. Dotted lines show the respect of DQLs. A: transparency; B: discoloration.*

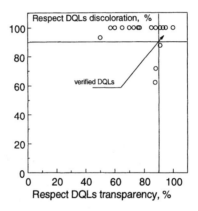

*Fig. 3 - Distribution of plots according to the achievement of DQLs. Lines show the limits of DQLs.*

## 4.2 LEVEL II

Overall, parallel evaluations were carried out on 5 permanent plots covering a total of 140 comparable trees. The overall findings for each of the indicators considered are illustrated in Figure 4. The fact that quality requirements were not met in assessments of non-explainable transparency is particularly important, although the results are better than those representative of the Level I plots. Figure 5 shows in detail the deviations in transparency evaluation, both explainable and non-explainable.

As can be seen, the majority of assessments do meet DQL requirements in explicable transparency, although there is an obvious trend to underestimate this index. As far as non-explicable transparency is concerned, there are no obvious systematic deviations, although in some cases the difference between surveyor and control reaches 30%.

The results provided by the controls suggest that the situation varies in relation to several different factors: surveyor, visibility of the trees, and meteorological conditions at the time of assessment. The main elements revealed by the study indicate that findings relating to those indices with a scoring system with fewer rating classes, and those which are usually absent and/or displaying low variability within a sample plot, are those which are fully reproducible.

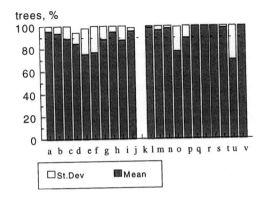

*Fig. 4 - Mean and standard deviation of values in DQLs with respect to the total of controls on level II plots.*

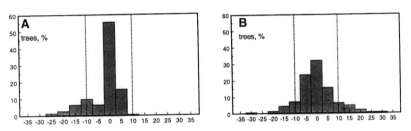

*Fig. 5 - Frequency and value of deviations from evaluations of explicable (A) and unexplainable (B) transparency. Data are expressed as differences in field crew - control (no: 140). Dotted lines show DQLs:*

## 5. Discussion and Conclusions

The data collected by the QC campaign allow the identification of some problems concerning the reproducibility of the field data in forest condition surveys. First, a full QA program was adopted in Italy for the first time since the beginning of the surveys in 1988. The introduction of SOPs at a national level may cause an abrupt change in the time series of the data, and - in principle - the comparability of the data collected before 1996 with the new data appearing to be problematic. Several important indices of tree condition (namely: crown transparency, branching type, colour deviation type, leaf curling and the effects of insects) do not meet the DQLs. This may be due to several reasons, including the need for additional training (with special regard to the effects of biotic agents) and the need for more suitable scoring systems and manuals. A consequence of this poor internal consistency is that great care must be placed in the interpretation phase. It is most important to note that crown transparency, the most used tree condition index in the international reports of forest condition, does not satisfy the MQOs. This places the suitability of the 1996 crown transparency data for mapping the condition of the Italian forests in serious question. When more than one index of tree condition is used, the reproducibility of the assessment should be checked for each index used in the survey if we want to derive combined indices. For example, in the past, crown transparency and discoloration were combined into a "damage" index. This is problematic for a variety of reasons (Ferretti, 1998). In view of data quality, it is obvious that such a damage index can be properly calculated only when DQLs are met for both transparency and discoloration, a circumstance that was seldom achieved in forest condition surveys in Italy in 1996.

At European level, no data about Quality Assurance within each country is provided in the EC- UN/ECE reports (cfr. Müller-Edzards et al., 1997). This leaves us unable to evaluate the consistency of results and the meaning of yearly fluctuations.

## Acknowledgements

The program CON.ECO.FOR. and the level I network are co-ordinated by S. Allavena, R. Isopi, B. Petriccione e E. Pompei, belonging to the Division V (National Focal Center) of the General Direction of Forests of the Italian Ministry for Agricultural Policy. The Authors are also grateful to many colleagues from the Italian Ministry for Agricultural Policy for the collection of field data.

## References

Cenni, E., Cozzi, A., Ferretti, M. and Bussotti F. (1995) *Valutazione delle condizioni degli alberi*. Regione Toscana, Firenze: 126 pp.

Cline, S.P. and Burkman, W.G. (1989) The Role Of Quality Assurance In Ecological Programs. In: *Air Pollution And Forest Decline*. eds. J.B. Bucher and I. Bucher-Wallin. Proc. 14th Int. Meeting for Specialists on Air Pollution Effects on Forest Ecosystems, Interlaken, Switzerland, Oct. 2-8, 1988. Birmensdorf: pp. 361-365.

Ferretti, M. (1997) *Environmental Monitoring and Assessment*, **48**, 45-72.

Ferretti, M., Cenni, E. and Cozzi A. (1994) *Monti e Boschi*, **2**, 5-13.

Ferretti, M, Cenni, E. and Cozzi A. (1995) *Assessment of tree crown transparency and crown discoloration as performed by surveyors from six european countries during the 7th International EC/ECE Mediterranean Intercalibration Course, Cagliari, Italy, 1994*. Regione Autonoma della Sardegna: 32 pp. + Annexes.

Innes, J. L. (1988) *Canadian Journal of Forest Researches*, **18**, 560-565.

Innes J.L. (1993) *Forest Health: Its Assessment And Status*. Commonwealth Agricultural Bureau, Wallingford: 677 pp.

Millers, I., Lachance, D., Burkman, W. G. and Allen, D. C. (1994) *North American Sugar Maple Project - cooperative field manual*, USDA, Canadian Forest Service: 51 pp.

Müller-Edzards, C., De Vries, W. and Erisman J.W. (1997) Ten years of monitoring forest condition in Europe. EC-UN/ECE, Brussels, Geneva: 386 pp.

Müller, E. and Stierlin, H.R. (1990) *Sanasilva - Kronenbilder; Couronnes d'arbres; Le chiome degli alberi*. Eidgenössische Anstalt für das Forstliche Vesuchswesen, Birmensdorf: 128 p.

Skelly, J. M. (1992) A Closer Look at Forest Decline: a Need for more accurate Diagnostics. In: *Forest Decline Concepts*. eds. P.D. Manion and D. Lachance. APS Press, St. Paul, Minnesota: pp. 85-107.

Tallent-Halsell N.G. ed. (1994) *Forest health monitoring 1994. Field methods guide*. EPA/620/R-94/027. U.S. Environmental Protection Agency, Washington, D.C.

# BIOINDICATION OF DIFFERENT STRESSES IN FOREST DECLINE STUDIES IN SLOVENIA

F. BATIČ[1], P. KALAN[2], H. KRAIGHER[2], H. ŠIRCELJ[1], P. SIMONČIČ[2], N. VIDERGAR-GORJUP[3] and B. TURK[1]

[1]*Agronomy Department, Biotechnical Faculty, University of Ljubljana, Jamnikarjeva 101, 1000 Ljubljana, Slovenia.* [2]*Slovenian Forestry Institute, Večna pot 2, 1000 Ljubljana, Slovenia.* [3]*Community Zagorje ob Savi, 1410 Zagorje ob Savi, Slovenia.*

Received : 25th September 1998; accepted 25th February 1999

**Abstract.** Determination of the stresses operating on a forest ecosystem demands the use of several bioindication methods. Air quality indicators were assessed from an inventory of forest decline based on the assessment of tree crowns and epiphytic lichens. Photosynthetic pigments, ascorbic acid and major macronutrients were studied in Norway spruce needles as indicators of physiological and biochemical stress. Analyses were carried out on selected forest plots in polluted areas (Zasavje district, vicinity of thermal power plants) and predominantly unpolluted areas (Triglav National Park, Julian Alps, Slovenia). For some bioindication methods, there was good agreement with measured air quality and climatological parameters. The best agreement was found between total foliar sulphur in needles and epiphytic lichens, especially in more polluted areas. Agreement with forest decline inventories and analyses of some needle stress physiological/biochemical parameters was less convincing. The strength of agreement was further decreased by soil characteristics and climatic parameters, influenced also by biotic parameters and forest stand history and management. It was concluded that there is no simple bioindication method available to evaluate the vitality of a forest.

**Keywords:** forest decline, stress, air pollution, bioindication, Norway spruce, photosynthetic pigments, ascorbic acid, macro nutrients, epiphytic lichens, Slovenia

## 1. Introduction

The concept of using bioindication and stress physiology to study forest decline has been promoted in Slovenia, and culminated in an international cooperative Tempus project dedicated specifically to bioindicators of forest site pollution (Kraigher *et al.*, 1996). In this paper two investigation areas are briefly presented, describing forest decline inventories which were later extended to include stress factors. Triglav National Park, representing a rather unpolluted area, was contrasted with the Zasavje district, historically one of the most air polluted regions in Slovenia due to mining and thermal power plant activity.

Photosynthetic pigments, ascorbic acid, thiols and plant hormone are often used as stress indicators for early diagnosis of disturbances in forest trees (Grill and Esterbauer, 1973; Darrall and Jäger, 1984; Wolfenden *et al.*, 1988; Pfeifhofer and Grill, 1989; Bermadinger *et al.*, 1989; Tausz *et al.*, 1996a). Combination of analyses of different stress physiological parameters in forest trees with air pollution measurements, analysis of mineral nutrition of forest trees and use of some other sensitive bioindicators like epiphytic lichens provide a more complete explanation of the process of forest decline (Tausz *et al.*, 1996b; Batič and Mayrhofer, 1996). The pattern of photosynthetic pigments in Norway spruce needles is widely used as an unspecific vitality indicator (Wolfenden *et al.*, 1988; Pfeifhofer and Grill, 1989). Analyses of thiols and macronutrients in combination with ascorbic acid analysis in needles can distinguish between stresses caused by air pollutants and those caused by natural phenomena like frost, drought, UV radiation etc. (Bermadinger *et al*,. 1989; Tausz *et al.,* 1996 b).

## 2. Material and methods

The Triglav National Park (TNP) is situated in the Julian Alps, in the western part of Slovenia. All forests are natural, or close to natural, due to forest management practice and inaccessible terrain. Norway spruce (*Picea abies* (L.) Karst.) is the commonest tree species, followed by the common beech (*Fagus sylvatica* L.), European larch (*Larix decidua* Mill.) and mountain pine (*Pinus mugo* Turra), which represents a shrub belt of "Krumholz" above the tree and forest limit.

The forest decline inventories were carried out in 1987, 1993 and 1995 (Šolar, 1989; Šolar, 1991; Gomišček, 1997) using the methodology developed by the ICP-Forest (Anonymous 1994), slightly modified by Slovenian foresters (Šolar 1991). Air quality was assessed at all forest decline inventory plots, using a very simple mapping methodology based on assessment of cover and frequency of major lichen thalli types, e.g. crustose, foliose and fructicose lichens (Batič and Kralj, 1989; Batič and Mayrhofe,r 1996). Subsequently, epiphytic lichen species were mapped along defined transects (Primožič, 1998; Surina, 1998).

Three transects, Trenta, Pokljuka and Kranjska gora were chosen, along which air pollutant concentrations were measured, and Norway spruce needles were sampled for physiological stress analyses. The transects were chosen in the south-western, eastern and northern parts of the park, representing major park forest types. Only results from the transects Kranjska gora (TNP Kranjska gora, 1A) and Trenta (TNP Trenta, 1B) are presented in this paper. Concentrations of $SO_2$ and $NO_x$ were measured for two week periods using passive samplers during the growing season, from March to November. On the high plateau, Pokljuka, $O_3$, $SO_2$, $NO_x$, VOC and climatological parameters were analysed during campaigns lasting for one week in spring, summer and autumn. Foliar macro nutrients were measured (Simončič and Kalan, 1996), and photosynthetic pigments and ascorbic acid were analysed by HPLC (Pfeifhofer, 1989; Bermadinger *et al.,* 1990).

The contrasting, polluted area, the Zasavje district is situated in the middle of Slovenia. The main air pollutor is the thermal power plant in Trbovlje-TET (125 MW), which began operation in 1968. Due to its location in a deep and narrow valley, it caused serious air pollution in the 1960s when half hour concentrations of $SO_2$ reached as high as 20 mg m$^{-3}$ in the most exposed localities. At that time forest damage was 10 fold higher than in other parts of Slovenia, and more than 84% of all damage was ascribed to air pollution (Šolar, 1989). Although Norway spruce is not an indigenous species in this district of Slovenia, it is very common due to historic forest practices, and after silver fir was most affected. Typical symptoms of damage were defoliation, chlorotic and necrotic needles, and changes in the branching system, all of which were included in the methodology of the assessment of forest decline (Šolar, 1989; Vidergar-Gorjup, 1998). Within this area 10 forest stands were selected according to forest damage level, topography and climatic circumstances. The forest decline inventory was carried out at the same time as in the TNP and the same measurements were made.

## 3. Results and discussion

Data obtained by measurements of climatological parameters are not presented and only selected air pollution values are presented. Further selected results include Norway spruce needle anlyses and epiphytic lichen mapping from both areas. Defoliation at the clean site (TNP) was lower than the national average 13 cf 22% respectively, indicating the healthy state of the park forests (Gomišček, 1997).

TABLE I
Concentrations of N, P and S (mg g$^{-1}$ DW), N/P ratio and N/S ratio and common S content class in current year Norway spruce (*Picea abies* (L.) Karst.) needles sampled on the profiles TNP Kranjska gora (**1A**) and Trenta (**1B**) in autumn 1994 and Zasavje (**2**) district in autumn 1995. Data are the average of two sampled trees (K. Gora, Trenta) and three trees (Zasavje).

| (1A) TNP Kranjska gora | altitude | N mg g$^{-1}$ | P mg g$^{-1}$ | N/P | S$_{94}$ mg g$^{-1}$ | N/S$_{94}$ | S$_{93}$ mg g$^{-1}$ | S conent class S$_{94+93}$ | No. of lichen species |
|---|---|---|---|---|---|---|---|---|---|
| Za žago | 920 | 11.9 | 0.90 | 13.0 | 1.10 | 10.8 | 0.96 | 2 | 19 |
| Greben 1 | 1020 | 9.5 | 0.77 | 12.0 | 0.89 | 10.7 | 1.12 | 1 | 28 |
| Greben 2 | 1130 | 9.9 | 0.80 | 12.0 | 0.91 | 10.9 | 0.94 | 1 | 31 |
| Plato | 1250 | 10.5 | 0.79 | 13.0 | 0.85 | 12.4 | 0.83 | 1 | 42 |
| Skalna glava | 1360 | 9.9 | 0.84 | 12.0 | 0.86 | 11.5 | 0.94 | 1 | 37 |
| Gozdna jasa | 1450 | 11.6 | 1.38 | 8.4 | | | 0.98 | 1 | 37 |
| Pod kurjim vrhom | 1540 | 12.0 | 1.50 | 8.0 | 0.85 | 14.1 | 0.82 | 1 | 37 |

| (1B) TNP Trenta | altitude | N mg g$^{-1}$ | P mg g$^{-1}$ | N/P | S$_{94}$ mg g$^{-1}$ | N/S$_{94}$ | S$_{93}$ mg g$^{-1}$ | S content class S$_{94+93}$ | No. of lichen species |
|---|---|---|---|---|---|---|---|---|---|
| Kamp korita | 520 | 11.4 | 1.80 | 6.3 | 1.10 | 10.8 | 0.96 | 2 | 29 |
| Pod vršičem | 610 | 10.6 | 1.24 | 8.5 | 0.89 | 10.7 | 1.12 | 1 | 29 |
| Za vršičem | 700 | 11.7 | 1.36 | 8.6 | 0.91 | 10.9 | 0.94 | 1 | 75 |
| Nad vršičem | 800 | 11.6 | 1.00 | 11.6 | 0.85 | 12.4 | 0.83 | 1 | 40 |
| Skokar | 910 | 10.5 | 0.90 | 11.7 | 0.86 | 11.5 | 0.94 | 1 | 62 |
| Ovinek | 1000 | 13.1 | 1.61 | 11.3 | | | 0.98 | 1 | 16 |
| Ob cesti | 1100 | 9.5 | 0.99 | 9.6 | 0.85 | 14.1 | 0.82 | 1 | 16 |
| Polog | 1215 | 13.3 | 1.39 | 9.6 | | | | | 16 |
| Glava | 1215 | 11.6 | 0.92 | 12.6 | | | | | 16 |
| Za glavo | 1400 | 12.7 | 1.86 | 6.8 | | | | | 29 |
| V Plazeh | 1620 | 12.2 | 1.71 | 7.1 | | | | | 40 |

| (2) Zasavje | altitude | N mg g$^{-1}$ | P mg g$^{-1}$ | N/P | S$_{95}$ mg g$^{-1}$ | N/S$_{95}$ | S$_{94}$ mg g$^{-1}$ | S content class S$_{95+94}$ | No. of lichen species |
|---|---|---|---|---|---|---|---|---|---|
| Ostenk | 390 | 14.8 | 1.21 | 12.3 | 1.43 | 10.3 | 1.52 | 3 | 6 |
| Marno | 390 | 12.5 | 1.13 | 11.1 | 1.72 | 6.8 | 1.50 | 4 | 7 |
| Ravenska | 480 | 11.7 | 1.16 | 10.1 | 1.53 | 8.2 | 1.87 | 3 | 7 |
| Retje | 490 | 13.3 | 1.47 | 9.0 | 1.30 | 10.2 | 1.44 | 3 | 7 |
| Kovk | 590 | 14.3 | 1.55 | 9.2 | 1.64 | 8.7 | 1.90 | 4 | 5 |
| Jelenca | 750 | 13.2 | 0.85 | 15.5 | 1.07 | 12.3 | 1.10 | 2 | 23 |
| Dobovec | 790 | 12.4 | 1.06 | 11.7 | 1.65 | 7.5 | 1.61 | 4 | 5 |
| Gore | 880 | 13.5 | 1.03 | 13.1 | 1.31 | 10.3 | 1.24 | 3 | 21 |
| Kljucevca | 1070 | 14.4 | 1.57 | 9.2 | 1.13 | 12.7 | 1.19 | 2 | 19 |
| Javor | 1070 | 14.0 | 1.26 | 11.1 | 1.24 | 11.3 | 1.23 | 3 | 21 |

Concentrations of $SO_2$ and $NO_x$ were low, especially in summer. At Trenta $SO_2$ reduces from 3.34 μg m$^{-3}$ in the valley bottom (520m) to 2.2. μg m$^{-3}$ at 1620m. Measured values for $NO_2$ for the same transect and period were from 2.7 μg m$^{-3}$ in the valley and 0.80 μg m$^{-3}$ at the highest sampling site. At Kranjska gora values for $SO_2$ decreased from 6.93 μg m$^{-3}$ in the valley (920 m) to 1.10 μg m$^{-3}$ at the highest sampling site (1540 m). Values for $NO_2$ were between 3.32 μg m$^{-3}$ and 0.72 μg m$^{-3}$ (Gomišček, 1997). Bulk N deposition ($NH_4$-N, $NO_3$-N) and S deposition ($SO_4$-S) based on measurements on Pokljuka plateau between 1996 and 1998 were between 10 and 12 kg N ha$^{-1}$ yr$^{-1}$ and 12 and 13 kg S ha$^{-1}$ yr$^{-1}$ (unpublished data, Simončič, Slovenian Forestry Institute), too small to damage the forest. However, measurements in open areas at low altitudes underestimate inputs to forests via occult deposition as can be seen from S analyses and the epiphytic lichens (Table 1 (1A)). $O_3$ concentrations (campaigns, Pokljuka) were relatively high (maxima 80-100 ppb), exceeding the threshold causing damage to sensitive crops and natural vegetation (40 ppb) (Gomišček, 1997). With the exception of some sites at lower altitudes at Trenta and Kranjska gora, due to pollution from the valley and transboundary sources, foliar S was low. N concentrations were just sufficient, and slightly increased in the needles from Trenta (Table 1 (IB)). The higher $SO_2$ concentration prevailing on the lower slopes could be seen in reduced number of epiphytic lichens particulary in Kranjska gora.

The age pattern of chlorophylls, the carotenoid content and their ratio reflect the $SO_2$ pollution and/or photooxidants and sampling altitude (Batič *et al.*, 1995; Bermadinger *et al.*, 1990; Grill *et al.*, 1979). Ascorbic acid content is a good indicator of oxidative stress in needles and also their reductive capability (Bermadinger *et al*, 1990). Its content in non polluted areas increases with altitude (UV light, $O_3$, drought, etc.) and is decreased by very high $SO_2$ pollution (Bermadinger *et al*,. 1989; Tausz *et al.* ,1996). The results from Triglav area are consistent with a non polluted situation. There was a slight decrease in chlorophyll, especially chlorophyll a, with sampling altitude (Figure 1), and an increase in ascorbic acid content, especially at Trenta (Figure 2), and in the Kranjska gora (Figure 2) in the TNP. Epihytic lichen populations reflect air pollutants earlier than spruce needles. The number of epiphytic lichen species in the south-west part of Julian Alps (transect Trenta) between 1000 and 1500 m, were drastically reduced and although no air pollution data is available (Primožič, 1998), occult deposition may be responsible (Gomišček, 1997).

In the Zasavje district the concentrations of $SO_2$ were very high; $NO_x$ and $O_3$ measured at Kovk exceeded the threshold values for sensitive vegetation. Bulk deposition in the open was 13 kg N ha$^{-1}$ yr$^{-1}$ and 25 kg S ha$^{-1}$ yr$^{-1}$ (Anon., 1998). At other sites in this area S deposition can exceed 30 kg S ha$^{-1}$ yr$^{-1}$ (Anon., 1998) and 83-100% of the assessed Norway spruce in damage classes 2-4 (Šolar, 1989). This damage can be largely ascribed to pollution from the thermal power plant and to the impact of $SO_2$ because almost all trees are grouped within classes 4 and 3 according to total foliar S concentrations (Table 1; II). Analyses of photosynthetic pigments and ascorbic acid gave quite different results in comparison with TNP (Figs. 1 and 2). Due to very high $SO_2$ concentrations in the valley the content of photosynthetic pigment increases with altitude while the ascorbic acid content decreases.

Fig.1 Concentration of Pigments in Needles (Average for 1992-95 Needle Age Classes)

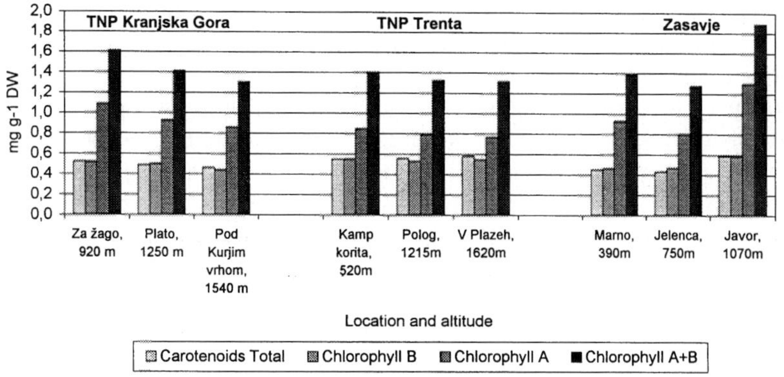

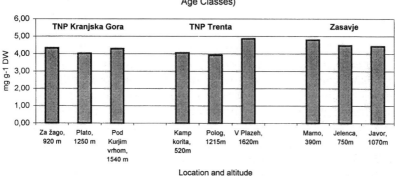

The number of epiphytic lichen species found on all sites in Zasavje clearly decreased with decreasing altitude (Tables I and II). This trend was less obvious in TNP, although expressed in the same way.

Fig. 2 Concentration of Ascorbic Acid in Needles (Average for 1992-95 Needle Age Classes)

The forest decline inventories, epiphytic lichen mapping, air pollutant measurements and analyses of Norway spruce needles confirm that forests in the Triglav National Park are not very polluted by comparison with forests in the Zasavje region. The impact of air pollution in the TNP is small and of two kinds: the first, observed especially in the east-northly part (Kranjska gora) has come from the steel plant in Jesenice, manifest as elevated foliar S concentration and in lower photosysnthetic pigment and ascorbic acid content, especially in the valley bottom. The second is the impact of eutrophication, expressed as higher foliar N concentration, especially at Trenta, in the south-western part of the park. Between 1000 and 1500m die-back of sensitive macro lichens is common and increasing cover of algae and the pollution resistant lichen species *Scoliciosporum chlorococcum* is observed. Eutrophication most probably originates from long range transport of pollutants from the south-west, very probably from industrial north Italy, which is the prevailing wind direction.

At the Zasavje the small number of epiphytic lichens present are primarily represented by species higly resistant to air pollutants (*Scoliciosporum chlorococcum, Lecanora conizaeoides, Lepraria* species) (Vidergar-Gorjup, 1998). Both, the epiphytic lichens and spruce needles have a high S content consistent with a low content of photosynthetic pigments and ascorbic acid. By comparison with the less polluted park the effects of altitude were negligible.

Comparison of the different bioindicator methods used in these two case studies (forest tree-defoliation, epiphytic lichens, S concentration, macronutrients, photosynthetic pigments, ascorbic acid) indicates that some of the responses could be explained and ascribed to the impact of $SO_2$ and to a lesser extent N deposition. However natural stresses (drought, light, biotic stress) may also affect forest health and their contribution could not be assessed by these methods.

## References

Anonymous: 1994, Manual on methods and criteria for harmonize sampling, assessment, monitoring and analysis of the effects of air pollution on forests, 1-177, PCC WEST, BFH Hamburg, PCC East, Jiloviste-Strnady.
Anonymous: 1998, Air Pollution in Slovenia. Year 1996. Hydrometeorologica Institute of Slovenia, p 146, Ljubljana.
Batič, F. and Kralj, T: 1989, *Zbornik gozdarstva in lesarstva* **34**: 51-70.
Batič, F., Grill, D., Kalan, P. and Ribarič-Lasnik, C.: 1995, *Acta Pharm.* **45**, 191-197.
Batič, F. and Mayrhofer, H.: 1996, *Phyton* **36**: (85)-(96).
Bermadinger, E., Grill, D. and Guttenberger, H.: 1989, *Phyton* **29**: (163)-(185).
Bermadinger, E., Guttenberger, H. and Grill, D.: 1990, *Environ. Pollut.* **68**, 319-330.
Darrall, N. M. and Jager, H. J.: 1984, *Gaseous Air Pollutants and Plant Metabolism* (Ed. by M.J.Koziol and F.R. Whatley), 333-351. Butterworths, London.
Gomišček, S.: 1997, Onesnaženje zraka v Triglavskem narodnem parku in njen vpliv na gozd in vode.- Atmosfera in nacionalni park. Final Report, Contract No.: S2-6241, pp., 1-133. Ministrstvo za znanost in tehnologijo Republike Slovenije. Ljubljana.
Grill, D. and Esterbauer, H.: 1973, *Eur. J. For. Path.* **3**: 65-71.
Grill, D., Esterbauer, H. and Klosch, U.: 1979, *Environmental Pollution (Series A)* **17**, 187-194.
Kraigher, H., Batič, F., Guttenberger, H., Grill, D., Agerer, R. and Hanke, D. E.: 1996, *Phyton* **36**: 1-188.
Pfeifhofer, H., and Grill, D.: 1988, In: Fuhrer E., Neuhuber, F., *FIW- Symposium 1988*, Wien, 290-291.
Pfeifhofer, H.: 1989, *Eur.J. For. Path.* **19**: 363-369.
Primožič, K.: 1998, *Graduation Thesis*, 81 University of Ljubljana, Biotechnical Faculty, Department of Biology, Ljubljana.
Simončič, P. and Kalan, P.: 1996. *Phyton* 36(3): 81-84.
Surina, B.: 1998, *Graduation thesis*, 1-52. University of Ljubljana, Biotechnical Faculty, Department of Biology, Ljubljana.
Šolar, M.: 1989, *Zbornik gozdarstva in lesarstva (Ljubljana)* **34**, 71-97.
Šolar, M.: 1991, *Graduation thesis*, 1-150, Sveučilište u Zagrebu, Hrvatska.
Tausz, M., Batič, F., and Grill, D.: 1996a, *Phyton* **36**: 7-14.
Tausz, M., Grill, D., Guttenberger, H.: 1996b, *Phyton* **36**: 31-34.
Vidergar-Gorjup, N.: 1998, *Master of Science Thesis*, 170. University of Ljubljana, Biotechnical Faculty, Agronomy Department, Ljubljana
Wolfenden, J., Robinson, D., Cape, J. N., Paterson, I. S., Francis, B. J., Mehlhorn, H., Wellburn A. R.: 1988, *New Phytol.* **109**: 85-95.

# CAN GRAPHICAL VECTOR ANALYSIS BE USED TO IDENTIFY MICRO NUTRIENT DEFICIENCY?

GUNNAR THELIN, ULRIKA ROSENGREN-BRINCK, and BENGT NIHLGÅRD

*Forest Ecology, Dept. of Plant Ecology, Ecology Building, Lund University,
S-223 62 Lund, Sweden*

(Received 25 September 1998; accepted 25 February 1999)

**Abstract.** Initial analysis of needle nutrient status and visible symptoms suggested that a declining Norway spruce (*Picea abies* L. Karst.) stand in south Sweden was suffering from B and Cu deficiency. Effects of addition of 1000 kg ha$^{-1}$ of a commercial N-free fertilizer containing macro nutrients plus B, Cu, and Zn were investigated. B concentrations and B/N ratios were increased significantly in needles from fertilized trees compared to the control, to levels well above critical levels of deficiency and optimum nutrient ratios. In contrast, the treatment did not change the low needle Cu status and there was no effect on growth measured as needle mass after one growing season. Graphical vector analysis together with the absence of change in needle mass suggested a luxury uptake of B. Application of the concepts for critical levels of deficiency and optimum nutrient ratios in relation to N suggested B and Cu deficiency were contributing factors to the observed forest damage in the area. Graphical vector analysis requires a rapid response in needle mass, in combination with improved needle nutrient status in order to identify nutrient deficiencies. The applicability of the technique appears to be limited to primarily growth-promoting nutrients.

Keywords: Boron, copper, deficiency, graphical vector analysis, n-free fertilisation, *Picea abies*.

## 1. Introduction

The forest at Kvarntorp, close to Örebro, in south Sweden was severely damaged as a result of mining and extraction of shale-oil from 1941 to 1966, during which period the emission of $SO_2$ was high (Eriksson, 1979). In addition, between the 1960s and the early 1980s large amounts of $NH_3$ were released from a fertilizer company in the area. Today the forest, within 1-2 km of the industrial site, is dominated by birch.

In March 1994 several trees in a Norway spruce (*Picea abies* L. Karst.) stand 3 km northeast of the industrial site showed symptoms of damage including abnormal needle loss and crown yellowing. Branch growth structure was often irregular showing dieback of leading shoots that could be traced several years back. Such symptoms typify B deficiency (Möller, 1982; Braekke, 1983), and possibly Cu deficiency (Turvey and Grant, 1990). Pre-treatment needle analysis indicated absolute and relative deficiencies of B, Cu, and K.

Investigations of nutrient imbalance and the connection to forest decline have usually focused on macro nutrients. Several studies have demonstrated the positive effects of N-free fertilization on macro nutrient status and vitality in declining stands (e.g. Hüttl, 1990; Boxman *et al.*, 1994; Katzensteiner *et al.*, 1995). A role for micro nutrient deficiencies in forest decline has, so far, received little attention.

Needle nutrient status is often evaluated using critical levels of deficiency (Nihlgård, 1990; Braekke, 1996), but by using nutrient ratios problems with annual and spatial variations, and dilution effects are reduced (Rosengren-Brinck and Nihlgård, 1995). The basis for using nutrient ratios as a diagnostic tool is that the proportion of a nutrient relative to N may be at least as important to tree vitality as the concentration of the nutrient (Linder, 1995).

A more complex method, graphical vector analysis (Timmer and Stone, 1978; Valentine and Allen, 1990), can be used to evaluate the pre-treatment status of nutrients by examining tree response to a treatment, e. g. fertilization, in terms of effects on needle mass and needle nutrient concentrations relative to a control. This technique assumes the addition of a deficient nutrient results in both increased concentration of the nutrient in the needles, and increased growth measured as current year needle mass.

The objectives of this study were to a) confirm the observed tree decline in a Norway spruce stand in south Sweden was due to micro nutrient deficiency, and b) see if graphical vector analysis provides a suitable tool for interpretation of the effects of N-free fertilization on needle nutrient, especially micro nutrient, status.

## 2. Material and methods

The study site lies 20 km south-east of Örebro in South Sweden, (lat. 59° 08′, long. 15° 19′, 40 m a.s.l). Mean annual precipitation is 614 mm and mean annual temperature is 5.7 °C. In 1994 estimated total N deposition was ~ 19 kg ha$^{-1}$ yr$^{-1}$ and total S deposition ~ 15 kg ha$^{-1}$ yr$^{-1}$ (Hallgren-Larsson and Westling, 1994). The stand consists of naturally regenerated 60-year old Norway spruce. The soil is a transition between a dystric cambisol and a ferric podzol (FAO 1987, 1988) with sandy to silty texture. The soil pH (H$_2$O) is 4.1 at 5-10 cm depth in the mineral horizon.

In May 1994 the stand, except for three control plots of 30 * 30 m, was treated with 1000 kg ha$^{-1}$ of an N-free fertilizer, which included B, Cu, and Zn, in solid phase (Supra Vital, SUPRA, Sweden) (Table I).

TABLE I

Composition of the fertilizer (kg 1000 kg$^{-1}$). Water soluble amounts are shown in brackets.

| P | K | Mg | Ca | S | B | Cu | Zn |
|---|---|----|----|----|----|----|----|
| 42 | 76 | 82 | 133 | 50 | 1.8 | 1.1 | 1.1 |
| [7.9] | [47] | | | | | | |

In March 1995 current year needles were sampled from branch whorl 7 - 12 on 4 co-dominant trees in each control plot and on 4 co-dominant trees 5 - 10 m outside each plot. Needles were dried at 40 °C for 48 h and 100 needles per sample weighed. Needle samples were digested in concentrated HNO$_3$. N was analyzed by the Kjeldahl-method. B, Ca, Cu, K, Mg, Mn, P, S, and Zn were analyzed using an ICP-ES (Perkin Elmer, Connecticut, USA).

Differences between means were evaluated using the Students t-test and the Mann-Whitney U-test (Sokal and Rohlf, 1995). The graphical vector analysis used was a modification by Valentine and Allen (1990) and Swift and Brockley (1994) (Figure 1).

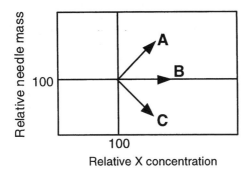

| Shift | Needle X concentration | Needle mass | Diagnosis |
|-------|------------------------|-------------|-----------|
| A | Increases | Increases | X is growthlimiting (deficiency) |
| B | Increases | Unchanged | X is stored (luxury consumption) |
| C | Increases | Decreases | X is toxic |

Fig. 1. Graphical representation of diagnoses arising from changes in needle nutrient concentration and needle mass after treatment. Response vectors radiate out from untreated condition (control; 100, 100) to treated conditions, plotted as percentages of the control. The length of the vectors shows the magnitude of the foliar response. (adapted from Valentine and Allen, 1990; and Swift and Brockley, 1994).

## 3. Results and discussion

In the control plots foliar concentrations of N, B, and Cu were in the deficiency range (Table II). The low B concentrations confirmed the visible symptoms and pre-treatment concentrations which all testified of B deficiency. B concentrations and the B/N ratio were significantly higher in the fertilized trees than in the control and well above critical levels of deficiency and optimum proportions respectively. There was no difference in needle mass between the fertilized trees and the control, and in the absence of a greater needle mass the graphical vector analysis indicated luxury uptake of B (Figure 2).

No significant differences between the treatment and the control were found for Ca, Cu, K, Mg, Mn, N, P, S, or Zn (Table II). Except for B and Cu, nutrient concentrations and ratios to N exceeded critical levels of deficiency and optimum ratios in both treated and control trees. N-concentrations in the control were on the low side of optimal. Cu-concentrations and Cu/N remained low in both the fertilized trees after treatment and the control (Table II). The lack of response does not rule out Cu deficiency. Cu was applied as $CuSO_4$, but the applied Cu must first be adsorbed to organic complexes before it can be taken up by the roots (McLaren et al., 1990). Limited time for complexation may explain why fertilized trees have not yet responded to the added Cu

(Table II). Another possibility is antagonistic effects from Ca and Mn on Cu uptake (Turvey and Grant, 1990; Kabata-Pendias and Pendias 1992). Concentrations of Mn and Ca were high (Table II), and Ca was added with the fertilizer (Table 1).

The findings point to the conclusion that the suboptimal status of B and Cu in the trees was a contributing factor to the damage symptoms observed. The former high levels of air pollution in the area (Eriksson, 1979) have, most likely, lead to increased acidification of the soil with subsequent leaching of nutrients (Ulrich, 1983; Falkengren-Grerup et al., 1987). B is especially easily leached (Brady, 1990). In addition, N deposition has probably lead to increased growth increasing the demand for other nutrients (Nihlgård, 1985). The combination of leaching and enhanced demand for nutrients may have resulted in the B and Cu deficiencies found in this study.

TABLE II

Mean nutrient concentrations and ratios to N in needles from control and treatment. Stars indicate difference between means.

|   | Concentration (mg g$^{-1}$ d.m.) (n=12) | | | Ratio to N (%) (n=12) | | |
|---|---|---|---|---|---|---|
|   | Ctrl. | Treat. | Def.[1] | Ctrl. | Treat. | Opt.[2] |
| N | 12.8 | 13.8 | 12–13 | 100 | 100 | 100 |
| K | 4.73 | 4.91 | 3.5 | 37.9 | 36.9 | 35–40 |
| P | 1.83 | 1.81 | 1.2–1.3 | 14.6 | 13.5 | 10–12 |
| Ca | 3.80 | 4.10 | 0.4–1.0 | 29.8 | 29.4 | 2.5–5 |
| S | 0.83 | 0.93 | 0.5–0.6 | 8.39 | 8.13 | 5 |
| Mg | 0.89 | 0.98 | 0.4–0.7 | 6.95 | 7.24 | 4–5 |
| Mn | 1.08 | 1.16 | 0.01–0.02 | 8.39 | 8.13 | 0.05 |
| Zn | 0.0272 | 0.0258 | 0.008–0.020 | 0.211 | 0.186 | 0.05 |
| B | 0.0047 | 0.0114*** | 0.004–0.005 | 0.037 | 0.085*** | 0.05 |
| Cu | 0.0019 | 0.0020 | 0.002–0.003 | 0.015 | 0.015 | 0.03 |

[1] Critical levels of deficiency (Nihlgård, 1990; Braekke, 1994)
[2] Optimum ratios (Nihlgård, 1990; Linder, 1995)

Graphical vector analysis, which uses the absence of change in needle mass in response to the treatment, suggests that B and Cu were not growth-limiting prior to fertilization. However, this does not rule out B and Cu deficiency. The connection between the amelioration of a deficiency and an immediate growth response is not self-evident as indicated by the multiple resource limitation hypothesis (Chapin et al., 1987; Gleeson and Tilman, 1992). Increased concentrations of one of two growth-limiting nutrients may not affect growth, but if they are both increased the effect on growth may be synergistic (Gleeson and Tilman, 1992). Also, after a change in nutrient availability there may be a time lag until the trees are able to respond optimally since they first must adjust their allocation of internal resources to the new environment (Gleeson and Tilman, 1992). According to Carter and Brockley (1990), applications of B alone would

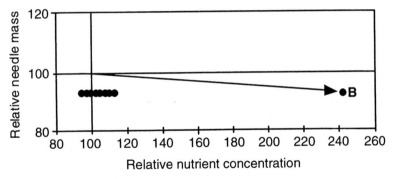

Fig. 2. Effects of the fertilization, shown as responses relative to the control (100, 100) on nutrient concentrations and mean needle mass.

not improve growth rate, but would prevent the occurrence of deficiency symptoms. They found increased branch elongation only when B and N were applied together. In a study of NPK and Cu applications to young Sitka spruce, Holstener-Jørgensen and Lauritsen (1983) found a positive growth response only when NPK and Cu were applied together. In our study, increased B concentrations in the needles might affect growth in the future, but without the amelioration of other constraints on growth the effect may not be realised in the short-term.

Graphical vector analysis has been shown to work well in fertilization trials with macro nutrients, but it appears from this study that the application of deficient elements, which are less closely related to growth, may yield inappropriate interpretations. Similar conclusions emerged from a study on Cu and Sr (Rosengren-Brinck, unpubl.) indicating that the technique does not work as well with micro nutrients as with macro nutrients. In addition, the graphical vector technique does not permit a possible time lag between the addition of a nutrient and a subsequent growth response. Interpretations based on the technique may be more reliable in single nutrient fertilization trials than when combinations of nutrients are added (Valentine and Allen, 1990).

## 4. Conclusions

Application of graphical vector analysis appears to be limited to growth-promoting nutrients. Without a quick response in needle mass in combination with improved needle nutrient status the technique cannot be used to identify nutrient deficiencies. This study suggests that micro nutrient deficiencies may be corrected without accompanying effects on growth. Use of graphical vector analysis for identification of micro nutrient deficiencies is recommended only in combination with other diagnostic tools such as critical levels of deficiency and optimum nutrient ratios.

## Acknowledgements

This study was financially supported by Falla-stiftelsen and the Swedish Environmental Protection Board. We thank A. Tapper and E. Westlund, Local Board of Forestry, for help in the field and landowner K-E. Dahm for access to the field site.

## References

Boxman, A. W., Cobben, P. L. W., and Roelofs, J. G. M. 1994. Environ. Pollut. **85**: 297-303.
Brady, N. C. 1990. *The nature and properties of soils - 10th ed.* Collier Macmillan Canada Inc., 621 pp.
Braekke, F. H. 1983. Commun. Inst. For. Fenn. **116**: 159-169.
Braekke, F. H. 1996. Trees. **11**: 23-33.
Carter, R. E., and Brockley, R. P. 1990. For. Ecol. Manage. **37**: 83-94.
Chapin, F. S. III, Bloom, A. J., Field, C. B., and Waring, R. H. 1987. BioScience **37**:49-57.
Eriksson, S. 1979. Kumla kommun och länstyrelsen i Örebro län, Sweden.
Falkengren-Grerup, U., Linnermark, N., and Tyler, G., 1987. Chemosphere **16**: 2239-2248
Gleeson, S. K., and Tilman, D. 1992. Am. Nat. **139**: 1322-1343.
Hallgren-Larsson, E., and Westling, O. 1994. IVL rapport B 1139, Sweden.
Holstener-Jørgensen, H., and Lauritsen, G. 1983. Det forstlige forsøgsvæsen i Danmark. Beretning nr. 328, bd. XXXIX, h. 1: 83-91.
Hüttl, R. F. 1990. Plant Soil. **128**: 45-58.
Kabata-Pendias, A., and Pendias, H. 1992. *Trace elements in soil and plants.* CRC Press Inc., London, 363 pp.
Katzensteiner, K., Eckmuellner, O., Jandl, R., Glatzel, G., Sterba, H., Wessely, A., and Hüttl, R. F. 1995. Plant Soil. **168-169**: 489-500.
Linder, S. 1995. Ecol. Bull. **44**: 178-196.
Marschner, H.,1986. *Mineral nutrition of higher plants.* Academic Press, London. 674 pp.
McLaren, R. G., Hogg, D. S., and Swift, R. S. 1990. For. Ecol. Manage. **37**: 131-142.
Möller, G. 1982. Institutet för skogsförbättring, Årsbok 1982, Uppsala, Sweden.
Nihlgård, B. 1985. Ambio, **14**: 2-8.
Nihlgård, B. 1990. *In* Liljelund, L.-E. (*Ed.*) *Svenska skogsträds vitalitet och näringstillstånd - stressymptom och orsakssamband.* SNV PM 3813, Sweden.
Rosengren-Brinck, U., and Nihlgård, B. 1995. Ecol. Bull. **44**: 168-177.
Sokal, R. R., and Rohlf, F. J. 1995. *Biometry.* W H Freeman and Company, New York, 887 pp.
SMHI. 1990-1994. Väder och vatten. SMHI, Norrköping, Sweden.
Swift, K. I., and Brockley, R. P. 1994. Can. J. For. Res. **24**: 594-602.
Timmer, V. R., and Stone, E. L. 1978. Soil. Sci. Soc. Am. J. **42**: 125-130.
Turvey, N. D., and Grant, B. R. 1990. For. Ecol. Manage. **37**: 95-122.
Ulrich, B. 1983. *In* Ulrich, B. and Pankrath, J. (*Eds.*), *Effects of accumulation of air pollutants in forest ecosystems.* D. Reidel Publishing Company, pp 127-146.
Valentine, D. W., and Allen, H. L. 1990. Can. J. For. Res. **20**: 144-151.

# RELATIONSHIPS BETWEEN CROWN CONDITION OF BEECH
(*Fagus sylvatica* L.) AND THROUGHFALL CHEMISTRY

J.NEIRYNCK[1] and P.ROSKAMS[2]

[1] *Laboratory of Forestry, Geraardsbergse Steenweg 267, 9090 Gontrode, Belgium.*
[2] *Institute for Forestry and Game Management, Gaverstraat 4, 9500 Geraardsbergen, Belgium.*

(Received 25 September 1998; accepted 26 February 1999)

**Abstract.** A two-step regression procedure was used to predict the impact of throughfall chemistry on the defoliation of *Fagus sylvatica* L. over a 10-year-monitoring period at three plots in northern Belgium. The impact of throughfall chemistry on crown condition was examined after accounting for influences of site, stand, climate and diseases. In a first step, defoliation was regressed on site, stand, climate and disease parameters. The residual defoliation of this standard set was correlated with 27 throughfall variables.

Climatic variables of the year preceding the year of crown assessment accounted for 79% of the variation in current defoliation. Site, stand and disease factors were not included and were still part of the residual defoliation. The study of the residual defoliation revealed that high throughfall depositions of sulphate and ammonium and high throughfall ratios of $SO_4/Mg$ enhanced defoliation.

**Keywords:** climate, crown condition, *Fagus sylvatica* L., residual defoliation, throughfall chemistry

## 1. Introduction

Forest health has been one of the major concerns in sylviculture for many years. As a consequence of the occurrence of new types of forest decline in Europe and North America in the early 1980s (Olson *et al.*,1992; Innes,1993; Hüttl and Mueller-Dombois, 1994), emphasis has been put on recording the visual assessment of crown condition as an indicator of forest health. The crown condition of sample trees has been recorded using different indices to monitor tree vitality. The acquisition of such a long time series can help to reveal if forest health is liable to a slowly developing long-term decline or if changes in vitality are merely due to natural (short or long term) fluctuations.

The data from such surveys have often been used to assess the impacts of pollution stress factors on stand nutrition or on crown condition (Freer-Smith and Read, 1995; Solberg and Tørseth, 1997; Audley *et al.*, 1998; Thomas and Büttner, 1998).

Leaf losses can be attributed to several biotic as well as abiotic stress factors. Climate stresses have become increasingly accepted as key factors influencing the crown condition of trees, although they were initially ignored (Innes, 1993; Landmann, 1995). Innes (1993) emphasised the importance of climate and stand management factors which may mask possible chronic pollution effects. Burton *et al.* (1991) stated that statistically verifiable effects of pollution on leaf area might be lost in the "noise" created by the vagaries of nature. Changes in leaf area resulting from acid deposition may be quite small compared to 20 to 30% leaf area reductions during heavy seed years. Effects of (air pollution) stress factors can, in principle, only be estimated or tested properly if all other relevant effects are taken into account (De Vries *et al.*, 1998). Because of the predominant role of climate in forest health, climate variables should be involved in pollution effect studies. This study aimed to explore relationships between throughfall chemistry and leaf losses of three *Fagus sylvatica* L. plots over a 10-year monitoring period in northern Belgium. The impact of throughfall chemistry was examined after accounting for the effects of site, stand, climate and diseases.

## 2. Materials and Methods

The data were collected in three plots which are part of the 'EU-UN/ECE intensive monitoring of the forest ecosystem' study in the Flemish Region (level II). The first plot is located at the forest of Wijnendale (280 ha, 51°04'10" N; 3°02'10" E) which is situated in the low plain of sandy Flanders. The soil is a moderately wet loam/sand with a partly destroyed textural B-horizon, with a sandy clay substrate beginning at moderate depth. (FAO classification = Gleyic Distric Cambisol). The second plot is located at Gontrode (30 ha, 50°58'35" N; 3°48'19" E) which is situated in the northern low plain of Flanders. The soil is a strongly gleyed (poorly drained) sandy loam with a partly destroyed textural B-horizon, with a clay substrate beginning at a depth of 60 cm (FAO classification = Dystric Podzoluvisol). The third plot is situated at the forest of Zoniën (4400 ha, 50°44'50" N, 4°24'53" E) located on a gently undulating plateau south-east of Brussels. The moderately to well drained loamy soil has a partly destroyed textural B-horizon beginning at a depth of ca. 40 cm (FAO = Dystric Podzoluvisol).

Soils at all sites are acidic with $pH_{CaCl2}$ not exceeding 4 throughout the soil column. Base saturation is lower than 25% except at the deeper horizons in Gontrode and Zoniën.

The climate of the sites is characterised as moderate Atlantic. Data from neighbouring meteorological weather stations display little differences in climatic features between the plots. Mean annual temperature and precipitation are 9.7°C and 754 mm, respectively. Precipitation during the growing season (May-October) averages 414 mm. There are on average 66 frost days each year.

The stands of the plots at Wijnendale and Zonien are homogeneous beech (*Fagus sylvatica* L.), age 75 and 89 years, respectively, with a basal area of 37.7 and 31.7 $m^2$ $ha^{-1}$, respectively. The standing volume amounts to 425 and 536 $m^3$ $ha^{-1}$, respectively. The stand in Gontrode is a 89-year-old mixed common oak (*Quercus robur* L.)-beech stand with a basal area and standing volume of 31.3 $m^2ha^{-1}$ and 391 $m^3ha^{-1}$, respectively.

Average annual defoliation rates over a 10-year-monitoring period are 24% in Wijnendale (range: 16 to 30%) and 15% in Gontrode (range: 12 to 22%) and Zoniën (range: 11 to 23%). Tree mortality has not been observed.

All plots are part of the intensive 'level II' monitoring network in Flanders. Crown condition is assessed on 29 (Wij.), 8 (Gon.) and 25 (Zon.) sample trees situated along the N-S and E-W axes of the circular plots (area 0.25 ha). Crown condition is assessed by measuring defoliation (5% classes), discoloration and the presence and intensity of insects and fungi (UN/ECE, 1994). Assessments are carried out in July-August. The evaluation of crown condition in Wijnendale and Gontrode started in 1988 and continued until 1997 (with the exception of 1989). Crown condition assessment in Zoniën started in 1991. Bulk rainfall and throughfall has been sampled since 1992 every two weeks. The water samples have been analysed for pH, Ca, Mg, K, Na, $HCO_3$, $SO_4$, $NH_4$, $NO_3$ and Cl. Deposition estimates were also available for the plots of Gontrode and Wijnendale for 1988.

The meteorological predictor variables were derived from an extensive data set of daily variables of rainfall, maximum and minimum temperature, relative humidity, wind speed and sunshine duration, from three neighbouring meteorological weather stations at

TABLE I
Set of meteorological predictor variables derived from meteorological weather stations

| Meteorological predictor variable | abbreviation |
|---|---|
| average daily rainfall in month $i$ (mm day$^{-1}$) | $R_i$ |
| average monthly minimum temperature in month $i$ (°C) | $Tmin_i$ |
| average monthly minimum temperature in month $i$ (°C) | $Tmax_i$ |
| average monthly relative humidity in month $i$ (%) | $RH_i$ |
| average monthly wind speed in month $i$ (km h$^{-1}$) | $V_i$ |
| average daily sunshine duration in month $i$ (min day$^{-1}$) | $SUN_i$ |
| number of rainless days in month $i$ (days month$^{-1}$) | $Ra_i$ |
| number of frost days in month $i$ (days month$^{-1}$) | $Tmina_i$ |
| number of days with Tmax >= 25 °C in month $i$ (days month$^{-1}$) | $Tmaxa_i$ |
| number of days with RH < 65% in month $i$ (days month$^{-1}$) | $RHa_i$ |
| number of days with V > 90 km h$^{-1}$ in month $i$ (days month$^{-1}$) | $Va_i$ |
| number of days with sunshine duration > 700 min day$^{-1}$ in month $i$ (days month$^{-1}$) | $SUNa_i$ |
| $\sum$ Tmin of frost days in month $i$ (degree days) | frost-index$_i$ |
| $\sum$ Tmax of days with Tmax >= 25 °C in month $i$ (degree days) | heat-index$_i$ |

Middelkerke, Munte and Zaventem (locations respectively 51°11'53"N; 2°52'04"E, 50°56'00"N; 3°44'00"E and 50°53'46"N; 5°28'00"E; the distance between each plot and the nearest weather station was 11, 4 and 6 km, respectively). From this database a set of monthly climatic variables from 20 months preceding the month of crown condition assessment was calculated (Table I).

The annual radial growth indices from each stand were included as a stand factor. The yearly transpiration stress (calculated as the difference between potential and actual transpiration) and drainage flux at 50 and 240 cm were obtained from a hydrological model (Hubrechts et al., 1997) and were introduced as site variables. The annual average intensity of insect attack and the percentage of sample trees affected by fungi were recorded at the same time as the crown condition assessment. The two-weekly deposition fluxes and ratios have been aggregated over a whole year (Table II).

The impact of throughfall chemistry was tested after making an allowance for the effects of site, stand, climate and pathogens. A two-step regression procedure was used to analyse the influence of deposition on average defoliation (Brooks, 1994). In a first step the annual average defoliation values from the three plots over 9, 9 and 7 years, respectively ($n = 25$) were regressed on a standard set of time-dependent site, stand, climate and disease predictor variables using a stepwise regression with a forward selection procedure. The problems related to multi-collinearity were avoided by omitting variables that were correlated with the variables included in the best-fit model. Therefore, prior to the regression analysis, a correlation study was performed. Positive residuals from this regression model implied that defoliation is underestimated by the set of predictor variables; it meant that there was a defoliation stimulating factor which had not been taken into account. Negative residuals indicated the presence of a factor which suppressed defoliation. The residual defoliation obtained from this multiple regression model was correlated individually with throughfall variables.

## 3. Results

A standard set of 4 monthly climatic variables from the year preceding the crown condition assessment explained 79% of the variability in defoliation (Table III). A low heat-index in May, high number of rainless days in June, high wind speed in August and

## TABLE II
Annual means of throughfall from 3 beech plots involved in the regression analysis (coefficient of variation in parentheses)

| Throughfall variable | Wijnendale (n = 7) | Gontrode (n = 7) | Zoniën (n = 6) |
|---|---|---|---|
| *throughfall fluxes* | | | |
| H (kg ha$^{-1}$ year$^{-1}$) | 0.003 (61%) | 0.011 (55%) | 0.044 (99%) |
| HCO$_3$ (kg ha$^{-1}$ year$^{-1}$) | 77.5 (39%) | 84.2 (33%) | 72.6 (35%) |
| SO$_4$-S (kg ha$^{-1}$ year$^{-1}$) | 19.4 (44%) | 26.2 (31%) | 18.5 (38%) |
| NO$_3$-N (kg ha$^{-1}$ year$^{-1}$) | 6.3 (37%) | 7.8 (32%) | 7.6 (19%) |
| NH$_4$-N (kg ha$^{-1}$ year$^{-1}$) | 20.8 (28%) | 21.2 (23%) | 14.6 (13%) |
| Cl (kg ha$^{-1}$ year$^{-1}$) | 41.2 (25%) | 40.0 (17%) | 32.3 (13%) |
| Ca (kg ha$^{-1}$ year$^{-1}$) | 10.9 (46%) | 17.2 (29%) | 18.5 (40%) |
| Mg (kg ha$^{-1}$ year$^{-1}$) | 3.7 (26%) | 5.1 (16%) | 4.3 (21%) |
| K (kg ha$^{-1}$ year$^{-1}$) | 23.5 (26%) | 33.0 (33%) | 25.8 (25%) |
| Na (kg ha$^{-1}$ year$^{-1}$) | 21.9 (26%) | 22.2 (16%) | 17.6 (24%) |
| *Throughfall ratios* | | | |
| NO$_3$/Cl (mg mg$^{-1}$) | 1.17 (44%) | 1.82 (58%) | 1.95 (44%) |
| SO$_4$/Cl (mg mg$^{-1}$) | 2.16 (27%) | 3.27 (35%) | 2.61 (35%) |
| Ca/K (mg mg$^{-1}$) | 0.73 (77%) | 0.79 (89%) | 1.20 (32%) |
| Mg/K (mg mg$^{-1}$) | 0.28 (99%) | 0.24 (97%) | 0.29 (23%) |
| Ca/Mg (mg mg$^{-1}$) | 3.19 (28%) | 3.53 (29%) | 4.53 (32%) |
| NH$_4$/K (mg mg$^{-1}$) | 2.36 (35%) | 1.73 (46%) | 1.39 (25%) |
| NH$_4$/Mg (mg mg$^{-1}$) | 9.52 (21%) | 6.38 (15%) | 5.28 (30%) |
| NH$_4$/Ca (mg mg$^{-1}$) | 3.43 (25%) | 2.05 (11%) | 1.35 (26%) |
| NO$_3$/K (mg mg$^{-1}$) | 2.82 (49%) | 2.16 (53%) | 2.53 (43%) |
| NO$_3$/Mg (mg mg$^{-1}$) | 10.1 (32%) | 7.90 (28%) | 9.5 (28%) |
| NO$_3$/Ca (mg mg$^{-1}$) | 4.14 (49%) | 2.59 (30%) | 2.33 (19%) |
| NO$_3$/NH$_4$ (mg mg$^{-1}$) | 1.24 (24%) | 1.44 (29%) | 2.01 (27%) |
| SO$_4$/K (mg mg$^{-1}$) | 4.92 (35%) | 4.94 (52%) | 3.68 (16%) |
| SO$_4$/Mg (mg mg$^{-1}$) | 18.9 (24%) | 17.7 (22%) | 14.2 (21%) |
| SO$_4$/Ca (mg mg$^{-1}$) | 7.20 (22%) | 5.78 (18%) | 3.79 (24%) |
| SO$_4$/NH$_4$ (mg mg$^{-1}$) | 2.36 (21%) | 3.21 (18%) | 3.49 (21%) |
| SO$_4$/NO$_3$ (mg mg$^{-1}$) | 3.12 (57%) | 3.58 (56%) | 2.50 (29%) |

a low relative humidity in November resulted in large leaf losses the subsequent year at all three beech plots. The relative humidity in November explained only a small portion of the total variability (5%). Site, stand and pathogens were not included in the model. Their minor components are still part of the residual.

The study of residual defoliation revealed that defoliation was generally underestimated for Wijnendale and Gontrode (Σ residuals resp. 2.8% and 3.9%) and overestimated for Zoniën (Σ residuals = -6.7%) (Figure 1). The residuals were regressed individually on throughfall data. Three out of 27 tested variables were found to be significantly correlated with the residual defoliation. Throughfall fluxes of NH$_4$-N and SO$_4$-S and the SO$_4$/Mg ratio in the throughfall ($p<0.01$, $<0.01$ and $<0.05$, respectively) were positively correlated with the residual defoliation (Table III; Figure 2). This implied that leaf loss rates were increasingly underestimated by the standard set of climatic variables in the presence of increasing deposition of sulphate and ammonium and increasing ratios of SO$_4$/Mg in throughfall.

## 4. Discussion

Changes in defoliation could be largely attributed to a combination of climatic variables from the year preceding the crown condition assessment. Summer droughts are

TABLE III
Statistics of two-step regression procedure on defoliation and residual defoliation

| Independent variable | regression coefficient | standard error | t-value | P | F-ratio | $R^2_{adj}$ |
|---|---|---|---|---|---|---|
| **defoliation (n =25)** | | | | | 23.62 | 0.79 |
| Constant | 60.585 | 24.240 | 2.52 | 0.0202 | | |
| Heat-index5* | -1.519 | 0.125 | -7.41 | 0.0000 | | |
| V8* | 0.566 | 0.008 | 5.13 | 0.0001 | | |
| Ra6* | 0.562 | 0.111 | 4.56 | 0.0002 | | |
| RH11* | -0.640 | 0.268 | -2.41 | 0.0257 | | |
| **residual defoliation** | | | | | | |
| $NH_4$ deposition (n=17) | | | | | 9.65 | 0.56 |
| Constant | -4.156 | 1.379 | -3.01 | 0.0087 | | |
| Slope | 0.206 | 0.066 | 3.11 | 0.0072 | | |
| $SO_4$ deposition (n =16) | | | | | 10.63 | 0.45 |
| Constant | -1.343 | 0.606 | -2.22 | 0.0451 | | |
| Slope | 0.086 | 0.026 | 3.26 | 0.0062 | | |
| $SO_4/Mg$ (n =18) | | | | | 5.14 | 0.26 |
| Constant | -2.226 | 1.447 | -1.538 | 0.1448 | | |
| Slope | 0.188 | 0.083 | 2.268 | 0.0385 | | |

\* meteorological variables from year preceding year of crown condition assessment

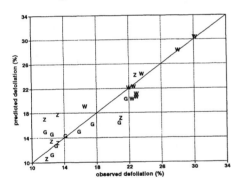

Fig. 1. Predicted (met. data) vs observed defoliation.

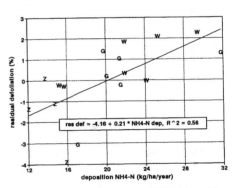

Fig. 2. Residual defoliation vs. $NH_4$-N deposition

reducing the twig growth (Roloff, 1984; Power, 1994) of beech in the following year. The relationship between previous summer drought and leaf loss is often connected with the occurrence of heavy mast years (Schröck, 1995; Hilton and Packham, 1997). This can lead to decreases in foliation due to allocation of assimilates to reproductive tissues (Burton et al., 1991). High cupule production in the 16 x 16 km network of Flanders was noticed in 1991 and 1995 (Sioen and Roskams, 1998).

The study of the residual defoliation suggests a decoupling of the relationship between foliation and climate in the presence of high throughfall deposition of ammonium, sulphate and high $SO_4$ to Mg ratios in throughfall (Figure 2). The results suggest adverse effects of ammonium and sulphate deposition on crown condition. Ammonium and sulphate deposition ranged from 12 to 32 kg N ha$^{-1}$ yr$^{-1}$ and from 13 to 44 kg S ha$^{-1}$ yr$^{-1}$ and declined during the monitoring period. Ammonium and sulphate were highly correlated in throughfall (r = 0.68, p < 0.001). Residual defoliation was set to zero at $NH_4$-N and $SO_4$-S deposition of 20.2 and 15.6 kg ha$^{-1}$ yr$^{-1}$ respectively. Values exceeding these rates lead to positive values of residual defoliation, implying additional leaf losses superimposed on the natural, climatic induced defoliation. Forest health could have been affected indirectly by the depletion of base cations from soils (nutrient deficiency and imbalances), increasing mobilisation of aluminium (increasing toxicity), inhibition of soil biological processes or increased bioavailability of nitrogen (Lucier and Haines, 1990).

Increasing ratios of $SO_4/Mg$ in throughfall were also associated with defoliation. High ratios of $SO_4$ to Mg may result from decreased canopy leaching rates of magnesium owing to depletion of mineral and forest floor magnesium pools or decreased dry deposition of Mg-bearing particles relative to gaseous or particulate interception of sulphur. Deposition of Mg and other base cations, on the contrary, did not contribute to the changes in residual defoliation. Base cation deposition in these regions of Europe is declining (Hedin *et al.*, 1994). It is not clear to what extent future decline will exacerbate stand nutrition and tree health at these base cation-poor sites.

## Acknowledgements

This research was financed by the Ministry of the Flemish Community and was executed in charge of the Minister of Environment of Flanders. The monitoring of the level II plots was co-ordinated by the Institute for Forestry and Game Management.

## References

Audley, D.E., Skelly, J.M., Mccormick, L.H.and Jackson, W.A.: 1998, *Water, Air, Soil Pollut.* **102**, 177-199.
Brooks, R.T.: 1994, *For. Sci.* **40**, 543-557.
Burton, A.J., Pregitzer, K.S. and Reed, D.D.: 1991, *For. Sci.* **37**, 1041-1059.
De Vries,W., Reinds, G.J., Deelstra, H.F., Klap, J.M. and Vel, E.: 1998, *Intensive Monitoring of Forest Ecosystems in Europe*, Technical Report FIMCI 1998, Herenveen, 170 pp.
Freer-Smith, P.H. and Read, D.B.: 1995, *For. Ecol. Manage.* **79**, 185-195.
Hedin, L.O., Granar, L., Likens, G.E., Buishand, T.A., Galloway, J.N., Butler, T.J. and Rodhe H.: 1994, *Nature* **367**, 351-354.
Hilton, G.M. and Packham, J.R.: 1997, *Forestry* **70**, 7-16.
Hubrechts, L., Van de Moortel R. and Feyen J.: 1997, *Berekening van de bodemwaterbalans onder bos voor de proefvlakken van het "Bosbodemetnet Vlaanderen"*, ILW, Leuven, 122 pp.
Hüttl, R.F. and Mueller-Dombois, D. (eds.): 1993, *Forest Decline in the Atlantic and Pacific Regions*, Springer, Berlin, 366 pp.
Innes, J.L.: 1992, *Forestry* **65**, 35 -60.
Innes, J.L.: 1993, *Forest Health: Its Assessment and Status*, CAB International, Wallingford, UK, 677 pp.
Landmann, G.: 1995, in *Forest decline and atmospheric deposition effects in the French mountains*, Landmann, G. and Bonneau, M. (eds.), pp. 407-461, Springer, New York.
Lucier, A.A. and Haines, S.G.: 1990, in *Mechanisms of Forest Response to Acidic Deposition*, Lucier, A.A. and Haines, S.G. (eds.), pp.1-9, Springer, New York.
Olson, R.K., Binkley, D. and Böhm, M. (eds.): 1992, *The response of Western forests to Air Pollution*, Springer, New York, 532 pp.
Power, S.A.: 1994, *Forestry* **67**, 13-30.
Roloff, A.: 1984, *Forst u. Holzwirt.* **14-15**, 364-366.
Schröck, H.W.: 1995, *Allge. Forstzeitschrift* **50**, 91-94.
Sioen, G. and Roskams, P.: 1998, *Bosvitaliteitsinventaris. Resultaten van het level I meetnet*, IBW, 43 pp.
Solberg, S. and Tørseth, K.: 1997, *Environ. Pollut.* **96**, 19-27.
Thomas, F.M. and Büttner, G.: 1998, *Forstw. Cbl.* **117**, 115-128.
UN/ECE (1994). *Manual on methodologies and criteria for harmonised sampling, assessment, monitoring and analysis of the effects of air pollution on forests*. Hamburg/Geneva, UN/ECE: ICP Forest, 117 pp.

# INDUSTRIAL POLLUTANTS TEND TO INCREASE GENETIC DIVERSITY: EVIDENCE FROM FIELD-GROWN EUROPEAN SCOTS PINE POPULATIONS

W. PRUS-GLOWACKI[1*], A.WOJNICKA-POLTORAK[1], J.OLEKSYN[2,3], and P.B.REICH[3]
[1]*Department of Genetics, Adam Mickiewicz University, Miedzychodzka 5, 60-371 Poznan, Poland*
[2]*Institute of Dendrology, Polish Academy of Sciences, Parkowa 5, 62-035, Kornik, Poland*
[3]*University of Minnesota, Department of Forest Resources, 1530 Cleveland Avenue N., St. Paul, MN 55108, USA*

(Received 7 December 1998; accepted 25 February 1999)

**Abstract.** The effect of heavy metal and gaseous pollution on genetic make up due to selection and mortality was studied in eight European Scots pine (*Pinus sylvestris* L.) populations. Genetic structure of these populations was evaluated after 13 years of growth at two sites, one located in the vicinity of a copper smelter in Glogow (southern Poland) and another in a control area (Kornik, central Poland). Based on data from 10 enzymatic loci, studied by starch gel electrophoresis, the following genetic parameters were characterized: mean number of alleles and genotypes per locus, heterozygosity and genotype polymorphism index, genetic diversity among sites and genetic similarity according to Nei and Roychoudhury. Most of the parameters differed both between sites as well as between pairs of populations. Mean multilocus observed heterozygosity and genotype polymorphism indices were 8% and 6% lower respectively at the control site than in the polluted one. Populations from the site polluted by heavy metals demonstrate higher genetic diversity then those from the control. Genetic similarities showed differentiation of particular populations from both locations in response to industrial pollution. These results provide evidence that the general pattern of adaptation strategy of populations to pollution stress is connected with increasing genetic variation. The best isoenzymatic markers for monitoring changes in the genetic structure of Scots pine populations under the influence of industrial pollution are fluorescent esterase (FEST), glutamic-oxaloacetatic transaminase (GOT-A and GOT-B), alcohol dehydrogenase (ADH), malate dehydrogenase (MDH-A) and shikimate dehydrogenase (ShDH-A) loci.

**Keywords:** Genetic structure, Scots pine, provenance, environmental pollution, allozyme

## 1. Introduction

Heavy metal pollution is a known factor inducing differentiation in mortality and tolerance among plant populations (Meerts and Van Isacker, 1997; Vekemans and Lefebvre, 1997; Koch *et al.*, 1998). There is evidence that both existence of heavy metal tolerant individuals within a population and differential mortality among populations can trigger microevolutionary processes leading towards changes in their genetic structure and adaptation to a polluted environment (Gartside and Neilly, 1974; Wu *et al.*, 1975; Bradshaw, 1984; Symeonides *et al.*, 1985; Al-Hiyaly *et al.*, 1988; Shaw, 1988; Nieminen, 1998; Bell *et al.*, 1998).

Effects of heavy metal and gaseous pollution on genetic structure of tree species reveal that it is not always possible to show whether tree mortality is associated with microevolutionary changes in genetic structure of population or with stochastic processes (Prus-Glowacki and Godzik, 1991; Prus-Glowacki and Nowak-Bzowy, 1992; Geburek *et al.*, 1987). Some authors are of the opinion that changes in allozymes are neutral for adaptation and that the relationships between heterozygosity, genetic multiplicity and fitness are generally poor (Kimura, 1983; Nei, 1987; Savolainen and Hedrick, 1995).

The aim of this study was to examine the genetic variation of eight geographically diverse Scots pine populations of 15-yr-old trees grown for 13 years in two experimental provenance sites of contrasting pollution intensity in Poland. The homogeneity of environmental conditions within each site and the presence in both sites of populations originating from the same seed sources makes it possible to answer the following questions: Is pollution inflicted mortality within populations related to changes in their genetic structure? To what extent can pollution pressure affect genetic diversity of Scots pine populations? What allozyme markers are most suitable as markers of changes in the genetic structure of Scots pine populations under the influence of industrial pollution stress? Thus this study addresses several key questions concerning long-term effects of pollution on genetic structure of diverse Scots pine provenances and helps to improve our understanding of the nature of selective processes and adaptation.

## 2. Materials and Methods

2.1. PLANT MATERIAL AND SITE CHARACTERISTICS

In 1984 twenty European provenances of 2-yr-old Scots pine (*Pinus sylvestris* L.) seedlings grown from seed were planted at two sites in western and southwestern Poland as a part of an experiment established under the auspices of IUFRO Working Party "Scots pine provenances and breeding" (Oleksyn *et al.*, 1988). One site was about 3.5 km from a copper smelter in Głogów (51°40'N and 16°06'E). The experiment consisted of 3 blocks, each with 18 plots. In each plot 20 plants were planted of each provenance. Due to high mortality among populations observed in the first six years only 33 to 55% of plants survived until the age of 16. Therefore, only eight populations, represented by more than 20 trees, were suitable for this study. The climate of the region is intermediate between maritime and continental. The average annual precipitation is 540 mm and average temperature 8.5 °C, with a mean growing season length, counted as the number of days with average temperature $> 5$ °C, of 225 days (Seniczak *et al.*, 1997).

A control site was located near Kórnik (52°15'N and 17°04'E) in a region free of acute pollution levels (Reich *et al.*, 1994; Oleksyn *et al.*, 1999a). This site consists of 7 blocks. Each provenance was planted in 3 to 7 replicated blocks with 48 plants (4 rows x 12 plants). The average annual precipitation is 526 mm and average temperature 7.7 °C, with a mean growing season length of 220 days. A total of eight populations with more than 20 trees on both sites were selected for this study (Table I, Figure 1).

According to the Main Statistical Office in 1985 the emission of $SO_2$ from the copper smelter in Głogów was near 57,000 t $y^{-1}$ and dusts more than 3,000 t $y^{-1}$ (Oleksyn *et al.*, 1988). The total magnitude of and dust air pollutants (including heavy metals) ranks this smelter in 6th place among the most noxious industrial plants in Poland. The mean annual concentration of $SO_2$ in the years 1975-1982 measured near the experimental area varied between 53 and 187 µg $m^{-3}$ (Oleksyn *et al.*, 1988). However, more recently annual emission rates and concentrations of gaseous pollutants have declined significantly to 22-44 µg $m^{-3}$ reflecting changes in industrial purification processes (Zwolinski, 1995).

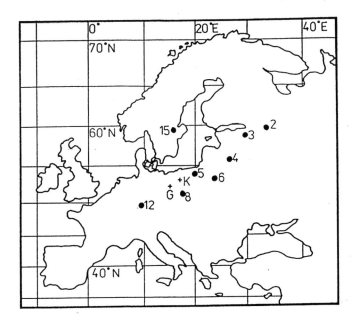

*Fig.1.* Seed origin of the Scots pine (*Pinus sylvestris* L.) provenanes (black circles). 2, Kondezhskoe, Russia (59°58'N, 33°30'E); 3, Serebryanskoe, Russia (58°50'N, 29°07'E); 4, Silene, Latvia (55°45'N, 26°40'E); 5, Milomlyn, Poland (53°34'N, 20°00'E); 6, Suprasl, Poland (53°12'N, 23°22'E); 8, Rychtal, Poland (51°08'N, 17°55'E); 12, Lampertheim, Germany (50°00'N, 10°00'E); 15, Sumpberget, Sweden (60°11'N, 15°52'E). G - Glogow (polluted site); K - Kornik (control site).

The soil type at both sites consists of light sands. Soil chemical properties measured 0 to 20 cm depth after removal of the $A_O$ horizon (Oleksyn *et al.*, unpublished) revealed no significant differences ($p>0.1$) between sites in concentrations of soluble $N-NH_4$, $N-NH_3$, P, K, Ca, Mg, Cl, S, Mn, and Zn. However, concentrations of soluble Cu, Pb and Cd were 56, 7 and 4 times higher, respectively in polluted vs. control sites ($p < 0.1$).

2.2. FOLIAR SAMPLING PROCEDURE AND CHEMICAL ANALYSES

For all provenances current-year needles were collected from the sunny upper part of tree crowns, from 4 to 15 trees per block in two blocks on both sites in November of 1996. Data are means of equally weighted composite samples of the individual trees sampled in each of two blocks. Needles were not washed.

Analyses of foliar concentrations of Cu, Pb, Cd, Zn and Ni were done simultaneously with an Inductively Coupled Plasma Emission Spectrometer (ICP) at the University of Minnesota Research Analytical Laboratory, St. Paul, MN, USA.

2.3 ISOZYME ANALYSES

Dormant winter buds were used for isozymatic analyses. Plant material was collected from 20 to 30 randomly chosen trees from each provenance. When less than 30 individuals were available, the samples from all living trees were collected (Table I).

Variation in the following polymorphic enzymatic loci was studied: fluorescent esterase, FEST (E.C. 3.1.1.1.); glutamic-oxaloacetatic transaminase, GOT=AAT, two loci (E.C. 2.6.1.1.), glutamate dehydrogenese, GDH (E.C. 1.4.1.3.), alcohol dehydrogenase, ADH (E.C.1.1.1.1.), shikimate dehydrogenase, ShDH, two loci, (E.C. 1.1.1.25), malate dehodrogenase, two loci MDH (E.C. 1.1.1.37). The separation of isozymes on starch gels and the genetic interpretation of the results were performed as described by Rudin and Ekberg, 1978; Szmidt and Yazdani, 1984; Prus-Glowacki, 1986; Muona and Szmidt, 1985. The following genetic parameters were calculated for each population: mean number of alleles (A/L) and genotypes (G/L) per locus, heterozygosity observed $H_O$ and expected $H_e$, genotype polymorphism index ($P_g$) genetic diversity index ($G_{ST}$) and genetic similarities and distances (Nei and Roychoudhury, 1974; Nei, 1975; El-Kassaby, 1991).

## 3. Results

The average concentrations of Cu, Pb, Cd, Zn and Ni in needles of eight Scots pine populations from two sites are shown in Table I. For all elements studied (Cu, Pb, Cd, Zn, Ni) needles from plants in the polluted site contained higher concentrations than those from the control site. For example concentrations of Pb in the control site were below the detection limit of 1.7 $\mu g\ g^{-1}$, whereas those in the polluted site varied between 33 and 67 $\mu g\ g^{-1}$. The concentration of Cu was on average 23 times higher in the polluted site than in the control one. Observed Cu concentrations from 50 to 71 $\mu g\ g^{-1}$ are higher than the critical toxicity level for most plants (20-30 $\mu g\ g^{-1}$, Marschner, 1995). In general, variation of needle heavy metal concentrations corresponds well with those observed in soil.

Data on gene diversity of Scots pine provenances are presented in Table I. On average the number of alleles and genotypes per locus were slightly higher in the polluted site than in the control one. However this difference was not significant statistically. The values of expected and observed heterozygosity were 6 and 8% higher in the polluted than in the control site, respectively. On average for all populations genotype polymorphism index was significantly higher in the polluted than in the control site (0.419 vs. 0.437, Table I).

The comparison of Nei's (1975) genetic diversity index ($G_{ST}$) indicates that not all enzymatic loci differ among populations in both sites (Figure 2). The most pronounced differences were observed for the following loci: FEST, GOT A, GOT B, ADH, MDH A and ShDH A. Little differentiation were noted for GDH, 6PGD, MDH C and ShDH B. Overall, mean values of $G_{ST}$ for both sites are similar, masking existing differentiation among individual loci in the polluted and control sites.

TABLE I

The origin of seeds of *Pinus sylvestris* used in the study, number of trees sampled, mean number of alleles per locus (A/L), genotypes per locus (G/L), heterozygosity expected ($H_e$), heterozygosity observed ($H_o$), genotype polymorphism index ($P_g$), fixation index (F) and mineral elements concentration. Provenances are ordered by latitude of origin.

| Population no. and country | Site | No. of trees sampled | Genetic parameter | | | | | | Needle mineral element, $\mu g\,g^{-1}$ | | | | |
|---|---|---|---|---|---|---|---|---|---|---|---|---|---|
| | | | A/L | G/L | $H_e$ | $H_o$ | $P_g$ | Cu | Pb | Cd | Zn | Ni |
| 15. Sweden | Control | 30 | 2.5 | 3.2 | 0.309 | 0.297 | 0.435 | 2.8 | - | - | 33.5 | 0.8 |
| | Polluted | 25 | 2.4 | 3.3 | 0.324 | 0.292 | 0.427 | 55.9 | 46.1 | 0.16 | 71.6 | 4.0 |
| 2. Russia | Control | 30 | 2.3 | 2.8 | 0.297 | 0.290 | 0.420 | 2.8 | - | - | 34.9 | 0.8 |
| | Polluted | 28 | 2.3 | 2.9 | 0.324 | 0.350 | 0.435 | 56.3 | 45.4 | 0.23 | 85.1 | 6.6 |
| 3. Russia | Control | 30 | 2.4 | 3.1 | 0.303 | 0.330 | 0.415 | 2.8 | - | 0.12 | 33.8 | 1.2 |
| | Polluted | 21 | 2.3 | 2.9 | 0.327 | 0.362 | 0.449 | 50.3 | 38.4 | 0.20 | 72.0 | 4.0 |
| 4. Latvia | Control | 30 | 2.4 | 3.0 | 0.314 | 0.337 | 0.436 | 2.8 | - | - | 39.6 | 1.2 |
| | Polluted | 27 | 2.6 | 3.3 | 0.326 | 0.337 | 0.446 | 65.9 | 61.4 | 0.23 | 82.9 | 4.8 |
| 5. Poland | Control | 30 | 2.3 | 2.9 | 0.301 | 0.333 | 0.411 | 2.3 | - | - | 26.5 | 0.8 |
| | Polluted | 29 | 2.7 | 3.3 | 0.312 | 0.324 | 0.435 | 55.1 | 46.5 | 0.17 | 72.7 | 6.6 |
| 6. Poland | Control | 30 | 2.4 | 3.1 | 0.269 | 0.250 | 0.394 | 2.5 | - | - | 34.8 | 0.8 |
| | Polluted | 30 | 2.4 | 3.1 | 0.325 | 0.347 | 0.430 | 71.1 | 66.9 | 0.13 | 68.3 | 2.9 |
| 8. Poland | Control | 29 | 2.3 | 3.0 | 0.302 | 0.335 | 0.409 | 2.4 | - | - | 28.5 | 0.9 |
| | Polluted | 20 | 2.2 | 2.8 | 0.309 | 0.325 | 0.412 | 65.9 | 62.6 | 0.20 | 58.8 | 3.6 |
| 12. Germany | Control | 30 | 2.3 | 3.2 | 0.319 | 0.260 | 0.436 | 2.6 | - | 0.16 | 34.5 | 0.8 |
| | Polluted | 23 | 2.7 | 3.4 | 0.322 | 0.300 | 0.461 | 54.9 | 32.9 | 0.14 | 58.8 | 4.7 |
| Mean | Control | 30 | 2.4 | 3.0 | 0.302 | 0.304 | 0.419 | 2.6 | - | - | 33.3 | 0.9 |
| | Polluted | 25 | 2.5 | 3.1 | 0.321 | 0.330 | 0.437 | 59.4 | 50.0 | 0.18 | 71.3 | 4.7 |
| Population effect $p > F$‡ | | | 0.25 | 0.38 | 0.01 | 0.1 | 0.04 | <0.001 | | | <0.001 | <0.001 |

†See Fig. 1 for origin; - Concentration below the detection limit. ‡Based on ANOVA of sites.

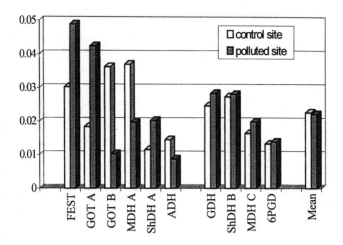

*Fig. 2.* Genetic diversity indices ($G_{ST}$) for 10 loci in eight *Pinus Sylvestris* populations from the control and polluted sites in Poland.

## 4. Discussion

These results indicate the existence of differences in genetic structure among diverse Scots pine populations grown in two sites with contrasting levels of pollution. Mortality among the populations at the control site was low (Oleksyn *et al.*, 1999b) while pollution-related mortality at the polluted site was very high, implying the observed differences in genetic structure resulted from this selection. Assuming the surviving trees are more tolerant and better adapted to pollution stress, we conclude that the adaptation processes are related to higher levels of genetic multiplicity or diversity. This is evident from significantly higher mean multilocus observed heterozygosity, genotype polymorphism indices and higher genetic diversity (Table I and Figure 2).

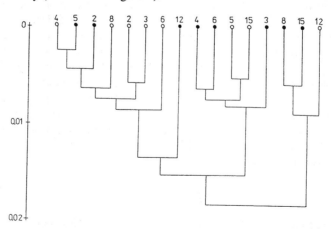

*Fig. 3.* Dendrogram based on genetic similarity coefficients (Nei and Roychoudhury, 1974). Open circle - K (Kornik, control site), solid circle - G (Glogow, polluted site).

Changes in genetic structure in trees as a result of environmental stress have also been noted in studies with *Pinus sylvestris* (Mejnartowicz, 1983; Prus-Glowacki and Nowak-Bzowy, 1992; Prus-Glowacki *et al.*, 1998), *Picea abies* (Bergmann and Scholz, 1984, 1987) or *Fagus sylvatica* (Müller-Starck, 1985). Studies with coniferous species revealed that some genetic parameters are related to quantitative traits such as growth rate or survival (Ledig *et al.*, 1983; Oleksyn *et al.*, 1994). However, these relationship often depend on population genetic parameters, age of population, their developmental homeostasis or pollution type (Prus-Glowacki and Godzik, 1991; Mitton and Grant, 1984).

Results of this study as well as other experiments conducted with diverse common-garden grown Scots pine populations subjected to environmental pollution stress (Oleksyn *et al.*, 1994; Prus-Glowacki *et al.*, 1998) indicate the usefulness of studies with isozymes in the assessment of pollution effects and forest decline. These findings may indicate that the expected neutral pattern of allozyme variation (Kimura, 1983; Nei, 1987) is not a universal phenomenon. It is especially apparent that for diverse plant populations exposure to rapid environmental alterations due to anthropogenic industrial pollution results in selective mortality. For example Swedish population no. 15 which shows relatively high survival in a polluted environment (Oleksyn, 1988) displayed fewer changes in genetic structure than more vulnerable populations from Germany (2), Poland (5) or Russia (3).

Not all isozymes are good markers of changes in Scots pine genetic structure. In this study where mortality of plants was inflicted by pollution associated with copper smelting (mainly heavy metals and $SO_2$) the most indicative were the following loci: FEST, GOT A, GOT B, ADH, MDH A and ShDH A (see $G_{ST}$ values in Figure 3). The value of these loci for detection of genetic effects of heavy metal pollution on Scots pine is supported by our earlier studies where we found that the most indicative were ADH, ShDH and MDH (Prus-Glowacki and Nowak-Bzowy, 1992). It is possible that some of these marker loci are associated with detoxification processes. However, the usefulness of any particular loci as markers of change in genetic structure of populations depends on the type of pollution. Studies conducted with the same Scots pine populations grown in a site polluted by a phosphate fertilizer factory ($SO_2$, $NO_X$, F, Al) indicated a different set of useful loci (Wojnicka-Poltorak, 1997). The usefulness of only some loci as markers of changes in genetic structure of forest tree populations was also pointed out by Geburek *et al.*, (1987), Bergmann and Scholz, (1989), Muller-Starck, (1989).

In summary, the results of this study showed that the differential mortality and selection caused by industrial pollution acts against less polymorphic, less genetically diverse and less heterozygous plants. The most useful marker loci for monitoring changes in the genetic structure of Scots pine populations affected by industrial pollutants such as heavy metals and $SO_2$ were FEST, GOT-A, GOT-B, ADH, MDH-A and ShDH-A.

## Acknowledgments

This study was made possible by support from the State Committee for Scientific Research (Poland) grant no. 6PO4C 08208).

## References

Al-Hiyaly, S.A., McNeilly, T. and Bradshaw, A.D.: 1988, *New Phytol.* **110**, 571-580.
Bell, R.L., Graham, A.K. and Roy, D.N.: 1998, *Forestry Chronicle* **74**, 530-532.
Bergmann, F. and Scholz, F.: 1984, *Lect. Notes Biomath.* **60**, 267-275.
Bergmann, F. and Scholz, F.: 1987, *Silvae Genet.* **36**, 80-83.
Bergmann, F. and Scholz, F.: 1989, in F. Scholz, H.R. Gregorius, and D. Rudin, (eds). *Genetics effects of air pollutants in forest tree populations*. Springer-Verlag Berlin, 143-160.
Bradshaw, A.D.: 1984, In: Origin and developement of adaptation (eds. D.Evered and G.M. Collins), CIBA Foundation Symposium 102. Pitman, London, 4-19.
El-Kassaby, Y.A.: 1991, in S. Fineschi, M.E. Malvotti, F. Cannata, and H. Hattemer, (eds). *Biochemical markers in forest genetics*. Academic Publ. BV, The Hague, 61-76.
Gartside, D.W. and McNeilly, T.: 1974,. *Heredity* **32**, 287-297.
Geburek, Th., Scholz, F., Knabe, W. and Vornweg, A.: 1987, *Silvae Genet.* **36**, 49-52.
Kimura, M.: 1983, *The neutral theory of molecular evolution*. Cambridge University Press, Cambridge.
Koch, M., Mummenhoff, K. and Hurka, H.: 1998, *Biochem. Syst.Ecol.* **26**, 823-838.
Ledig, F.T., Guries, R.P. and Bonefield, B.A.: 1983, *Evolution* **37**, 1227-1238.
Marschner, H.: 1995, *Mineral Nutrition of Higher Plants*. Second Edition. Academic Press, London, San Diego, New York, Boston, Sydney, Tokyo, Toronto.
Meerts, P. and Van Isacker, N.: 1997, *Plant Ecol.* **133**, 221-231.
Mejnartowicz, L.: 1983, *Genetica Pol.* **24**, 41-50.
Mitton, J.B. and Grant, M.C.: 1984, *Ann. Rev. Ecol. Syst.* **15**, 470-499.
Muona, O. and Szmidt, A.E.: 1985, *Lect. Notes Biomath.* **60**, 226-240.
Müller-Starck, G.: 1985, *Silvae Genet.* **34**, 241-247.
Muller-Starck, G.:1989, in F. Scholz, HR. Gregorius, and D. Rudin (eds). Genetic effects of air pollutants in forest tree populations. Springer Verlag Berlin, 127-142.
Nei, M. and Roychoudhury, A.K.: 1974, *Genetics* **76**, 379-390.
Nei, M.: 1975, *Molecular population, genetics and evolution*. North Holland, Amsterdam.
Nei, M.: 1987, *Molecular evolutionary genetics*. Columbia University Press, New York.
Nieminen, T.M.: 1998, *Chemosphere* **36**, 745-750.
Oleksyn, J.: 1988, *Environ. Pollut.*, **55**: 289-299.
Oleksyn, J., Karolewski, P. and Rachwal, L.: 1988, *Acta. Soc. Bot. Pol.* **57**, 107-115.
Oleksyn, J., Prus-Głowacki, W., Giertych, M. and Reich, P.B.: 1994, *Can. J. For. Res.* **24**: 2390-2344.
Oleksyn, J., Reich, P.B., Karolewski, P., Tjoelker, M.G. and Chalupka, W.: 1999a, *Water, Air, Soil Pollut.* **110**, 195-212.
Oleksyn, J., Reich, P.B., Chalupka, W. and Tjoelker, M.G.: 1999b, *Scan. J. For. Res.* **14**, (in press)
Prus-Glowacki, W.: 1986, in A. Kormutak and D. Uzak (eds). *Proc., Genpool of Forest Woody Species, Its Conservation and Utilization*. Nitra, Czechoslovakia. 107-122.
Prus-Glowacki, W. and Godzik, S.: 1991, *Silvae Genet.* **40**, 184-188.
Prus-Glowacki, W. and Nowak-Bzowy, R.: 1992, *Water, Air, Soil Pollut.* **62**, 249-259.
Prus-Glowacki, W., Oleksyn, J. and Reich, P.B.: 1998, *Chemosphere* **36**, 4-5, 813-818.
Reich, P.B., Oleksyn, J. and Tjoelker, M.G.: 1994, *Oecologia* **97**: 82-92.
Rudin, D. and Ekberg, I.: 1978, *Silvae Genet.* **27**, 1-12.
Savolainen, O. and Hedrick, P.: 1995, *Genetics* **140**, 755-842.
Seniczak, S., Dabrowski, J. and Dlugosz, J. 1997, *Water, Air, Soil Pollut.* **94**, 71-84.
Shaw, J.: 1988, *New Phytol.*, **109**, 211-222.
Symeonides, L., Mc Neily, T. and Bradshaw, A.D.: 1985, *New Phytol.* **101**, 309-315.
Szmidt, A.E. and Yazdani, R.: 1984, *Electrophoretic studies of genetic polymorphism of shikimate and 6-phospholuconate dehydogenase in Scots pine (Pinus sylvestrisL.) in Genetic studies of Scots pine domestication by means of isozyme analysis*. PhD dissertation SUAS, Umea, Sweden, 6-13.
Vekemans, X. and Lefebvre, C.: 1997, *J. Evol. Biol.*, **10**, 175-191.
Wojnicka-Póltorak, A.: 1997, *Microevolutionary processes in Pinus sylvestris L. populations affected by industrial pollution*. PhD dissertation, Adam Mickiewicz University, Poznan, Poland (in Polish).
Wu, L., Bradshaw, A.D. and Thurman, D.A.: 1975, *Heredity* **34**, 165-187.
Zwolinski J.:1995, *Prace IBL, Ser. A*, **809**, 1-85 (in Polish with English summary).

# AMONG-PROVENANCE VARIABILITY OF GAS EXCHANGE AND GROWTH IN RESPONSE TO LONG-TERM ELEVATED $CO_2$ EXPOSURE

JAMES L.J. HOUPIS[1], PAUL D. ANDERSON[2], JAMES C. PUSHNIK[3], AND DAVID J. ANSCHEL[4]

[1]Department of Biological Sciences, Southern Illinois University, Box 1099, Edwardsville, IL 62026, USA.
[2]USDA USFS, North Central Forest Experiment Station, 5985 Hwy K, Rhinelander, WI 54501, USA.
[3]Department of Biological Sciences, California State University, Chico, CA 95929, USA. [4]Health and Ecological Assessment, Lawrence Livermore National Lab. Box 808, Livermore, CA 94551, USA.

(Received 30 November 1998; accepted 25 February 1999)

**Abstract**: Genetic variability can have profound effects on the interpretation of results from elevated $CO_2$ studies, and future forest management decisions. Information on which varieties are best suited to future atmospheric conditions is needed to develop future forest management practices. A large-scale screening study of the effects of elevated $CO_2$ on 15 half-sibling sources of genetically superior ponderosa pine (*Pinus ponderosa* Dougl ex P. Laws.) is presented. These sources represent multiple elevations and latitudes throughout California. Among-provenance variability in the effects of elevated $CO_2$ on gas exchange and growth, and their correlation with geographic origin were investigated in ponderosa pine seedlings subjected to ambient or elevated $CO_2$ concentrations (525 µmol mol$^{-1}$ $CO_2$, and 700 µmol mol$^{-1}$ $CO_2$) for more than two years in open-top chambers. Substantial among-provenance variability in growth response to elevated $CO_2$ was evident, with 8 sources demonstrating no significant growth response to elevated $CO_2$ while 7 sources responded positively. For all sources, elevated $CO_2$ increased photosynthesis (ranging from 19% increase at 525 µmol mol$^{-1}$ $CO_2$ to 49% increase at 700 µmol mol$^{-1}$ $CO_2$). A modest correlation existed between geographic origin and above ground growth response to elevated $CO_2$.

**Keywords**: Genetic variability, elevated $CO_2$, climate change, pine, conifer, photosynthesis, growth.

## 1. Introduction

The projected rise in atmospheric $CO_2$ concentration and the consequential effects on climate will have profound effects upon terrestrial ecosystems (Cramer and Leemans, 1993; and Rind *et al.*, 1990). Globally, forest ecosystems represent the most persistent standing biomass of all terrestrial ecosystems and therefore play a significant role in the present and future global carbon (C) cycle. The importance of forests and their interactions with climate are considerable. Forests occupy 22 % of the earth's land area (excluding Polar Regions), and account for two-thirds of terrestrial photosynthetic C assimilation (Waring and Schlesinger, 1985). In addition to photosynthesis, forests contribute significantly to the hydrological cycle through evapotranspiration, and nutrient cycles that occur within the forest biosphere. The extent of the impact associated with elevated $CO_2$ concentration on future forest management practices, however, remains uncertain.

For tree species, a general trend has been reported of increased C assimilation that leads to increased growth of young trees. Increased growth is the result of biochemical and physiological adaptations that optimize C acquisition and allocation (Acock and Allen, 1985; Eamus and Jarvis, 1989). These growth responses vary widely between genera (Tolley and Strain, 1984), species within the same genera (Rogers *et al.*, 1994) and intraspecifically (Houpis *et al.*, 1995).

Variability in the observed difference in the response to elevated $CO_2$ will have substantial impact as land managers and plant breeders begin to focus attention on the use of forests as C sinks, on the desirability of establishing vast areas of new plantations, and on the development of new forest management practices in response to increasing $CO_2$ (Kellison and Weir, 1987; Sedjo, 1989). Forest plantations can be used in mitigation strategies for the reduction of atmospheric $CO_2$. If atmospheric $CO_2$ continues to rise, plantations can be designed to optimize the enhanced productive capacity associated with increasing $CO_2$ concentrations. Since there are species and genotype specific responses to elevated $CO_2$, tree breeding can have a substantial impact on the performance of these plantations (Ceulemans *et al.*, 1996; Ceulemans *et al.*, 1995; Callaway *et al.*, 1994). In tree breeding, it is desirable to develop genotypes that can grow well in a wide range of environments (Weber *et al.*, 1996). Genotypic differences in performance are also important in the establishment of plantations used for existing forest practices and future $CO_2$ sequestration strategies. This is due to the fact that genotypes must be well adapted to a wide range of $CO_2$ concentrations, as they will be planted under current ambient conditions, and over the course of their rotation, experience significant increases in atmospheric $CO_2$.

There have been a few studies on the intraspecific variability in the $CO_2$ response of tree species (e.g. *Populus* hybirds, Ceulemans *et al.*, 1996). Inter-population variation in *P. ponderosa*, (Callaway *et al.*, 1994; DeLucia *et al.*, 1994) and intra-population variation in the deciduous *Betula alleghaniensis* (Wayne and Bazzaz, 1995) have been examined. Thus, little information is known about intraspecific (among populations) tree growth response to elevated $CO_2$. In our study, we investigated among population (half-sib seed source) growth and gas exchange differences of a major commercial conifer species, *P. ponderosa*, to elevated $CO_2$, and whether the observed differences could be correlated to geographic origin. Understanding these underlying physiological and/or growth responses to elevated $CO_2$, and the extent of population variability is the first step in tree breeding.

## 2. Material and methods

### 2.1 PLANT MATERIAL AND GROWTH CONDITIONS

*Pinus ponderosa* seedlings from 15 half-sibling seed sources of different geographic origin (Table I) were grown at various atmospheric $CO_2$ concentrations in standard outdoor open-top chambers (OTC, a cylindrical shaped outdoor chamber, 3 m diameter and 3 m height; Rogers *et al.*, 1983) at the Lawrence Livermore National Laboratory exposure facility (Livermore, California). The 15 sources represent among-population variability throughout California. The seedlings were grown from seed and in 3 L containers containing a soil medium consisting of 2 parts clay, 3 parts red lava rock, 1 part colma sand, and 3 parts bark. All seedlings were well-watered and fertilized using a one-half strength Hoagland's solution.

TABLE I

Geographic origin of ponderosa pine half-sib sources from California included in this study. Source identification number, and classification of breeding zone and seed zone are based on the US Forest Service, Region 5 Tree Improvement Program.

| Source | Source Location of Origin | | | USFS Breeding Zone | USFS Seed Zone |
| --- | --- | --- | --- | --- | --- |
| | Elevation (m) | Latitude (deg.) | Longitude (deg.) | | |
| 3088 | 1330 | 38.8 | 120.4 | 3 | 526 |
| 3089 | 1800 | 38.8 | 120.0 | 3 | 526 |
| 3287 | 1730 | 39.8 | 120.6 | 2 | 523 |
| 3306 | 1570 | 34.8 | 119.9 | 3 | 531 |
| 3363 | 1570 | 37.3 | 119.3 | 3 | 532 |
| 3382 | 1770 | 39.4 | 120.7 | 2 | 525 |
| 3435 | 1930 | 37.2 | 119.3 | 4 | 533 |
| 3589 | 1870 | 35.6 | 118.6 | 4 | 540 |
| 4266 | 1500 | 40.6 | 121.4 | 7 | 732 |
| 4304 | 1500 | 40.3 | 120.6 | 7 | 771 |
| 4315 | 1730 | 39.8 | 120.6 | 7 | 732 |
| 4731 | 830 | 41.2 | 121.4 | 5 | 521 |
| 11953 | 1170 | 40.7 | 123.3 | 5 | 312 |
| 13005 | 1170 | 41.2 | 121.5 | 7 | 742 |
| 17520 | 570 | 41.0 | 122.4 | 5 | 521 |

2.2 $CO_2$ TREATMENTS

Seedlings of each source were exposed to three $CO_2$ exposure concentrations consisting of ambient (approximately 350 µmol mol$^{-1}$ $CO_2$), 525 µmol mol$^{-1}$ $CO_2$, and 700 µmol mol$^{-1}$ $CO_2$. The concentration of $CO_2$ in the OTCs was monitored using dedicated $CO_2$ analyzers (Horiba Model PIR-2000; all analyzers were zero and span checked daily and underwent a complete multipoint calibration every month). Chamber air was sampled twelve times per hour at canopy height, at the center of the chamber, for approximately one minute, and the values of $CO_2$ concentration were averaged over the one-minute sampling period. Over the entire length of the study, the chamber $CO_2$ concentrations were maintained 24 hr day$^{-1}$ within ±5% of the treatment concentration. Exposure began in April 1993 and continued through to October 1995.

2.3 GROWTH MEASUREMENTS

Seedling growth during the 1995 growing season was evaluated based on height and diameter of the main stem measured in August 1995. Height was measured to the nearest 0.5 cm using a tape measure. Diameter was measured to the nearest 0.1 mm at the cotyledon whorl using a vernier caliper. Measures of total height and diameter were used to estimate main stem volume ($\pi$ * radius$^2$ * height).

## 2.4 GAS-EXCHANGE MEASUREMENTS

Net photosynthesis ($P_n$), stomatal conductance ($g_s$), and leaf internal $CO_2$ concentration ($C_i$) were measured in August 1995, on current-year-old foliage using an open-loop photosynthesis system consisting of a portable infrared gas analyzer and microprocessor controller (model LI-6400, Licor Inc., Lincoln, NB). Values for $P_n$, $g_s$, and $C_i$ were calculated according to equations derived by von Caemmerer and Farquhar (1981). Gas exchange measurements were taken on a single fascicle, which was placed across the width of the LI-6400 cuvette. Gas exchange results are expressed on a total leaf surface area basis. Total leaf surface area was geometrically determined by measuring the radius and length of the needle inside the cuvette. All measurements were taken mid-morning (0900-1030) and light intensity (PAR) within the cuvette was maintained constant at 1000 µmol m$^{-2}$ s$^{-1}$ (saturating light intensity for *P. ponderosa*). Cuvette temperature was set at a leaf temperature of 25°C. The mean leaf temperature over all gas exchange measurements was 25 ± 0.1°C (leaf temperature ranged from 24.2 to 25.9°C). The mean leaf-air VPD over all gas exchange measurements was 2.0 ± 0.2 kPa (leaf-air VPD ranged from 1.8 to 2.2 kPa). All measurements were made at the treatment $CO_2$ concentration.

## 2.5 EXPERIMENTAL DESIGN

The study was conducted using a split-plot design. The three levels of atmospheric $CO_2$ concentration, the main plot factor, were randomly assigned to 18 OTCs to provide six replicates. Within each chamber, 15 sources of *P. ponderosa* were grown. Thus, source represented a sub-plot factor. Within each chamber, one seedling from each of the 15 sources of *P. ponderosa* was selected for measurement. The effects of $CO_2$ concentration and source on gas exchange were determined using analysis of covariance (ANACOV) with leaf-to-air vapour pressure deficit and air temperature as covariates. Treatment and source effects on seedling size were evaluated by analysis of variance (ANOVA) using log-transformed dependent variables to stabilize variance. Significance of $CO_2$ and $CO_2$-within-source effects was tested using either linear or quadratic orthogonal contrasts. Source differences were tested using the least significant difference (LSD) criterion if source effects were significant as indicated by ANOVA or ANACOV.

The relationship between geographic location of origin and physiological or growth performance was evaluated by means of regression. Each response variable was independently regressed on latitude, longitude and elevation. Interaction effects of location variables were not included due to high multicollinearity. Separate models were fitted for each $CO_2$ treatment.

## 3. Results

### 3.1 $CO_2$ EFFECTS

Net photosynthesis ($P_n$) and intercellular $CO_2$ concentration ($C_i$) were significantly influenced by $CO_2$ concentration (p=0.0033 and p<0.0001, respectively). Averaged over all half-sib sources, $P_n$ was increased by 19% and 49% at 525 and 700 µmol mol$^{-1}$ $CO_2$, relative to 350 µmol mol$^{-1}$ $CO_2$ (Figure 1). Increased $P_n$ was associated with increased $C_i$ of 32% at 525 µmol mol$^{-1}$ $CO_2$ and 88% at 700 µmol mol$^{-1}$ $CO_2$ (Figure 1). In contrast, stomatal conductance ($g_s$) did not change significantly with $CO_2$ concentration (p=0.434) although mean $g_s$ at 700 µmol mol$^{-1}$ $CO_2$ was almost 18 percent less than that at 350 µmol mol$^{-1}$ $CO_2$ (Figure 1).

### 3.2 SOURCE EFFECTS

Among sources, $P_n$ varied from 5.54 µmol m$^{-2}$ s$^{-1}$ in source 3589 to 6.91 µmol m$^{-2}$ s$^{-1}$ in source 4266, with the source effect being significant (p=0.0168, Figure 2). As with $CO_2$ response, $g_s$ did not vary significantly among sources (p=0.216). In contrast to the $CO_2$ response of $P_n$, significant source variation in $C_i$ was not evident (p=0.291).

### 3.3 $CO_2$ X SOURCE INTERACTION EFFECTS

Growth response to elevated $CO_2$ differed markedly among sources leading to significant $CO_2$ x source interaction effects on seedling height (p=0.0443), diameter (p=0.0193), and stem volume (p=0.0072). Of the 15 sources studied, only 7 sources had a positive growth response to elevated $CO_2$. Of the 7 sources that showed a positive response, 4 of these sources demonstrated a significant height response, 6 sources demonstrated a significant diameter response, and 7 sources demonstrated significant volume response to $CO_2$ concentration. No significant growth response was evident for 8 of the 15 sources. Of those sources having a significant response to elevated $CO_2$, the height response tended to be at a maximum at 525 µmol mol$^{-1}$ $CO_2$ and decreased with a further increase to 700 µmol mol$^{-1}$ $CO_2$ (Figure 3); diameter tended to increase linearly with increasing $CO_2$ concentration (Figure 3). Depending on the relative strength of diameter and height responses, the stem volume response was either linear (5 sources) or quadratic (2 sources; Figure 3).

### 3.4 SOURCE PERFORMANCE IN RELATION TO GEOGRAPHIC ORIGIN

There was evidence for a geographic pattern in physiological performance that differed with $CO_2$ concentration. At 350 and 525 µmol mol$^{-1}$ $CO_2$, $P_n$ was not significantly influenced by elevation, latitude or longitude of source origin (Table II). However, at 700 µmol mol$^{-1}$ $CO_2$, $P_n$ increased with elevation and with latitude (Table II). Even with this significant response, geographic origin only accounted for 14% of the variation in observed $P_n$ (Table II). There were no significant origin effects on $g_s$ or $C_i$.

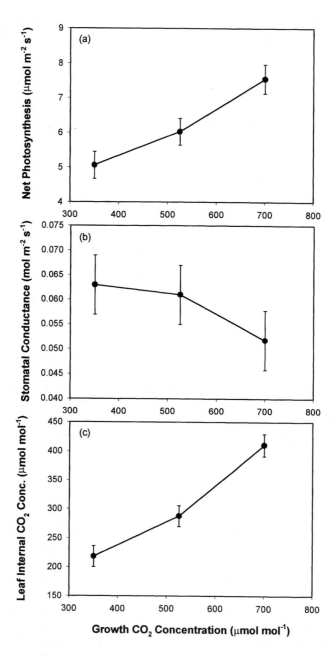

Fig. 1. Physiological responses of ponderosa pine seedlings to $CO_2$ concentration: a) net photosynthesis, b) stomatal conductance, and c) internal leaf $CO_2$ concentration. Values are means and standard errors of n=89-90 observations. Gas exchange measurements were made at saturating light conditions, ambient temperature, and growth $CO_2$ concentrations. Net photosynthesis and internal $CO_2$ concentration responses to $CO_2$ were linearly significant (p=0.001 and p=0.0001, respectively) as determined by orthogonal contrasts.

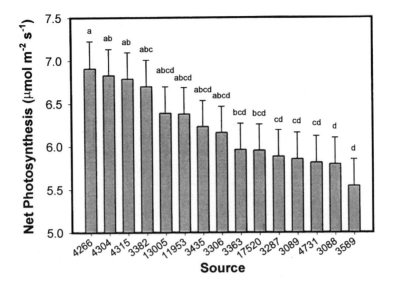

Fig. 2. Net photosynthesis by half-sib sources of ponderosa pine seedlings. Values are means and standard errors of n=17-18 observations. Bars denoted by a common letter do not differ at the p=0.05 level of significance as determined by the L.S.D. criterion. Overall source effect is significant at the p=0.017 level.

In contrast to physiological performance, growth was more strongly and consistently influenced by source geographic origin. Seedling size tended to decrease with increasing elevation, latitude, and longitude. As illustrated for stem volume, geographic variables accounted for 32 to 35% of the observed variation in growth regardless of $CO_2$ concentration (Table II). Under elevated $CO_2$, elevation and latitude were the strongest predictors of stem volume; at 350 µmol mol$^{-1}$ $CO_2$, longitude accounted for a substantial portion of observed variation in stem volume.

## 4. Discussion

The substantial among-population variation in the growth response to elevated $CO_2$ is consistent with earlier long-term studies with *P. ponderosa* (Houpis et al., 1988; Betsche, 1994; Ceulemans et al., 1996), although our data were measured over a greater number of sources and the extent of variability was greater. The variability in growth response to elevated $CO_2$ cannot be attributed to geographic location as indicated by trends for increasing growth with decreasing latitude and elevation that were common to seedlings at all $CO_2$ concentrations. Furthermore, those sources that did demonstrate a positive growth response included representatives of all USDA Forest Service breeding zones, covered the full range of latitudes included in the experiment and covered all but the lowest elevations included in the experiment (Table I and Figure 3).

In contrast to the growth response, seedlings from all sources demonstrated significantly increased photosynthesis and minor decreases in stomatal conductance at elevated $CO_2$ concentrations. Geographic origin had little effect on gas exchange rates.

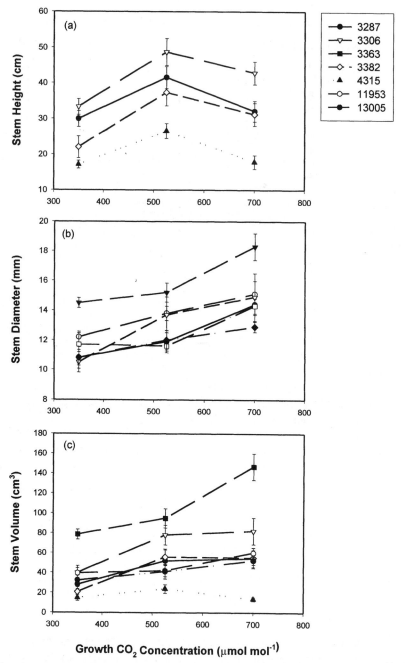

Fig. 3. Growth response to $CO_2$ concentration by half-sib ponderosa pine sources demonstrating significant a) stem height, b) stem diameter, or c) stem volume response. Values are means and standard errors of n=5-6 observations. Stem height responses were significantly quadratic and stem diameter response were significantly linear as determined by orthogonal contrasts at the p=0.05 level of significance. Stem value responses were either linear (sources 3287, 3306, 3363, 11953, 13005) or quadratic (sources 3382, 4315).

TABLE II.

Regressions of net photosynthesis and stem volume on geographic location of origin for 350, 525 and 700 µmol mol⁻¹ $CO_2$ treatments. Values are parameter coefficients (coeff.) and the probability that coefficients differ from zero (Prob.>$H_o$). Estimates of model fit include overall model significance (p), coefficient of determination ($R^2$), and standard error of estimation ($S_{y \cdot x}$).

| | Regression Models | | | | | |
|---|---|---|---|---|---|---|
| | Net Photosynthesis (µmol m⁻² s⁻¹) $CO_2$ Treatment | | | | | |
| | 350 µmol m⁻³ | | 525 µmol m⁻³ | | 700 µmol m⁻³ | |
| Source | Coeff. | Prob>$H_o$ | Coeff. | Prob>$H_o$ | Coeff. | Prob>$H_o$ |
| Intercept | 13.74 | 0.682 | -26.14 | 0.564 | -42.08 | 0.275 |
| Elevation | 0.00013 | 0.580 | 0.00024 | 0.440 | 0.00068 | 0.011 |
| Latitude | 0.155 | 0.286 | -0.075 | 0.704 | 0.360 | 0.033 |
| Longitude | -0.129 | 0.672 | 0.279 | 0.498 | 0.274 | 0.433 |
| | Model Fit | | | | | |
| | Model Signif. | | | | | |
| | p=0.615 | | p=0.883 | | p=0.0012 | |
| $R^2$ | 0.021 | | 0.008 | | 0.138 | |
| $S_{y \cdot x}$ | 1.561 | | 2.116 | | 1.794 | |
| | Stem Volume (cm³) $CO_2$ Treatment | | | | | |
| | 350 µmol m⁻³ | | 525 µmol m⁻³ | | 700 µmol m⁻³ | |
| Source | Coeff. | Prob>$H_o$ | Coeff. | Prob>$H_o$ | Coeff. | Prob>$H_o$ |
| Intercept | 39.53 | 0.0006 | 25.58 | 0.022 | 25.17 | 0.043 |
| Elevation | -0.00044 | 0.0001 | -0.00035 | 0.0001 | -0.00037 | 0.0001 |
| Latitude | -0.092 | 0.057 | -0.163 | 0.001 | -0.221 | 0.0001 |
| Longitude | -0.251 | 0.014 | -0.114 | 0.255 | -0.091 | 0.415 |
| | Model Fit | | | | | |
| | Model Signif. | | | | | |
| | p=0.0001 | | p=0.0001 | | p=0.0001 | |
| $R^2$ | 0.326 | | 0.319 | | 0.355 | |
| $S_{y \cdot x}$ | 0.517 | | 0.515 | | 0.573 | |

with the exception of increasing photosynthesis with increasing elevation of origin when grown at 700 µmol mol⁻¹ $CO_2$. At this point we speculate that in sources from higher elevations, stomatal regulation and water-use efficiencies may not be well-adapted to the high evaporative demand at the low-elevation experimental site. Conversely, low-elevation sources may be more adapted to high evaporative demand, and therefore may have evolved more conservative gas-exchange characteristics. Since all seedlings were given an adequate water supply, less conservative stomatal regulation by high elevation sources may lead to increased assimilation rates compared to low elevation sources. Grulke *et al.* (1993) demonstrated photosynthetic acclimation, while others have recently demonstrated that ponderosa pine of desert and montane origins have distinctly different water-relations, C assimilation, and C allocation characteristics that result in intraspecific differences in productivity (DeLucia *et al.*, 1988; Callaway *et al.*, 1994; Carey *et al.*, 1997; Carey *et al.*, 1998).

Together, the observed growth and photosynthesis responses suggest that with elevated $CO_2$, increased C acquisition rates are possible throughout the California ponderosa pine genome. However, increased C acquisition will inconsistently translate to increased growth due to intraspecific variation in allocation of C to structural and transient sinks. Furthermore, intraspecific variation in allocation patterns may be associated with variation in the metabolic regulation of assimilate translocation (Pushnik *et al.*, 1998). Thus, in order to capitalize on future growth potential associated with elevated $CO_2$, genotypic differences must be considered.

## Acknowledgements

In loving memory of a Great Father, Louis H. Houpis.

## References

Acock, B. and Allen, L.H.: 1985, in B. Strain and J. Cure (eds.), *U.S. Direct effects of increasing carbon dioxide on vegetation*, Department of Energy, Washington, D.C., USA, pp 53-97.
Betsche, T.: 1994, *Environ. Exper. Bot.* 34,75-86.
Callaway, R.M., DeLucia, E.H., Thomas, E.M., and Schlesinger, W.H.: 1994, *Oecologia* 98,159-166.
Carey, E.V., Callaway, R.M., and DeLucia, E.H.: 1997, *Oecologia* 111, 19-25.
Carey, E.V., Callaway, R.M., and DeLucia, E.H.: 1998, *Ecology* 79, 2281-2291.
Ceulemans, R., Jiang, X.N., and Shao, B.Y.: 1995, *Biogeogr.* 22, 261-268.
Ceulemans, R., Shao, B.Y., Jiang, X.N., and Kalina, J.: 1996, *Tree Phys.* 16, 61-68.
Cramer, W.P. and Leemans, R.: 1993, in A.M. Solomon and H.H. Shugart (eds.) *Vegetation Dynamics and Global Change*, Chapman and Hall, New York, USA, pp.190-217.
DeLucia, E.H., Schlesinger, W.H., and Billings, W.D.: 1988, *Ecology* 69, 303-311.
DeLucia, E.H., Callaway, R.M., and Schlesinger, W.H.: 1994, *Tree Phys.* 14, 669-676.
Eamus, D. and Jarvis, P.G.: 1989, *Adv. In Ecol. Res.* 19, 1-55.
Grulke, N.E., Hom, J.L., and Roberts, S.W.: 1993, *Tree Phys.* 12, 391-401.
Houpis, J.L.J., Surano, K.A., Cowles, S., and Shinn J.H.: 1988, *Tree Phys.* 4, 187-193.
Houpis, J.L.J., Pushnik, J.C., Anschel, D.J., and Anderson, P.D.: 1995, Air and Waste Manage Association, Pittsburgh, PA, USA, 16p.
Kellison, R.C. and Weir, R.J.: 1987, in W.E. Shands and J.S. Hoffman (eds.), *The Greenhouse Effect, Climate Change, and U.S. Forests*, The Conservation Foundation, Washington D.C., USA, pp.285-293.
Pushnik, J.C., Garcia-Ibilcieta, D., Bauer, S., Anderson P.D., Bell, J., Houpis, J.L.J.: 1999, *Water, Air, and Soil Pollut.* (in press).
Rind, D., Goldberg, R., Hansen, J., Rosenzweig, C., and Ruedy, R.: 1990, *J. Geophys. Res.* 95, 9983-10004
Rogers, H.H., Heck, W.W., and Heagle, A.S.: 1983, *Air Poll. Control Assoc.* 33, 42-44.
Rogers, H.H., Reunion, G.B., and Krupa, S.V.: 1994, *Environ. Poll.* 83, 155-189.
Sedjo, R.A.: 1989, *Environment* 31,15-20.
Tolley, L.C. and Strain, B.R.: 1984, *Can. Bot.* 62, 2135-2139.
von Caemmerer, S. and Farquhar, G.D.: 1981, *Planta* 153, 376-387.
Waring, R.H. and Schlesinger, W.H.: 1985, Academic Press Inc. Orlando, Fl., USA, 340pp.
Wayne, P.M. and Bazzaz, F.A.: 1995, *Global Change Biol.* 1, 315-324.
Weber, W.E., Wricke, G., and Westermann, T.: 1996, in M.S. Kang and H.G. Gauch, Jr. (eds.) *Genotype by Environment Interaction*, CRC Press, Inc. Boca Raton, FL, USA, pp. 354-371.

# BIOCHEMICAL RESPONSES AND ALTERED GENETIC EXPRESSION PATTERNS IN PONDEROSA PINE (*Pinus ponderosa* Doug ex P. Laws) GROWN UNDER ELEVATED $CO_2$.

J. C. PUSHNIK[1], D. GARCIA-IBILCIETA[1], S. BAUER[1],
P. D. ANDERSON[2], J. BELL[1] and J.L.J. HOUPIS[3]

[1]*Department of Biological Science, California State University, Chico CA, USA 95929-0515, USFS Forest Science Laboratory, Rhinelander WI, USA 54501,* [3] *Director of Environmental Studies, Southern Illinois University, Edwardsville, IL, USA 62026-1651*

(Received 30 November 1998; accepted 25 February 1999)

**Abstract.** Biochemical and gene expression changes in response to elevated atmospheric $CO_2$ were investigated in five maternal half-sibling breeding families of Ponderosa pine. Seedlings were grown in a common garden located at Lawrence Livermore National Laboratory, in open-topped chambers (OTC) for two years. Chamber atmospheres were maintained at ambient, ambient + 175 $\mu L\ L^{-1}\ CO_2$, or ambient + 350 $\mu L\ L^{-1} CO_2$. Growth measurements showed significant increases in stem volumes and volume enhancement ratios in three of the five families studied when grown under elevated $CO_2$. Biochemical and gene expression studies were undertaken to gain a mechanistic understanding of these phenotypic responses. Biochemical studies focused on sucrose phosphate synthase (SPS) specific activities at increased $CO_2$ levels. Kinetic evaluations of SPS showed an increase in $V_{Max}$. Specific SPS probes revealed increases in the transcriptional levels of one SPS gene with exposure to increasing $CO_2$. RT-PCR differential gene displays showed that overall only a small fraction of visualized gene transcripts responded to elevated $CO_2$ (8-10%). There were also significant differences between the gene expression patterns of the different families, some of which correlated with alterations in growth at elevated $CO_2$ levels.

**Keywords**: Ponderosa pine, elevated $CO_2$, growth, gene expression

## 1. Introduction

It has been widely accepted that atmospheric $CO_2$ concentration will increase from our present level of 360 $\mu L\ L^{-1}$ to a level of 550 $\mu L\ L^{-1}\ CO_2$ by the middle of the next century (Levine, 1992; Watson *et al.*, 1990). The effects on the biosphere of this global $CO_2$ increase and related climatic changes are of great concern.

Since the early-1990's a concerted effort has been made to measure the response of long-lived plant species, specifically long-lived forest trees, to increased levels of $CO_2$. Forests occupy 20 to 25 % of the Earth's surface and provide approximately 70% of terrestrial carbon (C) fixation (Waring and Schlesinger, 1985), which makes them a significant C sequestration sink on a planet-wide scale. The response of forest trees to increasing $CO_2$ is, therefore, of major importance for not only forest ecosystem structure but also for large scale feedback interactions with the atmosphere.

The direct linkage between forests and the atmosphere is through the photosynthetic process of C acquisition. Although stimulation of photosynthetic rates by elevated $CO_2$ has been widely observed, photosynthetic acclimation, and down-regulation of C assimilation, in response to elevated atmospheric $CO_2$ has also been demonstrated in no less than 47 different long-lived forest tree species (Gunderson and Wullschleger, 1994). A range of acclimation responses to elevated $CO_2$ have been observed. The crown leaf area has been shown to decrease in several tree species including *Castanea sativa* (Mousseau and Enoch, 1989), *Pinus ponderosa* (Pushnik et al., 1996) and *Liriodendron tulipifera*, in both growth chamber and field experiments (Norby and O'Neill, 1991). A significant linear decline has also been seen in leaf area ratios (LAR) in *Liriodendron tulipifera* (Norby et al., 1992) and *Maranthes corymbosa* (Berryman et al., 1993). Loss of apical dominance (Conroy et al., 1990) and changes in leaf anatomy and morphology, including increases in leaf thickness, are also responses to elevated atmospheric $CO_2$ (Thomas and Harvey 1988, Conroy et al., 1986; Mousseau and Enoch, 1989; Radoglou and Jarvis, 1990, Pushnik et al., 1995).

Only recently have researchers begun to look at the biochemical mechanisms underlying the acclimation process. C allocation within plants is highly regulated by both photosynthetic and cytosolic biochemical pathways (Stitt, 1989; Stitt et al., 1987; Van Oosten et al., 1992; Huber and Huber, 1992, 1991; Huber, 1983). The few biochemical studies that have been conducted under elevated $CO_2$ conditions have focused on the initial C reduction function of ribulose-1,5-bisphosphate carboxylase (Rubisco). The response of Rubisco to elevated $CO_2$ is unclear. Decreases in Rubisco activity have been documented by several research groups (Van Oosten et al., 1992; Tissue et al., 1993; Wilkins et al., 1993). In contrast, increases in Rubisco activity have also been demonstrated (Koch et al., 1986). Tissue et al. (1993), Pushnik et al. (1995), and Moore et al. (1997) observed that, although Rubisco content decreased per unit leaf area, the enzymatic activation state increased under elevated $CO_2$ conditions. Further, Tissue et al. (1993) demonstrated that these plants had less nitrogen (N) invested in Rubisco and more invested in the light-harvesting mechanisms of the plant. This suggests that the allocation of resources had moved from Rubisco towards other rate limiting processes in the C allocation pathway.

As yet no significant research has been aimed at elucidating the response of sucrose phosphate synthase (SPS) and C flow in trees exposed to elevated $CO_2$. SPS has a major role in the biosynthesis of sucrose and therefore has an immense impact on C allocation (Bruneau et al., 1991; Guy et al., 1992; Hampp et al., 1994; Huber 1983; Huber and Huber,1992; Kerr et al., 1987; Stitt, 1989; Stitt et al., 1987). Photoassimilates are translocated through the plant, primarily in the form of sucrose. The translocation occurs from the source of fixation to the utilization and storage sinks, and is dependent on the rate of sucrose synthesis and export. If photosynthetic C fixation rates exceed the capacity of the leaf to export sucrose, a shift toward chloroplastic starch storage occurs with a resultant negative feedback on the assimilation rate. Worrell et al. (1991) have cloned and over-expressed the SPS enzyme in tomato. These plant showed a six fold increase in SPS activity, stored less starch in the leaves and increased sucrose content two fold. However, the effect of increased sucrose synthesis on the development of other sinks has received little attention.

The objective of this research was to evaluate growth enhancement, as well as *in vivo* SPS performance and C allocation in *Pinus ponderosa* seedlings exposed to elevated levels of atmospheric $CO_2$. Ponderosa pine is a canopy dominant species in many Western U.S. coniferous forest ecosystems, and is therefore a significant C sequestering species in North America. Data gathered during these studies further elucidate the effects of increased $CO_2$ on the protein content and responsive gene expression patterns within these trees.

## 2. Materials and Methods

Three-year-old *Pinus ponderosa* seedlings representing five distinct open-pollinated breeding families were used in this study. The parent populations for these breeding families are geographically distributed longitudinally across the Sierra Nevada of California at a mid-elevation (*ca.* 1500 m). Seeds were obtained from the tree improvement program of the United States Forest Service Genetic Resource Center in Chico, California. Seedlings were grown from seed at Lawrence Livermore National Laboratory (Livermore, California). After one year of cultivation in the greenhouse under ambient conditions, these seedlings were transferred to 3 L pots containing a standard potting medium (45% peat, 45% vermiculite and 10% perlite by volume).

Seedlings were then distributed to eighteen standard outdoor open-top exposure chambers (3 m diameter and 3 m in height). They were fertilized using one-half strength Hoagland's solution and well watered throughout the study. Seedlings were continuously exposed to three $CO_2$ concentrations including ambient (*ca.* 350 µL $L^{-1}$ $CO_2$), ambient + 175 µL $L^{-1}$ $CO_2$ (*ca.* 525 µL $L^{-1}$ $CO_2$), and ambient + 350 µL $L^{-1}$ $CO_2$ (*ca.* 700 µL $L^{-1}$ $CO_2$) for two and half years (April 1993 until August of 1995). The concentrations of $CO_2$ in the open-top chambers were monitored using dedicated $CO_2$ analyzers (Horiba Model PIR-2000; all analyzers were zero and span checked daily and underwent a complete multi-point calibration monthly). Chamber atmospheres were sampled twelve times per hour at canopy height, at the center of the chamber. The chamber $CO_2$ concentrations were maintained within 5% of the treatment concentrations, 24 hr per day.

Growth measurements including determination of height, stem diameter and main stem volume were taken at the conclusion of each growing season and at the time of harvest.

Biochemical assessments of SPS (E.C. 2.3.1.14) were accomplished by the procedures of Kerr *et al.* (1987). Enzymatic activity was measured spectrophotometerically as described by Harbron *et al.* (1980). Kinetic parameters including velocity maximum ($V_{max}$) and $K_M$ were determined as described by Stitt *et al.*, (1988) and modified by Huber *et al.*, (1989) and Weiner (1995). $K_M$ values were derived from a Hanes-Woolf plot. Percent SPS activation (%A) state as determined by the ratio of SPS activity under limiting substrate concentrations (2 mM fructose-6-P, 10 mM glucose-6-P and 3 mM UDP-glucose) in the presence of the inhibitory influence of 5 mM $P_i$ versus SPS activity with saturating substrates (4 mM fructose-6-P, 20 mM glucose-6-P and 3 mM UDP-glucose). Aldolase activity was used to standardize the efficiency of enzyme extractions. Protein concentrations in the needle extracts were estimated by the bicinchronic acid method.

At the point of biomass harvest (August, 1995), total RNA was extracted and used from needle tissue of three replicate individuals per family at each of the three $CO_2$ exposure levels. Extraction of total RNA was accomplished by the procedures of Chang *et al.* (1993) followed by a DNase I digestion to remove contamination. Integrity of the

RNA was assessed by formaldehyde and formamide denaturing gel electrophoresis (Farrell,1993). Complementary DNA (cDNA) synthesis was performed by reverse-transcriptase polymerase chains reaction (RT-PCR) using the "Delta RNA Fingerprint Kit" from Clonetec (Palo Alto, Ca) according to the manufactures instructions. Combinations of arbitrary differential display primers (Clonetec) were used in the fingerprinting reactions. PCR programming was 1 cycle: 94°C for 5 min, 40°C for 5 min, 68°C for 5 min; 2 cycle: 94°C for 2 min, 40°C for 2 min, 68°C for 2 min; and 25 cycles: 94°C for 1 min, 60°C for 1min, 40°C for 2 min, 68°C for an additional 7 mins. Reaction products were used for the differential gene displays that were separated on a 6% polyacrylamide sequencing gel and visualized by silver staining.

Specific SPS degenerative oligonucleotide probes were constructed using a conserved amino acid sequence identified in seven known SPS clones (corn, rice, banana, spinach, sugar beet, potato, broad bean). The initial SPS selection from the RT-PCR amplified cDNA was conducted with degenerative oligonucleotide having the following sequences:5'..ACNGGNGGNCAYAGNCTYAG..3'and 5'..GGNCGRCARTCNCCYAANGC..3'; where N represents any nucleotide, Y is a pyrimidines and R is a purine. Sequencing of possible SPS PCR amplified products was conducted at California State University, San Diego Micro Core Facility. Results of the sequencing revealed that the 935 bp cDNA had a high degree similarity (70-82%) with all SPS sequences within the GenBank and EMBL databases. Specific SPS high stringency oligonucleotides primers (5'..TTGGATTAGAGGATTT CGGA and 5'..AAGTTTTATTACGCGGCAT..3') were constructed from a conserved 470bp region and employed to assess mRNA abundance. Specific SPS PCR amplification was conducted as follows: 5 min 95°C, 25 cycles of 1 min at 95°C, 1 min 46°C, 2 min at 70°C, all followed by 7 min extension at 70°C. Reaction products were separated electrophoretically and visualized as described above. Product quantification was determined as the relative change in banding intensity between $CO_2$ treatment levels by densiotometric analysis.

The growth data were analyzed as a split-plot complete factorial design. $CO_2$ concentration was the main plot factor and half-sib family was the split-plot factor. Significance of $CO_2$, family, and $CO_2$ x family interaction effects were determined by analysis of variance. For significant effects, differences among factor level means were determined by Tukey's HSD test at the $p=0.05$ level of significance. Biochemical data were analyzed by a post-hoc ANOVA Tukey's test between treatments.

## 3. Results and Discussion

Enhancement of seedling growth by elevated $CO_2$, expressed as the mean volume at elevated $CO_2$ relative to the mean stem volume at ambient $CO_2$, is summarized in Table I. Enhancement ratios for half-sib *Pinus ponderosa* families 3287 and 3306 indicated significant ($p<0.05$) increases of volume growth of 85 and 96% at 500 $\mu L\ L^{-1}$ $CO_2$ and 94 and 106%, respectively at 700 $\mu L\ L^{-1}$ $CO_2$. In contrast, elevated $CO_2$, at either 525 or 700 $\mu L\ L^{-1}$, had no significant affect on volume growth in families 4266 and 4731. Family 3089 demonstrated an intermediate response, with no significant change in volume when grown at 525 $\mu L\ L^{-1}$, but increased by 33% at 700 $\mu L\ L^{-1}$ $CO_2$.

Statistical analysis of total soluble protein and SPS specific activity (Table II), in general, show significant increases with exposure to elevated $CO_2$. At the 700 µL $L^{-1}$ $CO_2$, a significant increase in total soluble protein ($p<0.014$) was observed. Significant increases in SPS specific activity ($p<0.009$) were also seen for the 525 and 700 µL $L^{-1}$ treatment levels compared with seedlings which were kept under ambient $CO_2$ concentration.

Sucrose phosphate synthase $V_{max}$ differed significantly ($p<0.0$) among all $CO_2$ treatment levels. In detail, $V_{max}$ significantly increased at the 525 µL $L^{-1}$ treatment level compared to the 350 treatment, and the 700 µL $L^{-1}$ treatment significantly increased compared to the 525 µL $L^{-1}$ treatment level. The activation state of SPS also differed significantly ($p<0.02$) among all treatment concentrations. The activation state of the

TABLE I

Mean seedling volume and volume growth enhancement by five half-sib *Pinus ponderosa* families at ambient, 525 µL $L^{-1}$, 750 µL $L^{-1}$ atmospheric $CO_2$

| Family | Mean Stem Volume ($cm^3$) | | | Volume Enhancement Ratio | |
|---|---|---|---|---|---|
| | Ambient | 525 µL $L^{-1}$ | 700 µL $L^{-1}$ | 525 µL $L^{-1}$ | 700 µL $L^{-1}$ |
| 3089 | $47.9 \pm 4.7^a$ | $45.2 \pm 7.1^a$ | $62.1 \pm 17.4^b$ | 0.97 | 1.33 |
| 3306 | $28.0 \pm 4.4^a$ | $51.8 \pm 11.4^b$ | $54.3 \pm 8.8^b$ | 1.85 | 1.94 |
| 3287 | $39.8 \pm 4.8^a$ | $78.1 \pm 9.9^b$ | $82.1 \pm 13.3^b$ | 1.96 | 2.06 |
| 4266 | $27.1 \pm 3.4^a$ | $25.6 \pm 6.4^a$ | $22.3 \pm 3.7^a$ | 0.95 | 0.82 |
| 4731 | $73.0 \pm 7.5^a$ | $76.8 \pm 7.7^a$ | $66.6 \pm 9.1^a$ | 1.05 | 0.91 |

Mean volumes (n = 6) within families denoted by a common letter do not differ (p=0.05) among $CO_2$ treatments based on Tukey's HSD multiple comparison test.

TABLE II

Ponderosa pine biochemical responses to elevated $CO_2$

| $CO_2$ Concentration | Total soluble protein (mg/cm$^2$) | SPS specific activity (nmol mg$^{-1}$min$^{-1}$) | SPS $V_{max}^{app}$ (nmol mg$^{-1}$ min$^{-1}$) | SPS $K_m$ (mM) | Percent Activation |
|---|---|---|---|---|---|
| 350 µL L$^{-1}$ | 4.25± 1.10$^a$ | 2.71± 0.95$^a$ | 2.54± 0.85$^a$ | 10.0± 0.75$^a$ | 55.1±0.5$^a$ |
| 525 µL L$^{-1}$ | 4.63 ± 0.42$^a$ | 4.69± 0.68$^b$ | 6.58± 1.63$^b$ | 12.0± 0.89$^a$ | 69.5±0.8$^b$ |
| 700 µL L$^{-1}$ | 5.76 ± 0.35$^b$ | 5.05± 0.51$^b$ | 9.26± 0.89$^c$ | 11.4±01.0$^a$ | 45.4±1.0$^c$ |

Soluble protein and SPS specific activities represent mean ± standard error values of 30 pooled isolations of each $CO_2$ treatment level. SPS $V_{max}$, $K_m$ and percent activation data represent mean values of 6 isolations at each $CO_2$ treatment level. Means values denoted by a common letter do not differ significantly (p=0.05) among $CO_2$ treatments based on Tukey's HSD multiple comparison test.

525 µL L$^{-1}$ level was significantly higher than both the 350 and 700 µL L$^{-1}$ treatment levels, while the activation state of the 700 µL L$^{-1}$ level was significantly decreased compared to the 350 and 525 µL L$^{-1}$ treatments. SPS specific activity increased significantly at both the 525 and 700 µL L$^{-1}$ exposure levels (Table II). $V_{max}$ data at 525 µL L$^{-1}$ $CO_2$ correlates with the increase in specific activity. The increase in the activation state at 525 µL L$^{-1}$ $CO_2$ without a significant change in soluble protein content, suggests that the increase in both SPS specific activity and $V_{max}$ was probably caused by an unknown mechanism of enzyme activation, harboring no influence on $K_M$ value (See Table II). In contrast, SPS $V_{max}$ at 700 µL L$^{-1}$ was significantly greater but the enzyme activation state was significantly less at 700 µL L$^{-1}$ $CO_2$ compared to both the 350 and 525 µL L$^{-1}$ treatments. To maintain a high specific activity and high $V_{max}$ with a corresponding decrease in the enzyme activation state is likely to require increased SPS per needle tissue area, which is supported by the significant increase in total soluble protein content found at the 700µL L$^{-1}$ level. Taken as a whole, these data suggest that multiple SPS regulatory mechanisms not influencing enzyme substrate affinity are expressed in Ponderosa pine that maintain for the same alternative C allocation strategies responsive to concentration of atmospheric $CO_2$. The apparent allocation strategy at 525 µL L$^{-1}$ $CO_2$ is to increase SPS specific activity and $V_{max}$ by increasing the activation state of SPS, but without significant increases in the amount of enzyme present in the needle tissue. Conversely, due to the high inactivation at the 700 µL L$^{-1}$ level, increases in SPS activity and $V_{max}$ must be due to an increase in the amount of SPS present within the pathway.

The genetic basis of SPS response to elevated $CO_2$ was also examined. If genetic up-regulation was occurring (transcriptional), one would expect to find increased SPS mRNA levels within the plant and corresponding increased SPS protein. If the response was wholly translational, no changes in SPS mRNA level would occur, but alterations in SPS enzyme kinetics and amounts of proteins might result. To resolve this issue total RNA was isolated and used with specific oligonucleotide probes to examine the abundance of SPS mRNA under the contrasting $CO_2$ environments. The results of the

RT-PCR experiment are shown in Figure 1. The identity of these bands indicating SPS transcripts were confirmed by sequence analysis. Banding intensities for families that displayed positive volume enhancements to the elevated $CO_2$ atmospheres (3089 and 3287) indicated 3.0 and 2.25 fold increased mRNA abundance at the elevated $CO_2$ level. Family 3306 showed a slight decrease in banding intensity when the trees were grown at elevated $CO_2$ but still displayed increased volume enhancement. Family (3306) maintained the highest level of SPS expression under all experimental conditions tested. Family 4266 displayed a decrease in volume and did not exhibit changes in SPS gene expression at any of the $CO_2$ exposure levels. The one anomalous result was seen in family 4731 that displayed an increased abundance of the message but exhibited a decrease in stem volume growth under elevated $CO_2$. It is likely that growth response to elevated $CO_2$ may be regulated by several genetic mechanisms working in concert, such that a high level of SPS expression alone does not determine increased growth performance.

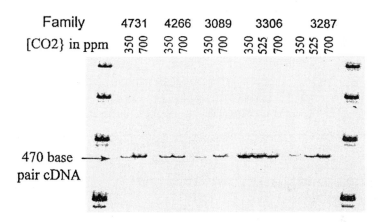

Fig. 1. Sucrose-phosphate specific expression in response to growth under elevated CO2. The bands represent RT-PCR amplified mRNAs using specific oligonucleotide probes constructed using known conserved sequence information from seven distinct plant species. All lanes are equally loaded with 5 µg cDNA quantified spectrophotometrically. The bands identity was confirmed by sequence analysis.

Differential gene displays employing generalized degenerative primers in the RT-PCR reaction were conducted to estimate the magnitude of changes in differential expressed genes associated with the various growth responses. A sample of these results are shown in Figure 2 and quantified in Table III . Approximately 90 % of the cDNA products are not affected by $CO_2$ exposure, the remaining 10% are equally split between gene products that are transcriptionally up-regulated and those that exhibit a decreased expression level. Some or all of these differential expressed genes may be involved as quantitatively expressed loci that modulate the differences in growth performance to elevated $CO_2$ observed within this species.

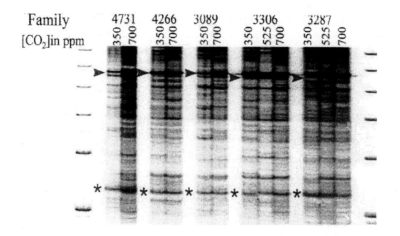

Fig. 2. RNA finger print showing evidence of differentially responsive gene expression patterns to elevated $CO_2$ exposure. These gene expression patterns are derived from geographically distinct breeding families. The arrow indicates cDNA which was up-regulated, while the asterisk points to a non-responsive cDNA.

TABLE III

Quantitative results of the differential gene display studies

|  | Family 4731 | Family 4266 | Family 3089 | Family 3306 | Family 3287 |
| --- | --- | --- | --- | --- | --- |
| Number of Bands Sampled | 131 | 124 | 114 | 114 | 109 |
| Percentage Remaining Constant | 89 | 92 | 91 | 87 | 89 |
| Percentage Up-regulated | 6 | 4 | 5 | 8 | 7 |
| Percentage Down-regulated | 5 | 4 | 4 | 5 | 4 |

Quantification of each single band obtained from the RNA fingerprint experiments depicted in Figure 2 was done as described in Methods and Materials. Up-regulated and down-regulated transcripts found under elevated $CO_2$ (700 µL L$^{-1}$) are indicated in percentage. Controls were performed at ambient atmospheric $CO_2$.

This research has demonstrated significant intraspecific variability in the responses of ponderosa pine half-sib families to changes in the atmospheric concentration of $CO_2$. These intraspecific differences correlate with biochemical responses and gene expression levels of sucrose-phosphate synthase, a key regulatory enzyme involved in carbon allocation. Increased growth response to elevated atmospheric $CO_2$ is most certainly a multigenic response. Differential gene displays suggest that about 10% of expressed genes are being differentially regulated in response to elevation of the $CO_2$ concentration in the atmosphere. Identification of the suite of differentially expressed genes responding to elevated $CO_2$ might serve as breeding markers in tree improvement programs or as a starting point for the bio-engineering of trees that will be grown under future atmospheres.

We have shown the existence of significant genetic variability in the response of a tree species to elevated $CO_2$. The genetic complexity and variability in the responses of long-lived species to elevated $CO_2$ will most likely be greater than that for annual species. Thus, using data solely obtained from annual crop species, with agronomically selected genetic backgrounds, will likely not be representative of the response of natural ecosystems to elevated $CO_2$. Further information on genetic variability residing in natural forest ecosystems is desperately needed, if we are to predict changes, and ultimately devise appropriate management practices.

## Acknowledgements

This research was supported by grant from the Western Regional Center of the National Institute of Global Environmental Change (WESGEC) 94-082.

## References

Berryman, C.A., Eamus, D. and Duff, G.A.,: 1993. *Australian J. Bot.* **41**: 195-209.
Bruneau, J.M., Worrell, A.C., Cambou, B., Lando, C. and Voelker, T.A.: 1991. *Plant Physiol* **96**: 473-478.
Chang, S., Puryear, J, and Cairney, J.:1993. *Plant Mol. Biol. Rep.* 11: 113-116.
Conroy, J.P., Barlow, E.W.R., and Bevege, D.I.: 1986. *Annals of Bot.* **57**: 165-177.
Conroy, J.P., Millham, P.J., Mazur, M. and Barlow, W.R.: 1990. *Plant, Cell Environ.* **13**: 329-337.
Doehlert, D.C. and Huber, S. C. :1983. *Plant Physiol* **73**: 989-994.
Farrell, R.E. Jr.: 1993. *In RNA Methodologies*, Acad. Press San Diego.
Gunderson, C.A. and Wullschleger, S. D.: 1994. *Photosyn Res* **39**: 369-388.
Guy, C.L., Huber, J.L. and Huber, S.C.: 1992. *Plant Physiol* **100**: 502-508.
Hampp, R., Egger, B., Effenburger, S. and Elnig, W. 1994. *Physiol Plant* **90**: 299-306.
Harbron, S., Woodrow, I..E., Kelly, G.J., Robinson, S.P., Latzko, E. and Walker, D.A. :1980. *Anal Biochem* **107**: 56-59.
Huber, J.L., Huber, S.C. and Neilsen, T.H.: 1989. *Arch Biochem Biophys* **270**: 681-690.
Huber, S.C.: 1983. *Plant Physiol* **71**: 818-821.
Huber, S.C. and Huber, J.L.: 1991. *Plant, Cell Physiol* **32**: 319-326.
Huber, S.C. and Huber, J.L.: 1992. *Plant Physiol* **99**: 1275-1278.
Kerr, P.S., Kalt-Torres, W. and Huber, S.C.: 1987. *Planta* **170**: 653-658.
Koch, K., Jones, P., Avigne, W. and Allen, L.: 1986. *Physiol. Plantarum* **68**: 477-484.
Levine, J.S.: 1992. Global climate change. P Firth, SC Fisher, eds, *Global Climate Change and Freshwater Ecosystems*. Springer-Verlag, Heidelberg, pp 1-25.
Moore, B.D., Palquist, D.E. and Seemann, J.R.: 1997. *Plant Physiol* **115**: 241-248
Mousseau, M. and Enoch, H.Z.,: 1989. *Plant, Cell Environ.* **12**: 927-934
Norby, R.J. and Oneil, E.G. : 1991. *New Phytologist* **117**: 515-528.

Norby, R.J., Gunderson, C.A., Wullscheger, S.D., O'Neill, E.G. and McCarken, M.K.:1992. *Nature* **327**:322-324.
Pushnik, J.C., Demaree, R.S., Houpis, J.L., Flory, W.B., Bauer, S.M. and Anderson, P.D.: 1995. *J. of Biogeography* **22**: 249-254
Pushnik,J.C., Bauer,S., Demaree, R.S., Houpis, J.L.J., Anderson, P.D. and Smith, R.L.: 1996.*Proceedings of the 16$^{th}$ Meeting for Specialist in Air Pollution Effects on ForestEcosystems IUFRO.* pp.69-81
Radoglou, K.M. and Jarvis, P.G.: 1990. *Annals of Bot.* **65**: 617-625.
Stitt, M.: 1989. *Phil Trans R Soc Lond* **323**: 327-338.
Stitt, M., Huber, S.C. and Kerr, P.S.: 1987. *In* Akazawa, Okamoto, eds, *The Biochemistry of Plants*, Vol. 10. Academic Press, NY, pp327-409
Stitt, M., Wilke, I., Feil, R. and Heldt, H.W.:1988. *Planta* **72**: 217-230
Stuiver, M.: 1978. *Science* **199**: 253-258
Tissue, D.T., Thomas, R.B. and Strain, B.R.: 1993. *Plant, Cell Environ* **16**: 859-865
Thomas, J.F. and Harvey, C.N.: 1988. *Bot. Gaz.* **144**: 303-309.
Van Oosten, J.J., Afif, D. and Dizengreme,l P.: 1992. *Plant Physiol Biochem* **30**: 541-547
Waring, R.H. and Schlesinger, W.H.: 1985. *Forest Ecosystems: Concepts and Management*. Academic Press Inc.
Watson, R.T., Rodhe, H., Oescheger, H. and Siegenthaler, U.: 1990. *In* J Houghton, G Jenkins, J Ephraums, eds, *Climate Change: The IPCC Scientific Assessment*, Cambridge University Press, Cambridge, pp 1-40
Weiner, H.: 1995. *Plant Physiol.* **108**: 219-225.
Worrell, A.C., Bruneau, J.M., Summerfelt, K., Boersig, M. and Voelker, T.A.: 1991. *Plant Cell* **3**: 1121-1130.

# TREATMENT OF YOUNG SPRUCE SHOOTS WITH $SO_2$ AND $H_2S$: EFFECTS ON FINE ROOT CHROMOSOMES IN RELATION TO CHANGES IN THE THIOL CONTENT AND REDOX STATE

ASTRID WONISCH[1], MICHAEL TAUSZ[1], MARIA MÜLLER[1], WILFRIED WEIDNER[1], LUIT J. DE KOK[2] and DIETER GRILL[1]

[1] *Institute of Plant Physiology, University of Graz, Schubertstraße 51, A-8010 Graz, Austria;* [2] *Department of Plant Biology, University of Groningen, P.O. Box 14, 9750 AA Haren, The Netherlands*

(Received 30 October 1998; accepted 25 February 1999)

**Abstract.** Three year old spruce trees (*Picea omorika*) were exposed to 100 and 225 nl $l^{-1}$ $SO_2$ and $H_2S$ for three weeks. The number of chromosomal aberrations and the mitotic index in the root tip meristems, and glutathione and cysteine contents in fine roots were determined twice weekly. An increase in glutathione content in fine roots of $H_2S$ exposed plants was only detectable after 13 days of fumigation. The number of chromosomal aberrations increased significantly after 9 days of exposure to 225 nl $l^{-1}$ $H_2S$ and after 13 days of exposure to 225 nl $l^{-1}$ $SO_2$ or 100 nl $l^{-1}$ $H_2S$. This increase in chromosomal damage persisted up to the end of the 3 week treatment. Neither $SO_2$ nor $H_2S$ exposure affected the cysteine content or the redox state of glutathione in fine roots. These results suggest that the development of chromosomal aberrations during $SO_2$ and $H_2S$ exposures does not directly reflect changes in thiol/glutathione content or redox state in the fine roots.

**Keywords:** sulfur, redox status, thiol, chromosome, glutathione, *Picea omorika*

## 1. Introduction

In today's environment plants may have to deal with elevated levels of atmospheric $SO_2$ and $H_2S$, both of which may cause significant changes in plant growth and development. The phytotoxicity and involvement of $SO_2$ in damaging forest trees has been well documented (De Kok, 1990). Plant shoots provide a sink for $SO_2$ and $H_2S$, which may be metabolized and used as a sulfur (S) source for plant growth (De Kok *et al.*, 1997; 1998; Tausz *et al.*, 1998). Prolonged exposure may result in increased S levels in the leaves, partially in the form of glutathione (Grill *et al.*, 1982, De Kok *et al.*, 1998). Many studies have dealt with these effects of sulfurous air pollutants on herbaceous plants and shoots of trees, however, few have studied effects of $H_2S$ on trees (De Kok *et al.*, 1989) and on tree roots (Tausz *et al.*, 1998). Recent investigations have shown that exposure of spruce seedlings to $SO_2$ pollution results in an increased amount of damage to chromosomes in dividing cells of the root meristems (Müller *et al.*, 1998a). The induction of such chromosomal aberrations in the roots by air pollutants is still poorly understood. It has been observed that exogenously applied glutathione induced an increase in the number of chromosomal aberrations in the root meristems of young spruce seedlings (Müller, unpublished results). Furthermore, glutathione was found to exert mutagenic effects on tissues (Glatt *et al.*, 1983). Glutathione accumulated in the foliage upon exposure to $SO_2$ and $H_2S$ may be transported to the roots via the phloem and may result in enhanced glutathione levels in the roots (De Kok *et al.*, 1997). It may therefore be assumed that chromosomal aberrations can be induced by enhanced levels of glutathione in the roots. The present study focused on the impact of $SO_2$ and $H_2S$ on fine roots of spruce seedlings and addressed the following questions: (1) How quickly do concentrations of glutathione and cysteine respond to $SO_2$ and $H_2S$ fumigation? (2) What is the time course of the development of chromosomal aberrations? (3) Do the occurrence of chromosomal aberrations coincide with changes in the thiol/glutathione content and redox state of fine roots?

## 2. Materials and Methods

### 2.1. PLANT MATERIAL

Three years old spruce trees (*Picea omorika* [Pancic] Pyrkyne) were obtained from Boomkwekerij Zundert B. V., Meirseweg 45, 4881 MJ Zundert, The Netherlands. Trees were potted using 14-cm diameter-pots in commercial potting soil without the addition of a fertilizer. Plants were kept in the greenhouse prior to the experiment.

### 2.2. FUMIGATION CONDITIONS

The $SO_2$ and $H_2S$ fumigations were carried out at the University of Groningen, The Netherlands (September 1995). Five treatments were applied: 1. Control treatment (ambient air, background level of both S gases below 2 nl $l^{-1}$), 2. $SO_2$ (low level, 100 nl $l^{-1}$), 3. $SO_2$ (high level, 225 nl $l^{-1}$), 4. $H_2S$ (low level, 100 nl $l^{-1}$), 5. $H_2S$ (high level, 225 nl $l^{-1}$). Plants were fumigated in 150 l cylindrical stainless steel cabinets with polycarbonate tops. Pressurized $H_2S$ and/or $SO_2$ diluted with $NO_2$ were injected into the incoming air stream at the desired concentrations by ASM electronic mass flow controllers (Bilthoven, The Netherlands). The air exchange rate in the cabinets was 40 l $min^{-1}$ and the air inside the cabinets was circulated by a fan (with an air movement capacity of 20 l $s^{-1}$) to reduce the boundary layer resistance of the leaves and facilitate gaseous uptake. The air temperature was maintained at $19 \pm 2$ °C and the relative humidity at $40 \pm 10$ %. The photoperiod was provided 14 h $day^{-1}$ and the photon flux density at plant height was 200-300 µmol $m^{-2}$ $s^{-1}$ (within the 400-700 nm range). A detailed description of the fumigation system is given in Tausz *et al.* (1996).

### 2.3. SAMPLING

Plant material was sampled on six occasions for chromosomal analysis, and five for biochemical analysis during the three week long exposure. Six trees were sampled from each treatment on each sampling occasion. Glutathione concentrations and chromosomal parameters were determined on fine roots (< 1 mm diameter).

### 2.4. BIOCHEMICAL ANALYSES

Fine roots were immersed in liquid nitrogen and lyophilized. The lyophilized material was powdered in a dismembrator, and thiols were determined according to the procedure of Kranner and Grill (1993). About 100 mg of plant material was extracted in 2 ml 0.1 mol $l^{-1}$ HCl prepared with 100 mg of PVPP (Polyvinylpolypyrrolidone) to remove phenolics. The derivation procedure and HPLC system were described previously (Kranner and Grill, 1993). Thiol concentrations in extracts of lyophilized plant powder were consistently higher (by about 10%) than in extracts prepared on fresh material according to the original procedure. This improved recovery could reflect a better extraction of the fine powder, compared to extraction using a mortar and pestle.

## 2.5. CHROMOSOMAL ANALYSES

Sample preparation and quantification of chromosomal aberrations and the mitotic index were performed according to Müller *et al.* (1998a). Root tips were cut and fixed in ethanol:glacial acetic acid (3:1, v/v). After fixation, the root tips were hydrolysed in 3 M HCl for 3 min at 63 °C, stained in freshly-prepared Schiff's reagent and squashed in a few drops of 45% acetic acid. Each aberrant meta- or anaphase was counted as one aberration and the percentage in total meta- and anaphases (a minimum of 200 cells per replicate tree) was calculated. Mitotic index was determined semi-automatically using an Optimas image analysis system, based on a minimum of 1000 cells per sample.

## 2.6. STATISTICS

At each sampling date, differences between treatments and control (n=6 replicates) were evaluated using the Mann-Whitney U-test. Calculations were performed with the help of Statistica (StatSoft, Tulsa, OK, USA) software package. Figures show medians and median deviations which are most suitable for small sample sizes (Sachs, 1992). Data points with less than six replicates were excluded from the analysis and the figures.

## 3. Results and Discussion

A three week exposure of *Picea omorika* to 100 and 225 nl $l^{-1}$ $SO_2$ and $H_2S$ did not affect tree growth (judged by tree height and root and shoot weight) and it did not result in symptoms of visible injury. Neither 100 nor 225 nl $l^{-1}$ $SO_2$ exposure affected the glutathione content of the fine roots (Figure 1). On the other hand, as reported previously, the glutathione content of fine roots of trees exposed to 225 nl $l^{-1}$ $H_2S$ was significantly increased after 20 days (Tausz *et al.*, 1998). The present time course analysis shows that both $H_2S$ concentrations induced a significant increase in glutathione after 13 days, but a further accumulation at prolonged exposure was only observed at 225 nl $l^{-1}$ $H_2S$ (Figure 1). The accumulation of glutathione in the roots of spruce trees exposed to $H_2S$ is likely to reflect a surplus of reduced S in the foliar tissue. Part of the accumulated glutathione in the needles following $H_2S$ exposure might be transported to the roots. In many herbaceous and woody plants such transport of reduced S to the roots in form of glutathione has been demonstrated (Rennenberg and Lamoureux, 1990). Alternatively, the roots might synthesize their own glutathione and export to the shoots might be reduced due to a diminished demand in the leaves, when an excess of reduced S from $H_2S$ is available. The latter possibility is supported by a lack of basipetal phloem transport of glutathione in spruce trees (Rennenberg and Herschbach, 1995). The cysteine content of the fine roots remained unaffected by $H_2S$ (and $SO_2$) exposure and was low compared to that of glutathione (maximum 10 nmol $g^{-1}$ dry weight). Glutathione is not only a central compound in S metabolism, but also an important antioxidant. Therefore, its redox state might be a significant factor for the regulation of several metabolic processes (Kunert and Foyer, 1993). Enhanced levels of oxidized glutathione in plant tissue can indicate oxidative stress (Polle and Rennenberg, 1994). However, upon $SO_2$ and $H_2S$ exposure glutathione was mainly present in its reduced form and the proportion of oxidized glutathione varied between 10 and 15%, which is comparable to other results on different plant tissues (Polle and Rennenberg, 1994).

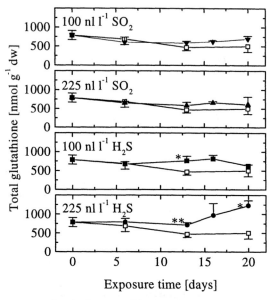

*Fig. 1.* Total glutathione concentrations in fine roots of fumigated spruce trees. Open squares indicate control, closed symbols exposure variant. Medians and median deviations of n=6 replicate trees. Significance levels for differences from control * $p<0.05$, ** $p<0.01$.

Due to recent studies (Sanchez-Fernandez et al., 1997; Müller, unpublished data) that pointed to an involvement of the glutathione system in cell division and chromosomal dynamics the mitotic index and chromosomal aberration rates of the root tip meristems were also measured. However, the mitotic indices (between 8 and 10 %) were not significantly affected by $SO_2$ and $H_2S$ exposure (data not shown), which suggests that the rate of cell division remained unaffected by S exposure. On the other hand, there was an increase in the number of chromosomal aberrations in the dividing cells of the fine roots upon prolonged exposure of the trees to 100 and 225 nl $l^{-1}$ $H_2S$ and to 225 nl $l^{-1}$ $SO_2$. This finding indicates that chromosomal damages are independent of effects on rates of cell division. Significant increases in chromosomal aberrations were observed after 9 days of exposure to 225 nl $l^{-1}$ $H_2S$ and after 13 days of exposure to 100 nl $l^{-1}$ $H_2S$ and 225 nl $l^{-1}$ $SO_2$. The increases (up to 8 % of aberrations) persisted to the end of the 3 week exposure (Figure 2). Trees from the control variant exhibited an aberration rate of 4 %, which is comparable to controls of other studies dealing with different stress impacts (Müller et al., 1998b; Wonisch et al., 1998).

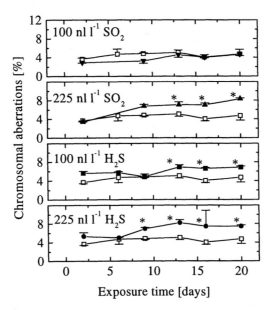

Fig. 2. Chromosomal aberrations in root tip meristems of spruce trees exposed to SO₂ and H₂S. Open squares indicate control, closed symbols exposure variant. Medians and median deviations of n=6 replicate trees. Significance levels for differences from control * p<0.05, **p<0.01.

Exposure to 100 nl l$^{-1}$ SO$_2$ did not significantly affect chromosomal structure. These results are consistent with previous experiments with different SO$_2$ (80 - 225 nl l$^{-1}$) levels applied to spruce trees for three months (Müller et al., 1998a). Effects of H$_2$S on the chromosomal aberrations in the meristematic cells of the roots of spruce trees have not been reported before. The accumulation of glutathione only occurred after 13 days of exposure to H$_2$S, whereas the increase in chromosomal aberrations was observed already after 9 to 13 days of exposure to H$_2$S and also at 225 nl l$^{-1}$ SO$_2$. Previous investigations revealed genotoxic effects (chromosomal aberrations) of glutathione fed via roots to spruce seedlings (Müller, unpublished data). Our results revealed that changes in glutathione content or in redox state of the whole fine root are not directly responsible for defects in the genetic material of spruce trees. On the other hand, there is a dearth of reliable information concerning the intercellular localization of glutathione in plants. Glutathione in plants is found in cytosol and chloroplast, but it is also present in the nucleoplasma (Sandstrom and Marklund, 1990; Thelander and Reichard, 1979; Müller, unpublished). Although shifts in the glutathione redox state in the nucleoplasma were not detectable by the method used such changes could not be excluded and might induce disturbances of the disulfide state of nuclear proteins and, consequently structual damage, because glutathione is an essential controlling factor in this respect. Therefore, this hypothesis requires further study, since our biochemical measurements of glutathione status in root tissues were focused on the whole tissue level and not on the subcellular level.

## Acknowledgments

This project was financially supported by the 'Fonds zur Förderung der Wissenschaftlichen Forschung', project No. P11836-Bio and by grants of NUFFIC and the Austrian Government. It was carried out in the framework of COST Action 829.

## References

De Kok, L.J.: 1990, *Sulfur Nutrition and Sulfur Assimilation in Higher Plants*, H. Rennenberg, C. Brunold, L. J. De Kok and I. Stulen (eds.), Academic Publishing bv, The Hague, The Netherlands, pp 111-130.
De Kok, L.J., Stahl, K. and Rennenberg, H.: 1989, *New Phytol.* **112**, 533-542.
De Kok, L.J., Stuiver C.E.E., Rubinigg, M., Westerman, S. and Grill, D.: 1997, *Bot. Acta* **110**, 1-9.
De Kok, L.J., Stuiver, C.E.E. and Stulen, I.: 1998, *Responses of Plant Metabolism to Air Pollution and Global Change*, L. J. De Kok and I. Stulen (eds.), Backhuys Publishers, Leiden, The Netherlands, pp. 51-63.
Glatt, H., Protic-Sabljic, M. and Oesch, F.: 1983, *Science* **220**, 961-963.
Grill, D., Esterbauer, H. and Hellig, K.: 1982, *Phytopath. Z.* **104**, 264-271.
Kranner, I. and Grill, D.: 1993, *Physiol. Plant.* **88**, 557-562.
Kunert, K.J. and Foyer, C.: 1993, *Sulfur Nutrition and Assimilation in Higher Plants*, L. J. De Kok, I. Stulen, H. Rennenberg, C. Brunold and W. E. Rauser (eds.), SPB Academic Publishing, The Netherlands, pp. 139-151.
Müller, M., Zellnig, G., Tausz, M., De Kok, L.J. and Grill, D.: 1998a, *Responses of Plant Metabolism to Air Pollution and Global Change*, L. J. De Kok and I. Stulen (eds.), Backhuys Publishers, The Netherlands, pp. 383-386.
Müller, M., Tausz, M., Guttenberger, H. and Grill, D.: 1998b, *ESPR - Environ. Sci. & Pollut. Res.*, Special Issue No.1, 101-104.
Polle, A. and Rennenberg, H.: 1994, *Causes of Photooxidative Stress and Amelioration of Defence Systems in Plants*, C. H. Foyer and P. M. Mullineaux (eds.), CRC Press, Boca Raton, pp. 199-218.
Rennenberg, H. and Lamoureux, G.L.: 1990, *Sulfur Nutrition and Sulfur Assimilation in Higher Plants*, H. Rennenberg, C. Brunold, L. J. De Kok and I. Stulen (eds.), Academic Publishing, The Netherlands, pp. 53-65.
Rennenberg, H. and Herschbach, C.: 1995, *J. Plant Nutr. Soil Sci.* **158**, 513-517.
Sachs, L.: 1992, *Angewandte Statistik*, Springer Verlag, Berlin.
Sanchez-Fernandez, R., Fricker, M., Corben, L.B., White, N.S., Sheard, N., Leaver, C.J., Van Montagu, M., Inze, D. and May M.J.: 1997, *Proc. Natl. Acad.Sci*, **94**, 2745-2750.
Sandstrom, B. E. and Marklund, S. L.: 1990, *Biochem. J.* **271**, 17-23.
Tausz, M., De Kok, L. J., Stulen, I. and Grill, D.: 1996, *J. Plant Physiol.* **148**, 362-367.
Tausz, M., Van der Kooij, T.A.W., Müller, M., De Kok, L.J. and Grill, D.: 1998, *Responses of Plant Metabolism to Air Pollution and Global Change*, L. J. De Kok and I. Stulen, Backhuys Publishers, The Netherlands, pp. 457-460.
Thelander, L. and Reichard, P.: 1979, *Ann. Rev. Biochem.* **48**, 133-158.
Wonisch, A., Müller, M., Tausz M., Soja, G. and Grill, D.: 1998, *Chemosphere* **36**, 709-714.

# EFFECT OF UV-B DOSE ON BIOSYTNTHESIS OF EPICUTICULAR WAXES IN BLUE SPRUCE (*PICEA PUNGENS* ENGELMANN.) PRIMARY NEEDLES: PRELIMINARY INVESTIGATION

## DAWN C. GORDON[1] AND KEVIN E. PERCY[2]

[1] Department of Biology, University of New Brunswick, Fredericton, NB, Canada E3B 6E1

[2] Natural Resources Canada, Canadian Forest Service,
Atlantic Forestry Centre, P.O. Box 4000, Fredericton, NB, Canada E3B 5P7

(Received 25 September 1998; accepted 25 February 1999)

**Abstract:** Blue spruce [*Picea pungens* Engelmann.] seedlings were reared from seed for 8 weeks under one of seven UV-B doses ranging from 0.0 to 9.2 kJ $m^{-2}$ $d^{-1}$. Emerging primary needles were chopped and incubated 48 h (22°C; 750 µmol $m^{-2}$ $s^{-1}$ PAR) with [1-$^{14}$C] $CH_3COONa$. Radioactivity incorporated into epicuticular waxes was measured using radio thin-layer chromatography. Biosynthesis of nonacosan-10-ol, the dominant constituent, was affected by the UV-B dose. The results suggest that, if no other factors are limiting, the optimum UV-B dose for wax biosynthesis in emerging primary needles of blue spruce is 6-7 kJ $m^{-2}$ $d^{-1}$. This dose is below that routinely measured in some northern temperate forests and well within the range of predicted values under stratospheric ozone depletion scenarios. UV-B dose levels above this threshold may cause changes to the wax composition that may predispose the tree to damage from other environmental stresses.

**Keywords:** blue spruce, epicuticular wax biosynthesis, nonacosan-10-ol, UV-B, primary needles.

## 1. Introduction

Incoming solar UV-B radiation (280 - 320 nm) is attenuated by the stratospheric ozone layer. Current predictions estimate that loading of certain ozone-destroying chemicals in the stratosphere will peak in 10 years, but the time line for ozone recovery is unknown (Pyle 1997). The relationship between stratospheric ozone decline and increasing radiation in the ultraviolet-B wavelengths is well established (Kerr and McElroy, 1993; Madronich, 1992; Pyle, 1997). It is predicted that over the next 30 years, ozone will decline ~15%, which corresponds to an increase in biologically effective UV-B of 30% over current levels (Lumsden, 1997).

Many studies have shown that UV-B radiation elicits stress responses in plants, manifest as reduced growth (Barnes *et al.*, 1990; Basiouny *et al.*, 1978; Biggs *et al.*, 1981), an overall reduction in net photosynthesis and $CO_2$ assimilation (Bornman, 1989; Musil and Wand, 1993; Sisson and Caldwell, 1977; Teramura, 1983) and DNA damage (Beggs *et al.*, 1986; Caldwell, 1979). Plant defensive mechanisms include attenuation of incident UV-B radiation. There is a series of UV-B screens in aerial plant parts, the first of which is the epicuticular wax layer (Clark and Lister, 1975; Edwards *et al*, 1996). For example, the epicuticular wax layer of blue spruce strongly reflects light in the blue and UV wavelengths, an adaptation of the plant to alpine environments where light in the visible and UV-B wavelengths is intense (Clark and Lister, 1975). Given that blue spruce epicuticular wax is comprised mainly (approximately 60%, data not shown) of tubular crystals of nonacosan-10-ol, this compound accounts for the main role of the epicuticular wax layer in the interaction with incident light.

In addition to its role in UV-B attenuation, the epicuticular wax layer minimizes cuticular transpiration; in maintaining a waterproof surface, the plant cuticle creates an environment inhospitable to fungal growth and reduces the residence time of pollutants on the leaf surface (Martin and Juniper, 1970). Biosynthesis of the wax layer is influenced in part by species, plant age, and environmental conditions such as light quality and quantity, humidity, and temperature (Baker, 1974; Macey, 1970). Thus, the final amount, chemical composition and structural arrangement of the epicuticular waxes depends to some extent on environmental conditions.

While the composition and quantity of surface wax changes with age, no new, compensatory epicuticular wax is produced after the first growing season (Percy et al., 1993). This, coupled with the fact that conifers retain needles for several years, means that any changes to the wax brought on by high levels of UV-B could significantly affect the functioning of the wax layer over the remaining lifetime of the needle (Bornman and Teramura, 1993).

This study reports significant effects of UV-B exposure during seedling development on epicuticular wax biosynthesis and subsequent synthesis of the dominant homologue, nonacosan-10-ol, in blue spruce (*Picea pungens*) primary needles. Interaction with PAR was not assessed in this preliminary investigation.

## 2. Materials and Methods

2.1 PLANT MATERIAL

Blue spruce [*Picea pungens* Engelmann.] (Nova Seed, range-wide collection) seeds were sown (250 per pot) in 7 pots (20.5 cm dia.) filled with peat/perlite (1:1). All seven pots were placed in a controlled environment chamber under the following conditions: 16 h photoperiod (300 ± 25µmol $m^{-2} s^{-1}$), temperature 22°/18° ± 1°C(day/night), relative humidity 70%. Each pot received one of seven UV-B treatments: 0.0, 1.1, 1.7, 4.3, 5.2, 8.1 or 9.2 kJ $m^{-2} d^{-1}$, delivered within the controlled environment chamber for 4.5 h each day, centered on the photoperiod. The details of the UV-B exposure system may be found in Gordon et al., 1998. Primary needles were collected for experimentation after approximately 8 weeks.
Ten replicates were collected for each UV-B treatment (ie., each pot); each replicate comprised primary needle material collected randomly from within a 3.5 cm x 3.5 cm area within each pot.

2.2 PROCEDURE

Primary needles were chopped (~1mm pieces) and transferred immediately to culture tubes (10 ml) containing 1 ml of [1-$^{14}$C] acetate, sodium salt, 2.5 µCi $ml^{-1}$ in 0.1 M phosphate buffer, pH 6.5. Tubes were incubated without UV-B, under a sodium vapour lamp (750 µmol $m^{-2} s^{-1}$ PAR, photoperiod 16 h) at 22 °C, for 48 h prior to lipid extraction.

Lipids, including epicuticular waxes, were extracted using a method modified from Bligh and Dyer (1959). The chloroform fraction was collected in glass scintillation vials (20 ml). The residue of the chloroform fraction was redissolved in chloroform (1 ml) and transferred to pre-weighed 1.8 ml vials. Residual solvent was removed under a stream of air at 60°C, and recovered lipid weights recorded. The amount of radioactivity incorporated into nonacosan-10-ol was determined by thin-layer chromatography. Radioactive samples were co-run with a Blue spruce cold wax sample. Wax (200 µg) was spotted on toluene-washed TLC plates (20 x 20 cm, silica gel 25 mm)(Aldrich) and run in toluene. Plates were counted on a BioScan System 200 Imaging Scanner (BioScan, Inc.). Peak areas were calculated using BioScan software, version 2.247. Following scanning, plates were developed with phosphomolybdic acid (10 % w/v) in 95% ethanol. The presence of nonacosan-10-ol was confirmed by GC/MS.

Treatment effects were assessed with non-linear regression analysis (SAS, 1997).

## 3. Results

The production of epicuticular wax and other chloroform soluble lipids measured at harvest time in blue spruce primary needles was affected by the UV-B dose applied during seedling rearing (Figure 1). The relationship was: $y = 58.3 - 30.4x + 7.31x^2 - 0.46x^3$ ($R^2=0.81$), where y = wax and lipid weight / oven dry needle weight (mg g$^{-1}$), and x = UV-B dose (kJ m$^{-2}$ d$^{-1}$). Wax and chloroform soluble lipid weight was greatest at the UV-B dose extremes, 0-1 and 8 kJ m$^{-2}$ d$^{-1}$, and least at ambient and near-ambient (2-5 kJ m$^{-2}$ d$^{-1}$) UV-B doses. Weights began to decrease again when UV-B doses increased above 8.1 kJ m$^{-2}$ d$^{-1}$ (Figure 1).

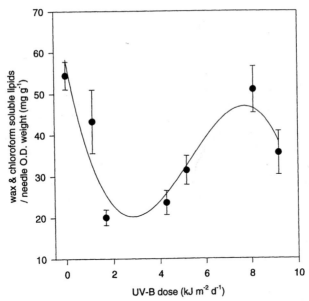

Fig. 1: Effect of UV-B dose (kJ m$^{-2}$ d$^{-1}$) on production of epicuticular wax and chloroform soluble lipids (mg g$^{-1}$ needle O.D. weight) in blue spruce primary needles. UV-B doses were delivered during seedling rearing (8 weeks). Data are means ± S.E. (n=10). Ambient UV-B range in Fredericton, NB, Canada in midsummer is ~4-5 kJ m$^{-2}$ d$^{-1}$.

Incorporation of [1-$^{14}$C] acetate into nonacosan-10-ol was significantly affected by UV-B dose applied concomitant with seedling development (Figure 2). The relationship was: $y = 0.034 - 6.23x + 3.136x^2 - 2.69x^3$ ($R^2=0.82$), where y = $^{14}$C activity / weight nonacosan-10-ol (cpm μg$^{-1}$), and x = UV-B dose (kJ m$^{-2}$ d$^{-1}$). Nonacosan-10-ol activity was reduced at the UV-B extremes: 0-2 and 9 kJ m$^{-2}$ d$^{-1}$. Ambient and near-ambient UV-B doses produced greater $^{14}$C incorporation into nonacosan-10-ol (Figure 2).

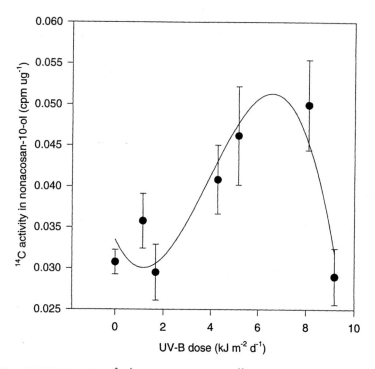

Fig. 2: Effect of UV-B dose (kJ $m^{-2}$ $d^{-1}$) on incorporation of [1-$^{14}$C] acetate into nonacosan-10-ol (cpm $\mu g^{-1}$ nonacosan-10-ol) of blue spruce chopped primary needles. UV-B doses were delivered during seedling rearing (8 weeks) and removed for incubation of chopped primary needle tissue. Incubations were 48 h, with [1-$^{14}$C]CH$_3$COONa. Data are means ± S.E.(n=10). Ambient UV-B range in Fredericton, NB, Canada in midsummer is ~4-5 kJ $m^{-2}$ $d^{-1}$.

## 4. Discussion

It is well established that epicuticular wax amounts and chemical composition vary between and within species, and are influenced by many environmental factors, including the quality (Tevini and Steinmuller, 1987) and energy of incident solar radiation (Baker 1974; Giese 1975). In this study we showed that the dose of UV-B radiation delivered during seedling development significantly affected epicuticular wax production, and subsequent incorporation of $^{14}$C acetate into the chloroform soluble wax fraction and into nonacosan-10-ol, which comprises up to 60% of the wax of blue spruce. Furthermore, the changes to lipid synthesis induced by UV-B were persistent: even after 48 h removed form UV-B exposure, the plants had not resumed normal metabolism. Previous workers have shown, through indirect means (analysis of waxes after exposure), that UV-B radiation has a direct and highly specific effect on wax biosynthesis (Barnes *et al.*, 1996; Barnes *et al.*, 1994; Gonzalez, 1996; Gordon *et al.*, 1998; Steinmuller and Tevini, 1985; Tevini and Steinmuller, 1987). These results provide direct evidence that the UV-B environment existing during seedling development affected nonacosan-10-ol production. Because the UV-B stress was absent during incubation of the needle tissue with $^{14}$C, it can be concluded that the effect was persistent, that is, exposure of developing primary needles to UV-B induced changes that remained even after the stress was removed.

Many studies investigating the effect of UV-B on various plant responses have employed extremely high UV-B dose levels which far exceed any predicted increases in UV-B (Fiscus and Booker, 1995). For example, in one study, the maximum dose applied in one experiment was 23.1 kJ m$^{-2}$ d$^{-1}$, (Sullivan et al., 1992). Furthermore, few studies have had more than 3 treatment levels, including a zero UV-B control. This study incorporated a range of seven doses, from 0 to 9.2 kJ m$^{-2}$ d$^{-1}$, and included doses close to the ambient PAS-weighted UV-B (4-5 kJ m$^{-2}$ d$^{-1}$) measured near Fredericton, N.B., Canada in midsummer (Rob Gordon, personal communication).

Wax and chloroform soluble lipid production measured in blue spruce primary needles, after 8 weeks' exposure, was affected by the UV-B dose (Fig.1). The peak of the regression curve occurred at a UV-B dose of about 8 kJ m$^{-2}$ d$^{-1}$, suggesting that the threshold for UV-B tolerance for wax and lipid production was 7-8 kJ m$^{-2}$ d$^{-1}$.

Studies involving agricultural crop plant species (Barnes et al. 1996; Steinmuller and Tevini, 1985) have indicated the response to UV-B radiation varied with species. Cucumber, bean and barley wax weight was increased, while tobacco wax weight was decreased, with enhanced UV-B. Clark and Lister (1975) showed that the epicuticular wax layer of blue spruce is highly reflective in the UV wavelengths, so it is not surprising to discover that total wax production is significantly affected by the UV-B environment during seedling development.

Exposure to UV-B during seedling development also affected the incorporation of [1-$^{14}$C] acetate into nonacosan-10-ol by blue spruce primary needles (Fig. 2). The peak of the regression curve occurred at a UV-B dose of about 7 kJ m$^{-2}$ d$^{-1}$, suggesting that the threshold for UV-B tolerance of nonacosan-10-ol production was 6-7 kJ m$^{-2}$ d$^{-1}$. Nonacosan-10-ol biosynthesis also appears to be responsive to air pollutants. The proportion of nonacosan-10-ol was reduced on exposure to O$_3$ and acid fog (pH 3.0) in *Picea rubens* (Percy et al., 1992), increased on exposure to acid precipitation in *Picea sitchensis* (Percy and Baker, 1990) and increased on exposure to a combination of O$_3$ and acid mist in *Picea abies* (Lütz et al., 1990).

It is important to note that the UV-B effects observed here were persistent. The 48 h incubation of the needle tissue with [1-$^{14}$C] acetate took place in an environment free from UV-B exposure. It is possible that the persistence of the UV-B effect may have implications for the functioning of UV-B repair mechanisms.

The integrity of the epicuticular wax layer is crucial for waterproofing, prevention of nutrient leaching and interaction with incident light. In the case of conifers, waxes are especially important because the investment in leaf tissue must be protected for several years (Chabot and Chabot, 1977). Given that nonacosan-10-ol is the major constituent of *Picea pungens* epicuticular wax, and that it, together with nonacosane-diols (Jetter and Riederer, 1994) forms the major structural feature of the needle surface (tubular crystals), its the sensitivity of its synthesis to UV-B may have significant implications for tree vitality in the future. Susceptibility to drought may increase if these UV-B induced changes have an effect on cuticular transpiration rates. Likewise, leaf surface topography may change, influencing susceptibility to microbial attack.

According to Bornman and Teramura (1993), conifer needles are particularly vulnerable to UV-B shortly after needle elongation and emergence from the protective bud scales. Conifer seedlings are particularly at risk because the greatest increases in UV-B exposure in the forest under storey are likely to occur in the spring, when new needles are emerging (Brown et al.,1994). In addition, conifer needles are retained for several years and UV-B damage is cumulative (McLeod and Newsham, 1997). Thus, subtle changes to leaf surface chemistry and topography may eventually affect competitive abilities, and the differential resistance to UV-B of competing plant species could lead to changes in plant community structure, in natural and managed forests (Brown et al., 1994; Caldwell, 1997; Gold and Caldwell, 1983).

The response of nonacosan-10-ol biosynthesis to UV-B dose seems to suggest that UV-B is an essential factor, that is, an environmental factor beneficial or essential for plant function. Plants normally have a defined optimum below or above which the response to the factor is inhibited (Godbold, 1998). This characteristic can be seen in Figure 2. The response of emerging blue spruce primary needles to increasing UV-B dose included increased production of the main wax constituent, nonacosan-10-ol. Furthermore, we suggest that if no other factors are limiting, the optimum / threshold dose for nonacosan-10-ol and total wax biosynthesis in emerging primary needles of blue spruce is 6-7 kJ $m^{-2}$ $d^{-1}$. One might predict, then, that the increasing UV-B levels projected for the future will significantly affect biosynthesis of the predominant wax constituent as well as *de novo* wax biosynthesis of blue spruce primary needles and possibly predispose affected trees to other environmental stresses. The hypothesis that increased nonacosan-10-ol synthesis is an adaptive response to increasing UV-B radiation could be further tested by analyzing epicuticular wax chemical composition of needles at various canopy locations and comparing those results to UV-B levels measured in the same locations during needle emergence.

## 5. Conclusion

The results suggest that the biosynthesis of nonacosan-10-ol is adaptive to the UV-B dose delivered during seedling rearing, but that the threshold dose for this response is 6-7 kJ $m^{-2}$ $d^{-1}$. The increased production of nonacosan-10-ol wax crystals in response to UV-B may result in an increased light reflective index of the needle surface. Given that the epicuticular wax layer performs vital plant functions such as waterproofing of the leaf surface, and that nonacosan-10-ol is the predominant wax constituent in blue spruce epicuticular wax, it may be proposed that seedling exposure to UV-B may predispose the plant to other environmental stresses and possibly affect the general health of the plant.

These results are the first to demonstrate a dose-response relationship between a full range of UV-B doses and an adaptive process with important physiological functions. The effect of the threshold dose recorded here was well within the range of ambient doses (6.9 - 8.5 kJ $m^{-2}$ $d^{-1}$) recorded (Naidu et al., 1993; Zeuthen et al., 1997), and well within levels predicted under current ozone depletion scenarios.

## Acknowledgements

Authors wish to thank Dr. Lucy Sheppard for her suggestions on this manuscript. Experiments were performed at the Canadian Forest Service, Atlantic Forestry Center, Fredericton, NB, Canada.

## References

Baker, E.A.: 1974, *New Phytol.* **73**:955.
Barnes, J., Flint, S.D. and Caldwell, M.M..: 1990, *Am. J. Bot.* **77**:1354.
Barnes, J., Paul, N., Percy, K., Broadbent, P., McLaughlin, C. and Wellburn, A.: 1994, 'Effects of UV-B radiation on wax biosynthesis' in: K.E. Percy, *et al.* (ed.), *Air Pollutants and the Leaf Cuticle. NATO ASI Series G, vol 36.* Springer-Verlag, Heidelberg. pp. 195-204.
Barnes, J.D., Percy, K.E., Paul, N.D., Broadbent, P., McLaughlin, C.K., Mullineaux, P.M., Criessen, G. and Wellburn, A.R.: 1996, *J. Exp. Bot.* **47**:99.
Basiouny, F.M., Van, T.K. and Biggs, R.H.: 1978, *Physiol. Plant.* **42**:29.
Beggs, C.J., Schneider-Ziebert, U. and Wellman, E.: 1986, 'UV-B radiation and adaptive mechanisms in plants' in: R.C. Worrest and M.M. Caldwell (eds), *Stratospheric ozone reduction, solar ultraviolet radiation and plant life*, Springer-Verlag, Berlin, pp 235-250.
Biggs, R.H., Kossuth, S.V. and Teramura, A.H.: 1981, *Physiol. Plant.* **53**:19.
Bligh, E.G. and Dyer, W.J.: 1959, *Canadian Journal of Biochemistry and Physiology* **37**:911.
Bornman, J.F.: 1989, *J. Photochem. Photobiol. B: Biol.* **4**:145.
Bornman, J.F. and Teramura, A.H.:1993, 'Effects of ultraviolet-B radiation on terrestrial plants, in: Young *et al.* (eds)., *Environmental UV Photobiology*. Plenum Press, New York.
Brown, M.J., Parker, G.G. and Posner, N.E.: 1994, *J. Ecol.* **82**:843.
Caldwell, M.M. 1979, *BioSci* **29**:520-525.
Caldwell, M.M., 1997.: 'Alterations in competitive balance', in Lumsden, P.J. (ed.) *Plants and UVB Responses to Environmental Change*, Cambridge University Press, Cambridge, pp 305-316.
Chabot, J.F. and Chabot, B.F.: 1977, *Can. J. Bot.* **55**:1064.
Clark, J.B. and Lister, G.R. 1975.: *Plant Physiol.* **55**:407.
Edwards, D., Abott, G.D. and Raven, J.A. 1996.: 'Cuticles of early land plants: a palaeoecophysiological evaluation', in G. Kersteins (ed.), *Plant Cuticles*. BIOS Scientific Publishers Ltd., Oxford. pp 1-31.
Fiscus, E.L. and Booker, F.L.: 1995, *Photosyn. Res.* **43**:81.
Giese, B.N.: 1975, *Phytochem.* **14**:921.
Godbold, D.L.: 1998, *Chemosphere* **36**: 859-864.
Gold, W.G. and Caldwell, M.M.: 1983, *Physiol. Plant.* **58**:435.
Gonzalez, R., Paul, N.D., Percy, K., Ambrose, M., McLaughlin, C.K., Barnes, J.D., Areses, M. and Wellburn, A.R.: 1996, *Physiol. Plant.* **98**:852.
Gordon, D.C., Percy, K.E. and Riding, R.T.: 1998, *Chemosphere* **36**:853.
Jetter, R. and Riederer, M.: 1994, *Planta* **195**:257.
Kerr, J.B. and McElroy, C.T.: 1993, *Science* **262**:1032.
Lumsden, P.J.: 1997, 'Preface', in: P.J. Lumsden (ed.), *Plants and UV-B: responses to environmental change. Society for Experimental Biology Seminar Series 64.* Cambridge University Press, Cambridge. pp. xiii-xx.
Lütz, C., Heinzmann, U. and Gulz, P-G.: 1990, . *Env. Poll.* **64**:313..
Macey, M.J.K.: 1970, *Phytochem.* **9**:757.
Madronich, S.: 1992, *Geophys. Res. Lett.* **19**:37.

Martin, J.T. and Juniper, B.E.: 1970, *The cuticles of plants*, Edward Arnold Publishers, Edinburgh.

McLeod, A.R. and Newsham, K.K.: 1997, 'Impacts of elevated UVB on forest ecosystems', in P.J. Lumsden (ed.), *Plants and UVB Responses to Environmental Change*. Cambridge University Press: Cambridge, pp247-282.

Musil, C.F. and Wand, S.J.: 1993, *Env. Exp. Bot.* **33**:233.

Naidu, S.L., Sullivan, J.H., Teramura, A.H. and DeLucia, E.H.: 1993, *Tree Physiol.* **12**:151.

Percy, K.E. and Baker, E.A.: 1990, *New Phytol.* **116**:79.

Percy, K.E., Jagels, R., Marden, S., McLaughlin, C.K. and Carlisle, J.: 1993, *Can. J. For. Res.* **23**:1472.

Percy KE, Jensen, K.F. and McQuattie, C.J.: 1992, *New Phytol.* **122**: 71.

Pyle, J.A.: 1997, 'Global ozone depletion: observations and theory', in: P.J. Lumsden (ed.), *Plants and UV-B Responses to environmental change*. Cambridge University Press, Cambridge, pp 3-12.

Sisson, W.B. and Caldwell, M.M.: 1977, *J. Exp. Bot.* **28**:691.

Steinmuller, D. and Tevini, M.: 1985, *Am. J. Bot.* **75**:225.

Sullivan, J.H., Teramura, A.H. and Ziska, L.H.: 1992. *Am. J. Bot.* **79**:737.

Teramura, A.H.: 1983, *Physiol. Plant.* **58**:415.

Tevini, M. and Steinmuller, D.: 1987, *J. Plant Physiol.* **131**:111.

Zeuthen, J., Mikkelsen T.N., Paludan-Müller, G. and Ro-Poulsen, H.: 1997, *Physiol. Plant* **100**:281.

# A NEW IMITATIVE MODEL TO PREDICT THE IMPACT OF AIR POLLUTANTS ON SCOTS PINE HEALTH AND RADIAL INCREMENT

A. AUGUSTAITIS

*Forest Monitoring Laboratory, Lithuanian University of Agriculture, Kaunas, Lithuania*

(Received 25 September 1998; accepted 25 February 1999)

**Abstract.** In order to address the problem of pollutant effects on trees against the background of environmental fluctuations a simulation model using radial increment as an indicator of tree productivity is proposed. Ring times series of large trees provide one of the most acceptable biological indicators for the assessment of pollutant impact. This research focused on the tallest severely damaged and healthy trees growing close to each other in the surroundings of Jonava Mineral Fertilizer plant, which primarily emits nitrogen (N) and sulphur (S) compunds. The annual increment was found to be strongly correlated with mean September air temperature of the previous year and S load. The correlation coefficient (k=0.961) between modelled and observed tree ring times series of severely damaged trees proves the impact of S load on growth and vitality of a tree and enables fulfillment of tree condition prognosis.

**Keywords:** simulative model, pollution impact, tree health, ring time series, prognosis

## 1. Introduction

The assessment of anthropogenic induced changes in various biological, as well as physical and chemical environmental indicators on the basis of their natural quasi-periodical fluctuations is a special and very complicated problem. The increasing anthropogenic impact on the environment stimulates the search for new methods to facilitate the assessment of air pollution impacts on forest ecosystems and to quantify the effects. Attempts to find regularities in such a complicated process simply by observation, collection and statistical analysis of data (monitoring) have been shown to be inadequate. The definition of factors limiting tree health and productivity helps to create the imitative model enabling problem solving and ecological standardization of forest ecosystems at the scientific level.

## 2. Materials and methods

Tree productivity represents the integration of genetic and environmental influences on tree condition. Therefore annual tree increment is one of the most acceptable biological indicators for the assessment of environmental impact. Methods and indicators must have high reproducibility to allow the use of retrospective information for a sufficiently long period. Despite its not very high sensitivity, the dendrochronological method provides an important long-term series of information from annual tree ring analysis (Juknys, 1993).

For assessing the anthropogenic induced changes in different biological, physical or chemical variables, the question of normality is most important. In our model the increment of a healthy tree is considered to be the norm. Scots pine (*Pinus sylvestris L.*) the most widespread and one of the most sensitive species to pollution in Lithuania was chosen for this research.

Investigations of the damaged Scots Pine stands in the surroundings of Jonava Mineral Fertilizer plant, where nitrogen (N) and sulphur (S) compounds are the main pollutants indicate that for tree growth simulation in polluted areas, it is reasonable to use ring time series of the largest and tallest trees. The largest changes in increment prove that the largest trees are most sensitive to pollution and other unfavorable environmental factors. Such trees also display great vitality and resilience and energetic losses (the carbon cost) of competition with other trees are minimal (Augustaitis, 1997).

High intraspecific diversity of Scots pines enables the development of an imitative model of severely damaged tree increment by comparing increment of damaged and healthy trees growing in the same locality. The main factors limiting tree growth and health are as follows: mean monthly air temperatures and pollutant loads. Analysis of the impact of different natural external factors indicate that the temperatures of late winter (February), early spring (March, April) and late summer (August) affect the growth of pines in Lithuania most strongly (Juknys, 1993). In our case the temperature in September of the last year was highly correlated to healthy tree increment ($r = 0.657$; $p < 0.05$). For the simulation of tree increment, the pollution load 1.5 km away from fertilizer plant was used (Armolaitis, 1998). Pollution load in the investigated stand, which was 3.5km away from plant, was not measured.

## 3. Results

3.1. IMITATIVE MODEL OF SEVERELY DAMAGED TREE INCREMENT

The main premise of this investigation is that early tree growth relies on nutrient and carbohydrate (CH) reserves created between September of the previous year and April of the current year. Thereafter assimiliation products (AP) accumulated from May to August are used for tree growth. Such a distribution of AP indicates that at the beginning of vegetation period, twigs and needles of trees grow very rapidly but CH export is minimal, because CH assimiliation is dependent on average monthly temperature (Larcher, 1978).

According to our investigation seasonal growth of Scots Pine twigs and needles, at the beginning of the vegetation period in control areas is rapid, ensured by AP accumulated in September-April month period (Augustaitis, 1989). In polluted areas, the beginning of twig and needle growth is delayed by 10-15 days and the intensity of growth within this period is significantly lower. Due to large energetic and AP losses incited neutralizing the impact of pollution, the remaining AP cannot ensure the beginning of growth at the same time as in control areas (see also Kramer, Kozlowsky, 1983).

The sequence of biological processes neutralizing the impact of pollution is as follows:

*1. Preservation of the tree crown; 2. Radial tree increment; 3. Crown increment.*

# A NEW IMITATIVE MODEL TO PREDICT THE IMPACT OF AIR POLLUTANTS

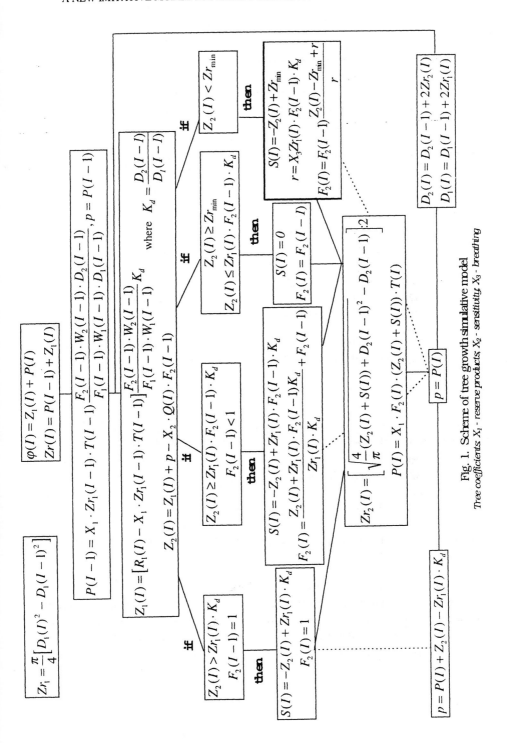

Fig. 1. Scheme of tree growth simulative model
Tree coefficients $X_1$ - reserve products, $X_2$ - sensitivity, $X_3$ - breathing

decreased respiration intensity. If the remaining AP is larger than needed for normal radial increment, crown phytomass increases (defoliation decreases). Assimilation products (AP) in this case are distributed between radial tree and crown increments. Figure 1 represents a simulation model of severely damaged trees.

Annual tree production was calculated according to the following equation:

$$\Psi(I) = Z_1(I) + P(I), \qquad (1)$$

We assume, the AP is expressed only by radial tree increment. In this case it is estimated according to the equation:

$$Zr(I) = Z_1(I) + P(I-1), \qquad (2)$$
$$P(I) = f(t), \qquad (3)$$

where: $\Psi(I)$ - annual assimilation production;
$Z_1$ - AP created for 05 - 08 month period;
$P(I)$ and $P(I-1)$ - AP created between 09 of the previous and 04 of the current year;
$f(t)$ - function of September mean temperature.

On the other hand annual AP depends directly on leaf area (phytomass), level of assimilation process and on tree parameters.

1. If AP for V-VIII month period is estimated according to the equation:

$$Z_1(I) = [Zr(I) - X_1 \cdot Zr(I-1) \cdot T(I-1)] \frac{F_2(I-1) \cdot W_2(I-1) \cdot D_2(I-1)}{F_1(I-1) \cdot W_1(I-1) \cdot D_1(I-1)}, \qquad (4)$$

and reserves of AP are estimated according to the equation:

$$P(I) = X_1 F(I)[Z_1(I) + P(I-1) - X_2 Q(I) F_2(I-1) + S(I)] T(I), \qquad (5)$$

where $T(I)$ - depends on an average temperature of September $t(I)$ and is equal:

$$T(I) = e^{0,1\, t(I)}, \qquad (6)$$

then radial increment of severely damaged trees is simulated according to the equation:

$$Zr(I) = \begin{cases} Z_1(I) + P(I-1) - X_2 \cdot Q(I) \cdot F_2(I-1), & \text{if } Z_1(I) + P(I-1) - X_2 \cdot Q(I) \cdot F_2(I-1) > Zr_{min} \\ Zr_{min}, & \text{if } Z_1(I) + P(I-1) - X_2 \cdot Q(I) \cdot F_2(I-1) < Zr_{min} \end{cases}, \qquad (7)$$

Where:  $Zr_1$ and $Zr_2$ - annual radial increment of healthy and severely damaged tree;
$F_1$ and $F_2$ - leaf area (phytomass) of healthy and severely damaged tree;
$W_1$ and $W_2$ - intensity of assimilation process of healthy and severely damaged tree;
$D_1$ and $D_2$ - diameter of healthy and severely damaged tree;
S - index of AP restructuring during neutralization of pollutant load;
Q - load of pollutant;
$X_1$ - coefficient of reserve products, $X_2$ - coefficient of tree sensitivity;
t - an average temperature of September.

The leaf area of severely damaged tree crown can change in such a way:

$$F_2(I) = \begin{cases} F_2(I), & \text{if } Z_1(I)+P(I-1)-[X_2 \cdot Q(I) \cdot F_2(I-1)] > Z_{min}, \\ \dfrac{F_2(I-1)\{Z_1(I)+P(I-1)-[X_2 \cdot Q(I) \cdot F_2(I-1)]-Z_{min}+r\}}{r}, & \text{if } Z_1(I)+P(I-1)-[X_2 \cdot Q(I) \cdot F_2(I-1)] < Z_{min}, \\ F_2(I-1)+\Delta F, & \text{if } Z_1(I)+P(I-1)-[X_2 \cdot Q(I) \cdot F_2(I-1)] > Z_{max} \cdot F_2(I-1). \end{cases}$$

(11)

where "s" - tree respiration which depends on tree increment and leaf area and is estimated using the formula:

$$rs = X_3 \, Zr_{max} \, F_2(I-1), \qquad (12)$$

where $X_3$ - respiration coefficient.

S(I) is restructuring of AP during the neutralization of pollutant load. This parameter can change in such a way:

$$0 < S(I) < P(I) + r, \qquad (13)$$

and is estimated according to the equation:

$$S(I) = \begin{cases} -\{Z_1(I)+P(I-1)-[X_2 \cdot Q(I) \cdot F_2(I-1)]+Z_{min}\}, & \text{if } Z_1(I)+P(I-1)-[X_2 \cdot Q(I) \cdot F_2(I-1)] < 0 \\ & \text{and } Z_1(I)+P(I-1)-[X_2 \cdot Q(I) \cdot F_2(I-1)]+r > 0 \\ 0, & \text{if } Z_1(I)+P(I-1)-[X_2 \cdot Q(I) \cdot F_2(I-1)] > 0 \end{cases}, \quad (14)$$

Correlation coefficient $r=0.961$ between modelled and observed tree ring times series of severely damaged trees proves the impact of sulphur (S) load on growth and vitality of a tree. Using the above model to assess the impact of air pollutants on increment and tree condition allows us to fulfil damaged tree condition prognosis according to different pollutant load.

3.2. DAMAGED TREE CONDITION PROGNOSIS

The average increment of a healthy tree and climatic conditions during the last 5 year period are used for the prognosis. As data on pollutant load in the investigated stand was not measured directly we used the relative S load based on S load near the plant which in the last year was 28 kg/ha for the tree health and increment prognosis.

The prognosis was fulfilled in 3 different tree damage groups:

Group I.   Tolerant trees. Healthy trees. Relative phytomass F=1.0 crown defoliation 0%;

Group II.  Moderately sensitive trees. Relative phytomass F=0.58, crown defoliation 42%;

Group III. Highly sensitive trees. Severely damaged trees. Relative phytomass F=0.28, defoliation 72%.

It is estimated that sensitive (severely damaged) trees can endure 1.5 times higher pollution than measured last year (28 kg S/ha), under the average climatic factors. Moderately sensitive trees endure 2 times higher S pollution loads than measured last year, whereas sensitive trees under such S loads would die after 5 years. For resistant individuals neutral S concentration can be 2.3 times higher in the last year of such pollutant loads (28 kg S/ha), however at such a concentration within a year all the sensitive trees for pollution would start dying and moderately sensitive trees would die within 5-6 years.

The investigation of tree condition recovery was carried out under the same climatic conditions without air pollution. It is estimated, that the most intensive recovery occurs in slightly damaged trees, where defoliation of the crown does not exceed 25%. Recovery indicated by nil defoliation takes 2-3 years. The recovery of more sensitive and moderately damaged trees is less pronounced, taking longer up to 7-9 years for full condition recovery. The recovery of the most sensitive and severely damaged trees, crown where defoliation exceeds 60%, should take theoretically 15 or more years. It is evident, that in nature, conditions favorable for growth will not last for such a long period. Thus we believe the health of severely damaged trees cannot recover.

## 4. Conclusion

A new method for estimation and prognosis of air pollution impact on tree condition and increment is suggested. We fully appreciate that this method could be improved but this is one of the first attempts to estimate in such a way the impact of pollutants on very complicated physiological processes and fulfill prognosis. This method could be applied to the estimation of regional pollutant load impact on tree condition and increment in National Forest Monitoring Programmes.

## References

Augustaitis, A.: 1997, Forestry. Lithuanian University of Agriculture. **40**, 5-16.
Augustaitis, A.: 1989, *Sankt Petersburg*. **12**, 32-51.
Armolaitis, K.: 1988, *Env. Pollut*. **102**, 55-60.
    Juknys, R.: 1993, Dendrochronological data applications at forest monitoring system. Climate and Atmospheric Deposition Studies in Forest. Conference Papers. Warszawa, 245-254.
Kramer, P.D. and Kozlowsky, T.T.: 1983, Physiology of Wooded Plants. Moscow, 462 pp.
Larcher, B.: 1978, Plant Ecology. Mockow. 388 pp.

# STATE OF SCIENCE AND KNOWLEDGE GAPS WITH RESPECT TO AIR POLLUTION IMPACTS ON FORESTS : REPORTS FROM CONCURRENT IUFRO 7.04.00 WORKING PARTY SESSIONS

K. PERCY[1], J. BUCHER[2], J. CAPE[3], M. FERRETTI[4], R. HEATH[5,] H.E. JONES[6], D. KARNOSKY[7], R. MATYSSEK[8], G. MULLER-STARCK[9], E. PAOLETTI[10], U. ROSENGREN-BRINCK[11], L. SHEPPARD[3], J. SKELLY[12] and G. WEETMAN[13]

[1]*Natural Resources Canada, Canadian Forestry Service-Atlantic, P.O. Box, 4000, Fredericton, NB, Canada E3B 5P7;* [2]*Swiss Federal Institute for Snow and Landscape Research, Zurcherstrasse 111, CH-8903 Birmensdorf, Switzerland;* [3]*Instutute of Terrestrial Ecology, Bush Estate, Penicuick, Midlothian EH26 OQB, UK;* [4]*LINNAEA ambient Srl, Via G. Sirtori 37, I-50137Firenze, Italy;* [5]*Department of Botany and Plant Sciences, University of California, Riverside, CA, 92521-0124, USA;* [6]*Institute of Terrestrial Ecology, Merlewood Research Station, Grange-over-Sands, Cumbria LA11 6JU, UK;* [7]*School of Forestry and Wood Products, Michigan Technological University, 1400 Townsend Drive, Houghton, MI, 49931-1295, USA;* [8]*Department of Forest Botany, Ludwig-Maximilians-Universitat, Am Hochanger 13, D-85354 Freising, Germany;* [9]*Lehrbereich Forstgenetik, Ludwig-Maximilians-Universitat Munchen,, LMU Munchen, Am Hochanger 13, D-85354 Freising, Germany;* [10]*IPAF-CNR, Piazzale delle Cascine 28, 50144 Firenze, Italy;* [11]*Department of Plant Ecology, University of Lund, Ecology Bldg., S-223 62 Lund, Sweden;* [12]*Department of Plant Pathology, 210 Buckhout Laboratory, Pennsylvania State University, University Park, PA, 16802, USA;* [13]*Forest Sciences Department, Faculty of Forestry, University of British Columbia, Vancouver, BC, Canada V6T 1Z4.*

Concurrent IUFRO Working Party (WP) sessions were convened and structured to address the state of science within each subject area. Oral summary presentations were made by rapporteurs during the closing plenary. The written output from the five sessions convened is included here in order of IUFRO WP designation.

## 1. Diagnosis, Monitoring and Evaluation (WP 7.04.01)

1.1 STATE OF SCIENCE

Presentations focused on four main topics. Nitrogen was discussed with reference to deficiency, enrichment and exceedance of critical loads, drought-induced reduction in nitrification, consequence on growth, and impact on water quality. Ozone presentations concentrated on visible foliar symptoms, exceedance of critical levels, and "memory" effects for both symptoms and morphological parameters. Sulfur dioxide was considered in the context of point sources. Finally, indicators and indices were discussed in terms of objective/measurable indicators, non-quantitative or subjective indicators, symptom description, and relationships between different indicators and indices.

1.2 KNOWLEDGE GAPS

Gaps were identified in diagnosis, monitoring and evaluation. These comprise: the absence of realistic maps for ozone exposure (e.g. Level II assessment); the poor understanding of interactions between abiotic and biotic agents; the poor understanding of time-delayed effects of ozone and nitrogen; the poor integration between monitoring and research; the lack of adequate planning, communication and cooperation between scientists, program managers, resource managers and field crews. This last item was considered to have major effects on QA/QC, an issue particularly important in diagnosis and monitoring.

Four major areas requiring increased attention were identified as: 1) the monitoring of air pollution effects in developing countries (SE Asia, South America, Africa); 2) the current lack of identification of ozone symptoms by the large-scale surveys of forest condition in Europe; 3) risk analysis and flux models for critical levels/loads assessment and the effects of nighttime exposure; and, 4) a clear assessment and measurement endpoints to facilitate accurate risk assessment.

Currently, different terminology is used in the context of forest health monitoring (eg. health, vitality, condition, vigor, stability, and integrity). However, each has its own scientific meaning, relevant indicator(s) and indices that are not explicitly defined. There needs to be a linking of monitoring strategy with ecosystem processes. There is also a clear need for adequate design and validation of field results.

## 2. Biochemical and Physiological Aspects (WP 7.04.02)

Discussions were initiated after consensus was attained on the need to broaden the context of the discussions from merely biochemical and physiological aspects of response, to include consideration of processes under the theme "biogeochemical and ecophysiological aspects".

### 2.1 STATE OF SCIENCE

Investigations regarding changes in the biochemistry and physiology of trees induced by stress from air pollutants have changed their focus from individual to ecosystem, and from differential or biochemical to a holistic or whole system approach. These changes have been favored by a realization that different pollution regimes (from primary to secondary pollutants, and from single to multiple sources of those multiple pollutants) induce different kinds of tree injuries (acute versus chronic, local versus regional). As a result, the regulatory standards set by governments have changed from atmospheric concentrations to critical loads/levels. Often, however, the actual regulation appears to be not well-fitted to vegetation. For example, questions still exist about the correspondence between the AOT40 exposure index (current European standard) and actual injury induced by the ozone dose absorbed by vegetation. Further improvements are offered by a calculation of the "physiologically effective ozone dose" and its vegetation response.

The literature dealing with biochemical and physiological aspects in natural or artificially stressed trees and published over the last few decades is impressive. However, most is still descriptive, linking visual observations and chemical analysis with little regard for any model or hypothesis.

### 2.2 KNOWLEDGE GAPS

Future investigations should address the explanatory or mechanistic approach which focuses upon systems analysis, using ecophysiological and molecular-biological technology, and should include such parameters as growth, allocation, biodiversity, competition, progeny fitness and soil leachates. In particular, the physiology of the root system in stressed trees has been largely ignored and requires more intensive investigation.

The question to be asked is: does pollutant A cause effect B on site C? By focusing our attention on the effects caused by pollutants, including the mechanisms of damage or injury, we should strive to predict plant and ecosystem responses. Models can be most useful as prediction tools, if the input parameters, functional relationships within the model and validation are well defined and accessible to all researchers.

A balanced research program is suggested comprising three components: 1) controlled experiments using artificial protocols, designed to define basic mechanisms including the varied effects of different pollutants; 2) key studies in the field based upon defined hypotheses, to monitor representative stands and link chronic effects observed under controlled experiments; and 3) models, well-structured to link data from 1) and 2).

The first two steps have to be strongly integrated through literature or specific research programs, even if they cannot be carried out in the same system by the same researchers. A combined approach with an interdisciplinary team is needed to attain mechanistic and integrated answers. The main result should be the ability to carry out risk prediction, which is essential as a "scientifically sound" base for policy makers.

## 3. Soil Organisms, Rhizosphere and Nutrient Uptake (WP 7.04.03)

3.1 STATE OF SCIENCE

Discussion on the biological (mycorrhizas and pathogenic fungi) component of the soil was limited to mycorrhizas and pathogenic fungi. For mycorrhizas, it is particularly important to identify the function of different types, and their relative sensitivities to air pollutants. If different species have different functions (enhancement of surface area for nutrient uptake, nutrient mobilization of organic material, enhancement of ion uptake, nutrient storage, detoxification mechanisms and weathering through organic acid excretion), then their different sensitivities to tree stresses from air pollutants will have unpredictable results. It is well-known that mycorrhizal infection of fine roots declines in conditions of high N input, with an accompanying shift in the community structure.

Ozone has been shown to have a similar effect to N deposition on mycorrhizas, as it impairs C allocation to the roots. Ambient levels of ozone cause damage to species associated with deciduous and coniferous trees, more so at higher altitudes.

In oak stands across Europe, several species of the root pathogen, *Phytophthora*, have been shown to be associated with oak decline. Higher numbers of sporangia produced at high experimental levels of N suggest that N deposition might exacerbate the extent of the effect of the pathogen in soils with a pH above 3.5. An indirect effect could also result from high N deposition, as mycorrhizal roots are known to have better protection against pathogen attacks.

Discussion on soil chemical indices focused on C:N and Ca:Al ratios. In terms of the former, it was felt that past emphasis had been on the N part of the ratio at the expense of possible changes in C. The ratio is a useful index to identify N saturation, although difficult to interpret on disturbed sites, such as clear-felled ones. However, in interpreting it, we need to know how stable the organic layer will be under air pollution stresses, what mechanisms are underlying measured changes (for example, what triggers changes in the rate of nitrification and nitrate leaching) and the role of the fine root/mycorrhizal interaction.

The critical substrate Ca:Al ratio is often quoted as 1, although this is based on values derived for Norway spruce (often seedlings in pot experiments or in hydroponic culture), which may be a factor of 10 lower than those for agricultural crops and deciduous forest floor species. While the ratio is more useful than absolute concentrations of either in the soil solution, the role of other base cations is often overlooked. There is wide variation in toxicity of different forms of Al in the soil, and the ratio would be more meaningful if confined to the most toxic, inorganic $Al^{3+}$.

Although $NH_4^+$ can be chemically fixed, only N immobilization in the microbial biomass was considered. The measured mineralization rate is a net value, representing gross mineralization minus immobilization and leaching. Seasonal studies have some use, but temporal variation is overlooked in assessing critical loads.

### 3.2 KNOWLEDGE GAPS

With respect to mycorrhizas, there is limited knowledge of how different mycorrhizal species vary in sensitivity (for example, some rhizomorph types are particularly sensitive to N deposition). Also, although the role of the extramatrical mycelium is still not clear, it is highly likely that interactions between the hyphal mats of saprotrophic and mycorrhizal fungi will be influenced by pollutant effects on the mycorrhizas via the trees.

The Ca:Al ratio is calculated on bulk soil samples, so we require better techniques to characterize the rhizosphere environment, with its often deviating soil chemistry. There is also a need for studies of the variations in the soil solution, both spatially and temporally.

Although the N in fungal fruiting bodies, dead hyphae and bacteria can be remobilized, knowledge of the relative time scales for this process is lacking.

## 4. Genetic Aspects of Air Pollution and Climate Change (WP 7.04.04)

### 4.1 STATE OF SCIENCE

Presentations focused on the genetic response of individual trees and forest tree populations, respectively, to various forms of environmental stress. Corresponding biochemical and physiological aspects were addressed with the main emphasis on the response to elevated $O_3$, $CO_2$, heavy metals and temperature change. The severe impact of long-term air pollution from smelters on the population structure of Scots pine was clearly demonstrated. Similarly, clear evidence of adverse impacts of $O_3$ on epicuticular waxes and a genetic difference in response were shown. Finally, the potential impacts of $CO_2$ on the genetic structure of ponderosa pine was outlined. The role of antioxidants in controlling the response of trees to air pollutants was demonstrated for aspen clones differing in tolerance to $O_3$.

A case study on changes in flushing periods of European beech provenances to defined environmental conditions revealed strong dependencies on heat and other climatic parameters. These data can be utilized for simulations on effects of climatic change on growth parameters. The great efficiency of molecular genetic markers for the verification of biotic stress components in forest ecosystems was highlighted by means of the species specific detection and quantification of *Phytophthora* species which cause severe root-rot diseases in deciduous trees.

### 4.2 KNOWLEDGE GAPS

Despite increasing knowledge about genetic dynamics in populations related to air pollution, there is little information on actual competition between genotypes differing in tolerance to air pollution. The establishment of competition interaction studies along pollution gradients or in Free Air $CO_2$ Exposure (FACE) systems are required to allow for more definitive estimates of genetic change under the influence of normal stand dynamics. Generally, more information is required on the genetic response of individuals and populations to specific stress factors. In addition, there are knowledge gaps with respect to interactions between air pollution and other forms of biotic/abiotic environmental stress. As well, there is little information on the genetics of host-parasite-systems in stress response.

Until recently, isozymes were mainly used to estimate genetic parameters from forest trees. Now, geneticists have an array of molecular tools at their disposal offering unparalleled power for detecting genetic changes in forest tree populations or their associated pests. Linkage between isozyme and molecular approaches are likely to prove exceptionally useful in studies of population genetics of forest trees. In particular, there is a strong need for the study of correlations between genetic markers and phenotypic traits such as tolerance or sensitivity of individuals with respect to certain stress factors. Such information contributes much to the maintenance of forest ecosystems because it supplies diagnostic tools having inclusive bioindicators for forest ecosystems exposed to long-term stress.

"*Ex situ*" preservation of forest tree germplasm has long been promoted as a means of preserving genetic diversity for future generations. Continual changing pollutant environments, along with concomitant global climate change means that "*in situ*" preservation and tree breeding must pay special attention to adaptability and adaptiveness through the maintenance of high levels of genetic diversity.

## 5. Silviculture in Polluted Areas (WP 7.04.05 )

5.1 STATE OF SCIENCE
Three case studies were discussed as surrogates for the state of science with respect to silviculture and forest management in polluted areas. The WP discussions centered on forest sustainablility and the silvicultural implications of air pollution. In Lithuania nitrogen and sulphur pollution associated with N fertilizer plants have induced changes in cutover response, manifest as a grass (*Calamagrostis* sp) invasion. Herbicide use was recommended but a switch in tree species composition may be required. Silvicultural treatments suggested included: denser stands, mixed species and belts of deciduous trees. Mitigation is needed if insect and disease effects, especially *Ips typographis*, are to be avoided.

In southern Sweden, the SUFOR project is examining long-term effects of acid and N deposition at the landscape level. Programme components include: production management, biogeochemistry, biodiversity, forest health and the economy. It is suspected that the artificial Norway spruce monocultures may not be sustainable due to acidification, buttrots and nutrient imbalances including water quality problems. Alternatives include more deciduous and mixed species stands with lower canopy filtration of N and S, use of less soil volume due to different rooting patterns, mixed litter and nutrient cycling. Models are being developed based on weathering and growth rates to predict these effects.

In the UK, questions are arising on the sustainability of Sitka spruce plantations on degraded upland soils which require fertilization with P and sometimes N. Nevertheless, second rotation productivity initially seems to be satisfactory. A critical loads approach has been used for acidic deposition. At present, only very limited areas are being affected and there is little evidence for growth reductions. Nevertheless, the question remains as to whether forest management policies should favor mixed species, which may be more resistant over the longer-term. It was agreed that leaching of nitrate N in standing crops may be a potential indicator of pollution impact.

In North America, there is a long history of repeated, sustainable rotations of spruce and pine plantations in the absence of air pollution and human impacts. Natural forests are arranged on landscapes in a very orderly and systematic way. Pollution stress when added to endogenous stress due to climate, insects, disease, windthrow and old age may cause the death of forests. Local examples have been attributed to acid rain in old eastern red spruce forests. Notions and concepts about artificial European spruce monoculture sustainability in polluted environments are not applicable to natural monocultures in North America.

## 5.2 KNOWLEDGE GAPS

The questions posed and solutions proposed depend upon the forest systems under investigation. Generalizations and application of results to systems other than those under study are not possible with current science.

A major knowledge gap exists when the combined effects of both natural and man-made stress factors are taken into consideration. There is a world-wide requirement for scientific answers to questions about the long-term sustainability of forest ecosystems under air pollution stress.

# 6. Conclusions

The interdisciplinary and multinational composition of the WP sessions resulted in very lively and productive scientific discussions. Clearly, a comprehensive review of the state of science in these five areas of air pollution effects on forest ecosystems was not feasible within the time available and tended to be based on the specialities represented by those present. From these discussions and the summaries presented above, however, several strategic points emerged which are of immediate concern to research and monitoring activities and of direct relevance to forest managers and policy makers.

Changes in forest health due to air pollutants are routinely monitored at national/multinational scales in North America and Europe. Current detection methods and indicator designs are likely to be inadequate for early detection and the final attribution of cause-effect. Measurement endpoints do not often relate to growth and indicators are generally not specific for air pollutants under study. We cannot assume that because symptoms are not reported, there are no effects. Only where monitoring is linked temporally/spatially with process-based research, have linkages been made.

A balanced, interdisciplinary approach is needed in the future. Research and monitoring must be better integrated so that air pollution issues affecting forestry should be investigated using the multiple stress concept, where abiotic/biotic factors are considered along with air pollutants. Major gaps remain in our knowledge of below-ground processes, calculation of critical levels/loads and their application across landscapes, and transfer of ozone air concentrations into effective pollutant dose. Assessment of risk is becoming an increasing priority.

While case studies have documented significant air pollutant impacts on forests in North America and Europe, it is clear that patterns of decline are different, due in part to different ecophysiological/biogeochemical processes and forest management histories/practices.